Cell Biology
of the Secretory
Process

Editor: M. Cantin, Montreal, Que.

184 figures and 19 tables, 1984

 KARGER

S. Karger · Basel · München · Paris · London · New York · Tokyo · Sydney

National Library of Medicine, Cataloging in Publication
  Cell Biology of the Secretory Process /
  Editor, M. Cantin. – Basel; New York: Karger, 1984
  1. Endocrine Glands – Physiology 2. Endocrine Glands – Secretion 3. Exocrine Glands – Physiology
  4. Exocrine Glands – Secretion I. Cantin, Marc
  WK 102 C392
  ISBN 3–8055–3619–4

Drug Dosage
  The authors and publisher have exerted every effort to ensure that drug selection and dosage set forth in this text are in accord with current recommendations and practice at the time of publication. However, in view of ongoing research, changes in government regulations, and the constant flow of information relating to drug therapy and drug reactions, the reader is urged to check the package insert for each drug for any change in indications and dosage and for added warnings and precautions. This is particularly important when the recommended agent is a new and/or infrequently employed drug.

Contents

Foreword

Investigations on the cell biology of the secretory process in various endocrine and exocrine glands have progressed so fast and the variety of approaches is now so vast that the resulting literature is both extremely abundant and widely dispersed in specialized journals. This book has been devised so that investigators and graduate students may find, under a single cover, as much of the pertinent information as possible. The first chapters deal with generalities applicable to all glands: ligand-receptor interactions, stimulus-secretion coupling and the participation of key organelles in the synthesis, processing and secretion of products. The other chapters deal with the formation and secretion of hormonal products in a wide variety of glands in all their intricate details. While some contributions are mainly morphological and others are almost exclusively biochemical, the resulting integration will, it is hoped, prove helpful.

I am very grateful to the various contributors for their friendly cooperation and to *Vivianne Jodoin* for her invaluable editorial assistance.

Marc Cantin

Cell Biology of the Secretory Process, pp. 1–51 (Karger, Basel 1984)

Membrane Receptors and Hormone Action

Karen A. Valentine, Morley D. Hollenberg

Department of Pharmacology and Therapeutics, Endocrine Research Group, Faculty of Medicine, University of Calgary, Calgary, Alta., Canada

Introduction

Specific receptors localized in the cell membrane are integrally involved in the action of many hormones. In the absence of a functional receptor, hormone responsiveness is not possible. In contrast, the presence of a receptor is permissive for hormone action, but does not guarantee hormone responsiveness. Since the development of the receptor concept at the turn of the century, knowledge of receptor structure and function has advanced to the point where the biochemical details of several of the oligomeric protein moieties involved in hormone and neurotransmitter recognition and action are now known. In addition, for a number of hormones, the sequences of events leading from hormone recognition and binding to the alteration of cellular function are being elucidated.

It would be impossible to document all of the recent advances in the field of hormone receptors in a single chapter. Rather, we will present some of the basic concepts involved in the study of hormone receptors, and we will describe several specific examples that are illustrative of our present understanding of receptor structure, function, and regulation. The chapter will deal, for the most part, with membrane receptors for peptide hormones and neurotransmitters. The structure and function of cytosolic steroid hormone receptors will not be covered here; the interested reader is directed to a number of review articles dealing with this subject [23, 24, 68, 113, 143, 187].

Pharmacologic and Biochemical Characterization of Receptors

Development of the Receptor Concept

The term receptor is usually used to describe that cellular component which recognizes a specific ligand (e.g., hormone, neurotransmitter, drug) and sets into motion a sequence of events which culminate in a cellular response characteristic of that ligand. The genesis of the receptor concept can be traced to the early 1900s in the work of *Ehrlich* [53], *Langly* [120], and *Clark* [27, 28]. *Ehrlich* [53], investigating the action of tetanus toxin, concluded that the toxin must unite with some receptive substance in the cells in order for it to have its effect. Further studies by *Ehrlich* [53] on the antimicrobial action of triphenylmethane dyes led him to conclude that the receptors for these substances had very strict stereochemical requirements and that families of compounds worked through their own specific receptors. At about the same time, *Langly* [120] was studying the actions of nicotine and curare at the neuromuscular junction. He observed that: (1) nicotine stimulated the contraction of denervated muscle; (2) curare could block the stimulatory action of nicotine, and (3) denervated, curare-treated muscle could still contract when electrically stimulated. These observations led *Langly* [120] to conclude that both curare and nicotine acted through a receptive substance on the muscle. He reasoned that nicotine combined with the receptive substance so as to trigger contraction, while curare combined with the receptive substance in a manner which did not trigger a response yet blocked the combination of nicotine with the substance. Further, he postulated that 'whether curare-muscle complexes or nicotine-muscle complexes were formed depended on the amounts of each compound present and on their relative affinities for the receptive substance'.

A focus on the quantitative aspects of receptor function was stimulated by the work of *Clark* [27–29]. Using a bioassay for acetylcholine (contraction of frog heart muscle), *Clark* [28] estimated that only 10^{-14} mol or 20,000 molecules of acetylcholine were required to produce a response, and thus it became apparent that the numbers of receptors present on a responsive cell were infinitesimal. In his remarkable treatise on pharmacology *Clark* [30] went on to address many of the same questions that are still challenging us today.

In summary, the early work of these investigators defined many of the fundamental aspects of the receptor concept; i.e., that receptors are highly specific recognition entities present in or on cells in small numbers and which, in combination with specific series of compounds, alter some cellular biochemical process (probably in a rate-limiting way). It is important to remember that the concept of a specific receptor for a given ligand is inextricably linked to the response of that system to the ligand. Ligand-binding studies have for the most part focused on the recognition function of receptors; recently, advances have also been made in elucidating some of the biochemical reactions that occur as a consequence of ligand binding. However, the in-depth evaluation of both of these aspects of receptor function must ultimately be integrated to provide a complete reaction sequence for hormone-modulated cellular responsiveness.

Relation of Binding to Biological Response

It is a fundamental assumption of receptor theories that for a hormone to act it must first be bound. The initial binding reaction may be thought of in simple terms as a bimolecular chemical reaction:

$$R + H \underset{k_2}{\overset{k_1}{\rightleftharpoons}} RH \rightarrow \rightarrow response, \tag{1}$$

where R, H, and RH represent receptor, hormone, and receptor-hormone complex, respectively, and the equilibrium dissociation constant K_D is given by the quotient of the rate constants, k_2/k_1. This relatively simple equilibrium forms the basis of a number of theories of drug and hormone action that focus either on the equilibrium concentration of RH (occupancy theory) as developed by *Clark* [27–30], *Gaddum* [57, 58], *Ariens* [5], *Ariens and Simonis* [6], *Stephenson* [179], and *Stephenson and Barlow* [180] or on the rates of formation and dissociation of the RH complex (rate theory) as developed by *Paton* [145], *Paton and Rang* [146], and *Paton and Wand* [147].

The two major theories differ with respect to the immediate function of the receptor-hormone complex. However, as details of the sequence of events following ligand binding emerge, these apparent differences may become less meaningful [88]. The details of these receptor theories have been reviewed extensively elsewhere [25, 34, 40, 74, 88, 94, 156] and as such will not be dealt with here.

As indicated earlier, the binding of a hormone to its receptor is a prerequisite for a biological response. Prior to the development of high specific activity radiolabeled ligands, analysis of the bioresponse characteristics provided the only source of information about the ligand recognition function of receptors. The dose-response relationship indicated by equation 1 can be evaluated in a way analogous to the evaluation of substrate-velocity relationships for enzymes [40, 88]. In the case of enzyme kinetics, one has the advantage of knowing a great deal about the reacting species (enzyme and substrate) and about the end products of the enzymatic reaction. The evaluation of dose-response relationships is hampered by: (1) chemical information is usually available for only one of the reacting species (hormone); (2) the end product is not a metabolite but a biological response to the receptor-ligand interaction, and (3), as is illustrated by equation 1, there are a great many uncertainties about the coupling of the response to receptor occupation [40, 88, 94]. Thus, the information that one can gain from dose-response data about the affinities and mechanisms of interaction of stimulatory ligands (agonists) with receptors is rather limited.

As a first approximation, the concentration of an agonist producing a half-maximal response (ED_{50}) can be taken to reflect the affinity of the binding site for the agonist. In addition, the relative ED_{50} values of a series (family) of agonists may be expected to indicate the relative affinities of these compounds for the receptor. The ED_{50} for a ligand can be taken as identical to the equilibrium dissociation constant (K_D) only in cases where the receptor concentration is relatively low (with respect to the K_D) and where response is directly related to receptor occupation [40]. In actuality, the above situation may well represent only an 'idealized' situation, and in experimental systems the ED_{50} may be less than, equal to, or greater than the true dissociation constant of the receptor for the ligand, as expressed by K_D [40, 156]. The relation of the ED_{60} to the K_D will be dependent on the nature of the coupling reactions linking receptor occupation to response. Therefore, the shape of dose-response curves cannot prove or disprove any proposed model of ligand receptor interaction, but are useful only in the assessment of the relative affinities of agonist ligands for recognition sites. Dose-response relationships are, however, of great value in determining receptor characteristics using chemical and enzymatic probes. The use of these methods to study receptors is based on the assumption that a perturbation of responsiveness in a characteristic manner (e.g., a shift of the dose-response curve to the right indicating a lowered affinity) reflects a similar change in ligand-receptor interaction.

On the other hand, for antagonists, dose-response curves can give accurate information about the recognition site, when analyzed in conjuction with a series of active compounds. The dose-ratio method originally developed by *Arunlakshana and Schild* [7] and *Schlessinger* et al. [168] for determining antagonist equilibrium dissociation constants (K_D) is based on a null hypothesis. It is assumed that when the response to a particular concentration of agonist in the absence of inhibitor is the same as the response to a higher concentration of agonist in the presence of competitive inhibitor, then the amount of agonist reaching the receptors in both cases must be the same. This assumption simplifies the mathematical analysis of dose-response data and allows one to estimate antagonist affinities without making any assumption about the mechanisms coupling agonist occupation of the receptor to response. Thus, one can use bioassay systems alone to characterize antagonist interactions with receptors. These studies are directly comparable to the measurement of receptor affinities with the use of ligand-binding studies with radiolabeled probes. The bioassay approach employing antagonists and the dose-ratio method has been used profitably in the characterization of both nicotinic and muscarinic cholinergic receptors and for identifying both α- and β-adrenergic receptors. A close agreement has been found between the receptor affinities as measured by ligand binding and by the dose-ratio methods [15, 31, 146].

It is evident from the above discussion that the analysis of biological responsiveness provides important but somewhat limited information concerning agonist and antagonist recognition sites. The dose-response data are critical, nonetheless, for the interpretation of ligand binding studies.

Direct Binding Studies

A major advance in the study of receptors has come from the development of methods for the direct analysis of the interaction of radiolabeled ligands with receptors. Although much work has been done with the use of ^{3}H-labeled ligand probes, it has been the adaptation of immunoassay technology that has permitted the explosive advance in the studies of receptors for polypeptide hormones. The radioimmunoassay technology provided for the preparation of high specific activity radiolabeled ligands which retained both biological and immunological activity. In addition, the immunoassay methods were directly applicable for the rapid separation of receptor-bound from unbound labeled ligand [for the methodologies involved see references 40, 94, 95].

In ligand-binding studies it is often a problem to determine whether or not the binding site detected represents a pharmacologically relevant receptor. To aid in distinguishing receptor-related binding from nonspecific ligand interactions, a number of criteria of hormone binding have been developed. The fulfilment of which criteria is considered necessary for the demonstration of a specific receptor site. The initial studies of ligand binding should include: (1) an assessment of the biological activity of the radiolabeled hormone probe; (2) measurement of high affinity binding of the labeled hormone, in keeping the low physiological concentrations of most hormones; (3) observation of saturability of binding over a concentration range corresponding to the concentration range of biological activity of the hormone studied; (4) an evaluation of binding in the initial instance in organs known to be target tissues for the hormone of interest; (5) an assessment of the reversibility of binding, in keeping with the reversibility of hormone action, and (6) an evaluation of the relative binding affinity of hormone analogues with known relative biological activities [40, 94].

When the information indicated above has been obtained, it should be possible to determine if the binding measured is receptor related, i.e., 'specific'. It is essential that the binding data be consistent with the bioassay data. If the binding data are out of keeping with the bioassay data, then it will become necessary to question the interpretation of the ligand-binding studies and consider that a nonreceptor (albeit chemically specific) interaction may be under study [40, 89, 93, 94].

Over the past decade or so, the study of the binding of a variety of polypeptide hormones to putative membrane receptors has met with considerable success. As indicated above, there are by now a number of comprehensive reviews that summarize the techniques used, the kind of progress that has been made, and the kinds of concerns and pitfalls inherent in such studies [16, 40, 60, 93, 94, 108, 160]. It is possible to generalize somewhat on the properties of receptor-binding interactions. For most peptide hormones, the affinities are remarkably high, with dissociation constants usually less than $10^{-8}\ M$. In addition, the receptors found on any given cell are present in vanishingly small number, usually less than 10^5 per cell. Because of the high affinities and small numbers of receptors present on responsive cells, ligand-binding studies necessitate the use of radioactively labeled compounds of very high specific activity (e.g., 1,000–3,000 Ci/mmol). To achieve sufficiently radioactive derivatives, peptides are usually substituted with ^{125}I or ^{131}I to the extent of one atom of iodine per peptide molecule. Polypeptides are particularly attractive compounds for such studies, since it is frequently possible to introduce the iodine atom at a position that is not critical for the biological activity of the peptide of interest. As discussed elsewhere [40], there are a variety of methods for the preparation of highly radioactive peptide derivatives.

As already indicated, it is important that the radioactive peptide be evaluated for its biological activity. Ideally, the derivative should be fully biologically active, as has been established for insulin molecules substituted to the extent of one iodine atom per molecule. In contrast, the substitution of iodine into the tyrosine residue of oxytocin or vasopressin abolishes biological activity. If it can be established that the substituted polypeptide derivative is fully active, it can be logically assumed that the binding of the derivative reflects the binding of the unlabeled compound to the receptor recognition site.

Most measurements of ligand-receptor interactions are done under equilibrium conditions. Aliquots of the cell, membrane, or soluble receptor preparation of interest are incubated with increasing concentrations of radioactive ligand both in the absence and in the presence of an amount of unlabeled ligand (100- to 1,000-fold excess) sufficient to saturate all of the high-affinity binding sites. The preparation is allowed to equilibrate (usually 30–60 min at 24 °C), and the receptor-bound radioactivity is then rapidly separated from the free ligand in the supernatant (e.g., cells or membranes are rapidly collected and washed on Millipore filters, or cultured cell monolayers are rapidly rinsed free of supernatant). It is usually the case that even in the presence of a large excess of unlabeled ligand, an appreciable amount of radioactivity is still bound to the membrane or cell preparation. The amount of bound radioactivity for which the unlabeled parent compound cannot compete is assumed to represent 'nonspecific' or 'nonreceptor' binding. The 'specific' binding, usually reported in receptor studies, represents the total binding minus the 'nonspecific' binding. With the use of the above experimental paradigm, a binding isotherm for the ligand of interest can be obtained, as indicated for urogastrone (fig. 1a). The concentration of radioactive ligand at which binding is half-maximal provides a preliminary estimate of the equilibrium dissociation constant. The binding data can be analyzed further, mathematically, in a variety of ways to yield the affinity constant and maximum binding capacity [e.g., see discussion in ref. 157].

As an alternative to measuring the entire binding isotherm with radioactively labeled ligand, it is possible to determine binding-competition curves using a constant amount of radioactive ligand and increasing amounts of unlabeled parent ligand or ligand analogues (fig. 1b). Such

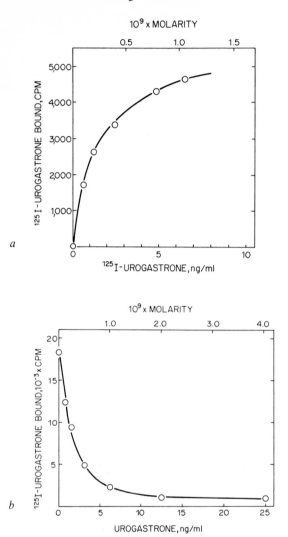

Fig. 1. Binding of epidermal growth factor-urogastrone to human fibroblasts. *a* Intact fibroblast monolayers were incubated at room temperature with increasing concentrations of ^{125}I-labeled human urogastrone, and the specific binding of radioligand was determined by previously described methods [40]. *b* Fibroblast monolayers were incubated with a fixed amount of ^{125}I-labeled human urogastrone and with increasing amounts of unlabeled peptide. After equilibration for 1 h at room temperature, the amount of bound radioligand was determined.

binding-competition curves are analogous to those used for radioimmu-noassay, where the binding agent is an antibody rather than a receptor. For receptor binding, it is expected that the relative order of potency of ana-logues to compete for binding will reflect the relative biological potency of the analogues in a variety of test systems. For insulin, the binding data have agreed remarkably well with observed biological data, as studied in detail by *Gliemann and Gammeltoft* [62]. The use of peptide analogues for binding-competition experiments thus provides essential information to establish the 'receptor' nature of the observed binding.

In addition to equilibrium data, measurements of the rates of binding (both on and off rates) provide useful information concerning the receptor-ligand interaction. In the case of insulin, a half-life of the receptor-ligand complex of approximately 16 min can be observed in membrane prepa-rations at 24 °C; at higher temperatures, the half-life is much shorter. The rate of formation of the receptor-ligand complex can be analyzed in terms of a simple bimolecular reaction, so as to yield an on rate constant. The quotient of the off rate (k_{-1}) and on rate (k_1) constants can be used to calculate the equilibrium constant: $K_D = k_{-1}/k_1$. In general, remarkably good agreement has been observed between the dissociation constant determined from equilibrium-binding measurements and the constant calculated from the rate constants. Particularly illustrative analyses of the rates of formation and dissociation of the insulin-receptor complex have been provided by *Pollet* et al. [150] and by *DeLean and Rodbard* [46].

It is now realized that upon binding to their receptors in intact cells many ligands can become rapidly internalized and degraded, not only at 37 °C, but at room temperature as well. Thus, in studies of intact cells, it is important to account for internalized and/or degraded ligand, when the 'binding' data are analyzed. In some cases [73], it is possible to dissociate the surface-bound ligand in acidic isotonic solutions, so as to obtain a measurement of internalized peptide. Alternatively, measurements at 0 °C, at which temperature the internalization process does not proceed, can yield an estimate for the maximum binding capacity of intact cell preparations. Unfortunately, because of the internalization process in intact cells, ligand affinity calculation based on data obtained at room temperature or above must be interpreted with some caution.

Probing Receptor Structure with Enyzmes and Lectins

The use of enzyme and lectin probes has served to complement mea-surements of ligand binding, thus adding considerably to the accepted models of receptor structures. The ligand recognition site of a number of receptors (e.g., for insulin) has been found to be remarkably sensitive to proteolytic enzymes. Such data not only indicate the protein nature of the receptors, but suggest that the ligand recognition sites are located in a relatively exposed portion of the plasma membrane. In addition to the exposed sites for ligand recognition, it is evident from experiments with phospholipases that a certain portion of the recognition sites may not be exposed to the external environment, but may be masked by membrane lipids. Other data with enzymes that hydrolyze glycosidic bonds (e.g., neuraminidase, β-galactosidase) suggest that sugar residues may play a role not only in the coupling of receptor occupation to cellular activation, but also in the specific recognition of the ligand. For example, neuraminidase abolishes the action of insulin on adipocytes without affecting insulin binding, whereas the simultaneous treatment of adipocytes with neuraminidase and β-galactosidase markedly affects insulin binding. The use of enzyme probes for the study of receptor structure has been well documented in studies with the insulin receptor [37].

Studies with plant lectins suggest further that peptide hormone receptors are glycoproteins. Not only can lectins such as concanavalin A and wheat germ agglutinin block the binding of polypeptides such as insulin and epidermal growth factor-urogastrone (EGF-URO) to receptors, but the lectins themselves can be observed to cause hormonelike effects. For example, both concanavalin A and wheat germ agglutinin have insulinlike effects in fat cells. In addition to serving as probes of receptor structure, the lectins have proved useful, as insolubilized derivatives, for the isolation of receptors by affinity chromatographic techniques.

Solubilization, Characterization, and Isolation of Receptors
Certain peptide hormone receptors (e.g., for insulin) are able to retain the ligand recognition property subsequent to solubilization with nonionic detergents like Triton X-100, whereas some receptors (e.g., the placental receptor for EGF-URO) lose the ligand recognition property upon solubilization. In the soluble state, the receptor for insulin is amenable to analysis by a variety of physicochemical methods, including velocity sedimentation, column chromatography, isoelectric focusing, and gel electrophoresis. In the majority of the studies with the insulin receptor, use has been made of the receptor's insulin-binding property to detect the presence of receptor (e.g., in sucrose gradients or subsequent to chromatogra-

phy). Only relatively recently has sufficient receptor been isolated to detect the receptor by more conventional means [106].

Based on the data so far accumulated for the insulin receptor, the recognition macromolecule appears to be a glycoprotein that in nonionic detergent solutions (Triton X-100) behaves as a large molecule (apparent stokes radius about 72 Å) with an estimated molecular weight of 300,000 [36]. Recent work indicates that even in nonionic detergents the receptor may be able to dissociate into subunits of smaller size (about 40 Å) [61, 132]. It is likely that the properties of the insulin receptor will not prove too dissimilar from the characteristics of receptors for other polypeptide hormones.

In the event that solubilization abolishes the ligand recognition property, other approaches are necessary for receptor characterization. Polypeptides provide the advantage that a variety of covalent coupling methods may be used to 'affinity-label' the receptor prior to solubilization. For the polypeptide, EGF-URO, both photoaffinity labeling [44, 86] and affinity cross-linking with glutaraldehyde [161] or with bis-succinimidyl substrate [142, our unpublished data] have succeeded in identifying a specific recognition macromolecule. On gel electrophoresis, the recognition site exhibits an apparent molecular weight of about 200,000. As with the insulin receptor, there is evidence that the EGF-URO receptor is a glycoprotein [21, 85, 161], and that it may possess a complex structure in the membrane [21, 161].

For the EGF-URO receptor, it was necessary to develop a new method for detection of the soluble receptor [95, 161]. The assay is based on the ability of the receptor to bind to lectin-agarose derivatives and on the success of measuring the binding of radiolabeled EGF to the lectin-insolubilized receptor (fig. 2). In principle, this method should work for any glycoprotein receptor that binds a nonglycosylated ligand and that does not require oligosaccharide for its binding activity. The use of this method has permitted a study of the solubilized EGF-URO receptor from human placenta [87].

Photolabeling [126] and crosslink-labeling methods have proved instrumental in analyzing the oligomeric structure of receptors like the one for insulin [131, 191]. The principle of the photolabeling procedure is outlined in figure 3. In essence, the radiolabeled photoprobe is first allowed to bind to the receptor and is then activated by an intense light. The highly chemically reactive species generated by light activation become covalently linked to the protein molecules to which the photo-

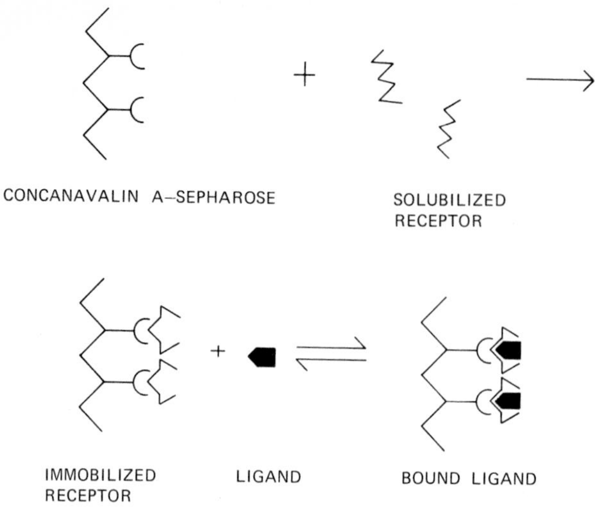

Fig. 2. Scheme for the lectin immobilization assay. The soluble receptor is first adsorbed to concanavalin A-Sepharose, and the binding assay is performed on the bead-bound receptor that has been washed with an appropriate binding buffer.

probe has bound. The use by *Yip* et al. [191] of insulin photoprobes substituted at position B-1 and B-29 of the insulin molecule has permitted the identification in intact adipocytes of four glycoprotein components present in the insulin receptor oligomer (α, molecular weight about 135,000; β, 90,000, and γ, δ, molecular weights about 40,000). It is, as yet, unknown if all of these subunits play a role in the insulin-binding process or if some of the peptide species (e.g., γ and δ) represent closely associated proteins that do not participate in binding but which may play a role in cell activation (see figure 4). The complementary work by *Massague* et al. [131] and by *Jacobs* et al. [104, 105] using either crosslink-labeled [131] or radiolabeled purified receptor [105] has employed limited reduction and gel electrophoretic analysis of the radiolabeled insulin receptor to deduce a tentative oligomeric model for the insulin receptor (fig. 4b).

Receptor isolation has depended not only on conventional methods of column chromatography (e.g., molecular sieve, ion exchange, hydrophobic adsorption) but also on affinity chromatography methods using both lectin-agarose and hormone-agarose derivatives. These methods, used initially for the isolation of the insulin receptor [36, 37] have now

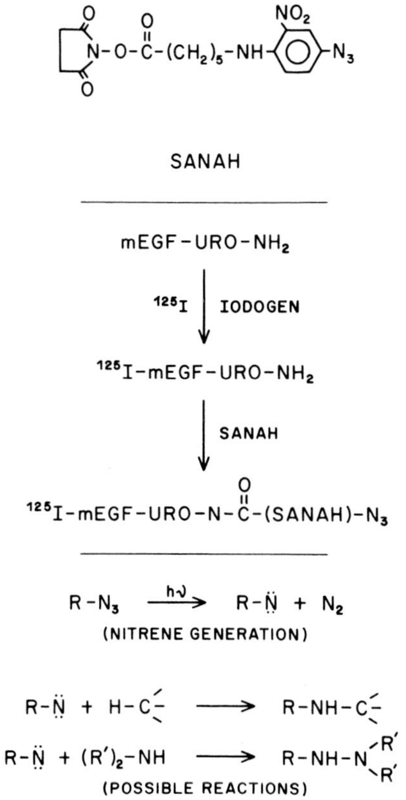

SANAH

---

mEGF-URO-NH$_2$

^{125}I $\downarrow$ IODOGEN

^{125}I-mEGF-URO-NH$_2$

$\downarrow$ SANAH

^{125}I-mEGF-URO-N-$\overset{\text{O}}{\overset{\|}{\text{C}}}$-(SANAH)-N$_3$

R-N$_3$ $\xrightarrow{h\nu}$ R-N̈ + N$_2$
(NITRENE GENERATION)

R-N̈ + H-C$\overset{\diagup}{\underset{\diagdown}{}}$ $\longrightarrow$ R-NH-C$\overset{\diagup}{\underset{\diagdown}{}}$

R-N̈ + (R')$_2$-NH $\longrightarrow$ R-NH-N$\overset{\diagup\text{R}'}{\underset{\diagdown\text{R}'}{}}$
(POSSIBLE REACTIONS)

Fig. 3. Procedure for photolabeling the receptor for EGF-URO with N-succinimidyl-6(4'-azido-2'-nitrophenylamino)-hexanoate (SANAH). The ^{125}I-labeled peptide is first prepared using the iodogen procedure and is incubated directly, without further purification, with SANAH to form the photolabeling derivative (R-N$_3$). Upon binding to the receptor (usually done at 0 °C to avoid nonspecific coupling reactions), the preparation is irradiated to activate the photoprobe, that then couples covalently to the receptor via a number of possible reactions.

been applied to several receptor systems, including the β-adrenergic receptor [174] and the receptor for EGF-URO [32].

Receptors versus 'Acceptors'

It is becoming increasingly evident that, aside from 'true' receptors and other chemically specific nonreceptor binding sites, there are also membrane recognition sites ('acceptors') that, in addition to the familiar ion or metabolite transport sites, have to do with the communication of chemical information from the cell exterior to the cytoplasm.

The term 'receptor' may, therefore, require a more restricted and precise definition. In pharmacologic terms, the membrane receptor for agents such as neurotransmitters can be thought of as a macromolecule (probably an oligomer) that has the dual function of both recognizing a ligand of interest in a chemically specific manner (recognition function) and causing an immediate perturbation of membrane function (i.e., the action function) that in some manner leads to a biological response. This recognition-action function of a receptor may be distinguished from a membrane 'acceptor' site that may function solely as a recognition molecule for the selective cellular uptake of certain serum-borne constituents.

An example of a cellular 'acceptor' can be seen in the function of transcobalamin II (TC II) [129, 137], a protein that serves as a transport agent for cobalamin (vitamin B_{12}) in the circulation and subsequently delivers cobalamin to the cell interior [129]. In the case of cobalamin, the TC II-cobalamin complex can bind to a specific cellular acceptor site, leading to the translocation of the complex and the subsequent intracellular release of cobalamin for further metabolic processes. In this instance, cobalamin can be thought of as the pharmacological agent active at an intracellular (enzyme) receptor; the membrane constituent that recognizes the TC II-cobalamin complex, in a highly specific manner (the TC II-cobalamin complex, but neither free TC II nor free cobalamin binds with high affinity to the acceptor site), clearly functions in a manner different from the one envisioned for neurotransmitter receptors and may, therefore, be termed an 'acceptor'. The cellular binding site for low-density lipoprotein (LDL) [67] can be thought of in similar terms, wherein the feedback regulator, cholesterol, is the pharmacologically active ligand in the cell interior subsequent to internalization via the LDL acceptor. Importantly, for acceptors as well as receptors, it is to be expected that a strict chemical specificity for ligand binding will be observed along with other criteria that are consistent with a high-affinity recognition function. In the context of the above discussion, steroid hormone receptors, which reside in the cytoplasm, can be seen to play a role separate from either that of the recognition-action function of the neurotransmitter receptor or that of the passive translocation function of the TC II-cobalamin or LDL acceptor.

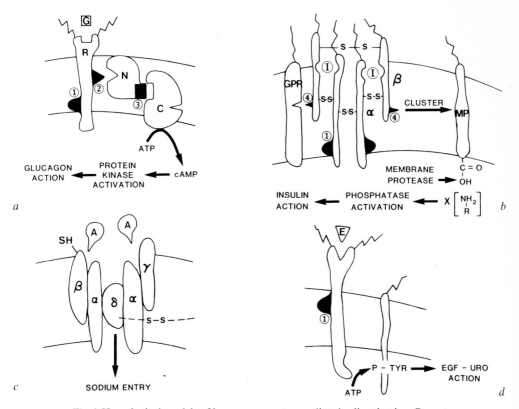

Fig. 4. Hypothetical models of hormone receptor mediated cell activation. Receptors for glucagon *(a)*, insulin *(b)*, acetylcholine *(c)* and EGF-URO (E; *d*) are shown as integral membrane glycoproteins (oligosaccharide constituents; waved lines) that traverse the plasma membrane. Common receptor regions that relate to receptor function are depicted schematically: region 1 is involved in receptor internalization; regions 2 and 3 are related to the coupling of hormone receptor occupation to cyclase activation. Region 4 on the insulin receptor is shown to interact with a ligand affinity regulator glycoprotein (GPR); an analogous region may possibly interact with the catalytic subunit of cyclase. The four distinct activation mechanisms are discussed in the text. *a* G = Glucagon; R = glucagon receptor; N = cyclase regulatory subunit; C = cyclase catalytic subunit; ATP = adenosine-5′-triphosphate; cAMP = cyclic adenosine 3′,5′-monophosphate. *b* I = Insulin; α, β = subunits of insulin receptor; GPR = glycoprotein regulator of insulin receptor affinity; MP = membrane protein substrate of insulin-mediated proteolytic activity; X = peptide mediator of insulin action. *c* A = acetylcholine; α, β, γ, δ = subunits of nicotinic acetylcholine receptor. *d* E = EGF-URO; P-Tyr = phosphotyrosine.

Disulphide bonds that link receptor oligomers are shown – S-S – as is a free sulfhydryl (SH) on the acetylcholine receptor. As pointed out in the text dealing with the mobile receptor model (fig. 5), individual receptors could modulate a number of membrane-localized processes.

Receptors, Acceptors, Ligand Internalization, and the
Action of Polypeptide Hormones

As already indicated, the term receptor can be regarded in a restricted sense to mean that membrane-localized macromolecule that both recognizes the polypeptide hormone in a specific manner and then causes a membrane perturbation leading to a biological response. It is as yet not certain whether or not the 'recognition function' and 'action function' reside in the same polypeptide chain, or whether the receptor exists as an oligomeric structure, only part of which functions as a recognition site, and part of which may play a role in the activation process. According to this model, it is the entire oligomeric structure that forms the functional activating species and that becomes triggered by the binding of the specific hormone.

Despite the above model of the receptor, and despite the now incontrovertible evidence that the cell membrane constitutes the *primary* site of action for most polypeptide hormones, it is still reasonable to ask: does the cell surface constitute the *only* site of action for polypeptide hormones? Recent ligand-binding studies with radioactively labeled polypeptides are directing a close look at this question. It is now evident from a number of studies that, subsequent to the binding of a radioactively labeled polypeptide at the receptor site, ligand internalization can occur. In the case of mouse EGF-URO, the disappearance in intact cells of available receptor sites (receptor 'down-regulation') observed subsequent to the binding of ligand is associated with the appearance in the medium (at 37 °C) of ligand degradation products [20]. It appears that the EGF-URO receptor complex once formed undergoes pinocytosis and lysosomal degradation. As alluded to above, an analogous mechanism is thought to liberate cobalamin into the cell interior from the TC II-cobalamin complex and to free up cholesterol from the LDL-cholesterol complex. It is, therefore, not unreasonable to hypothesize that a degradation fragment of EGF-URO or of its receptor released in the cell interior subsequent to receptor binding, and internalization may play a role in the well-known mitogenic action of this polypeptide. Since it is the rule, rather than the exception, that at least some proportion of an active ligand present in the external medium can be found in the cellular cytoplasm, proving or disproving of the above hypothesis may be much more difficult than establishing the cell surface as the primary point of hormone action. Even in the case of well-controlled studies with polypeptide-agarose derivatives, it could be argued that at the cell surface a portion of the peptide might be cleaved for local cell uptake or that a triggering of receptor internalization might be achieved so as to cause cell activation without the release of appreciable ligand from the inert support into the medium. The detection of nuclear binding sites for EGF-URO [164] and the observation of the binding of insulin to nuclear sites that appear to be distinct from the membrane-localized insulin-binding sites [63–65] while controversial [10] make it difficult to rule out a potential role for internalized hormone fragments.

The above arguments are not to be construed as opposing the main tenet that has developed, implicating membrane-localized reactions in the action of perhaps the majority of neurotransmitters and polypeptide hormones. Indeed one might predict that all agents bringing about rapid (i.e., seconds to minutes) cellular events (e.g., membrane depolarization, stimulation of glucose transport, activation of adenylate cyclase, and modulation of cyclic AMP dependent processes) would act solely at the plasma membrane. In this context, the process of ligand internalization and lysosomal degradation may prove to be an important aspect of receptor regulation per se that is linked to but separate from the 'action function' of the receptor as envisioned by the mobile receptor paradigm (see below). The observations that anti-insulin receptor antibodies exhibit insulinlike actions in isolated adipocytes [55, 103, 109] argue convincingly, if not unequivocally, in favor of a membrane-localized site of action for insulin. On the other hand, it is important not to rule out the possibility that ligand and/or receptor internalization may play a role for agents causing a relatively slow (hours to days) cellular response (e.g., nerve cell differentiation in the case of nerve growth factor). In this regard, insulin, which can cause both rapid (glucose transport, antilypolysis) and delayed (fibroblast growth) cellular effects could potentially act both at the cell surface and via an internalization process, depending on the cell type affected.

To date, the best data related to polypeptide action and ligand internalization come from studies with mouse EGF-URO, an agent that can also cause both rapid (inhibition of gastric acid secretion, stimulation of amino acid transport) and delayed (cell growth) cellular effects. Highly fluorescent analogues of both EGF-URO and insulin can be observed to bind initially to highly mobile receptor sites in viable fibroblasts [72, 167, 170]; subsequently, the receptors can be observed to cluster rapidly in small discrete extracellular patches which within 10 min at 37°C become internalized. Other experiments using antibodies to monitor the disappearance of cell surface EGF-URO [20], using ^{125}I-labeled EGF-URO to monitor the 'down-regulation' of the receptor [2] and using photoaffinity-labeled EGF-URO receptor to follow receptor processing [43], document further the internalization of the EGF-URO-receptor complex.

The relationship of EGF-URO receptor internalization to cellular stimulation is as yet uncertain. Strikingly, the addition of anti-EGF-URO antibody to cells stimulated by EGF-URO can reverse the mitogenic response as long as 6–8 h after the initial stimulus [20, 166]. At such time, a large proportion of receptors would have become internalized and degraded so as to liberate fragments both of the receptor and the polypeptide in the cytoplasm. Further, it can be demonstated that a very brief exposure to EGF-URO (30 min) followed by removal of EGF-URO from the medium is sufficient to trigger a mitogenic response [166]; however, the addition of anti-EGF-URO antibody to these primed cells, as long as 6 h after the brief exposure, reverses the mitogenic effect. Taken together, the above data make it difficult to attribute the mitogenic process to internalized receptor or hormone fragments. Clearly, further work will be necessary to clarify the role of receptor internalization (and shedding) in the stimulation of cells by hormones like EGF-URO and insulin.

Cooperativity of Receptor Binding and Hormone Action

It is often the case that an analysis of ligand-binding data – e.g., by the method of Scatchard [165] – suggests either the presence of ligand-receptor cooperativity or the presence of more than one ligand-binding site. While there are a number of possible factors that can result in nonlinear Scatchard plots of the data [discussed at some length in references 156 and 157], recent discussions in the literature have favored a negative cooperativity model for the interaction of a number of hormones with specific receptors. The interpretations rest on two principal kinds of data: (1) equilibrium-binding data yielding Scatchard plots that are concave up and (2) a kinetic analysis of ligand receptor dissociation kinetics done in either the absence or the presence of an excess of unlabeled ligand.

It should be noted at the outset that the interpretation of both kinds of data is fraught with difficulty. First, even if the equilibrium-binding data are interpreted in terms of multiple binding sites, it is often very difficult to establish the ligand specificity of each binding site according to the criteria outlined above and discussed elsewhere [94]. Furthermore, the occurrence of anomalous dissociation kinetics in nonreceptor preparations, such as talc [39], indicates that the unequivocal analysis of similar data in biological systems may prove difficult. As orginally demonstrated, the talc-binding data for insulin can be seen to be of a nonreceptor character, distinct from data obtained in membrane preparations.

The most extensively documented data supporting a negative cooperativity model have come from the work with insulin [47–49], based primarily on an assay measuring the accelerated dissociation rate of ^{125}I-insulin from cultured IM-9 lymphocytes and from mouse liver membranes, caused by unlabeled insulin and by a variety of insulin analogues. It has been concluded that the insulin molecule possesses a receptor-binding region as well as a distinct region responsible for causing an increased insulin receptor dissociation rate. Despite these detailed studies with insulin analogues, which are consistent with an agonist-mediated acceleration of the dissociation of previously bound hormone, an alternative careful kinetic analysis of insulin binding to either unsaturated or partially saturated receptor preparations reveals no difference in the intrinsic receptor affinity constant for insulin [150]. It is evident that different methods of kinetic analysis yield apparently conflicting results. The kinetic experiments of both *DeMeyts* et al. [48, 49] and *Pollet* et al. [150] have been recently reevaluated by *DeLean and Rodbard* [46]. It should be apparent from the above discussion that, in the absence of evidence at least as extensive as that obtained by *DeMeyts* et al. [48, 49] for insulin, the simple demonstration that a ligand enhances its own dissociation rate is insufficient to confirm or disprove receptor cooperativity.

A consideration of insulin action, in terms of the mobile or floating receptor hypotheses – to be discussed below [38, 45, 100] –, predicts that insulin binding by a single receptor macromolecule could readily exhibit multiple affinities, as well as negative cooperativity [100]. Recent work suggests that whereas the soluble insulin receptor isolated by affinity chromatography does not exhibit negative cooperativity [132], other membrane-localized constituents can interact with the insulin receptor so as to increase the receptor's apparent molecular size and confer upon the receptor the complicated equilibrium-binding kinetics observed either in membrane preparations or in crude soluble receptor preparations.

Thus, while the receptor binding of insulin may indeed exhibit cooperativity, as suggested by *DeMeyts* et al. [48], a model akin to the mobile receptor paradigm wherein receptor-effector interactions lead to alterations in ligand affinity is suggested, rather than a model comprising site-site interactions between receptors, as was originally proposed. Future work should provide data to distinguish between the models not only for insulin, but also for other polypeptide hormones for which cooperative receptor interactions are suspected. It should be noted that the role of receptor cooperativity in terms of biological response is readily accommodated by the mobile receptor model.

Membrane Fluidity and the Mobile Receptor Paradigm of Hormone Action

Development of the Model

As reviewed elsewhere [130, 175, 176], there has been considerable progress in understanding the organization of the cell membrane. It has been realized for some time now, based on fluorescence and electron microscopic observations, that certain membrane proteins are free to diffuse in the plane of the plasma membrane. It was a logical extension of such observations to propose that receptors for hormones would also be mobile constituents that could interact with other elements in the plane of the membrane. The 'mobile' or 'floating' receptor model, developed separately by *Bennett* et al. [9], *Cuatrecasas* [38], *Cuatrecasas and Hollenberg* [40], *Jacobs and Cuatrecasas* [100, 102], *DeHaen* [45], and by *Boeynaems and Dumont* [17], proposes that a hormone-receptor complex may interact with a number of 'effector' substituents in the plane of the membrane, as depicted in figure 5. In large part, the genesis for the theory centers around the observation that a number of receptor-specific agonists (catecholamines, prostaglandins, glucagon, ACTH, etc.) can independently stimulate adenylate cyclase in a cell such as the adipocyte, in a manner indicative of a unique adenylate cyclase enzyme responding in a complex way to various hormonal stimuli. Rather than supposing that all of the receptors are clustered about the cyclase-enzyme complex or that each receptor is physically associated with its own cyclase, it is proposed that each independent hormone-receptor moiety can freely compete for the effector (adenylate cyclase) in the plane of the membrane. The model, as depicted in figure 5, does not restrict the number of effectors with which the hormone-receptor complex may interact (e.g., an ion transport channel may be perturbed as well as a membrane enzyme complex), and the model is generally applicable to the modulation of any membrane process by a variety of hormones. Additionally, the model does not preclude the self-

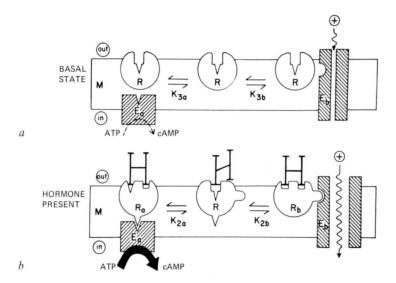

Fig. 5. The mobile receptor model of hormone action. The ability of the hormone-receptor complex to interact in the plane of the membrane is shown for two effector macro-molecules: E_a = Adenylate cyclase, converting adenosine 5′-triphosphate to adenosine 3′,5′-cyclic monophosphate (cAMP); E_b = a transport system for a cation (depicted as a closed or open membrane pore). The equilibria are portrayed schematically in terms of the equilibrium constants that are expressed both in the absence *(a)* and presence *(b)* of a hormone agonist (H). The differences in hormone affinities between the uncomplexed and complexed states (receptor conformations R_a and R_b) are indicated by different 'lock and key' configurations; the possible (weak) association between receptor and effectors in the absence of hormone is also shown. Only the ternary complex (HRE) is fully active, either converting ATP to cyclic AMP at an accelerated rate (HR_aE_a) or opening an ion channel (HR_bE_b). A receptor site for the binding of an inhibitor to form a nonassociating complex (IR, not shown) is also depicted. 'out' = Outside of cell; 'in' = inside of cell; M = cell membrane. See text for further explanation.

association of one hormone-receptor complex with another to form a receptor cluster.

Predictions Based on the Mathematical Analysis of the Model

In a simplified version, the equilibrium involved in the mobile receptor model can be expressed:

$$H + R \rightleftarrows HR \quad K_1 \tag{2}$$

$$HR + E \rightleftarrows HRE \quad K_2 \tag{3}$$

where H, R, and E represent hormone, receptor, and effector (e.g., membrane-bound adenyl cyclase), respectively. It is evident that the equilibria could readily be made more complex, so as to account for cooperative phenomena, by varying the stoichiometry of the reacting species. For example, either the receptor or effector may well represent oligomeric macromolecular species. Indeed, should receptors exist as clusters within the plane of the membrane, negative cooperativity between receptor molecules alone could account for 'spare' receptors as discussed by *Levitzki* [124], whereby low-ligand occupancy would lead to a large configural change of the receptor cluster. In equations 1 and 2 the values K_1 and K_2 represent the microscopic equilibrium association constants for the reactions with overall forward and reverse rate constants, k_1, k_{-1}, k_2, and k_{-2}. It is a fundamental hypothesis of the mobile receptor model that the affinity of the hormone-receptor complex for the effector, as expressed by the equilibrium constant, K_2, is greater than the affinity of the uncomplexed receptor for the effector, as given by the following equation:

$$R + E \rightleftarrows RE \ K_3 \tag{4}$$

and expressed by the equilibrium constant K_3. It is thus proposed that for a hormone inhibitor, I, it would be expected that the ternary complex (IRE; depicted as HRE in figure 5) would be biologically inactive and that for the inhibitor, K_2, would be equal to or less than K_3. A fourth equilibrium that can occur is the dissociation of the hormone-receptor-effector complex according to the equation:

$$H + RE \rightleftarrows HRE \ K_4. \tag{5}$$

The above simple equilibria can be rendered more complex, for instance, by supposing that at least two hormone-receptor complexes must cluster before an effector molecule can interact, for example

$$2H + 2R \rightleftarrows (HR)_2 \ K_5 \tag{6}$$

$$(HR)_2 + E \rightleftarrows (HR)_2E \ K_6. \tag{7}$$

A more generalized scheme than the one outlined above has been described by *DeHaen* [45] for the action of hormones that stimulate adenylate cyclase.

It can readily be demonstrated from the above equilibria (equations 2–5) that:

$$K_3 \cdot K_4 = K_1 \cdot K_2. \tag{8}$$

Given the above hypothesis, that $K_2 > K_3$ (i.e., the affinity of the receptor for the effector is greater in the presence of hormone), it is evident that $K_4 > K_1$. Thus, the binding of a hormone by a homogeneous population of receptor molecules can, as suggested by the mobile receptor model, lead to complicated binding kinetics; the complicated binding kinetics predicted by the mobile receptor model (for example, nonlinear Scatchard plots) might otherwise (incorrectly) be interpreted in terms of a heterogeneity of receptor molecules.

The mobile receptor model introduces enormous flexibility into the possible actions of hormones and permits a complexity of kinetics that potentially may account for many observed phenomena in connection with hormone effects. It would, as suggested above and depicted in figure 5, rationalize the differential modulation of independent membrane processes by a single hormone-receptor complex, if the complex exhibited different affinities toward two or more membrane-localized effectors. Furthermore, as developed by *DeHaen* [45], the differential maximal activation of an effector (e.g., adenyl cyclase) common to several hormones could be explained in terms of distinct affinities of the various hormone receptors for the common enzymatic unit. Additionally, the complex α- and β-adrenergic effects in biological systems may be explained in terms of the interactions of multiple receptors with multiple effectors.

An important aspect of the mobile receptor model concerns both the levels of effectors and receptors present in a given cell, and the receptor/effector ratio, which may well vary under different physiological conditions. For example, denervation supersensitivity might be rationalized in terms of variations in the receptor-effector ratio. Alternatively, the ratio of adenyl cyclase effector molecules relative to the total number of receptors for distinct hormone activators would determine whether 'additivity' of enzymatic activation might be observed for separate hormones.

The detailed mathematical analyses of the mobile receptor model by *Jacobs and Cuatrecasas* [100] and by *DeHaen* [45] indicate that the binding of a homogeneous ligand with a unique receptor molecule can exhibit nonlinear Scatchard plots, Hill plots consistent with 'negative' cooperativity, and increased ligand off-rates determined with radiolabeled derivatives in the presence of high concentrations of hormone. By supposing a situation in which there is an excess of receptors (e.g., 10- to 20-fold) compared with effectors, the model can also predict the effect on the ligand-binding data of 'spare' but equivalent receptors. In such a case, at least two 'affinity' states for the ligand would be detected by binding studies, only one of which 'sites' would appear to coincide with the ED_{50} for the biological dose-response curve. All of the receptors would, nonetheless, be equivalent and would contribute to the overall responsiveness of the system.

Several studies have now confirmed that receptors for a number of active agents are mobile in the plane of the membrane. Fluorescent cholera toxin, which binds to membrane ganglioside, GM_1 and activates adenylate cyclase, can be observed to patch and cap in lymphocytes [35]. As mentioned above, studies with fluorescent derivative of insulin and epidermal growth factor have confirmed that the receptors for these two polypeptides are also mobile [167, 170]. Using image intensification photomicrography, the fluorescent hormones can be observed to bind initially in a diffuse pattern on the cell surface; subsequently the fluorescence can be observed first to aggregate in discrete patches and then to become internalized. The formation of receptor aggregates appears to depend on the presence of hormone. Strikingly, the internalization of both hormone-receptor complexes and ligand 'acceptor' complexes appear to proceed via the same cell surface structures, the so-called 'coated pits' [66, 67, 134]. The hormone-receptor complexes are remarkably mobile, as estimated by photobleaching recovery techniques, such that the microscopic aggregation of hormone-receptor oligomers could occur within milliseconds [169].

Receptor Aggregation and Hormone Action

Given that the macroscopic aggregation of receptors can be visualized by fluorescence photomicrography, it is reasonable to anticipate that receptor microaggregation at the molecular level may play a role in cellular activation. Studies with antibodies directed against the receptor for insulin appear to implicate receptor microaggregation in insulin action [110]. Remarkably, antibodies directed against the insulin receptor can be shown to possess insulinlike activity both in adipocytes and in muscle [103, 109, 110, 123]. However, monovalent antibodies, prepared by enzymatic digestion of the intact molecules, are not only devoid of insulinlike activity in adipocytes, but behave as competitive inhibitors of insulin binding and insulin action in these cells; furthermore, the monovalent human antibodies, once bound, can be rendered active in the presence of antihuman IgG which would presumably cause crosslinking and aggregation of the monovalent antibody-receptor complex [110]. Thus, for insulin and possibly for many other hormones, receptor clustering may prove to be a prerequisite for hormone action. In terms of the mobile receptor paradigm, the simplest initial situation may be the one outlined by equation 6, where two hormone-receptor complexes dimerize before interacting with an effector.

The Mobile Receptor Model and New Directions for Receptor Research

The above discussion should serve to indicate that the mobile receptor model is sufficiently versatile to accommodate many of the phenomena observed in connection with hormone action. To date, the model has proved of considerable value in stimulating new experiments. The predicted mobility of receptors can now be visualized experimentally [167, 170]. There is also now evidence that a receptor such as the one for insulin can interact with other nonrecognition macromolecules present in the plasma membrane [132, 133]. It will thus be of interest in future work to determine the factors that control the mobility and clustering of hormone-receptor complexes and to characterize those membrane-localized macromolecules (possibly effectors) with which receptors can interact. It is to these ends that more recent studies are now being directed.

Receptor Regulation

The previous discussion has indicated that hormone receptors can no longer be thought of as static elements in the cell membrane. Upon combination with a hormone, receptors can cluster, aggregate in supramolecular patches at an internalization site, and can then undergo endocytosis with subsequent proteolytic receptor processing. Added to this dynamic state of receptor mobility caused by specific hormone-receptor interactions, there are a number of other factors that must be considered with respect to the control of receptor number at the cell surface.

Receptor Synthesis and Turnover

The rates of synthesis and turnover of hormone receptors can be measured by a combination of ligand-binding methods, with the heavy isotope technique originally developed by *Devreotes and Fambrough* [50, 51], *Devreotes* et al. [52], and *Hartzell and Fambrough* [78] for the study of nicotinic acetylcholine receptors in cultured chick skeletal muscle cells. More recently, this method has been used to study the insulin receptor in 3T3-L1 preadipocyte cells [154, 155]. The technique depends on the synthesis of new, 'heavy' receptor produced in the presence of ^{2}H-, ^{13}C-, and ^{15}N-labeled 'heavy' amino acids and on the subsequent separation of heavy and light receptors by density gradient ultracentrifugation.

In cultured chick skeletal cells it was observed that the newly synthesized, 'heavy' nicotinic receptors appeared in the membrane within 3–3.5 h after exposure of the cells to heavy amino acids [50, 51, 52, 78]. The precursor-product relationship between the intracellular pool of receptors and surface receptors was found to be somewhere between a strictly linear assembly line process and a random selection of intracellular receptors to be inserted into the cell surface.

Similar experiments using the heavy isotope technique in 3T3-L1-differentiating preadipocytes [154, 155] have shown that 'heavy' insulin receptors appear in the plasma membrane of differentiated 3T3-L1 cells within 2–3 h of exposure to the heavy isotope-labeled amino acids. It was further noted that within 24 h the heavy receptor replaces the light receptor completely in the plasma membrane. The half-life of the insulin receptor in differentiated 3T3-L1 cells was estimated to be 6.7 h [154]. The differentiation process in these cells involves a 10- to 20-fold increase in the numbers of insulin receptors present on the cell surface. Based on their data with the heavy receptor technique, *Reed* et al. [155] concluded that the increase in receptor was due to an increase in the synthesis of the receptor, rather than decreased degradation of receptors or the unmasking of cryptic membrane receptors. The heavy receptor technique could well be applied to any receptor type. Very likely the dynamics of the insulin and acetylcholine receptors are representative of the dynamics of other macromolecular glycoprotein receptors.

In addition to changes in the rates of receptor synthesis or turnover, one must consider the de novo synthesis of a specific receptor, as has been reported for insulin receptors in lymphocytes, activated by a number of natural and artificial stimuli [81, 82, 92, 117]. Neither cell surface nor cytoplasmic binding sites for insulin can be detected in T or B cell lymphocytes before stimulation. After activation of the cells either with plant lectins or by immunologic stimuli, binding sites for insulin can be detected. The appearance of insulin-binding sites is independent of DNA synthesis and cell division, but can be inhibited by inhibitors of RNA synthesis. The evidence suggests that the receptors appear de novo, dependent on the state of differentiation of the lymphocyte induced by the mitogenic stimulus.

Developmental Aspects of Receptor Regulation

Regulation of hormone action can be accomplished by developing systems by regulation of the appearance of receptors as well as by regulating the receptor-effector coupling mechanisms. Receptors for EGF-URO have been observed to increase with age in mouse embryos, with pronounced changes evident in target tissues such as maxilla [139]. Similarly chick dorsal root ganglia have been observed to be highly responsive to β-nerve growth factor (β-NGF), as indicated by the neurite outgrowth bioassay and by binding studies with ^{125}I-βNGF during early development (8–14 days). However, after 18–21 days both the response to β-NGF and ^{125}I-β-NGF binding are reduced markedly [84].

Hormone binding and hormone responsiveness need not parallel each other. For example, the maturation of rat reticulocytes into erythrocytes results in a marked reduction of response (activation of adenylate cyclase) to the adrenergic agonist isoproterenol. However, the reduction in response is not paralleled by a concomitant decrease in the number of β-adrenergic receptors on the plasma membrane [116]. In this case, the reduction of response is attributed to a change in the coupling between receptor occupation and adenylate cyclase activation [8, 13, 14].

Growth Control and Receptor Regulation

Cells grown in culture exhibit a variety of changes in receptor number depending on the stage of the cell cycle. For example, BSC-1 cells at low-density culture possess eight times as many receptors for EGF-URO as do cells at confluency [97]. Conversely, BALB/3T3 fibroblasts exhibit greater insulin binding either at confluency or under reduced serum concentration (i.e., cells that have stopped growing) compared with cells that are rapidly growing [184].

A final example of receptor regulation during cell growth has been observed in melanoma cell cultures [188]. In this case, receptors for melanocyte-stimulating hormone have been detected only in the G_2 phase of the cell cycle of a melanoma cell line. The biological response to melanocyte-stimulating hormone parallels the appearance and disappearance of melanocyte-stimulating hormone receptors.

Hormonal Regulation of Receptors

The ability of hormones to regulate their own receptors (homospecific regulation) has already been mentioned. Early studies by *Gavin* et al. [59] indicated that the number of insulin-binding sites in cultured IM-9 lymphocytes could be reduced by preincubating the cells with insulin, albeit at concentrations markedly above those that could be considered physiologic. The original data with insulin in lymphocytes were not entirely confirmed in studies with fibroblasts [98]. However, data obtained with short-term adipocyte cultures indicate that elevated insulin concentrations within the physiologic range can indeed cause a diminution in receptor binding and a predictable shift in the dose-response curve to the right [128]. Similar observations have now been made in a number of systems containing receptors for EGF-URO and for β-adrenergic agonists [20, 122]. The best data demonstrating the hormone-specific nature of this homospecific down-regulatory process have come from these latter two studies. As with any receptor-related process, it is essential to demonstrate an appropriate ligand specificity (e.g., potency of isoproterenol > epinephrine > norepinephrine, for the β-adrenergic system) in order to characterize completely the 'homospecific' nature of the down-regulation phenomenon.

Homospecific receptor regulation can be in the positive as well as negative direction. For instance, prolactin administration results in the appearance of its own receptors in the liver of hypophysectomized rats [151, 152]. In addition, angiotensin II administration in rats has been observed to cause initially (within 24 h) an increase both in receptor number and affinity and subsequently (36 h) an increase only in adrenal receptor number [79]. In parallel with the increase in receptor number, an increase in adrenal cell responsiveness was oberved [1]. In the studies with prolactin and angiotensin, it remains to be determined if the effects are due to a direct action of each hormone on the receptor-bearing cells or via an indirect process, possibly mediated by a second hormone signal generated in vivo. Presumably, studies with cultured cells will be able to resolve this issue.

The regulation of the numbers of receptors for one hormone by a second hormone (so-called heterospecific receptor regulation) has been observed in a number of instances for both steroid and polypeptide hormones. A particularly intriguing example of this kind of regulation relates to the action of the gonadotropins. In cultured rat ovarian fragments [140], follicle-stimulating hormone causes the appearance of binding sites for ^{125}I-labeled human chorionic gonadotropin (these sites are presumably intended for luteinizing hormone, LH). In contrast, the hypothalamic modulatory factor that stimulates LH release (LHRH) causes a reduction in gonadotropin receptors, seemingly via a direct effect on the Leydig cells [118, 121]. This intriguing example of heterospecific receptor regulation merits further investigation to determine the mechanism involved.

The influence of steroid hormones on the numbers of receptors for polypeptide hormones has been examined in a number of instances. For example, estrogens can cause an increase in the number of oxytocin receptors in rat uterus [177] and can augment receptors for prolactin and growth hormone in rat liver [152]. Similarly, corticosteroids increase the binding of EGF-URO to cultured fibroblasts. To date, this kind of hete-rospecific receptor regulation has not been sufficiently explored vis-à-vis the influence of polypeptide hormones on the numbers of receptors for steroid hormones; such studies appear more than warranted.

The above examples of heterospecific receptor regulation presumably involve mechanisms that relate to the rates of synthesis and turnover of receptor. Recent data have brought to light examples of heterospecific receptor regulation that very likely involve mechanisms located solely in the plasma membrane. For instance, the tumor promoter, 12-o-tetradeca-noyl-phorbol-13-acetate can interact with its own receptor in fibroblasts to cause a rapid (tens of minutes) time- and temperature-dependent selective decrease in the binding of EGF-URO [96]; the receptor for insulin is unaffected. Given the data available for the time course of receptor synthesis and turnover alluded to above (hours), one must conclude that a novel (membrane-localized) mechanism (induction of receptor internalization?) is probably involved in this process. Another example of hete-rospecific control that probably occurs at the level of the plasma membrane can be observed in the complex reciprocal relationship between the binding of γ-aminobutyric acid (GABA) and benzodiazepines in isolated membrane preparations. The binding data complement electrophysiological studies indicating that diazepam augments the GABA-mediated inhibitory effects on neurons. An increase in ^{3}H-GABA binding caused by diazepam and an increase in ^{3}H-methyldiazepam binding in the presence of GABA can be seen within minutes in isolated membrane preparations. The effect of diazepam on GABA binding is attributed to its ability to complex with a distinct nonreceptor membrane protein (GABA modulin; molecular weight about 15,000) that reduces the receptor's affinity for GABA. The mechanism whereby GABA alters diazepam binding is unknown [19, 69, 135].

Nonspecific Factors Affecting Receptors

In addition to being hormonally regulated, receptors have also been shown to be regulated by a wide variety of compounds and agents acting via nonreceptor mechanisms. For example, butyric acid can induce the appearance of β-adrenergic receptors in cultured HeLa cells [183] and cyclic AMP increases the numbers of insulin receptors in cultured fibroblasts and lymphocytes [184]. Murine or feline sarcoma virus-mediated transformation of cells causes a marked reduction of EGF-URO receptors [185] as does the chemical transformation of cultured hamster fibroblasts by chemicals [91] or by DNA tumor virus [11]. In the case of the RNA tumor virus mediated transformation, the reduction in EGF-URO binding can be attributed to the production of substances (so-called tumor growth factors) that are distinct from EGF-URO, but that are able to occupy the EGF-URO receptors and mimic the effects of EGF-URO in cells [178, 185]. In addition to changes in receptor numbers, changes in the specificity of receptors have also been noted. For example, simian virus transformation of mouse 3T3 cells causes a change in the catecholamine receptors from a $β_1$ specificity to a $β_2$ specificity [173].

It is evident from the above discussion that hormone receptors are regulated by a wide variety of factors related both to intracellular events such as turnover, cell cycle, cell differentiation, and to extracellular stimuli caused by hormones and other agents.

Effects of Receptor Regulation on Cell Response

The regulation of the numbers of cell receptors provides an intriguing mechanism for the control of cell responsiveness. Since the initial interaction of a hormone with a cell is governed by the equilibrium described in equation 1, it is possible to predict the consequences of receptor regulation, irrespective of the mechanisms that couple receptor occupation to cell response (fig. 6). The response is in some manner a function of the concentration of hormone receptor complexes, HR [i.e., response α (HR)].

It is evident from equation (1) that

$$HR = R_t/(1 + \frac{K}{H_f}),$$

where R_t is the total receptor concentration, K the equilibrium constant, and H_f the free concentration of hormone. Thus, if R_t decreases (e.g., downregulation), there must be an increase in the concentration of hormone (H_f) to achieve the same concentration of occupied receptor com-

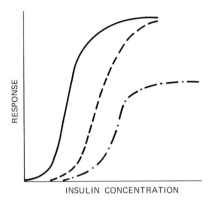

RESPONSE

INSULIN CONCENTRATION

Fig. 6. Schematic dose-response curves for insulin. The response curves depict insulin acting on normal cells or on target cells in which receptor numbers and/or postreceptor responses have diminished, as discussed in the text. — = normal; ––– = reduced receptors or endogenous competitor; –·– = receptor + postreceptor defect.

plexes (HR) and thereby the same cell response; thus, the dose-response curve will be shifted to the right (fig. 6). Since in most systems so far studied, there are more receptors present on cells than are required for a maximum cell response (i.e., 'spare' receptors are present) even in a 'down-regulated' system, a sufficiently high concentration of hormone will still be able to occupy the number of receptors that must be activated for a maximum cell response. Thus, in many receptor systems, a decrease in receptors will result in a shift of the dose-response curve to the right, without a change in the maximum response (middle curve, fig. 6). Similarly, an increase in receptor number would result in a leftward shift in the dose-response curve. In the situation where postreceptor mechanisms reduce cell responsiveness, the maximum response will not be realized even if all receptors be occupied; thus, the combined effect of a reduction of receptors with a postreceptor defect would be to shift the dose-response curve to the right, with a simultaneous reduction in the maximum achievable response (curve on the far right, fig. 6). In practice, dose-response curves that have been measured for the action of insulin in obese and diabetic subjects [115, 116] have been observed to be consistent with both of the curves shown on the right in figure 6.

It is evident that the situation with receptors differs somewhat from the situation with the action of enzymes on substrate molecules. In contrast to the situation with enzymes, where the net velocity is proportional the *fraction* of enzyme molecules activated, in the cellular receptor system it is the total *number* rather than the fraction of available receptor entities occupied that governs net cellular responsiveness. Thus, in enzyme systems, reducing the enzyme concentration does not alter the concentration of substrate at which reaction velocity is half maximal, whereas in receptor systems, a reduction in receptor number causes a rightward shift in the dose-response curve.

Receptor Regulation and Tachyphylaxis

A long-recognized pharmacological phenomenon concerns the diminution in response to a system upon repeated exposure to an agonist: so-called tachyphylaxis or desensitization. In many cases the phenomenon is specific, in that the response (for example, muscle contraction) to one agent is markedly diminished, whereas the same response to a second agent is unaffected. The receptor-related mechanisms for tachyphylaxis are as yet poorly understood. Nonetheless, the downregulation of receptors observed by ligand-binding studies provides one possible mechanism. In the limited number of studies that have been done to date using ligand-binding methods, it appears that the reduction in ligand binding and the corresponding tachyphylaxis appear to be caused by agonists, but not antagonists. It is as yet unclear whether receptors are selectively lost from responsive cells, for example, by shedding or by internalization, or whether the receptor remains inaccessible in the membrane as a high-affinity hormone receptor complex. It should be mentioned, however, that receptor downregulation can occur in some systems (e.g., β-adrenergic receptor of turkey erythrocytes) in which desensitization is not observed [75]. In the case of the nicotinic receptor for acetylcholine, it is believed that the 'high-affinity' form of the receptor that can be detected in detergent extracts of electroplax membranes represents the desensitized form of the receptor [189]. In view of this possibility, further equilibria might be considered in terms of the mobile receptor paradigm discussed above:

$$HRE \rightleftharpoons HR' + E \tag{9}$$

$$HR^1 \rightleftharpoons H + R^1 \rightleftharpoons R \tag{10}$$

$$HR' \rightleftharpoons HR, \tag{11}$$

where HR′ represents an altered high-affinity state of the hormone-receptor complex that might be produced only consequent to the formation of the agonist (but not antagonist) complex, HRE. In a sense, the agonist, in terms of the rate theory of drug action [20], becomes its own ideal antagonist, with a markedly reduced off rate. A conformational change of the kind outlined by equations 9–11 has often been suggested as a mechanism for desensitization, as discussed by *Colquhoun and Rang* [34]. The fundamental difference between the interactions of agonists and antagonists with the β-adrenergic receptor has been discussed in terms of the equilibria shown in equations 9–11 [190].

A provocative example of receptor desensitization comes from work with angiotensin analogues that are full agonists in causing ileal contraction, but, unlike native angiotensin II, do not lead to desensitization of the preparation [148]. Thus, the conformation of the hormone-receptor complex that leads to cellular activation may differ from the conformation that leads to receptor-specific densensitization (or downregulation). It is evident that the agonist property of compounds may be a necessary but not sufficient condition for the production of tachyphylaxis. It will thus be of interest to ascertain the role of receptor mobility in connection with the phenomenon of desensitization.

Hormonal Modulation of Adenylate Cyclase

The coupling mechanism, whereby receptor occupation leads to cellular activation, has been studied most thoroughly for those hormones that act via cyclic adenosine monophosphate, subsequent to the stimulation of membrane-localized adenylate cyclase. Indeed, this aspect of study is in such an active state that the following can be considered only as an interim progress report.

Distinction of the Ligand Recognition Site from the
Enzymatic Catalytic Site

Although the receptors for those agents activating the cyclase appear to be intimately linked to the enzymatic activity [107, 136, 158, 182], evidence from a number of sources now convincingly demonstrates that the ligand recognition site and the catalytic site converting ATP to cyclic AMP reside in distinct molecular species. Part of the evidence is inferrential, based on a consideration of the number of independently acting agents that can activate adenylate cyclase in a single cell such as the adipocyte [38, 45]. Other more direct evidence comes from: (1) cell fusion experiments [144, 171, 172] wherein a receptor-deficient cell possessing an active adenylate cyclase can be rendered responsive to a hormone subsequent to fusion with a receptor-containing cell in which the cyclase has been inactivated by N-ethylmaleimide; (2) studies of cellular differentiation in which there are nonparallel changes in the number of β-adrenergic receptors, fluoride-sensitive cyclase, and catecholamine-sensitive cyclase – this nonparallelism can be observed during rat erythrocyte maturation [8, 13, 14, 26, 111], cultured HeLa cells subsequent to the induction of β-receptors by butyrate [83], and, most instructively, during the differentiation of 3T3-L1 proadipocytes to mature adipocytes [119]; (3) studies of membrane fractionation subsequent to cellular disruption in which the β-receptor and adenylate cyclase can be observed to partition with different distributions in sucrose density gradients [162]; (4) solubilization and chromatographic separation of the adenylate cyclase and the β-adrenergic receptor [70, 125], and (5) genetic evidence [18, 71, 99] indicating that the β-adrenergic receptor and adenylate cyclase are products of separate genes.

A very exciting aspect of the genetic studies with S-49 lymphoma cell lines was the discovery of a mutant possessing both a β-adrenergic recognition site and a functional adenylate cyclase, but in which cell receptor occupation does not lead to cyclase activation [71]. The use of the S49 mutant cells, along with complementation assays, has permitted the isolation of the G/F or N-subunit of the cyclase system, a dimer of two distinct proteins, that is responsible for nucleotide binding and for the guanosine triphosphate mediated modulation of hormone-stimulated activation of adenylate cyclase. There is the possibility that the catalytic activity may be modulated by more than one protein factor [76, 80, 141, 163, 181]. The new advances in the assay and detection of regulatory proteins in this complicated system may reveal positive as well as negative regulatory components that may be reconstituted into an active complex [22, 80, 108, 149, 159, 163]. At present the receptor-cyclase complex is thought to comprise at least two and possibly three [119] regulatory proteins forming the G/F or N subunit in addition to the catalytic and ligand recognition moieties. Clearly, the picture concerning the hormonal modulation of adenylate cyclase activity and the nonhormonal modulation of this enzyme, which is central to the action of a variety of neurotransmitters, hormones, and toxins, is changing and expanding at a rapid pace. Many new developments in this area are anticipated in the near future.

On the Question of Receptor-Adenylate Cyclase Coupling

From the above discussion, it is now evident that hormone receptors and adenylate cyclase are separate macromolecules whose function and physical state can be regulated and monitored independently, as predicted by the mobile receptor hypothesis. Further, the recent reconstitution data [76] indicate that the G/F or N subunit plays a critical role in governing the qualitative characteristics of the cyclase response (e.g., stimulation by nonhydrolyzable analogues of guanosine triphosphate in the presence or absence of β-agonist). However, it must be acknowledged that there is presently no evidence that the receptors and cyclase-regulatory complex are physically associated, even in the presence of hormones. It is possible that, as suggested by *Tolkovsky and Levitzki* [186], the intermediate, HRE, may have a very short half-life and may never accumulate as such. Nonetheless, now that these molecules have been proved to exist apart, positive proof must be provided in order to conclude that functional coupling (i.e., modification of enzyme activity) is related to direct physical coupling; the latter has been nearly universally assumed to occur, even in the 'collision coupling' version [186] or the mobile receptor hypothesis. It is perhaps ironic that, whereas early studies of the hormone-cyclase system have led to the 'mobile receptor model', more recent data suggest that alternative models may be necessary.

In the absence of evidence for direct receptor-effector (cyclase) association, the possibility of indirect coupling must be seriously entertained. In fact, it has been suggested, on the basis that adrenergic receptors and cyclase can be found in different membrane vesicle populations derived from erythrocyte ghosts (even when prepared in the presence of an agonist, isoproterenol), that the hormonal regulation of this enzyme could in principle be mediated indirectly [162]. In such a case, cyclic AMP would very possibly not be literally the true 'second' messenger of hormone action. Also, although not interpreted in this manner by the authors, the kinetics of coupling of hormonal reconstitution in cell fusion experiments [172] may be too rapid to be explained simply on the basis of protein intermixing to allow direct receptor-cyclase associations; the rate of the process of reconstitution appears to be almost coincident with that of the rate of fusion. Possible mechanisms by which hormone-receptor complexes could affect adenylate cyclase, other than by direct association, include intermediary, primary, hypothetical chemical substances (for example, resulting from membrane phospholipase protein kinase or protease activation), changes in ionic complexes, and/or electrochemical gradients across the membrane or through an interconnecting network of a submembranous protein mesh that may simultaneously alter distant proteins.

Receptor Structure

Except for the nicotinic receptor for acetylcholine, the amounts of receptor material available for study at present preclude the use of direct chemical analysis (amino acid analysis, sequence, etc.). Nonetheless, as indicated in previous sections of this chapter and also outlined schematically in figure 4, a great deal of information has been obtained with the use of the specific high-affinity ligands as receptor 'markers', along with photoaffinity and affinity-crosslinking labeling methods, enzymatic probe techniques, and lectin-probe methods. At the moment, structural data are available for the receptors for acetylcholine (nicotinic), insulin, and EGF-URO; data for other receptors (β-adrenergic, nerve growth factor) are rapidly becoming available. For illustrative purposes, this section will focus on the cholinergic, EGF-URO, and insulin receptors.

In general, receptors for polypeptides such as insulin or EGF-URO appear to be glycoproteins that are only partially embedded in the membrane lipid environment. It is interesting to note that important functional information resides both in the protein and nonprotein constituents of receptors. It is now evident, for instance, that the oligosaccharide portion of receptors may play a role both in the ligand recognition function (e.g., removal of sialic acid augments the binding of EGF-URO to its receptor) and in the signal-transduction process (e.g., neuraminidase abrogates insulin action in adipocytes without affecting insulin binding). A contribution of other nonprotein moieties (possibly tightly receptor associated via noncovalent mechanisms) to ligand recognition can be seen in the possible participation of gangliosides in the binding and action of agents such as thyroid-stimulating hormone and interferon [114].

As previously discussed, the insulin receptor contains two or more polypeptide chains that appear to be involved in ligand recognition [data summarized in several communications in ref. 4 and 41]. As outlined under 'Solubilization, Characterization and Isolation of Receptors', taken together, the data indicate an oligomeric structure for the insulin receptor recognition subunit with a two-chain structure [41] that probably exists in the membrane as a disulfide-linked multimer $(\alpha\beta)_2$. It remains to be seen if the two constituents in the 40,000-dalton range described by *Yip* et al. [191] form an integral part of the recognition molecule or if these substituents are noncovalently associated with the recognition subunit. Thus, the ligand recognition event may turn out to be a very complicated process, including not only the participation of a number of disulphide-linked polypeptide chains comprising the receptor recognition oligomer per se, but also including input from other 'non–recognition' or 'nonreceptor' glycoprotein moieties with which the recognition oligomer can interact.

Even more detailed information about receptor structure-function relationships can be anticipated from studies with the nicotinic-cholinergic receptor [112], for which substantial amounts of material have been obtained from Torpedo and electroplax species. In detergent solutions, the receptor behaves as a species with an apparent molecular weight of about 250,000; the entire oligomeric structure $(\alpha_2\beta\gamma\delta)$ comprises two alpha subunits (molecular weight about 40,000) and three other distinct, but chemically related [153] substituents with molecular weights of about 48,000 (β), 58,000 (γ), and 64,000 (δ) [112]. Although the exact function of each polypeptide chain has yet to be determined, it appears likely that the subunit is the recognition moiety and that the remainder of the oligomer may comprise the ion channel. As the detailed structures of the putative ligand recognition components and the ion channel species become available, it should be possible to improve our understanding of the complex interaction of agonists, partial agonists, antagonists, and ion channel specific agents that modulate the ion transport function of this complex oligomeric receptor.

In contrast to the receptors for insulin and acetylcholine, the receptor for EGF-URO appears to be a single-chain glycoprotein species, with a molecular weight in the 200,000-dalton range [44, 86]. The receptor-associated kinase site [32] is thought to reside on the inner aspect of the plasma membrane [126, 127] and appears to be particularly susceptible to endogenous cellular proteolytic activity. The membrane-associated receptor seems to be organized in proteolytic 'domains' that can be cleaved by a variety of enzymes [142]. In some cells (e.g., A431 epidermoid carcinoma) the receptor may be present as large aggregates, even in the absence of stimulating ligand. The schematic representation of the EGF-URO receptor along with the schemes for the receptors for insulin, glucagon, and acetylcholine are depicted in figure 4.

Models for Hormone-Mediated Cell Activation

From the information already presented, it is possible to construct tentative models for four distinct reaction mechanisms that can lead to receptor-modulated cell activation (fig. 4). As outlined above, the receptors (R) for those hormones like glucagon (G) acting by adenylate cyclase activation interact via a guanosine triphosphate binding regulatory subunit (G/F or N subunit) to modulate the activity of the catalytic (C) subunit. In figure 4 it is suggested that a receptor region (region 2), that may be similar in a variety of cyclase-related hormone receptors, interacts directly with the N subunit. Further, the receptor is illustrated as having an internalization-related region (region 1) in common with a variety of receptors (fig. 4b, d); interaction of region 1 with elements of the 'coated pit' area is presumed to be involved in the internalization process but not in the cell activation event. In figure 4a, the catalytic subunit of cyclase is shown to have a region (region 3) for interacting with a specific portion of the regulatory subunit; evidence suggests a conservation of this region between cell types and between species. In addition, a site on the catalytic subunit is shown (region 4) that might permit inhibitory regulation by a receptor like the one for insulin (fig. 4b) or by other cyclase regulatory proteins.

Very recent work suggests that insulin (I; fig. 4) may act, in part, via the liberation of a intracellular chemical mediators generated in concert with by proteolysis of a membrane protein [42]. In the model for insulin action (fig. 4b), essentially as proposed by *Czech* [42], the disulphide-stabilized oligomeric insulin receptor (α, β_2), as reviewed previously [42, 90], is shown to be bivalent; a receptor area related to receptor internalization (region 1) and a region (region 4) for the interaction with a membrane glycoprotein receptor affinity regulator are depicted. It can be hypothesized that glycoprotein receptor and membrane protein may account for the two distinct 40,000-dalton membrane proteins detected in the photolabeling experiments of *Yip* et al. [191]. Evidence for an affinity regulatory protein (glycoprotein receptor) has also been obtained by neutron bombardment experiments coupled with ligand-binding measurements [77]. As suggested above, an area like region 4 might also play a role in the modulation of the catalytic subunit of adenylate cyclase. It is suggested (fig. 4b) that receptor clustering (possibly involved in insulin-induced membrane potential changes) leads to the activation of a membrane protease that liberates a chemical mediator of insulin action (a phosphatase activator). The role and chemical nature of the mediator(s) released from membranes by insulin stimulation is presently under intensive study. In the case of the nicotinic receptor for acetylcholine (A; fig. 4c) cell stimulation is mediated by activation of an ion channel (fig. 4c). Interestingly, this cholinergic receptor is thought to be bivalent like the insulin receptor. It is believed that two receptor oligomers, linked in tandem by a disulphide bond, form a functional unit.

The EGF-URO receptor is shown as a single-chain transmembrane species. The receptor (fig. 4d) is depicted to have the common internalization site (region 1) alluded to above. In addition, it is suggested that the receptor possesses an intracellular catalytic site responsible for the phosphorylation of membane protein tyrosine residues. It is quite possible that the phosphorylation reaction represents a first step in EGF-URO action. Although the phosphorylation reaction can proceed in detergent solution, it is uncertain whether or not receptor clustering is a prerequisite for the EGF-URO-stimulated enzymatic activity.

Thus, four distinct reaction pathways (cyclase activation and cyclic AMP modulated kinase activity, activation of membrane enzymes to liberate chemical mediators, activation of ion transport, and stimulation of intrinsic receptor kinase activity, leading to membrane constituent phosphorylation) appear to lead to cell activation. It is important to reemphasize that, in keeping with the mobile receptor model [40, 89], any particular receptor oligomer could possibly modulate a number of such reaction pathways. For instance, the insulin receptor, in addition to activating a membrane protease, might also stimulate the production of cyclic inositol phosphate, a compound thought to be involved in the action of a large number of hormones [12]. Therefore, it may be inappropriate to attempt to identify a *single* reaction pathway as the *primary* event in the action of a particular hormone. Rather, the challenge in the future is to identify the matrix of concurrent receptor-modulated membrane-localized reactions that are involved in cell activation.

Receptors and Disease

It is most rewarding that advances in receptor-related research are leading to a better understanding of the pathophysiology of a variety of disease entities [42, 101]. In a number of instances, the pathology appears to reside not in the receptor per se, but rather in the presence of autoantibodies that can interact with the receptor either as a competitive antagonist or as an agonist. Diseases in which antireceptor antibodies play a role are: Graves' disease (receptor for thyroid-stimulating hormone), myasthenia gravis (nicotinic receptor for acetylcholine), the syndrome of acanthosis nigricans type B with insulin resistance (insulin receptor), and a population of patients exhibiting β-adrenergic hyporesponsiveness (β-receptor) [56]. It is of interest that this latter group of patients has been discovered as a result of a deliberate search for antibodies directed against the β-adrenergic receptor; evidence suggests that the circulating antibodies act as competitive inhibitors of isoproterenol action [56]. Thus, it would appear fruitful in an instance of a disease exhibiting either hormone resistance or the symptoms of hormone hyperresponsiveness to hunt for the presence of autoantibodies directed against the receptor for the hormone of interest.

Aside from antireceptor antibodies, defects in the receptor system itself can lead to cellular malfunction. For instance, in type II hypercholesterolemia, there is a genetically determined defect in the acceptor for LDL responsible for the cellular uptake of LDL-associated cholesterol [3]. At least two kinds of defects are known, one relating to an absolute deficiency of receptors and a second manifested by a defective internalization mechanism. In a variant of pseudohypoparathyroidism, it appears that the defect does not reside in the parathyroid hormone receptor itself, but in another component of the receptor-effector system. The lack of cellular response has been attributed to a defect in the subunit (guanine nucleotide binding, G/F or N subunit) discussed previously, responsible for coupling receptor occupation to the activation of adenylate cyclase [54]. The essence of the above discussion is, that as more details become available concerning the multiple membrane localized steps leading from receptor occupation to cell activation, the more likely it is that certain of the receptors system components will be found to play a role in the pathophysiology of certain diseases. It is thus pleasing to note that the in-depth study of receptor-related mechanisms is of particular interest and importance not only to the molecular biologist, but also the entire biomedical community at large.

Acknowledgements

A portion of the work described in this chapter, coming from the authors' laboratory, was supported by grants from the National Institutes of Health, USA, the Canadian Medical Research Council, the March of Dimes Birth Defects Foundation, and the Alberta Heritage

Foundation for Medical Research (to *K.A.V.*). *K.A.V.* is a recipient of a scholarship from the Medical Research Council of Canada.

References

1   Aguilera, C.; Hauger, R. L.; Catt, K. J.: Control of aldosterone secretion during sodium restriction: adrenal receptor regulation and increased adrenal sensitivity to angiotensin II. Proc. natn. Acad. Sci. USA *75:* 975–979 (1978).

2   Aharonov, A.; Pruss, R. M.; Herschman, H. R.: Epidermal growth factor. Relationship between receptor regulation and mitogenesis in 3T3 cells. J. biol. Chem. *253:* 3970–3977 (1978).

3   Anderson, R.G.W.; Goldstein, J. L.; Brown, M. S.: A mutation that impairs the ability of lipoprotein receptors to localize in coated pits on the cell surface of human fibroblasts. Nature, Lond. *270:* 695–699 (1977).

4   Andreani, D.; DePirro, R.; Lauro, R.; Olefsky, J. M.; Roth, J.: Current views of insulin receptors. Serano Symposium, vol. 41 (Academic Press, New York 1981).

5   Ariens, E. J.: Affinity and instrinsic activity in the theory of competitive inhibition. Archs int. Pharmacodyn. Thér. *99:* 32–49 (1954).

6   Ariens, E. J.; Simonis, A. M.: A molecular basis for drug action. J. Pharm. Pharmac. *16:* 137–157, 289–312 (1964).

7   Arunlakshana, O.; Schild, H. O.: Some quantitative uses of drug antagonists. Br. J. Pharmacol. *14:* 48–58 (1959).

8   Beckman, B. S.; Hollenberg, M. D.: Beta-adrenergic receptors and adenylate cyclase activity in rat reticulocytes and mature erythrocytes. Biochem. Pharmac. *28:* 239–248 (1979).

9   Bennett, G. V.; O'Keefe, E.; Cuatrecasas, P.: The mechanism of action of cholera toxin and the mobile theory of hormone-receptor-adenylate cyclase interactions. Proc. natn. Acad. Sci. USA *72:* 33–37 (1975).

10   Bergeron, J.J.M.; Posner, B. I.; Josefsberg, Z.; Sikstrom, R.: Intracellular polypeptide hormone receptors. The demonstration of specific binding sites for insulin and human growth hormone in Golgi fractions isolated from the liver of female rats. J. biol. Chem. *253:* 4058–4066 (1978).

11   Berhanu, P.; Hollenberg, M. D.: Epidermal growth factor-urogastrone receptor: selective alteration in simian virus 40 transformed mouse fibroblasts. Archs Biochem. Biophys. *203:* 134–144 (1980).

12   Berridge, M. J.: Phosphatidylinositol hydrolysis: a multifunctional transducing mechanism. Mol. cell. Endocrinol. *24:* 115–140 (1981).

13   Bilezikian, J. P.; Speigel, A. M.; Gammon, D. E.; Aurbach, G. D.: Identification and persistence of beta-adrenergic receptors during maturation of the rat reticulocyte. Molec. Pharmacol. *13:* 775–785 (1977).

14   Bilezikian, J. P.; Speigel, A. M.; Gammon, D. E.; Aurbach, G. D.: The role of guanyl nucleotides in the expresion of catecholamine-responsive adenylate cyclase during maturation of the rat reticulocyte. Molec. Pharmacol. *13:* 786–795 (1977).

15   Birdsall, N.J.M.; Hulme, E.C.: Biochemical studies on muscarinic acetylcholine receptors. J. Neurochem. *27:* 7–16 (1976).

16 Blecher, M.: Methods in receptor research; in Laskin, Last, Methods in molecular biology, vol. 9, parts I and II (Dekker, New York 1976).

17 Boeynaems, J. M.; Dumont, J. E.: The two-step model of ligand-receptor interaction. Mol. cell. Endocrinol. 7: 33–47 (1977).

18 Bourne, H. R.; Coffino, P.; Tomkins, G. M.: Selection of a variant lymphoma cell deficient in adenylate cyclase. Science 187: 750–751 (1975).

19 Braestrup, C.; Neilson, M.: Benzodiazapine receptors. Drug. Res. 30: 852–857 (1980).

20 Carpenter, G.; Cohen, S.: ^{125}I-labeled human epidermal growth factor. Binding, internalization and degradation in human fibroblasts. J. Cell Biol. 71: 159–171 (1976).

21 Carpenter, G.; Cohen, S.: Influence of lectins on the binding of ^{125}I-labeled EGF to human fibroblasts. Biochem. biophys. Res. Commun. 79: 545–552 (1977).

22 Cassel, D.; Zelinger, Z.: Mechanism of adenylate cyclase activation on cholera toxin: inhibition of GTP hydrolysis at the regulatory site. Proc. natn. Acad. Sci. USA 74: 3307–3311 (1977).

23 Chan, L.; O'Malley, B. W.: Mechanism of action of the sex steroid hormones. New Engl. J. Med. 294: 1322–1328 (1976).

24 Chan, L.; O'Malley, B. W.: Mechanism of action of the sex steroid hormones. New Engl. J. Med. 294: 1372–1381 (1976).

25 Changeux, J.-P.; Blumenthal, R.; Kasai, M.; Podleski, T.: Conformational transitions in the course of membrane excitation; in Porter, O'Connor, Molecular properties of drug receptors, pp. 197–214 (Churchill, London 1970).

26 Charness, M. G.; Bylund, D. B.; Beckman, B. S.; Hollenberg, M. D.; Snyder, S. H.: Independent variation of β-adrenergic receptor binding and catecholamine-stimulated adenylate cyclase activity in rat erythrocytes. Life Sci. 19: 243–250 (1976).

27 Clark, A. J.: The reaction between acetylcholine and muscle cells. J. Physiol. 61: 530–546 (1926).

28 Clark, A. J.: The antagonism of acetylcholine by atropine. J. Physiol. 61: 547–556 (1926).

29 Clark, A. J.: The reaction between acetylcholine and muscle cells. Part II. J. Physiol 64: 123–143 (1927).

30 Clark, A. J.: General pharmacology; in Handbook of experimental pharmacology, vol. 4 (Springer, Berlin 1937).

31 Cohen, J. B.; Changeux, J.-P.: The cholinergic receptor protein in its membrane environment. A. Rev. Pharmacol. 15: 83–103 (1975).

32 Cohen, S.; Carpenter, G.; King, L., Jr.: Epidermal growth factor-receptor-protein kinase interactions. Copurification of receptor and epidermal growth factor-enhanced phosphorylation activity. J. biol. Chem. 255: 4834–4842 (1980).

33 Colquhoun, D.: The relation between classical and cooperative models for drug action; in Rang, Drug receptors, pp. 149–182 (University Press, Baltimore 1973).

34 Colquhoun, D.; Rang, H. P.: Effects of inhibitors on the binding of iodinated α-bungarotoxin to acetylcholine receptors in rat muscle. Molec. Pharmacol. 12: 519–535 (1976).

35 Craig, S.; Cuatrecasas, P.: Mobility of cholera toxin receptor on rat lymphocyte membranes. Proc. natn. Acad. Sci. USA 72: 3844–3848 (1975).

36 Cuatrecasas, P.: Properties of the insulin receptor isolated from liver and rat cell membranes. J. biol. Chem. 247: 1980–1991 (1972).

37 Cuartrecasas, P.: Insulin receptor of liver and fat cell membranes. Fed. Proc. *32:* 1838–1846 (1973).

38 Cuartrecasas, P.: Membrane receptors. A. Rev. Biochem. *43:* 169–214 (1974).

39 Cuatrecasas, P.; Hollenberg, M.D.: Binding of insulin and other hormones to non-receptor materials: saturability, specificity and apparent negative cooperativity. Biochem. biophys. Res. Commun. *62:* 31–41 (1975).

40 Cuatrecasas, P.; Hollenberg, M.D.: Membrane receptors and hormone action. Adv. Protein Chem. *30:* 251–451 (1976).

41 Czech, M.P.: Molecular basis of insulin action. A. Rev. Biochem. *46:* 359–384 (1977).

42 Czech, M.: Insulin action. Am. J. Med. *70:* 142–150 (1981).

43 Das, M.; Fox, C.F.: Molecular mechanism of mitogen action: processing of receptor induced by epidermal growth factor. Proc. natn. Acad. Sci. USA *75:* 2644–2648 (1978).

44 Das, M.; Miyakawa, T.; Fox, C.F.; Pruss, R.M.; Aharonov, A.; Herschman, H.R.: Specific radiolabeling of a cell surface receptor for epidermal growth factor. Proc. natn. Acad. Sci. USA *74:* 2790–2794 (1977).

45 DeHaen, C.: The non-stoichiometric floating receptor model for hormone-sensitive adenylate cyclase. J. theor. Biol. *58:* 383–400 (1976).

46 DeLean, A.; Rodbard, D.: Kinetics of cooperative binding; in O'Brien, The receptors, a comprehensive treatise, vol. 1, pp. 143–192 (Plenum Press, New York 1978).

47 DeMeyts, P.: Insulin and growth hormone receptors in human cultured lymophocytes and peripheral blood monocytes; in Laskin, Last, Methods in molecular biology, vol. 9, part I, pp. 301–383 (Dekker, New York 1976).

48 DeMeyts, P.; Bianco, A.R.; Roth, J.: Site-site interaction among insulin receptors. Characterization of the negative cooperativity. J. biol. Chem. *251:* 1877–1888 (1976).

49 DeMeyts, P.; Roth, J.; Neville, D.M., Jr.; Gavin, J.R., III; Lesniak, M.A.: Insulin interaction with its receptors: experimental evidence for negative cooperativity. Biochem. biophys. Res. Commun. *54:* 154–161 (1973).

50 Devreotes, P.N.; Fambrough, D.M.: Acetylcholine receptor turnover in membranes of developing muscle fibres. J. Cell Biol. *65:* 335–358 (1975).

51 Devreotes, P.N.; Fambrough, D.M.: Synthesis of the acetylcholine receptor by cultured chick myotubes and denervated mouse extensor digitorum longus muscles. Proc. natn. Acad. Sci. USA *73:* 161–164 (1976).

52 Devreotes, P.N.; Gardner, J.M.; Fambrough, D.M.: Kinetics of Biosynthesis of acetylcholine receptor and subsequent incorporation into plasma membrane of cultured check skeletal muscle. Cell *10:* 365–373 (1977).

53 Ehrlich, P.: Nobel lecture (1908) on partial functions of the cell; in Himmelweit, Marquardt, Dale, The collected papers of P. Ehrlich, vol. III, p. 183 (Pergamon Press, Oxford 1956).

54 Farfel, Z.; Brickman, A.S.; Kaslow, H.R.; Brothers, V.M.; Bourne, H.R.: Defect of receptor-cyclase coupling protein in pseudohypoparathyroidism. New Engl. J. Med. *303:* 237–242 (1980).

55 Flier, J.S.; Kahn, C.R.; Jarrett, D.B.; Roth, J.: Characterization of antibodies to the insulin receptor. A cause of insulin-resistant diabetes in man. J. clin. Invest. *58:* 1442–1449 (1976).

56   Frazer, C.M.; Harrison, L.C.; Kaliner, M.C.; Venter, J.C.: Autoantibodies to the β-adrenergic receptor are associated with β-adrenergic hyporesponsiveness (Abstract). Clin. Res. *28:* 236 (1980).

57   Gaddum, J.H.: The action of adrenalin and ergotamine on the uterus of the rabbit. J. Physiol. *61:* 141–150 (1926).

58   Gaddum, J.H.: The quantitative effects of antagonistic drugs. J. physiol. *89:* 7P–9P (1937).

59   Gavin, J.R., III; Roth, J.; Neville, D.M., Jr; DeMeyts, P.; Buell, D.N.: Insulin-dependent regulation of insulin receptor concentration. A direct demonstration in cell culture. Proc. natn. Acad. Sci. USA *71:* 84–88 (1974).

60   Ginsberg, B.H.: The insulin receptor: properties and regulation; in Litwack, Biochemical actions of hormones, vol.IV, pp.313–349 (Academic Press, New York 1977).

61   Ginsberg, B.H.; Kahn, C.R.; Roth, J.; DeMeyts, P.: Insulin-induced dissociation of its receptor into subunits: possible molecular concomitant of negative cooperativity. Biochem. biophys. Res. Commun. *73:* 1068–1074 (1976).

62   Gliemann, J.; Gammeltoft, S.: The biological activity and the binding affinity of modified insulins determined on isolated rat fat cells. Diabetologia *10:* 105–113 (1974).

63   Goldfine, I.D.; Smith, G.J.: Binding of insulin to isolated nuclei. Proc. natn. Acad. Sci. USA *73:* 1427–1431 (1976).

64   Goldfine, I.D.; Smith, G.J.; Wong, K.Y.; Jones, A.L.: Cellular uptake and nuclear binding of insulin in human cultured lymphocytes: evidence for potential intracellular sites of insulin action. Proc. natn. Acad. Sci. USA *74:* 1368–1372 (1977).

65   Goldfine, I.D.; Vigneri, R.; Cohen, D.; Plaim, N.B.; Kahn, C.R.: Intracellular binding sites for insulin are immunologically distinct from those on the plasma membrane. Nature, Lond. *269:* 698–700 (1977).

66   Goldstein, J.L.; Brown, M.S.: Familial hypercholesterolemia. A genetic regulatory defect in cholesterol metabolism. Am. J. Med. *58:* 147–150 (1975).

67   Goldstein, J.L.; Brown, M.S.: Hyperlipidemia in coronary artery disease: a biochemical genetic approach. J. Lab. clin. Med. *85:* 15–28 (1975).

68   Gorski, J.; Gannon, F.: Current models of steroid hormone action: a critique. A. Rev. Physiol. *38:* 425–450 (1976).

69   Guidotti, A.; Baraldi, M.; Costa, E.: 1,4-Benzodiazepines and gamma-aminobutyric acid: pharmacological and biochemical correlates. Pharmacology *19:* 267–277 (1979).

70   Haga, T.; Haga, K.; Gilman, A.G.: Hydrodynamic properties of the β-adrenergic receptor and adenylate cyclase from wild type and variant S49 lymphoma cells. J. biol. Chem. *252:* 5776–5782 (1977).

71   Haga, T.; Ross, E.M.; Anderson, H.J.; Gilman, A.G.: Adenylate cyclase permanently uncoupled from hormone receptors in a novel variant of S49 mouse lymphoma cells. Proc. natn. Acad. Sci. USA *74:* 2016–2020 (1977).

72   Haigler, H.; Ash, J.F.; Slinger, S.J.; Cohen, S.: Visualization by fluorescence of the binding and internalization of epidermal growth factor in human cardinoma cells A-431. Proc. natn. Acad. Sci. USA *75:* 3317–3321 (1978).

73   Haigler, H.T.; Maxfield, F.R.; Willingham, M.C.; Pastan, I.: Dansylcadaverine inhibits internalization of ^{125}I-epidermal growth factor in BALB 3T3 cells. J. biol. Chem. *255:* 1239–1241 (1980).

74 Hall, Z.W.: Release of neurotransmitters and their interaction with receptors. A. Rev. Biochem. *41:* 925–952 (1972).

75 Hanski, E.; Levitzki, A.: The absence of desensitization in the beta-adrenergic receptors of turkey reticulocytes and erythrocytes and its possible origin. Life Sci. *22:* 53–60 (1978).

76 Hanski, E.; Sternweiss, P.C.; Northup, J.K.; Dromerick, A.W.; Gilman, A.G.: The regulatory component of adenylate cyclase. Purification and properties of the turkey erythrocyte protein. J. biol. Chem. *256:* 12911–12919 (1981).

77 Harmon, J.T.; Kahn, C.R.; Kempner, E.S.; Schlegel, W.: Characterization of the insulin receptor in its membrane environment by radiation inactivation. J. biol. Chem. *255:* 3412–3419 (1980).

78 Hartzell, H.C.; Fambrough, D.M.: Acetylcholine receptor production and incorporation into membranes of developing muscle fibers. Devl. Biol. *30:* 153–165 (1973).

79 Hauger, R.L.; Aguilera, G.; Catt, K.J.: Angiotensin II Regulates its receptor sites in the adrenal glomerulosa Zone. Nature, Lond. *271:* 176–177 (1978).

80 Hebdon, M.; Levine, H., III; Sahyoun, N.; Schmitges, C.J.; Cuatrecasas, P.: Properties of the interaction of fluoride- and Gpp(NH)p-regulatory proteins with adenylate cyclase. Proc. natn. Acad. Sci. USA *75:* 3693–3697 (1978).

81 Helderman, J.H.; Strom, T.B.: Emergence of insulin receptors upon alloimmune T cells in the rat. J. clin. Invest. *59:* 334–338 (1978).

82 Helderman, J.H.; Strom, T.B.: Specific insulin binding site on T and B lymphocytes as a marker of cell activation. Nature, Lond. *274:* 62–63 (1978).

83 Henneberry, R.C.; Smith, C.C.; Taliman, J.F.: Relationship between β-adrenergic receptors and adenylate cyclase in HeLa cells. Nature, Lond. *268:* 252–254 (1977).

84 Herrup, K.; Shooter, E.M.: Properties of the β-nerve growth factor receptor in development. J. cell. Biol. *67:* 118–125 (1975).

85 Hock, R.A.; Hollenberg, M.D.: Characterization of the receptor for epidermal growth factor-urogastrone in human placenta membrane. J. biol. Chem. *255:* 10731–10736 (1980).

86 Hock, R.A.; Nexo, E.; Hollenberg, M.D.: Isolation of the human placenta receptor for epidermal growth factor-urogastrone. Nature, Lond. *277:* 403–405 (1979).

87 Hock, R.A.; Nexo, E.; Hollenberg, M.D.: Solubilization and isolation of the human placenta receptor for epidermal growth factor-urogastrone. J. biol. Chem. *255:* 10737–10743 (1980).

88 Hollenberg, M.D.: Receptor models and the action of neurotransmitters and hormones; in Yamaura, Neurotransmitter receptor binding, pp. 13–39 (Raven Press, New York 1978).

89 Hollenberg, M.D.: Hormone receptor interactions at the cell membrane. Pharmac. Rev. *30:* 393–410 (1979).

90 Hollenberg, M.D.: Membrane receptors and hormone action I and II. Trends pharmacol. Sci. *2:* 320–323; *3:* 25–28 (1981).

91 Hollenberg, M.D.; Barrett, J.C.; Ts'o, P.O.P.; Berhanu, P.: Selective reduction in receptors for epidermal growth factor-urogastrone in chemically transformed tumorigenic Syrian hamster embryo fibroblasts. Cancer Res. *39:* 4166–4169 (1979).

92 Hollenberg, M.D.; Cuatrecasas, P.: Hormone receptors and membrane glycoproteins during in vitro transformation of lymphocytes; in Clarkson, Baserga, Control of prolife-

ration of animal cells, pp. 423–434 (Cold Spring Harbor Laboratory, Cold Spring Harbor 1974).

93 Hollenberg, M. D.; Cuatrecasas, P.: Membrane receptors and hormone action: recent developments. Prog. Neuro-Psychopharmacol. 2: 287–302 (1978).

94 Hollenberg, M. D.; Cuatrecasas, P.: Distinction of receptor from non-receptor interactions in binding studies: historical and practical perspectives; in O'Brien, The receptors, a comprehensive treatise, vol. I, pp. 193–214 (Plenum Press, New York 1979).

95 Hollenberg, M. D.; Nexo, E.: Receptor binding assays; in Jacobs, Cuatrecasas, Receptors and recognition, series B, vol. 11: membrane receptors: methods for purification and characterization, pp. 1–31 (Chapman & Hall, London 1981).

96 Hollenberg, M. D.; Nexo, E.; Berhanu, P.; Hock, R. A.: Phorbol ester and the selective modulation of receptors for epidermal growth factor-urogastrone; in Middlebrook, Kohn, Receptor mediated binding and internalization of Toxins and hormones, pp. 181–195 (Academic Press, New York 1981).

97 Holley, R. W.; Armour, R.; Baldwin, J. H.; Brown, K. D.; Yeh, Y.-C.: Density-dependent regulation of growth of BSC-1 cell culture: control of growth by serum factors. Proc. natn. Acad. Sci. USA 74: 5046–5050 (1977).

98 Huang, D.; Cuatrecasas, P.: Insulin-induced reduction of membrane receptor concentrations in isolated fat cells and lymphocytes: independence from receptor occupation and possible relation to proteolytic activity of insulin. J. biol. Chem. 250: 8251–8259 (1975).

99 Insel, P. A.; Maguire, M. E.; Gilman, A. G.; Bourne, H. R.; Coffino, P.; Melmon, K. L.: Beta-adrenergic receptors and adenylate cyclase. Products of separate genes? Molec. Pharmacol. 12: 1062–1069 (1976).

100 Jacobs, S.; Cuatrecasas, P.: The mobile receptor hypothesis and cooperativity of hormone binding. Appliction to insulin. Biochim. biophys. Acta 433: 482–495 (1976).

101 Jacobs, S.; Cuatrecasas, P.: Cell receptors in disease. New Engl. J. Med. 297: 1383–1386 (1977).

102 Jacobs, S.; Cuatrecasas, P.: The mobile receptor hypothesis for cell membrane receptor action. Trends biochem. Sci. 2: 280–282 (1977).

103 Jacobs, S.; Cuatrecasas, P.: Antibodies to purified insulin receptor have insulin-like activity. Science 200: 1283–1284 (1978).

104 Jacobs, S.; Hazum, E.; Cuatrecasas, P.: The subunit structure of rat liver insulin receptor. Antibodies directed against the insulin binding subunit. J. biol. Chem. 255: 6937–6940 (1980).

105 Jacobs, S.; Hazum, E.; Shechter, Y.; Cuatrecasas, P.: Insulin receptor: covalent labeling and identification of subunits. Proc. natn. Acad. Sci. USA 76: 4918–4921 (1979).

106 Jacobs, S.; Shechter, Y.; Bissel, K.; Cuatrecasas, P.: Purification and properties of insulin receptors from rat liver membranes. Biochem. biophys. Res. Commun. 77: 981–988 (1977).

107 Jard, S.; Roy, C.; Barth, T.; Rajerison, R.; Bockaert, J.: Antidiuretic hormone-sensitive kidney adenylate cyclase. Adv. cyclic Nucleotide Res. 5: 31–53 (1975).

108 Kahn, C. R.: Membrane receptors for hormones and neurotransmitters. J. Cell Biol. 70: 261–286 (1976).

109 Kahn, C. R.; Baird, K.; Flier, J. S.; Jarrett, D. B.: Effect of anti-insulin receptor anti-bodies on isolated adipocytes. Diabetes 25: suppl. 1, p. 322 (1976).

110 Kahn, C. R.; Baird, K. L.; Jarrett, D. B.; Flier, J. S.: Direct demonstration that receptor crosslinking or aggregation is important in insulin action. Proc. natn. Acad. Sci. USA 75: 4209–4213 (1978).

111 Kaiser, G.; Wiemer, G.; Kremer, G.; Dretz, J.; Hellwich, M.; Palm, D.: Correlation between isoprenaline stimulated synthesis of cyclic AMP and occurrence of β-adre-noreceptors in immature erythrocytes from rats. Eur. J. Pharmacol. 48: 255–262 (1978).

112 Karlin, A.; The acetylcholine receptor; in Cotman Poste, Nicolson, The cell surface and neuronal function, pp. 191–260 (Elsevier/North-Holland Biomedical press, Am-sterdam 1980).

113 Katzenellenbogn, B. S.: Dynamics of steroid hormone receptor action. A. Rev. Physiol. 42: 17–35 (1980).

114 Kohn, L. D.: The thyrotropin receptor in the exophthalmos of Graves' disease; in Melnechuk, Cell receptor disorders, pp. 28–38 (Western Behavioral Sciences Institute, La Jolla 1978).

115 Kolterman, O. G.; Gray, R. S.; Griffin, J.; Burstein, P.; Insel, J.; Scarlett, J. A.; Olefsky, J. M.: Receptor and postreceptor defects contribute to the insulin resistance in non-insulin-dependent diabetes mellitus. J. clin. Invest. 68: 957–969 (1981).

116 Kolterman, O. G.; Inse, J.; Saekow, M.; Olefsky, J. M.: Mechanisms of insulin resist-ance in human obesity. Evidence for receptor and postreceptor defects. J. clin. Invest. 65: 1272–1284 (1980).

117 Krug, U.; Kurg, F.; Cuatrecasas, P.: Emergence of insulin receptors on human lym-phocytes during in vitro transformation. Proc. natn. Acad. Sci. USA 69: 2604–2608 (1972).

118 Labrie, F.; Belanger, A.; Cusan, L.; Seguin, C.; Pelletier, G.; Kelly, P. A.; Reeves, J. J.; Lefebvre, F. A.; Lemay, A.; Gourdeau, Y.; Raynaud, J.-P.: Antifertility effects of LHRH agonists in the male. J. Andrology 1: 209–228 (1980).

119 Lai, E.; Rosen, O. M.; Rubin, C. S.: Differentiation-dependent expression of catechol-amine-stimulated adenylate cyclase. Roles of the β-receptor and G/F protein in dif-ferentiating 3T3-L1 adipocytes. J. biol. Chem. 256: 12866–12874 (1981).

120 Langly, J. N.: On nerve endings and on special excitable substances. Proc. R. Soc. B 78: 170–194 (1906)

121 Lefebvre, F. A.; Reeves, J. J.; Seguin, C.; Massicot, J.; Labrie, F.: Specific binding of a potent LHRH agonist in rat testis. Mol. cell. Endocrinol. 20: 127–134 (1980).

122 Lefkowitz, R.: Identification and regulation of alpha and beta-adrenergic receptors. Fed. Proc. 37: 123–129 (1978).

123 Lemarchand-Brustel, Y.; Gorden, P.; Flier, J. S.; Kahn, C. R.; Freychet, P.: Anti-insulin receptor antibodies inhibit insulin binding and stimulate glucose metabolism in skeletal muscle. Diabetologia 14: 311–318 (1978).

124 Levitzki, A.: Negative cooperativity in glustered receptors as a possible basis for mem-brane action. J. theor. Biol. 44: 367–372 (1974).

125 Limbird, L. E.; Lefkowitz, R. J.: Resolution of β-adrenergic receptor binding and ade-nylate cyclase activity by gel exclusion chromatography. J. biol. Chem. 252: 799–802 (1977).

126 Linsley, P. S.; Das, M.; Fox, C. F.: Affinity labeling of hormone receptors and other

ligand binding proteins; in Jacobs, Cuatrecasas, Receptors and recognition, series B, vol. 11: membrane receptors, methods for purification and characterization, pp. 87–113 (Chapman & Hall, London 1981).

127 Linsley, P. S.; Fox, C. F.: Controlled proteolysis of EGF receptors: evidence for trans-membrane distribution of the EGF binding and phosphate acceptor sites. J. supramol. Struct. *14:* 461–471 (1980).

128 Livingston, J. N.; Purvis, B. J.; Lockwood, D. H.: Insulin-dependent regulation of the insulin-sensitivity of adipocytes. Nature, Lond. *273:* 394–396 (1978).

129 Mahoney, M. S.; Rosenberg, L. E.: Inborn errors of cobalamine metabolism; in Babior, Cobalamine biochemistry and pathophysiology, pp. 369–402 (Wiley & Sons, New York 1975).

130 Marchesi, V. T.; Furthmayr, H.; Tomita, M.: The red cell membrane. A. Rev. Biochem. *45:* 667–698 (1976).

131 Massague, J.; Pilch, P. F.; Czech, M. P.: Electrophoretic resolution of three major insulin receptor structures with unique subunit stoichiometries. Proc. natn. Acad. Sci. USA *77:* 7137–7141 (1980).

132 Maturo, J. M., III; Hollenberg, M. D.: Insulin receptor: interaction with non-receptor glycoprotein from liver cell membranes. Proc. natn. Acad. Sci. USA *75:* 3070–3074 (1978).

133 Maturo, J. M., III; Shackelford, W. H.; Hollenberg, M. D.: Characteristics of the solubilized insulin receptor of human placenta. Life Sci. *23:* 2063–2072 (1978).

134 Maxfield, F. R.; Schlessinger, J.; Shechter, Y.; Pastan, I.; Willingham, M. C.: Insulin epidermal growth factor and α_2-macroglobulin rapidly collect in the same patches on the surface of cultured fibroblasts and are internalized together. Cell *14:* 805–810 (1978).

135 Muller, W. E.: The benzodiazepine receptor: an update. Pharmacology *22:* 153–161 (1981).

136 Murad, F.; Chi, Y.-M.; Rall, T. W.; Sutherland, E. W.: Adenyl cyclase. III. The effect of catecholamines and choline esters on the formation of adenosine 3′,5′-phosphate by preparations from cardiac muscle and liver. J. biol. Chem. *237:* 1233–1238 (1962).

137 Nexo, E.: Transcobalamine I and other receptor-binders: purification, structural, spectral and physiologic studies. Scand. J. Haematol. *20:* 221–236 (1978).

138 Nexo, E.; Hock, R. A.; Hollenberg, M. D.: Lectin-agarose immobilization. A new method for detecting soluble membrane receptors. J. biol. Chem. *254:* 8740–8743 (1980).

139 Nexo, E.; Hollenberg, M. D.; Figueroa, A.; Pratt, A. M.: Detection of epidermal growth factor-urogastrone and its receptor during fetal mouse development. Proc. natn. Acad. Sci. USA *77:* 2782–2785 (1980).

140 Nimrod, A.; Tsafriri, A.; Linder, H. R.: In vitro induction of binding sites for hCG in rat granulosa cells by FSH. Nature, Lond. *267:* 632–633 (1977).

141 Northup, J. K.; Sternweiss, P. C.; Smigel, M. D.; Schleifer, L. S.; Ross, E. M.; Gilman, A. G.: Purification of the regulatory component of adenylate cyclase. Proc. natn. Acad. Sci. USA *77:* 6516–6520 (1980).

142 O'Keefe, E. J.; Battin, T. K.; Bennett, V.: Proteolytic domains of the epidermal growth factor receptor of human placenta. J. supramol. Struct. *15:* 15–27 (1981).

143 O'Malley, B. W.; Schrader, W. T.: Steroid hormone action: structure and function of

receptor; in Bitensky, Collier, Steine, Fox, Transmembrane signalling, pp. 629–638 (Liss, New York 1979).

144 Orly, J.; Schramm, M.: Coupling of catecholamine receptor from one cell with adenylate cyclase from another cell by cell fusion. Proc. natn. Acad. Sci. USA *73:* 4410–4414 (1976).

145 Paton, W. D. M.: A theory of drug action based on the rate of drug-receptor combination. Proc. R. Soc. B *154:* 21–69 (1961).

146 Paton, W. D. M.; Rang, H. P.: The uptake of atropine and related drugs by intestinal smooth muscle of the guinea pig in relation to acetylcholine receptors. Proc. R. Soc. B *163:* 1–44 (1965).

147 Paton, W. D. M.; Waud, D. R.: A quantitative investigation of the relationship between rate of access of a drug to a receptor and the ratio of onset or offset of action. Naunyn-Schmiedebergs Arch. exp. Path. Pharmak. *248:* 124–143 (1964).

148 Pavia, A. C. M.; Pavia, T. B.; Juliand, L.; Olivera, M. C. F.: Structure-activity relationships in the renal pressor system (Abstract). PAABS-IUB Symp., No. 86, p. 26.

149 Pfeuffer, T.: GTP-binding proteins in membranes and the control of adenylate cyclase activity. J. biol. Chem. *252:* 7224–7234 (1977).

150 Pollet, R. J.; Standaert, M. L.; Haase, B. A.: Insulin binding to the human lymphocyte receptor. Evaluation of the negative cooperativity model. J. biol. Chem. *252:* 5828–5834 (1977).

151 Posner, B. I.; Kelley, P. A.; Friesen, H. G.: Prolactin receptor in rat liver: possible induction by prolactin. Science *188:* 57–59 (1978).

152 Posner, B. I.; Kelley, P. A.; Friesen, H. G.: Induction of lactogenic receptor in rat liver: influence of estrogen and the pituitary. Proc. natn. Acad. Sci. USA *71:* 2407–2410 (1974).

153 Raftery, M. A.; Hunkapillar, M. W.; Strader, C. D.; Hood, L. E.: Acetylcholine receptor: complex of homologous subunits. Science *208:* 1454–1456 (1980).

154 Reed, B. C.; Lane, M. D.: Insulin receptor synthesis and turnover in differentiating 3T3-L1 preadipocytes. Proc. natn. Acad. Sci. USA *77:* 285–289 (1980).

155 Reed, B. C.; Ronnett, G. V.; Clements, P. R.; Lane, M. D.: Regulation of insulin receptor metabolism. Differentiation-induced alteration of receptor synthesis and degredation. J. biol. Chem. *256:* 3917–3925 (1981).

156 Rodbard, D.; Mathematics of hormone-receptor interacton. I. Basic principles; in O'Malley, Means, Receptors for reproductive hormones, pp. 289–326 (Plenum Press, New York 1973).

157 Rodbard, D.; Bertino, R. E.: Theory of radioimmunoassays and hormone-receptor interactions. II. Simulation of antibody divalency, cooperativity and allosteric effects; in O'Malley, Means, Receptors for reproductive hormones, pp. 327–341 (Plenum Press, New York 1973).

158 Rodbell, M.; Kran, H. M. J.; Pohl, S.; Birnbaumer, L.: The glucagon-sensitive adenyl cyclase system in plasma membranes of rat liver. III. Binding of glucagon. Method of assay and specificity. J. biol. Chem. *246:* 1861–1871 (1971).

159 Ross, E. M.; Gilman, A. M.: Reconstitution of acetylcholine-sensitive adenylate cyclase activity: interaction of solubilized components with receptor-deplete membranes. Proc. natn. Acad. Sci. USA *74:* 3715–3719 (1977).

160 Roth, J.: Peptide hormone binding to receptors: a review of direct studies in vitro. Metabolism *22:* 1059–1073 (1973).

161 Sahyoun, N.; Hock, R. A.; Hollenberg, M. D.: Insulin and epidermal growth factor-urogastrone: affinity crosslinking to specific binding sites in rat liver membranes. Proc. natn. Acad. Sci. USA 75: 1675–1679 (1978).

162 Sahyoun, N.; Hollenberg, M. D.; Bennett, V.; Cuatrecasas, P.: Topographic separation of adenylate cyclase and hormone receptors in the plasma membrane of toad erythrocyte ghosts. Proc. natn. Acad. Sci. USA 74: 2860–2864 (1977).

163 Sahyoun, N.; Schmitges, C. J.; Levine, H., III; Cuatrecasas, P.: Molecular resolution and reconstruction of the Gpp(NH)p and NaF sensitive adenylate cyclase system. Life Sci. 21: 1857–1864 (1977).

164 Savion, N.; Vlodavsky, I.; Gospodarowicz, D.: Nuclear accumulation of epidermal growth factor in cultured bovine corneal endothelial and granulosa cells. J. biol. Chem. 256: 1149–1154 (1981).

165 Scatchard, G.: The attraction of proteins for small molecules and ions. Ann. N.Y. Acad. Sci. 51: 660–672 (1949).

166 Schild, H. O.: pAx and competitive drug antagonism. Br. J. Pharmacol. 4: 277–280 (1949).

167 Schlessinger, J.; Shechter, Y.; Cuatrecasas, P.; Willingham, M. C.; Pastan, I.: Quantitative determination of the lateral diffusion coefficients of the hormone receptor complexes of insulin and epidermal growth factor on the plasma membrane of cultured fibroblasts. Proc. natn. Acad. Sci. USA 75: 5353–5357 (1978).

168 Schlessinger, J.; Shechter, Y.; Willingham, M. C.; Pastan, I.; Direct visualization of binding, aggregation and internalization of insulin and epidermal growth factor on living fibroblastic cells. Proc. natn. Acad. Sci. USA 75: 2659–2663 (1978).

169 Schramm, M.; Orly, J.; Eimerl, S.; Korner, M.: Coupling of hormone receptors to adenylate cyclase of different cells by cell fusion. Nature, Lond. 268: 310–313 (1977).

170 Schulster, D.; Orly, J.; Seidel, G.; Schramm, M.: Intracellular cyclic AMP production enhanced by a hormone receptor transferred from a different cell. J. biol. Chem. 253: 1201–1206 (1978).

171 Shechter, Y.; Hernaez, L.; Cuatrecasas, P.: Epidermal growth factor: biological activity requires persistent occupation of high-affinity cell surface receptors. Proc. natn. Acad. Sci. USA 75: 5788–5791 (1978).

172 Shechter, Y.; Schlessinger, J.; Jacobs, S.; Chang, K.-J.; Cuatrecasas, P.; Fluorescent labeling of hormone receptors in viable cells: preparation and properties of highly fluorescent derivatives of epidermal growth factor and insulin. Proc. natn. Acad. Sci. USA 75: 2135–2139 (1978).

173 Sheppard, J.: Catecholamine hormone receptor differences identified on 3T3 and simian virus transformed 3T3 cells. Proc. natn. Acad. Sci. USA 74: 1091–1094 (1977).

174 Shorr, R. G. L.; Lefkowitz, R. J.; Caron, M. G.: Purification of the β-adrenergic receptor. Identification of the hormone binding subunit. J. biol. Chem. 256: 5820–5826 (1981).

175 Singer, S. J.: The fluid mosaic model of membrane structure. Some applications to ligand-receptor and cell-cell interactions; in Bradshaw, Frazier, Merrell, Gottlieb, Hogue, Surface membrane receptors, pp.1–24 (Plenum Press, New York 1976).

176 Singer, S. J.; Nicholson, G. L.: The fluid mosaic model of the structure of cell membranes. Science 175: 720–731 (1972).

177 Soloff, M.: Uterine receptor for oxytocin: effects of Estrogen. Biochem. biophys. Res. Commun. *65:* 205–212 (1975).

178 Sporn, M. B.; Todaro, G. J.: Autocrine secretion and malignant tansformation of cells. New Engl. J. Med. *303:* 878–880 (1980).

179 Stephenson, R. P.: A modification of receptor theory. Br. J. Pharmacol. *11:* 379–393 (1956).

180 Stephenson, R. P.; Barlow, R. B.: Concepts of drug action, quantitative pharmacology and biological assay; in Passmdore, Robson, A companion to medical studies, vol.2, pp. 3.1–3.19 (Blackwell, London 1970).

181 Sternweis, P. C.; Northup, J. K.; Smigel, M. D.; Gilman, A. G.: The regulatory component of adenylate cyclase. Purification and properties. J. biol. Chem. *256:* 11517–11526 (1981).

182 Sutherland, E. W.; Oye, I.; Butcher, R. W.: The action of epinephrine and the role of the adenyl cyclase system in hormone action. Recent Prog. Horm. Res. *21:* 623–642 (1965).

183 Tallman, J. F.; Smith, C. C.; Henneberry, R. C.: Induction of functional β-adrenergic receptors in HeLa cells. Proc. natn. Acad. Sci. USA *74:* 873–877 (1977).

184 Thompoulos, P.; Kosmakos, F. C.; Pastan, I.; Lovelace, E.: Cyclic AMP increases the concentration of insulin receptors in cultured fibroblasts and lymphocytes. Biochem. biophys. Res. Commun. *75:* 246–252 (1977).

185 Todaro, G. J.; Delarco, J. E.; Cohen, S.: Transformation by murine and feline sarcoma viruses specifically blocks binding of epidermal growth factor to cells. Nature, Lond. *264:* 26–31 (1976).

186 Tolkovsky, A. M.; Levitzki, A.: Collision coupling of the β-receptor with adenylate cyclase; jin Dumont, Nunez, Hormones and cell regulation, vol.2, pp. 89–105 (Elsevier/North-Holland Biomedical Press, Amsterdam 1977).

187 Tymoczko, J. L.; Liang, T.; Liao, S.: Androgen receptor interactions in target cells: biochemical evaluation; in O'Malley, Birnbaumer, Receptors and hormone Action, pp. 121–156 (Academic Press, New York 1978).

188 Varga, J. M.; Dipasquale, A.; Pawelek, J.; McGuire, J. S.; Lerner, A. B.: Regulation of melanocyte stimulating hormone action at the receptor level: discontinuous binding of hormone to synchronized mouse melanoma cells during the cell cycle. Proc. natn. Acad. Sci. USA *71:* 1590–1593 (1974).

189 Weber, M.; David-Pfeuty, T.; Changeux, J.-P.: Regulation of binding properties of the nicotinic receptor protein by cholinergic ligands in membrane fragments from *Torpedo marmorata.* Proc. natn. Acad. Sci. USA *72:* 3443–3447 (1975).

190 Weiland, G. A.; Minneman, K. P.; Molinoff, P. B.: Fundamental difference between the molecular interactions of agonists and antagonists with the β-adrenergic receptor. Nature, Lond. *281:* 114–117 (1979).

191 Yip, C. C.; Moule, M. L.; Yeung, C. W. T.: Subunit structure of insulin receptor of rat adipocytes as demonstrated by photoaffinity labeling. Biochemistry *21:* 2945–2950 (1982).

Karen A. Valentine, Department of Pharmacology and Therapeutics,
Endocrine Research Group, Faculty of Medicine, University of Calgary,
Calgary, Alta. T2N 4N1 (Canada)

Cell Biology of the Secretory Process, pp. 52–72 (Karger, Basel 1984)

Stimulus-Secretion Coupling

Ronald P. Rubin

Medical College of Virginia, Richmond, Va., USA

Historical Aspect

The role of calcium as a pivotal component of the secretory process has been unquestionably established, and there is now little doubt that calcium is critical for providing the link between the membrane signal and activation of the secretory apparatus. The basis for the concept of stimulus-secretion coupling, which encompasses 'all the events occurring in the cell exposed to its immediate stimulus that lead, finally, to the appearance of the characteristic secretory product in the extracellular environment' was provided by the work of *Douglas* [18]. The adrenal medulla was employed for the initial experiments in part because it had already been established – mainly through the endeavors of *Katz* [38] – that its physiological neurotransmitter, acetylcholine, acts on the plasma membrane of other responsive tissues to increase permeability to commonly occurring ions. In the light of previous work on neurotransmitter release from cholinergic nerve endings [32, 39], calcium was deemed the cation most likely to mediate the action of acetylcholine on medullary chromaffin cells.

The conclusion that calcium is the physiological mediator of stimulus-secretion coupling was based mainly upon the evidence that calcium was essential and sufficient for evoking the secretory response to acetylcholine and other secretagogues, including excess potassium and autacoids such as histamine, 5-hydroxytryptamine, and peptides (bradykinin, angiotensin) [18, 20]. In fact, calcium alone was sufficient to elicit an explosive discharge of catecholamines, if the chromaffin cells were made leaky by a previous period of calcium deprivation. The action of acetylcholine was associated with the depolarization of the chromaffin cell as a result of an increase in membrane permeability and an inward movement of sodium and calcium ions. Depolarization was, however, neither sufficient nor absolutely necessary to trigger the secretory response, since acetylcholine could still elicit secretion in the absence of sodium ions, or in the presence of depolarizing concentrations of potassium [20]. The theory was promulgated that the entry of calcium into the medullary chromaffin cell was the critical event in promoting secretion, in the same general manner as it promotes acetylcholine release from motor nerve terminals during depolarization elicited by the nerve action potential. This 'calcium hypothesis' was substantiated by the subsequent demonstration of a causal relationship between calcium uptake and evoked secretion by isolated chromaffin cells [42]. With the advent of the calcium ionophore A23187 – which proved to be a potent medullary secretagogue – the view that an increase in calcium availability was the key event was strengthened [26].

Membrane and Ionic Events

Inward transport of calcium across the plasma membrane of excitable cells includes a downhill (passive) movement, permitted by the large concentration difference between the extracellular and intracellular calcium compartments. The conductive nature of the voltage-dependent calcium entry during the activation of excitable cells reflects the existence of membrane channels specific for calcium ions; and the identification and characterization of these channels provided a greater understanding of the nature of the events associated with stimulus-secretion coupling.

Studies carried out on the intact adrenal gland and on isolated chromaffin cells revealed that calcium enters through two operationally distinct channels, a fast channel (through which sodium also enters) and a slow channel [75] (fig. 1). Tetrodotoxin, which selectively blocks the fast channel, exerts only a slight depression of acetylcholine-evoked catecholamine secretion [43]. By contrast, the ability of the organic calcium blocker D-600 (methoxy-verapamil) and certain divalent cations, including lanthanum, manganese, and cobalt to obtund evoked secretion implies that the slow calcium channel which regulates the major portion of the inward calcium current during membrane activation is responsible for triggering secretion [78].

Conventional neurons possess cation channels similar to those of medullary chromaffin cells [3]. Experiments utilizing tetrodotoxin and/or

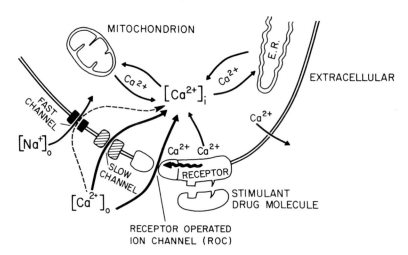

Fig. 1. Mechanisms for increasing ionized cellular calcium during activation of the secretory process. Following stimulation, calcium enters excitable cells through voltage-dependent 'fast' and 'slow' channels. Sodium enters through the fast channel, but it is the slow channel that is mainly responsible for conveying the inward calcium current during secretory activity. Non-excitable cells, such as exocrine glands, permit calcium entry through receptor-operated channels (ROC). Calcium may also be mobilized from the plasma membrane, or from the mitrochondrion and/or endoplasmic reticulum (ER). The ionic calcium concentration is restored to prestimulation levels by sequestration into intracellular organelles or by extrusion from the cell (pumps or counter transport mechanism) [reprinted with permission from ref. 79].

sodium deprivation confirm that it is the activation of the voltage-dependent slow calcium channel that is the critical event culminating in neurotransmitter release [40]. Not only do neuroendocrine organs such as the neurohypophysis also possess voltage-dependent calcium channels that are sensitive to organic calcium antagonists and divalent cations [94], but the behavior of endocrine glands of non-neuronal origin such as the pancreatic beta-cells and the adenohypophysis follows the same general pattern [52, 63]. Accordingly, secretagogues induce electrical activity which is a direct consequence of a voltage-dependent calcium influx, although like the chromaffin cell, there may also be a sodium component [10, 92]. Changes in potassium permeability may also serve to trigger the electrical events during the secretory response of certain cells [74]. But regardless of the ionic basis of electrical activity, membrane activation in excitable secretory cells is associated with the inward movement of calcium ions by way of channels in the membrane.

Douglas [18] had the perspicacity to employ the term stimulus-secretion coupling rather than excitation-secretion coupling to describe these events, not only to distinguish it from muscle where excitation can be defined as a propagated electrical event, but also to encompass the diverse ways by which an increase in ionic calcium may be brought about within the secretory cell. An enhanced influx of extracellular calcium is not the principal mechanism for triggering secretion in all secretory systems, but in fact there is convincing evidence that in certain secretory systems, stimulation brings about a mobilization of calcium from cellular stores (fig. 1). The endoplasmic reticulum, isolated as microsomes, accumulates and stores calcium in the cell. Energy-dependent calcium uptake activity has been found in microsomal fractions from a variety of secretory tissues, including adrenal glands, blood platelets, salivary glands, and the endocrine and exocrine pancreas [35, 37, 49, 70, 87]. These fractions represent stores of cellular calcium potentially available for release. The question as to whether the endoplasmic reticulum can release calcium has not been thoroughly answered, although *Feinstein* [23] using chlorotetracycline as a fluorescent calcium probe reported the release of calcium from the endoplasmic reticulum during the blood platelet release reaction. This finding supports the proposal that blood platelets behave similarly to muscle in that a dense tubular system (smooth endoplasmic reticulum) represents the calcium storage site from which calcium is released during stimulation [27]. Calcium is also sequestered by active transport in mitochondria, which may provide an important reservoir for calcium mobilization during secretory activity [14, 31, 90]. In such systems where cellular calcium is mobilized, the primary signal is usually generated at the level of the plasma membrane, so that intracellular messengers must transmit the signal to intracellular stores. Cyclic nucleotides (cyclic AMP or cyclic GMP) and/or arachidonic acid metabolites, including the prostaglandins, may act as positive mediators in these secretory organs to increase calcium availability.

The plasma membrane which is relatively rich in calcium may also provide the source of activator calcium for secretion (fig. 1), just as this site appears to be the source of activator calcium in certain types of smooth muscle [98]. The potential importance of calcium localized to the plasma membrane should not be minimized, particularly since it is the primary locus of action of many stimuli. This assertion is supported by studies on the exocrine pancreas suggesting that a releasable store of membrane-bound calcium, inaccessible to extracellular chelating agents, is mobilized

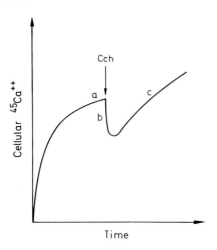

Fig. 2. Mobilization of calcium during activation of muscarinic (cholinergic) receptors in the exocrine pancreas. Isolated rat pancreatic acinar cells were preloaded with ^{45}Ca until a steady state was reached (a), and the acetylcholine analogue carbachol (Cch) added where indicated and cellular ^{45}Ca determined for periods up to 2–3 h. The fall (b) and rise (c) of the cellular calcium content are attributed to a release of calcium from a membrane-bound pool and an increase in the calcium permeability of the plasma membrane, respectively. For additional experimental details, see *Schulz* [86].

during an early stage of secretagogue action [48]. In addition, an increase in calcium entry from the extracellular fluid is indicated by the developing dependence of acetylcholine-evoked secretion on extracellular calcium. Moreover, calcium fluxes during secretagogue-induced pancreatic enzyme release are biphasic [86] (fig. 2). The initial process – utilized for short-term secretion – involves efflux and reflects the mobilization of calcium from the plasma membrane. The secondary process – required for sustained secretion – involves influx of extracellular calcium brought about by an increase in membrane permeability. Designating the source of 'trigger' calcium as the plasma membrane is intriguing, because the release of calcium bound to the plasma membrane during the actions of secretagogues could likely lead to an increase in membrane permeability promoting further increases in free cytoplasmic calcium levels [20, 86].

The findings in the exocrine pancreas that calcium associated with the plasma membrane is not easily depleted by calcium deprivation cannot be summarily extrapolated to other secretory organs. In fact, the seemingly contradictory findings that the action of corticotropin to initiate corticosteroid synthesis and release depends upon extracellular calcium, but is only modestly impaired by the calcium channel blocker D-600, may be interpreted on the basis of a critical pool in a superficial membrane compartment that is readily exchangeable with the extracellular fluid [78]. Alternatively, such results may also be explained by the existence of receptor-operated channels in the cell membrane that are not voltage dependent (fig. 1). Cations such as lanthanum, manganese, cobalt, and nickel will impede calcium entry through this pathway, while organic calcium antagonists inhibit only at high concentrations [72]. These receptor-operated channels even exist in systems that are excitable and possess voltage-dependent channels such as the adrenal chromaffin cell [41].

There are other examples that confirm the existence of multiple pathways to increase calcium availability within secretory cells. Most notably, alpha-adrenergic and cholinergic receptor stimulation mediates amylase release from mouse and rat parotid (salivary) glands by promoting calcium entry into the cell. By contrast, beta-adrenergic receptor stimulation has been suggested to bring about enzyme release through a cyclic-AMP-dependent mobilization of cellular calcium [13, 97]. The additional findings that stimulation of muscarinic receptors is associated with an increased formation of cyclic GMP [86] and that ionophore A23187 mimics the effects of muscarinic agents on cyclic GMP levels and enzyme release [12] prompt speculation that cyclic GMP formation might be a key calcium-mediated event in the salivary gland following alpha-adrenergic or cholinergic receptor stimulation. But whether or not cyclic GMP is a pivotal component of enzyme secretion by the salivary gland, it is clear from such studies that activation of receptors even on the same cell may generate diverse signals that result in the mobilization of distinct pools of calcium, as well as the formation of multiple putative messengers [7].

Nature of the Secretory Process – Exocytosis

It is generally agreed that cytoplasmic granules are the reservoir and the primary source of secretory product of the cell. The 'quantal' theory of acetylcholine release from motor nerve terminals first espoused by *Katz* [39] envisioned the release process as discrete all-or-none events caused by a collision between the vesicular and axonal membranes. Ultrastructural analysis of the secretory process revealed that fusion of the secretory granule membrane and the plasma membrane (termed exocytosis) was the mode of secretion [64], although the serial fusion of granule membranes may be an accompanying event [19]. Morphological studies have been extended and elaborated in a number of different secretory systems [19, 22, 29, 61] and confirmed by freeze-fracture techniques [61]. This latter technique enables exocytotic images to become more evident and correlates them temporally with enhanced secretion [34].

Biochemical evidence to support the concept of exocytosis has also been obtained by assaying the effluent of secretory tissues for cellular constituents. The reasoning behind such experiments rests with the premise that if the secretory organelle were to secrete its contents directly into the extracellular space, then the soluble components of the organelle should appear in the outflow while substances free in the cytoplasm should be retained. This biochemical approach, originally utilized to investigate the mechanism of catecholamine secretion [18, 96], has also been successfully applied to studies of secretion from the neurohypophysis and parathyroid gland [30, 56]. In these systems specific granular proteins sequestered within the secretory granules of each gland are concomitantly released with secretory product, while enzymes bound to the granule membrane (dopamine-beta hydroxylase) and those harbored in the cytoplasm (lactate dehydrogenase) are retained within the cell.

While a variety of secretory products are released by diverse cell types, the basic components of the secretory process, the plasma membrane and the granule (vesicle) membrane are functionally similar in different cells [18]. Moreover, the ubiquitous requirement for calcium reinforces a common mechanism of secretion and forges a link between the action of calcium and exocytotic secretion. In fact, critical insight into the molecular nature of the secretory process will probably be achieved by investigations of calcium-mediated membrane fusion phenomena [65, 66, 71]. This conclusion is supported by an almost inextricable linking of calcium with membrane fusion reactions and by the ability of barium and strontium to mimic this fusogenic action of calcium, just as they replace calcium in supporting secretory activity [50, 56, 71].

Biochemical Aspects of Stimulus-Secretion Coupling

Despite the indisputable importance of calcium in the secretory process, the action of this cation at the molecular level is not understood. The primary role of calcium is not simply to modulate cellular levels of other mediators, such as cyclic AMP, cyclic GMP, and/or prostaglandins or other arachidonic acid metabolites. While it is true that calcium controls · the formation of these putative mediators as a result of its well-established effects on adenylate and guanylate cyclase, as well as phospholipases, the actions of cyclic nucleotides and prostaglandins – when they are manifest – are in all probability exerted through mobilization of cellular calcium [79]. Moreover, in the adrenal medulla, the prototypical model of stimulus-secretion coupling, there is no convincing evidence that calcium requires the participation of any other known mediators to elicit its stimulatory effect on catecholamine release [6]. Finally, neither cyclic nucleotides nor known arachidonic acid metabolites universally act as agonists of the secretory process. In fact, in certain systems, they may be negative modulators, perhaps by virtue of their ability to promote the sequestration of free intracellular calcium [6, 24].

Activator Proteins

In order to elucidate the mechanism of the action of calcium on the secretory apparatus, the identity and chemical nature of the intracellular calcium 'receptor', which appears to be universally present in secretory tissue, must be established. The proposal by *Kretsinger* [44] that cellular functions are mediated by calcium-binding proteins was supported by the discovery of calmodulin, a cellular protein first identified as an activator of phosphodiesterase [15]. Recognition of the potential importance of calmodulin in the secretory process is afforded by its identification in the endocrine and exocrine pancreas, neurohypophysis, parathyroid gland, and in nerve endings [4, 11, 17, 91, 94]. Moreover, agents such as trifluoperazine, which inhibit calmodulin-induced stimulation of enzyme activity (e.g., adenylate cyclase), block evoked secretion [17, 25, 91].

Accepting the premise that calmodulin functions as the calcium recognition site in secretory cells then raises the question as to the mechanism by which the interaction of calcium with its protein receptor triggers the secretory response. The ability of the calcium-calmodulin complex to enhance adenylate cyclase, phosphodiesterase, and phospholipase activities [15, 100] provides a potentially important link between calcium and the other known putative mediators. On the other hand, calcium-dependent phosphorylation of certain proteins, through the activation of protein kinases that require calmodulin as a coenzyme, may regulate a number of biological processes in the cell, including the secretory process. In fact, secretory activity appears to be associated somehow with the calcium-dependent activation of protein kinase and the phosphorylation of cellular proteins [5, 17, 54, 85, 89]. Calcium-dependent neurotransmitter release evoked from synaptosomes (isolated nerve endings) by either depolarization or the ionophore A23187 involves a calmodulin-mediated activation of a calcium-dependent protein kinase that phosphorylates proteins in the vesicle membrane [17, 54]. The increase in acetylcholine release and phosphorylation occurs on a similar time scale; and strontium and barium which support transmitter release also sustain the phosphorylation mechanism, while magnesium lacks this ability to maintain the release response and phosphorylating activity.

While such studies implicate calcium-dependent phosphorylation as an important component of calcium-induced secretion, one must also be cognizant of the fact that there are protein kinase systems that are activated by cyclic AMP, cyclic GMP, as well as phospholipids [45, 60]; so, protein phosphorylating mechanisms can proceed through alternate pathways, not necessarily involving the calcium-calmodulin pathway. Additionally, we cannot ignore the possible roles of other calcium-binding proteins in the processes associated with secretion. The facilitatory actions of a protein isolated from the adrenal medulla called synexin on the calcium-induced fusion of medullary chromaffin granules may represent a valid model of exocytosis [68]; and there is the possibility that still other as yet undiscovered calcium-binding proteins will be identified in secretory cells.

Secretion as a Contractile Event

The isolation of actin and myosin from secretory organs [62, 95] raises the question as to the role of these contractile proteins in stimulus-secretion coupling. While we really do not understand how cytoplasmic motility is controlled in non-muscle cells, the particular association of actin with the microfilaments of the cell has implicated this component of the cytoskeletal system in the secretory process. An involvement of the tubulin-containing microtubular system has also received considerable attention [53]. Attachment of actin filaments to the plasma membrane could provide a mechanism for linking the microtubules in the cytoplasm to the cell surface. Calcium may bring about some alteration in the microtubular-microfilamentous system, thereby enabling this system to provide the framework and the motile force to convey the secretory granules to the plasma membrane. For example, the calcium-calmodulin complex may play a role in the assembly and disassembly of microtubules by regulating the activity of tubulin kinase [16]. The resulting phosphorylation of the microtubules may be a critical control mechanism in promoting granule-microtubule interactions. However, despite some intriguing evidence that contraction of microtubules may be somehow associated with secretory activity [36, 46], the inconsistent and variable effects of purported inhibitors of the microtubular-microfilamentous system on the secretory response do not convincingly substantiate the supposition that the cytoskeletal system plays a direct role in the discharge of secretory product [53, 59].

Nevertheless, the obvious parallels between the actions of calcium on stimulus-secretion coupling and excitation-contraction coupling aver that a basically similar molecular mechanism underlies both processes [20, 76]. So, to probe this issue further, one may consider the secretory cell as a modified contractile element apart from the cytoskeletal system. This perspective is justified because precise information concerning the subcellular localization of actin in secretory cells is still lacking [2, 95]. In this context, the similarity between these two calcium-dependent processes is strengthened by the knowledge that a calmodulin-dependent phosphorylation of the myosin light chain kinase appears to control the activity of the blood platelet by permitting actin to activate the myosin molecule [1, 24]. This biochemical reaction – which is likewise observed in smooth muscle [1] – may lead to the migration of secretory granules to the periphery of the cell.

But while secretion and contraction may share some basic characteristics, another property of actin, that is not utilized in muscle contraction, could have an important function in the events associated with secretory activity. Thus, actin and other proteins that regulate motility in non-muscle cells can exist in monomeric form as a diffuse meshwork, as well as bundles of microfilaments [69]. So, the rapid polymerization of actin into a fibrous form (F-actin) may be a pivotal control mechanism in the secretory cell, with or without the participation of myosin. Alternatively, systems have been described in non-muscle cells in which gel-sol transitions occur in preparations of actin and actin-binding proteins [93]. The gel consists of cross-linked filaments, and calcium dissolves the actin gel by dissociating these filaments. The calcium-mediated gel-sol transitions – which can explain granule migration through alterations in cytoplasmic structure – may also involve other as yet unidentified cytoplasmic proteins. There are, therefore, still insufficient grounds for tacitly regarding the secretory cell as a close functional analogue of muscle and viewing actomyosin as the site of calcium action. Convincing evidence has yet to be attained that actomyosin has a universal function to provide the force generation required for promoting either calcium-dependent membrane fusion or even the peripheral migration of secretory organelles.

Phospholipid Turnover

While the role of calcium in the secretory process should be viewed as being closely linked to specific cellular proteins, it is perhaps equally important to consider the phospholipids as another modality for expressing the actions of calcium in the secretory cell. These effects of calcium may be expressed either by a direct action on phospholipids or indirectly through effects on phospholipases, the enzymes that purportedly regulate phospholipid turnover. The basic element that makes these enzymes such an intriguing candidate for a critical role in the calcium-dependent events associated with secretion rests with the finding that phospholipases, particularly phospholipase A_2, require calcium for activity [80].

Studies using artificial membrane systems have been employed to identify specific biochemical components of each membrane system that are crucial for calcium-induced exocytosis. Thus, experiments utilizing liposomes have revealed that the role of phospholipids in membrane fusion is related to their ability to form complexes with calcium [66]. Calcium particularly influences the structure of membranes containing acidic phospholipids by causing a phase transition of the lipid bilayer. The lipid phase transitions induced by calcium involve a phase separation of mixed acidic and neutral phospholipids, which by altering the stability of the membranes may make them more susceptible to fusion [21, 65, 66].

The asymmetric distribution of phospholipids in biological membranes provides the potential for controlling fusion events. Calcium influx initiated by the primary stimulus would bring calcium in contact with the acidic phospholipid head group on the cytoplasmic side of the plasma membrane and the abutting secretory organelle. Calcium could then induce a phase change with the formation of crystalline domains and promote fusion at the domain boundaries [66]. In addition to the interaction of calcium with acidic phospholipids of the cell membrane, calcium-induced hydrolysis of membrane ATP is also a required event during exocytosis, perhaps to 'destabilize' apposing membranes [67, 71, 77].

The relevance of phospholipase activity to calcium-mediated exocytosis is apparent not only because these enzymes require calcium for

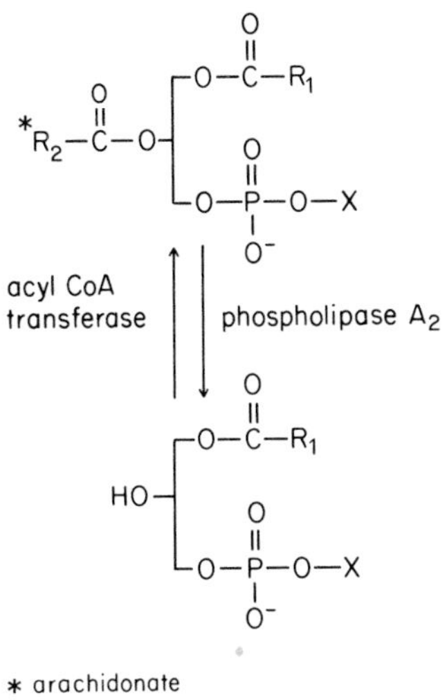

O
‖
─O─C─R₁ → $-O-C-R_1$

O
‖
*R₂─C─O─ → $*R_2-C-O-$

O
‖
─O─P─O─X → $-O-P-O-X$
|
O⁻

acyl CoA
transferase   phospholipase A₂

O
‖
─O─C─R₁ → $-O-C-R_1$

HO─

O
‖
─O─P─O─X → $-O-P-O-X$
|
O⁻

* arachidonate

Fig. 3. The deacylation-reacylation cycle. Stimulation of secretory cells causes activation of membranous phospholipase A_2 which brings about the release of arachidonic acid from position 2 of phospholipids (deacylation). The reincorporation of fatty acid into lysophospholipid (reacylation) is catalyzed by acyl CoA transferase. X-polar head group (inositol, choline, serine, or ethanolamine).

activity, but because of recent attention devoted to the possible role of arachidonic acid metabolites in cellular function, including secretion [78, 88]. The activation of a number of secretory organs, including the adrenal cortex, neutrophils, blood platelets, and the adenohypophysis involves the release of arachidonic acid from position 2 of phospholipids [9, 58, 80, 82, 84] (fig. 3). Arachidonic acid is metabolized through the cyclooxygenase and lipoxygenase pathways to prostaglandins and hydroxy fatty acids, which may express their effects by increasing calcium availability and elevating cyclic GMP levels [28, 57, 78].

It follows then that the calcium-induced activation of phospholipase A_2 may very well occupy a strategic position in the pathway leading to exocytotic secretion. Additional insight into this aspect of the action of calcium may be gleaned by analysis of the lysophospholipids, another product of the phospholipase A_2 reaction (fig. 3). These substances are of particular interest in the present context because they possess fusogenic properties [50]. It has been proposed that lysophosphatidylcholine and lysophosphatidylserine are involved in the secretion of catecholamines and histamine from chromaffin and mast cells, respectively [51, 99]. However, with regard to lysophosphatidylcholine, more recent evidence indicates that this lysophospholipid does not directly participate in the process of membrane fusion [50]. Additionally, our own studies have demonstrated in adrenocortical cells and neutrophils that the phospholipase A_2-mediated turnover of phosphatidylinositol and its conversion to lysophosphatidylinositol may be a pivotal event for activating the secretory process [81, 82, 84].

Alternatively, phospholipase-C-mediated reactions, involving the cleavage of polar head groups from phosphatidylinositol to produce diacylglycerol and phosphatidic acid by subsequent rephosphorylation, may also be crucial for mobilizing calcium or inducing membrane fusion [33, 55, 73]. It has even been proposed that phospholipase-C-mediated turnover of phosphatidylinositol is coupled to a phospholipase-A-mediated deacylation of phospholipids during platelet activation [8]. Thus, evidence that focuses on phospholipase A_2 as an important component of calcium-activated secretion must be considered in relation to other modifications in phospholipid turnover.

Conclusions

Since the role of calcium as an obligatory mediator of the secretory process is now unquestioned, studies are now concerned with (a) the details of the diverse mechanisms by which calcium availability is enhanced and (b) the nature of the cellular mechanisms involved in calcium-dependent secretion. In excitable secretory cells such as the adrenomedullary chromaffin cell, neurohypophysis, pancreatic beta-cell, and the neuron, during activation calcium traverses the cell membrane through voltage-dependent channels. Alternatively, receptor-operated channels that function independently of the membrane potential may also permit the influx of extracellular calcium in both excitable and non-excitable cells. Membrane activation is also associated with the release of calcium from the plasma membrane or from intracellular stores (endoplasmic reticulum or mitochondria). In non-excitable secretory organs such as exocrine glands, a relatively complex mechanism for regulating calcium availability exists, in that the pattern of calcium handling varies within a given organ depending upon the duration of stimulation, as well as the type of surface receptor activated. The signal is transmitted to the intracellular stores of calcium by such intracellular messengers as cyclic nucleotides and/or arachidonic acid metabolites.

Since the primary feature of stimulus-secretion coupling is an increase in the amount of free calcium in the cell, the interaction of calcium with various membranous and cytoplasmic components (phospholipases, phospholipids, and cyclic nucleotides) may serve to mediate the events that are necessary to convey the secretory organelle to the plasma membrane and the subsequent discharge of its contents from the cell. The complex, but coordinated, series of interactions between calcium, arachidonic acid metabolites, and cyclic nucleotides can be portrayed by employing the rabbit neutrophil to formulate a model that incorporates the observations that have been considered (fig. 4). The mechanism of stimulus-secretion coupling controlling the release of lysosomal enzymes from neutrophils is believed to be similar to that regulating secretion in other secretory cells, wherein calcium-evoked enzyme release occurs by exocytosis [79]. In this scheme, activation of plasmalemmal phospholipase A_2 is coupled to the increase in free or available calcium, resulting from the action of formyl methionyl peptides, phorbol esters, or ionophore A23187 acting as a mobile cation carrier [81, 82]. Free arachidonic acid is made available to form products that may modulate secretion either by mobilizing calcium [57] and/or modifying cyclic nucleotide levels [28]. The calcium-dependent accumulation of cyclic GMP might be a consequence of the calcium-induced release of arachidonic acid. On the other hand, lysophospholipids, another product of phospholipase activation, may act as fusogens, or may stimulate guanylate cyclase [79].

Thus, calcium not only plays a crucial role in arachidonate metabolism, as expressed through its regulation of phospholipase activity, or by a direct action on acidic phospholipids of cellular membranes; but arachidonic acid metabolites may participate in a feedback mechanism to regulate calcium metabolism in the cell. This concept implies the existence of multiple calcium-dependent pathways involved in stimulus-secretion coupling in the neutrophil. A similar sequence appears to exist in the adrenal cortex [78].

While the calcium-dependent deacylation of phospholipids may be instrumental in promoting events associated with exocytosis, the reacylating component may also subserve important functions (fig. 3). Reacylation of lysophospholipids will not only prevent these cytotoxic substances from accumulating to concentrations detrimental to the cell, but modification of the fatty acid profile in membrane phospholipids may be crucial

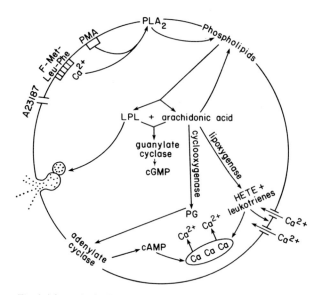

Fig. 4. A hypothetical model depicting the multiple sites of calcium mobilization during activation of lysosomal enzyme secretion from the rabbit neutrophil. The increase in calcium availability resulting from the interaction of the peptide formyl-methionyl-leucyl-phenyl-alanine (F-Met-Leu-Phe) or the tumor promoter phorbol myristate (PMA) with cell surface receptors, or by ionophore A23187 acting as a mobile carrier for the cation brings about the stimulation of membranous phospholipase A_2 (PLA_2) and the degradation (deacylation) of phospholipids. The resulting formation of lysophospholipids (LPL) and arachidonic acid can exert effects on enzyme secretion by enhancing the accumulation of cyclic GMP (cGMP) or in the case of the LPL by promoting membrane fusion reactions. The availability of precursor arachidonic acid enhances the production of prostaglandins (PG) and hydroxy fatty acids (HETE and leukotrienes) through the cyclooxygenase and lipoxygenase pathways, respectively. Prostaglandins and hydroxy fatty acids may express their actions by increasing the ionized calcium of the cell either by enhancing membrane permeability or by mobilizing calcium from cellular reservoirs, e.g. endoplasmic reticulum or mitochondria. Prostaglandins may also act by promoting cyclic AMP formation. A rapid reacylation of the lysophospholipid is deemed necessary to prevent the accumulation of this potentially cytotoxic substance(s). Moreover, the deacylation-reacylation sequence, which is thought to preferentially reincorporate arachidonate into phospholipids, may impart a higher degree of unsaturation to the cell membrane, thereby altering membrane function. For further details, see text.

for inducing permeability changes and/or regulating activity of membrane-bound enzymes [47, 83]. For example, an enrichment in polyunsaturated fatty acids (e.g., arachidonate) may bring about an increase in membrane fluidity with concomitant alterations in membrane function. But regardless of the particular role(s) that each of these reactions plays in the overall scheme, the action of calcium as an obligatory activator of phospholipid turnover is deemed pivotal. This interaction between calcium and phospholipids – and activator proteins as well – could in large measure account for the unique role of this cation in stimulus-secretion coupling.

References

1 Adelstein, R. S.; Conti, M. A.; Pato, M. D.: Regulation of myosin light chain kinase by reversible phosphorylation and calcium-calmodulin. Ann. N.Y. Acad. Sci. *356:* 142–150 (1980).

2 Aunis, D.; Guerold, B.; Bader, M.-F.; Cieselski-Treska, J.: Immunocytochemical and biochemical demonstration of contractile proteins in chromaffin cells in culture. Neuroscience *5:* 2261–2277 (1980).

3 Baker, P. F.: Transport and metabolism of calcium ions in nerve. Prog. Biophys. molec. Biol. *24:* 177–223 (1972).

4 Bartelt, D. C.; Scheele, G. A.: Calmodulin and calmodulin-binding proteins in canine pancreas. Ann. N.Y. Acad. Sci. *356:* 356–357 (1980).

5 Baum, B. J.; Freiburg, J. M.; Ito, H.; Roth, G. S.; Filburn, C. R.: β-Adrenergic regulation of protein phosphorylation and its relationship to exocrine secretion in dispersed rat parotid gland acinar cells. J. biol. Chem. *256:* 9731–9736 (1981).

6 Berridge, M.: The interaction of cyclic nucleotides and calcium in the control of cellular activity. Adv. cyclic Nucleotide Res. *6:* 1–98 (1975).

7 Berridge, M. J.: Phosphatidylinositol hydrolysis: a multifunctional transducing mechanism. Mol. cell. Endocrinol. *24:* 115–140 (1981).

8 Billah, M. M.; Lapetina, E. G.; Cuatrecasas, P.: Phospholipase A_2 and phospholipase C activities of platelets. J. biol. Chem. *255:* 10227–10231 (1980).

9 Bills, T. K.; Smith, J. B.; Silver, M. J.: Metabolism of [^{14}C] arachidonic acid by human platelets. Biochim. biophys. Acta *424:* 303–314 (1976).

10 Brandt, B. L.; Hagiwara, S.; Kidokoro, Y.; Miyazaki, S.: Action potentials in the rat chromaffin cell and effects of acetylcholine. J. Physiol., Lond. *263:* 417–439 (1976).

11 Brown, E. M.; Dawsonhughes, B. F.; Wilson, R. E.; Adragaz, N.: Calmodulin in dispersed human parathyroid cells. J. clin. Endocr. Metab. *53:* 1064–1071 (1981).

12 Butcher, F. R.: The role of calcium and cyclic nucleotides in α-amylase release from slices of rat parotid: studies with the divalent cation ionophore A23187. Metabolism *24:* 409–418 (1975).

13 Butcher, F. R.; Putney, J. W., Jr.: Regulation of parotid gland function by cyclic nucleotides and calcium. Adv. Cyclic Nucleotide Res. *13:* 215–249 (1980).

14 Carafoli, E.; Crompton, M.: The regulation of intracellular calcium by mitochondria. Ann. N.Y. Acad. Sci. *307:* 269–284 (1978).

15 Cheung, W. Y.: Calmodulin and the adenylate cyclase-phosphodiesterase system. Cell Calcium *2:* 263–280 (1981).

16 Dedman, J. R.; Brinkley, B. R.; Means, A. R.: Regulation of microfilaments and microtubules by calcium and cyclic AMP. Adv. Cyclic Nucleotide Res. *11:* 131–174 (1979).

17   DeLorenzo, R.J.: The calmodulin hypothesis of neurotransmission. Cell Calcium *2:* 365–385 (1981).

18   Douglas, W.W.: Stimulus-secretion coupling: the concept and clues from chromaffin and other cells. Br. J. Pharmacol. *34:* 451–474 (1968).

19   Douglas, W.W.: Involvement of calcium in exocytosis and the exocytosis-vesiculation sequence. Biochem. Soc. Symp. *39:* 1–28 (1974).

20   Douglas, W.W.: Secretomotor control of adrenal medullary secretion: synaptic, membrane, and ionic events in stimulus-secretion coupling; in Blaschko et al., Handbook of physiology, volume VI, sect. 7, pp. 367–388 (American Physiological Society, Washington 1975).

21   Duzgunes, N.; Wilschut, J.; Fraley, R.; Papahadjopoulas, D.: Studies on the mechanism of membrane fusion. Role of head-group composition in calcium- and magnesium-induced fusion of mixed phospholipid vesicles. Biochim. biophys. Acta *642:* 182–195 (1981).

22   Farquhar, M.G.: Processing of secretory products by cells of the anterior pituitary gland. Mem. Soc. Endocr. *19:* 79–122 (1971).

23   Feinstein, M.B.: Release of intracellular membrane-bound calcium precedes the onset of stimulus-induced exocytosis in platelets. Biochem. biophys. Res. Commun. *93:* 593–600 (1980).

24   Feinstein, M.B.; Rodan, G.A.; Cutler, L.S.: Cyclic AMP and calcium in platelet function; in Gordon, Platelets in biology and pathology, pp. 437–472 (Elsevier, Amsterdam 1981).

25   Fleischer, N.; Schubart, U.K.; Fleckman, A.; Erlichman, J.: Calcium antagonists and the secretion of insulin and thyrotropin; in Weiss, New perspectives on calcium antagonists, pp. 177–190 (Waverly Press, Baltimore 1981).

26   Garcia, A.G.; Kirpekar, S.M.; Prat, J.C.: A calcium ionophore stimulating the secretion of catecholamines from the cat adrenal. J. Physiol., Lond. *244:* 253–262 (1975).

27   Gerrard, J.M.; Peterson, D.A.; White, J.G.: Calcium mobilization; in Gordon, Platelets in biology and pathology, pp. 407–436 (Elsevier, Amsterdam 1981).

28   Goldberg, N.D.; Graff, G.; Haddox, M.K.; Stephenson, J.H.; Glass, D.B.; Moser, M.E.: Redox modulation of splenic cell soluble guanylate cyclase activity: activation by hydrophilic and hydrophobic oxidants represented by ascorbic and dehydroascorbic acids, fatty acid hydroperoxides and prostaglandin endoperoxides. Adv. Cyclic Nucleotide Res. *9:* 101–130 (1978).

29   Grynszpan-Winograd, O.: Ultrastructure of the chromaffin cell; in Blaschko et al., Handbook of physiology, vol. VI, sect. 7, pp. 295–308 (American Physiological Society, Washington 1975).

30   Habener, J.F.; Potts, J.T.: Subcellular distributions of parathyroid hormone, hormonal precursors and parathyroid secretory protein. Endocrinology *104:* 265–275 (1979).

31   Hahn, H.J.; Gylfe, E.; Hellman, B.: Calcium and pancreatic β-cell function. Evidence for cyclic AMP induced translocation of intracellular calcium. Biochim. biophys. Acta *630:* 425–432 (1980).

32   Harvey, A.M.; MacIntosh, F.C.: Calcium and synaptic transmission in a sympathetic ganglion. J. Physiol., Lond. *97:* 408–416 (1940).

33   Hawthorne, J.N.; Pickard, M.R.: Phospholipids in synaptic function. J. Neurochem. *32:* 5–14 (1979).

34 Heuser, J.E.; Reese, T.S.; Dennis, M.J.; Jan, Y.; Jan, L.; Evans, L.: Synaptic vesicle exocytosis captured by quick-freezing and correlated with quantal transmitter release. J. Cell Biol. *81:* 275–300 (1979).

35 Howell, S.L.; Montague, W.: Regulation by nucleotides of ^{45}calcium uptake in homogenate of rat islets of Langerhans. FEBS Lett. *52:* 48–52 (1975).

36 Kanazawa, Y.; Kawasu, S.; Ikeuchi, M.; Kosaka, K.: The relationship of intracytoplasmic movement of beta granules to insulin release in monolayer cultured pancreatic beta-cells. Diabetes *29:* 953–959 (1980).

37 Käser-Glanzmann, R.; Jakabova, M.; George, J.N.; Lüscher, E.F.: Further characterization of calcium-accumulating vesicles from human blood platelets. Biochim. biophys. Acta *512:* 1–12 (1978).

38 Katz, B.: Microphysiology of the neuromuscular junction. Bull. Johns Hopkins Hosp. *102:* 275–312 (1958).

39 Katz, B.: The release of neural transmitter substances (Thomas, Springfield 1969).

40 Katz, B.; Miledi, R.: Tetrodotoxin-resistant electrical activity in presynaptic terminals. J. Physiol., Lond. *203:* 459–487 (1969).

41 Kidokoro, Y.; Ritchie, A.K.: Chromaffin cell action potentials and their possible role in adrenaline secretion from rat adrenal medulla. J. Physiol., Lond. *307:* 199–216 (1980).

42 Kilpatrick, D.L.; Slepetis, R.J.; Corcoran, J.J.; Kirshner, N.: Calcium uptake and catecholamine secretion by cultured bovine adrenal medulla cells. J. Neurochem. *38:* 427–435 (1982).

43 Kirpekar, S.M.; Prat, J.C.: Release of catecholamines from perfused cat adrenal gland by veratridine. Proc. natn. Acad. Sci. USA *76:* 2081–2083 (1979).

44 Kretsinger, R.H.: Calcium-binding proteins. A. Rev. Biochem. *45:* 239–266 (1976).

45 Kuo, J.F.; Andersson, R.G.G.; Wise, B.C.; Mackerlova, L.; Salomonsson, I.; Brackett, N.L.; Katoh, N.; Shoji, M.; Wrenn, R.W.: Calcium-dependent protein kinase: widespread occurrence in various tissues and phyla of the animal kingdom and comparison of effects of phospholipid, calmodulin and trifluoperazine. Proc. natn. Acad. Sci. USA *77:* 7039–7043 (1980).

46 Lacy, P.E.; Finke, E.H.; Codilla, R.C.: Cinemicrographic studies on β-granule movement in monolayer culture of islet cells. Lab. Invest. *33:* 570–576 (1975).

47 Lands, W.E.M.; Crawford, C.G.: Enzymes of membrane phospholipid metabolism in animals; in Martonosi, The enzymes of biological membranes, vol.2, pp.3–85 (Plenum Press, New York 1976).

48 Laugier, R.; Petersen, O.H.: Pancreatic acinar cells: electrophysiological evidence for stimulant-evoked increase in membrane calcium permeability in the mouse. J. Physiol., Lond. *303:* 61–72 (1980).

49 Laychock, S.G.; Landon, E.J.; Hardman, J.G.: The effect of ACTH in adrenal cortical, microsomal vesicles. Endocrinology *103:* 2198–2206 (1978).

50 Lucy, J.A.: Mechanisms of chemically induced cell fusion; in Poste, Nicolson, Membrane fusion, pp.268–304 (North Holland, Amsterdam 1978).

51 Martin, T.W.; Lagunoff, D.: Interactions of lysophospholipids and mast cells. Nature, Lond. *279:* 250–252 (1979).

52 Matthews, E.K.: Insulin secretion; in Dumont, Nunez, First European Symposium on Hormones and Cell Regulation, pp.57–76 (Elsevier, Amsterdam 1977).

53 Meldolesi, J.; Borgese, N.; DeCamilli, P.; Ceccarelli, B.: Cytoplasmic membranes and the secretory process; in Poste, Nicolson, Membrane fusion, pp. 510–627 (North Holland, Amsterdam 1978).

54 Michaelson, D.M.; Avissar, S.: Ca^{2+}-dependent protein phosphorylation of purely cholinergic Torpedo synaptosomes. J. biol. Chem. 254: 12542–12546 (1979).

55 Michell, R.H.: Inositol phospholipids and cell surface receptor function. Biochim. biophys. Acta 415: 81–147 (1975).

56 Morris, J.F.; Nordmann, J.J.; Dyball, R.E.J.: Structure-function correlation in mammalian neuroscretion. Int. Rev. exp. Path. 18: 1–95 (1978).

57 Naccache, P.H.; Sha'afi, R.I.; Borgeat, P.; Goetzl, E.J.: Mono- and dihydroxyeicosatetraenoic acids alter calcium homeostasis in rabbit neutrophils. J. clin. Invest. 67: 1584–1587 (1981).

58 Naor, Z.; Catt, K.J.: Mechanism of action of gonadotropin-releasing hormone. Involvement of phospholipid turnover in luteinizing hormone release. J. biol. Chem. 256: 2226–2229 (1981).

59 Nemeth, E.F.; Douglas, W.W.: Effects of microfilament-active drugs, phalloidin and the cytochalasins A and B, on exocytosis in mast cells evoked by 48/80 or A23187. Arch. Pharmacol. 302: 153–156 (1978).

60 Nishizuka, Y.; Takai, Y.; Hashimoto, E.; Kishimoto, A.; Kuroda, Y.; Sakai, K.; Yamamura, H.: Regulatory and functional compartment of three multifunctional protein kinase systems. Mol. cell. Biochem. 23: 153–165 (1979).

61 Orci, L.; Perrelet, A.: Ultrastructural aspects of exocytotic membrane fusion. Cell Surface Rev. 5: 629–656 (1978).

62 Ostlund, R.E., Jr.; Leung, J.T.; Kipnis, D.M.: Myosins of secretory tissues. J. Cell Biol. 77: 827–837 (1978).

63 Ozawa, S.; Kimura, N.: Membrane potential changes caused by thyrotropin-releasing hormone in the clonal GH_3 cell and their relationship to secretion of pituitary hormone. Proc. natn. Acad. Sci. 76: 6017–6020 (1979).

64 Palade, G.E.: Functional changes in the structure of cell components; in Hayashi, Subcellular particles. Soc. of Gen. Physiologists Symp., pp. 64–80 (Ronald Press, New York 1959).

65 Papahadjopoulos, D.: Calcium-induced phase changes and fusion in natural and model membranes; in Poste, Nicolson, Membrane fusion, pp. 766–790 (North Holland Publishing Company, Amsterdam 1978).

66 Papahadjopoulos, D.; Portis, A.; Pangborn, W.: Calcium-induced lipid phase transitions and membrane fusion. Ann. N.Y. Acad. Sci. 308: 50–66 (1978).

67 Plattner, H.; Reichel, K.; Matt, H.; Beisson, J.; Lefort-Tran, M.; Pouphile, M.: Genetic dissection of the final exocytosis steps in Paramecium tetraurelia cells: cytochemical determination of Ca^{2+}-ATPase activity over preformed exocytosis sites. J. Cell Sci. 46: 17–40 (1980).

68 Pollard, H.B.; Creutz, C.E.; Pazoles, C.J.: Mechanisms of calcium action and release of vesicle-bound hormones during exocytosis. Recent Prog. Horm. Res. 37: 299–332 (1981).

69 Pollard, T.D.; Fujiwara, K.; Niederman, R.; Maupin-Szamier, P.: Evidence for the role of cytoplasmic actin and myosin in cellular structure and motility; in Goldman, Pollard, Rosenbaum, Cell motility, Book B, Actin, Myosin, and associated proteins, pp. 689–724 (Cold Spring Harbor Laboratory, Cold Spring Harbor 1976).

70   Ponnappa, B.C.; Dormer, R.L.; Williams, J.M.: Characterization of an ATP-dependent Ca^{2+} uptake system in mouse pancreatic microsomes. Am. J. Physiol. *240:* G122–129 (1981).

71   Poste, G.; Allison, A.C.: Membrane fusion. Biochim. biophys. Acta *300:* 421–465 (1973).

72   Putney, J.W., Jr.: Calcium antagonists and calcium-gating mechanisms in the exocrine glands; in Weiss, New perspectives on calcium antagonists, pp. 169–175 (Waverly Press, Baltimore 1981).

73   Putney, J.W., Jr.: Recent hypotheses regarding the phosphatidylinositol effect. Life Sci. *12:* 1183–1194 (1981).

74   Ribalet, B.; Beigelman, P.M.: Cyclic variation of K^+ conductance in pancreatic β-cells: Ca^{2+} and voltage dependence. Am. J. Physiol. *237:* C137–C146 (1979).

75   Ritchie, A.K.: Catecholamine secretion in a rat pheochromocytoma cell line: two pathways for calcium entry. J. Physiol., Lond. *286:* 541–561 (1979).

76   Rubin, P.P.: The role of calcium in the release of neutrotransmitter substances and hormones. Pharmac. Rev. *22:* 389–428 (1970).

77   Rubin, R.P.: Calcium and the secretory process (Plenum Press, New York 1974).

78   Rubin, R.P.: Actions of calcium antagonists on secretory cells; in Weiss, New perspectives on calcium antagonists, pp. 147–158 (Waverly Press, Baltimore 1981).

79   Rubin, R.P.: Calcium and cellular secretion (Plenum Publishing Corporation, New York 1982).

80   Rubin, R.P.; Laychock, S.G.: Prostaglandins and calcium-membrane interactions in secretory glands. Ann. N.Y. Acad. Sci. *307:* 377–390 (1978).

81   Rubin, R.P.; Sink, L.E.; Freer, R.J.: On the relationship between formyl methionyl-leucyl-phenylalanine stimulation of arachidonyl phosphatidylinositol turnover and lysosomal enzyme secretion by rabbit neutrophils. Molec. Pharmacol. *19:* 31–37 (1981).

82   Rubin, R.P.; Sink, L.E.; Freer, R.J.: Activation of arachidonyl phosphatidylinositol turnover in rabbit neutrophils by the calcium ionophore A23187. Biochem. J. *194:* 497–505 (1981).

83   Sandermann, H.: Regulation of membrane enzymes by lipids. Biochim. biophys. Acta *515:* 209–237 (1978).

84   Schrey, M.P.; Rubin, R.P.: Characterization of a calcium mediated activation of arachidonic acid turnover in adrenal phospholipids by corticotropin. J. biol. Chem. *254:* 11234–11241 (1979).

85   Schubart, U.K.; Fleischer, N.; Erlichman, J.: Ca^{2+}-dependent protein phosphorylation and insulin release in intact hamster insulinoma cells. J. biol. Chem. *255:* 11063–11066 (1980).

86   Schulz, I.: Messenger role of calcium ion function of pancreatic acinar cells. Am. J. Physiol. *239:* G335–347 (1980).

87   Selinger, Z.; Naim, E.; Lasser, M.: ATP-dependent calcium uptake by microsomal preparations from rat parotid and submaxillary glands. Biochim. biophys. Acta *203:* 326–334 (1970).

88   Sha'afi, R.I.; Naccache, P.H.: Ionic events in neutrophil chemotaxis and secretion; in Weissmann, Advances in inflammation research, vol. 3, pp. 115–148 (Raven Press, New York 1981).

89   Sieghart, W.; Theoharides, T.C.; Alper, S.L.; Douglas, W.W.; Greengard, P.: Cal-

cium-dependent protein phosphorylation during secretion by exocytosis in the mast cell. Nature, Lond. *275:* 329–331 (1978).

90 Sugden, M.C.; Ashcroft, S.J.H.: Effects of phosphoenolpyruvate, other glycolytic intermediates and methylxanthines on calcium uptake by a mitochondrial fraction from rat pancreatic islets. Diabetologia *15:* 173–180 (1978).

91 Sugden, M.C.; Christie, M.R.; Ashcroft, S.J.H.: Presence and possible role of calcium-dependent regulator (calmodulin) in rat islets of Langerhans. FEBS Lett. *105:* 95–100 (1979).

92 Taraskevich, P.S.; Douglas, W.W.: Catecholamines of supposed inhibitory hypophysiotrophic function suppress action potentials in prolactin cells. Nature, Lond. *276:* 832–834 (1978).

93 Taylor, D.L.; Hellelwell, S.B.; Virgin, H.W.; Heiple, J.: The solation-contraction coupling hypothesis of cell movements; in Hatano et al., Cell motility: molecules and organization, pp.363–377 (University Park Press, Baltimore 1978).

94 Thorn, N.A.; Russell, J.T.; Torp-Pedersen, C.; Treiman, M.: Calcium and neurosecretion. Ann. N.Y. Acad. Sci. *307:* 618–639 (1978).

95 Trifaro, J.: Contractile proteins in tissues originating in the neural crest. Neuroscience *3:* 1–24 (1978).

96 Viveros, O.H.: Mechanism of secretion of catecholamines from adrenal medulla; in Blaschko et al., Handbook of physiology, vol. VI, sect. 7, pp. 389–425 (American Physiological Society, Washington 1975).

97 Watson, E.L.; Williams, J.A.; Siegel, I.A.: Calcium mediation of cholinergic stimulated amylase release from mouse parotid gland. Am. J. Physiol. *236:* C233–237 (1979).

98 Weiss, G.B.: Sites of action of calcium antagonists in vascular smooth muscle; in Weiss, New perspectives on calcium antagonists, pp.83–94 (Waverly Press, Baltimore 1981).

99 Winkler, H.; Schneider, F.H.; Rufener, C.; Nakane, P.K.; Hörtnagl, H.: Membranes of adrenal medulla: their role in exocytosis; in Ceccarelli et al., Adv. in Cytopharmacology, vol.2, Cytopharmacology of secretion, pp.127–139 (Raven Press, New York 1974).

100 Wong, P.Y.K.; Lee, W.H.; Chao, P.H-W.: The role of calmodulin in prostaglandin metabolism. Ann. N.Y. Acad. Sci. *356:* 179–189 (1980).

Prof. Ronald P. Rubin, PhD, Medical College of Virginia, Department of Pharmacology, PO Box 613 – MCV Station, Richmond, VA 23298 (USA)

Cell Biology of the Secretory Process, pp. 73–101 (Karger, Basel 1984)

Role of Presecretory Proteins in the Secretory Process

Thomas Carne[a], George Scheele[b]

[a] Department of Chemistry, University of Calgary, Alta., Canada; [b] Department of
Cell Biology, Rockefeller University, New York, N.Y., USA

Introduction

Cellular and molecular mechanisms for the secretion of proteins from
cells have largely been elucidated over the past 25 years. The intracellular
pathway for protein secretion in eukaryotic cells includes the following 6
steps: (1) protein synthesis on ribosomes bound to the rough endoplasmic
reticulum [82]; (2) segregation of nascent proteins into the cisternal space
of the RER [40, 64, 65, 68, 78, 82]; (3) intracellular transport from the
RER to the Golgi apparatus [11, 39]; (4) condensation at the level of
immature secretory granules on the trans-face of the Golgi apparatus [40];
(5) storage of secretory product in mature secretory granules [27, 33, 40,
78], and (6) release of product to the extracellular environment through the
process of exocytosis [58, 91]. The intracellular organelles which participate in this process (RER, Golgi, secretory granules) are limited by membranes which define, for the purpose of discussion of the secretory process,
two spatial regions in the cell: the extraorganellar space or cytosol, which
contains the mRNA, ribosomes, activated amino acids, and factors for
protein synthesis, and the intraorganellar or cisternal space which channels the secretory products to the cell surface. It is in this latter space where
post-translational modifications (glycosylation, disulfide bonding, proteolytic reduction, hydroxylation, phosphorylation, and sulfation) occur in
secretory proteins prior to their release from the cell. Central to the
secretory process is the translocation of secretory proteins from their site of
synthesis in the cytosolic space across the membrane of the RER to the
cisternal space where they are channeled through a series of interconnect-
able membrane-bound compartments to the extracellular medium.

Work over the past 10 years has defined, in general terms, the mechanism of translocation of secretory proteins across the RER membrane. *Milstein* et al. [54] first demonstrated that a secretory protein, the immunoglobulin light chain of the murine myeloma cell, is synthesized with an amino terminal peptide extension, which they believed signaled attachment of the nascent polypeptide to the RER membrane. *Blobel and Dobberstein* [5] confirmed the precursor finding of *Milstein* et al. [54] for the light chain of MOPC 41 myeloma cell and, using RER membranes derived from dog pancreas [74], performed functional reconstitution studies which first demonstrated that the translocation process was cotranslational. Similar studies carried out by others demonstrated that a variety of secretory proteins, including enzymes and zymogens, nutritional or storage proteins, protease inhibitors, serum proteins, immunoglobulins, toxins, and hormones are synthesized with amino terminal extensions responsible for transport of their proteins into the cisternal secretory pathway (table I). In the most comprehensive and unambiguous study on the translocation of secretory proteins, *Scheele* et al. [77] followed simultaneously the synthesis, segregation and processing (proteolytic removal of peptide extensions) of 14 nonglycosylated pancreatic exocrine proteins in the presence of dog pancreatic microsomal membranes. Using an improved in vitro reconstitution system, they showed that processing and segregation of proteins were complete and demonstrated for the first time absolute coupling between these two events. They also showed that translocation of secretory proteins into microsomal vesicles was an irreversible process and more recently *Scheele and Jacoby* [75, 76] have demonstrated that proteins translocated by this mechanism show conformational properties and biological activities indistinguishable from authentic secreted proteins.

These studies have led to the development, over time, of the transport peptide hypothesis [5, 54, 73; for recent reviews, see 19, 37, 41, 46, 69, 87]. In this chapter we will outline in greater detail the present state of the transport peptide hypothesis and then focus on the amino acid sequences of a large number of transport peptides in an attempt to relate the structures of these peptides to the mechanism of protein translocation.

The Transport Peptide Hypothesis

Central to the hypothesis is the transport peptide, an amino terminal extension of 15–30 amino acid residues, which plays a prominent role in

Table I. The amino acid sequences of transport peptides associated with secretory and transmembrane proteins

(A.) Eukaryotic enzymes and zymogens

```
        -15        -10        -5                -1
         |          |          |                 |
X-X-F-F-L-L-L-X-V-I-G-F-X-V-A-Q-                CANINE PANCREATIC PREAMYLASE (10)
X-X-P-L-L-I-L-A-F-L-X-A-A-V-A-X-                CANINE PRETRYPSINOGEN 1 (10)
A-L-X-I-T-F-L-A-L-L-X-X-V-A-F-                  CANINE PRETRYPSINOGEN 2+3 (10)
A-F-L-I-L-V-X-A-F-A-L-X-X-G-A-F-X-X-            CANINE PRECHYMOTRYPSINOGEN 2 (10)
X-X-L-I-L-V-F-G-A-L-L-X-A-I-Y-X-Q-              CANINE PREPROCARBOXYPEPTIDASE A1 (10)
         -L-L-S-L-I-G-F-C-Y-A-Q-                RAT PANCREATIC PREAMYLASE (47)
M-X-F-F-L-L-L-X-L-I-X-L-                        RAT SALIVARY PREAMYLASE (26)
M-K-F-F-L-L-L-S-L-G-F-C-W-A-Q-                  RAT LIVER AND SALIVARY PREAMYLASE (29)
M-K-F-V-L-L-L-S-L-I-G-F-C-W-A-Q-                MOUSE PANCREATIC PREAMYLASE (30)
M-S-A-L-L-I-L-A-L-V-G-A-A-V-A-                  RAT PANCREATIC PRETRYPSINOGEN 1 (48)
M-S-A-L-L-I-L-A-L-V-G-A-A-V-A-                  RAT PANCREATIC PRETRYPSINOGEN II (48)
M-L-R-F-L-V-F-A-S-L-V-L-Y-G-H-S-                RAT PANCREATIC PREPROELASTASE I (48)
M-I-R-T-L-L-L-S-A-F-V-A-G-A-L-S-                RAT PANCREATIC PREPROELASTASE II (48)
M-R-S-L-L-I-L-V-L-C-F-L-P-L-A-A-L-G-K-          CHICKEN PRELYSOZYME (60)
```

(B.) Eukaryotic nutritional and storage proteins

```
    -20        -15        -10        -5              -1
     |          |          |          |               |
M-K-V-L-I-L-A-C-L-V-A-L-A-L-A-R-                OVINE PRE β CASEIN (51)
M-K-L-L-I-L-T-C-L-V-A-V-A-L-A-R-                OVINE PRE αS₁ CASEIN (51)
M-K-V-L-M-K-A-C-L-V-A-V-A-L-A-R-                OVINE PRE αS₂ CASEIN (51)
M-R-K-S-I-L-L-V-V-T-L-A-L-T-L-P-F-L-I-A-P-      OVINE PRE κ CASEIN (51)
M-M-S-E-F-V-S-L-L-V-G-I-L-F-X-A-T-P-A-E-        OVINE PRE α LACTALBUMIN (51)
M-K-C-L-L-L-A-L-G-L-A-L-A-C-G-V-Q-A-I-          OVINE PRE β LACTALBUMIN (51)
M-K-L-I-L-C-T-V-L-S-L-G-I-A-A-V-C-F-A-A-        CHICKEN PRECONALBUMIN (92)
```

Table I (continued)

(C.) Eukaryotic protease inhibitors

```
 -25      -20      -15      -10       -5       -1
  |        |        |        |        |        |
M-A-M-A-G-V-F-V-L-F-S-F-V-L-C-G-F-L-P-D-A-A-F-G-A-     CHICKEN PREOVOMUCOID (93)
```

(D.) Eukaryotic serum proteins

```
          -20      -15      -10       -5       -1
           |        |        |        |        |
        M-K-W-V-T-F-L-L-L-L-F-I-S-G-S-A-F-S-R-         RAT PREPROALBUMIN (86)
        M-K-L-L-L-L-C-L-G-L-T-L-V-C-G-H-A-E-           RAT LIVER PRE α2μ GLOBULIN (20)
```

(E.) Eukaryotic immunoglobulins

```
          -20      -15      -10       -5       -1
           |        |        |        |        |
  M-D-M-R-A-P-A-Q-I-F-G-F-L-L-L-L-F-P-G-T-R-C-D-    MOUSE PRELIGHT CHAIN IMMUNOGLOBULIN (MPC 41A) (9)
(M)M-X-A-P-A-X-I-F-X-P-L-L-L-L-F-P-X-T-X-C-         MOUSE PRELIGHT CHAIN IMMUNOGLOBULIN (MPC 41B) (8)
  M-X-T-X-T-L-L-L-W-V-L-L-L-W-V-P-X-X-T-W-          MOUSE PRELIGHT CHAIN IMMUNOGLOBULIN (MPC 321) (71)
  M-A-W-I-S-L-I-L-S-L-L-A-L-S-S-G-A-I-S-Z-          MOUSE PRELIGHT CHAIN IMMUNOGLOBULIN (MPC 104E) (8)
  M-A-W-T-S-L-I-L-S-L-L-A-L-C-S-G-A-S-S-Z-          MOUSE PRELIGHT CHAIN IMMUNOGLOBULIN (MPC 315) (21)
      E-T-D-T-L-L-L-W-V-L-L-L-W-V-P-G-              MOUSE PRE κ LIGHT CHAIN (9)
```

(F.) Eukaryotic toxins

```
          -20      -15      -10       -5       -1
           |        |        |        |        |
  M-K-F-L-X-N-V-A-L-V-F-M-V-V-Y-I-X-Y-I-Y-A-A-       HONEY-BEE PREPROMELITTIN (42)
```

(G.) *Eukaryotic hormones*

```
          -25        -20        -15        -10         -5         -1
           |          |          |          |          |          |
M- -P-G-S-R-T-S-L-L-A-F-A-L-L-C-L-P-W-L-Q-E-A-G-A-            HUMAN PREPLACENTAL LACTOGEN (80)
M-A-A-D-S-Q-T-P-W-L-L-T-F-S-L-L-C-L-L-W-P-Q-E-A-G-A-L-        RAT PREGROWTH HORMONE (79)
M-N-S-Q-V-S-A-R-L-A-G-T-L-L-L-L-M-M-S-N-L-L-F-L-Q-N-V-Q-T-L- RAT PREPROLACTIN (49)
M-P-R-L-C-S-S-R-S-G-A-L-L-L-A-L-L-L-Q-A-S-M-E-V-R-G-W-       BOVINE PREPRO-OPIOMELANOTROPIN (56)
M-M-S-A-K-D-M-V-K-V-M-I-V-M-L-A-I-C-F-L-A-R-S-D-G-K-         BOVINE PREPROPARATHYROID HORMONE (28)
    M-A-L-W-M-R-F-L-P-L-L-A-L-L-V-L-W-E-P-K-P-A-Q-A-F-       RAT PREPROINSULIN 1 (13)
    M-A-L-W-I-R-F-L-F-F-L-A-L-L-T-L-W-E-P-K-P-A-Q-A-F-       RAT PREPROINSULIN 2 (13)
    M-A-L-W-M-R-L-L-P-L-L-A-L-L-A-L-W-G-P-D-P-A-A-A-F-       HUMAN PREPROINSULIN (89)
```

(H.) *Eukaryotic membranes proteins*

```
M-E-F-S-L-L-L-L-L-A-F-L-A-G-L-L-L-L-L-F-    RABBIT CYTOCHROME P-450 LM₂ (31)
```

(I.) *Prokaryotic secretory proteins*

```
          -25        -20        -15        -10         -5         -1
           |          |          |          |          |          |
    M-S-I-Q-H-F-R-V-A-L-I-P-F-F-A-A-F-C-L-P-V-F-A-H-        E. COLI PREPENICILLINASE (βLACTAMASE) (2)
M-K-I-K-T-G-A-R-I-L-A-L-S-A-L-T-T-M-M-F-S-A-S-A-L-A-K-      E. COLI PREMALTOSE BINDING PROTEIN (4)
```

(J.) *Prokaryotic membrane proteins*

```
          -25        -20        -15        -10         -5         -1
           |          |          |          |          |          |
M-K-K-S-L-V-L-K-A-S-V-A-V-A-T-L-V-P-M-L-S-F-A-A-            E. COLI F₁ PRECOAT PROTEIN (88)
    M-K-A-T-K-L-V-L-G-A-V-I-L-G-S-T-L-L-A-G-C-             E. COLI PRELIPOPROTEIN (38)
M-M-I-T-L-R-K-L-P-L-A-V-A-V-A-A-C-V-M-S-A-Q-A-M-A-V-       E. COLI PRE λ RECEPTOR PROTEIN (32)
M-I-T-L-R-K-L-P-L-A-V-A-V-A-A-C-V-M-S-A-N-A-M-A-V-         E. COLI PRE LAM B (55)
```

Table I (continued)

(K.) Prokaryotic viral membrane proteins

```
         -20      -15      -10      -5       -1
         |        |        |        |        |
M-K-K-S-L-V-L-K-A-S-V-A-V-A-T-L-V-P-M-L-A-F-A-A-    E. COLI PRE Fd MAJOR PHAGE COAT PROTEIN (88)
        M-K-K-L-L-F-A-I-P-L-V-V-P-F-Y-S-H-S-A-      E. COLI PRE Fd MINOR PHAGE COAT PROTEIN (70)
```

(L.) Eukaryotic viral membrane proteins

```
               -15      -10      -5       -1
               |        |        |        |
M-K-C-L-L-Y-L-A-F-L-F-I-H-V-N-C-    VSV MEMBRANE GLYCOPROTEIN (44)
  M-A-I-Y-L-I-L-L-F-I-A-V-R-G-      INFLUENZA A HEMAGGLUTININ RI/5/57 (H2) (1)
M-A-I-I-Y-L-I-L-L-F-T-A-V-R-G-      INFLUENZA A HEMAGGLUTININ JAPAN 305/57 (H2) (99)
M-K-T-I-I-A-L-S-Y-I-F-C-L-V-L-G-    INFLUENZA A HEMAGGLUTININ MEMPHIS 102/72 (H3)(98)
```

A = Ala; B = Asx; C = Cys; D = Asp; E = Glu; F = Phe; G = Gly; H = His; I = Ile; K = Lys; M = Met; N = Asn; P = Pro; Q = Gln; R = Arg; S = Ser; T = Thr; V = Val; W = Trp; X = unknown; Y = Tyr. Proteolytic cleavage of the transport peptide occurs after residue −1. The membrane protein, cytochrome P450 LM$_2$, contains an uncleaved transport peptide.

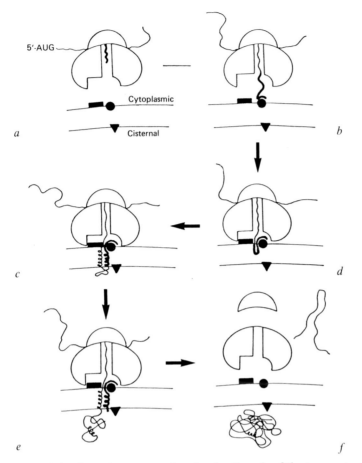

5′-AUG

Cytoplasmic

a

Cisternal

b

c

d

e

f

Fig. 1. A schematic representation showing the role of the transport peptide in the translocation of eukaryotic secretory proteins through the RER membrane. The six stages of translocation shown *(a–f)* are described in detail in the text. The membrane receptor complex responsible for the initial binding of the transport peptide is represented by (●), the membrane complex responsible for electrostatic ribosomal binding by (■), and the transport peptide hydrolase by (▼). Adaptation of the transport peptide hypothesis to the translocation of prokaryotic secretory proteins is described in the text.

the identification of a protein as 'secretory' and in the mechanism for the transfer of the protein across the RER membrane. The essential features of the transport peptide hypothesis are shown schematically in figure 1. As the transport peptide emerges from the tunnel in the large ribosomal subunit (fig. 1a), it is responsible for the attachment of the ribosomal-mRNA complex to a receptor complex in the membrane of the rough endoplasmic reticulum (fig. 1b). In addition, electrostatic interactions form between the large ribosomal subunit and the RER membrane (fig. 1c) [95]. The transport peptide inserts into the membrane as a loop structure with amino-terminal residues remaining on the cytoplasmic face of the membrane (fig. 1c) [19, 85]. As protein synthesis continues, the loop spans the membrane and the nascent chain is extruded into the cisternal space where it begins to fold, thus anchoring the protein within the cisternal space (fig. 1d). Some time after the transport peptide spans the membrane, but prior to chain termination, it is cleaved from the remainder of the nascent chain by a membrane-bound peptidase, believed to be located in the cisternal leaflet of the membrane (fig. 1e). The fate of the hydrophobic segment following cleavage is unknown although it appears to be rapidly degraded [62]. Following chain termination, the newly synthesized protein is released into the cisternal space and the synthetic apparatus dissociates from the membrane into its component parts (fig. 1f).

In prokaryotes, synthesis of secretory proteins occurs on ribosomes bound to the inner leaflet of the plasma membrane and secretion occurs directly through this membrane into the periplasmic space. Translocation of proteins across the plasma membrane of prokaryotes also requires the presence of amino-terminal transport peptides [cf. 90] and the translocation process appears, in several respects, to be similar to the mechanism observed at the level of the RER in eukaryotes [17]. However, in prokaryotes there is no evidence at present for either electrostatic interactions between the ribosomes and the membrane or the presence of a membrane-bound protein receptor that reacts with the transport peptide.

Functional Role of the Transport Peptide

During the translocation process, the transport peptide has several functions. In eukaryotes it must recognize a binding site on the rough endoplasmic reticulum. Recent evidence suggests that binding occurs on the cytoplasmic leaflet of the membrane and involves an interaction with a membrane protein complex [52, 53, 97]. This complex has been isolated and shown to be essential for the translocation process. In order to penetrate the hydrophobic layer of the membrane as a loop structure, the transport peptide must maintain its amino-terminus on the cytoplasmic face of the membrane and achieve a transverse orientation with respect to the plane of the membrane. Orientation of this type will require specific interactions between the transport peptide and protein or lipid components within the membrane. For example, the transport peptide may additionally interact with a protein complex located either in the hydrophobic layer or in the cisternal leaflet of the membrane although the existence of such complexes has not yet been established. Finally, the transport peptide must signal its own removal by the membrane-associated transport peptide hydrolase.

The translocation mechanism appears to be common among higher organisms, both invertebrates and vertebrates, since dog pancreatic microsomes translocate and correctly process proteins directed by a wide variety of eukaryotic mRNAs [77, 81], and nascent presecretory polypeptide chains from a number of eukaryotic sources [45] as well as synthetic transport peptides [67] will compete for the translocation machinery. However, the mechanisms by which the prokaryotic transport peptide and ribosome bind to the cellular membrane may not be identical to that by which the eukaryotic transport peptide and ribosome bind to the RER membrane (see below).

In the case of transmembrane proteins the transport peptide hypothesis also accounts for the transport of polypeptide segments across the lipid bilayer. If the transport peptide is not cleaved, and stop-transport sequences are not present (see below), the membrane protein will orient with the NH_2-terminus on the cytoplasmic leaflet and the COOH-terminus directed intracisternally. In this position the transport peptide may serve to anchor the protein to the membrane. If the transport peptide is removed but translocation is interrupted by a stop-transport sequence (a peptide segment of 20–30 hydrophobic residues followed by a sequence containing 2–8 charged, usually cationic, residues), the membrane protein will orient with the NH_2-terminus directed intracisternally and the COOH-terminus directed toward the cytoplasm. Insertion of the transport peptide into the membrane as a loop structure also allows for multiple translocations in polytopic transmembrane proteins, e.g. bacterial rhodopsin. Such proteins can be expected to show alternating transport and stop-transport sequences [12, 24].

Penetration of the transport peptide into the membrane in a loop configuration also allows for the possibility of an internal transport peptide in secretory proteins. Such may be the case with ovalbumin which shows no proteolytic reduction during its transport within the cell or secretion from the cell [61]. *Lingappa* et al. [45] have isolated an internal peptide comprising residues 229 to 276 that they believe contains the transport peptide sequence. This peptide has a similar sequence to the transport peptides of preovomucoid and prelysozyme (table I) and was shown to compete with preprolactin for the translocation machinery. However, using kinetic experiments to determine the ovalbumin chain lengths which, during synthesis, bind to dog pancreatic microsomal membranes, *Meek* et al. [50] found that 50% of maximum binding occurred when the nascent chains were only 50–65 residues in length. This binding is consistent with a functional transport peptide located at the amino-terminus, as has been described in detail for pancreatic presecretory proteins [77].

The term 'transport peptide' is used here and elsewhere [10, 72, 73, 77] in functional terms since it is responsible for the transport of polypeptide segments across the lipid bilayer of membranes. In the protein transport process described in this chapter, it functions cotranslationally since binding to the membrane and translocation of the polypeptide chain across the membrane are cotranslational events. Cotranslational transport mechanisms should be distinguished from posttranslational transport mechanisms, which are responsible for the transport of proteins across membranes following polypeptide chain termination [for recent discussions, see 16, 69].

The term 'transport peptide' is synonymous with 'leader peptide' used by *Meek* et al. [50] and *Palmiter* et al. [61], 'signal peptide' [5, 6], and 'cotranslational insertion peptide' [69]. We favor use of 'transport peptide' for the following reasons. The term 'leader sequence', while appropriate for presecretory proteins which show amino-terminal peptide extensions, does not correctly apply to the internal transport peptides described above. 'Signal peptide' refers, in the generic sense, to any peptide which acts as a recognition unit in nature [77]. *Sabatini* et al. [69], in their recent discussion on protein topogenesis, included transport, stop transport, and sorting (destination) sequences as examples of signal sequences. Finally, the term 'cotranslational insertion sequence' appropriately describes the insertion of the transport peptide in the membrane. However, this designation would also include 'stop transport' sequences which may also be cotranslationally inserted into membranes.

'Transport peptide' designates any peptide which inserts into the membrane in a loop configuration and initiates (or starts) translocation of subsequent polypeptide segments, located at the carboxyl-terminus of the transport peptide, across the lipid bilayer. While (start) transport and stop-transport sequences are structurally similar, their orientation in the polypeptide chain is in opposite directions. In the (start) transport peptide, the amino-terminus contains charged residues and the major segment of the peptide is nonpolar (see below). In the stop-transport peptide the major segment at the amino-terminus is nonpolar [69] and the carboxy-terminus contains the charged residues. Thus, despite their structural similarity, the orientation of these two peptides in the polypeptide chain leads to functionally dissimilar consequences during protein topogenesis. We believe the terms for these two peptides should adequately distinguish between their opposite functions. Considered in these functional terms, 'transport peptide' adequately describes both amino-terminal peptide extensions associated with the majority of secretory proteins and internal transport peptides associated with polytopic membrane proteins and possibly ovalbumin.

Primary Structure in Transport Peptides

Table I shows the sequence data of 53 transport peptides from both eukaryotic and prokaryotic preproteins. Although the majority of these transport peptides are associated with secretory proteins, several represent peptides associated with membrane proteins. With one possible exception, ovalbumin, the transport peptides are located at the amino-termini of the precursor proteins. With this same possible exception, the transport peptides for all secretory proteins are removed by a membrane-bound protease during translocation. The indicated transport peptides vary in size from 15 to 29 residues and show unusually high concentrations of hydrophobic amino acids. Sites for proteolytic cleavage occur between residues −1 and +1. A close examination of the data reveals little or no sequence homology among the various transport peptides that could represent the signal for translocation of the corresponding secretory proteins across the RER membrane. Observed homologies are more related to overall homologies observed throughout the remainder of the protein than with the function of secretion, even within a single secretory system. For example, in the canine pancreas, the transport peptides for two forms of trypsinogen (trypsinogen 1 with $IEP_u = 4.9$ and trypsinogen 2+3 with $IEP_u = 5.7–6.2$) show 40% homology [10]. In the mammary gland, the transport peptides for α and β casein [51] show strong sequence homology to each other, particularly in the carboxyl half of the peptide. Proteins synthesized and secreted by a single secretory cell type but lacking strong sequence homology show little sequence homology within their transport peptides. K casein, α-lactalbumin, and β-lactalbumin are also secreted by the mammary gland but none of their transport peptides show significant homology to either each other or to the transport peptides of α and β casein. Despite an earlier report [18], the exocrine pancreatic proteins demonstrating separate enzyme and zymogen activities also fail to show significant sequence homology among their transport peptides [10]. It is well established that these proteins are coprocessed within the same cisternal compartments in the pancreatic acinar cell [for a review of the secretory pathway in the exocrine pancreas, see 14, 59, 73].

Although there is little or no primary sequence homology, there are several features shared by many of the transport peptides. Based on regions of similarity, the transport peptide can be divided into several domains, each domain of which might play a role in one or more of the functions of the peptide.

The amino-terminal regions of a large number of the transport peptides, particularly those from prokaryotes, contain one or more charged residues. This charged domain can be expected to anchor the amino-terminal portion of the transport peptide at the cytoplasmic face of the membrane. Such an interaction would ensure that the remainder of the transport peptide would enter the membrane as a loop structure (fig. 1). The insertion of the transport peptide into the rough endoplasmic reticulum membrane as a loop is important since the complete insertion of the hydrophobic peptide across the lipid bilayer should provide sufficient energy to insert a similar length of peptide from the amino-terminus of the authentic protein such that the second peptide will also span the lipid bilayer. If the insertion of the transport peptide occurred in a linear fashion with the amino-terminus penetrating the membrane, as postulated by *Blobel and Dobberstein* [5], the affinity of the hydrophobic residues for the lipid bilayer would prevent the extrusion of the peptide through the cisternal face of the membrane and, hence, would not allow translocation.

An examination of the amino-terminals of the eukaryotic transport peptides shows that several of these sequences do not contain charged residues (cf. prechymotrypsinogen 2 and α-lactalbumin). Eukaryotic proteins are synthesized with unblocked and therefore positively charged amino termini. The presence of a single charged α amino group may be sufficient to perform the function (orientation of the transport peptide in a loop structure) of charged residues observed in this region of many of the transport peptides. Prokaryotic secretory membrane and viral proteins contain exclusively positively charged residues at the amino-terminus of their transport peptide. Positive charged residues in this region may be more important in bacterial proteins since not only are they synthesized with an amino-terminal N-formyl methionine, which is uncharged, but they may assist in the initial attachment of the nascent bacterial preprotein to the negatively charged cytoplasmic membrane through ionic interactions [19]. Eukaryotic cells contain a number of diverse intracellular membrane-bound organelles and cotranslational transport of nonmitochondrial proteins across membranes occurs exclusively at the level of only one of these membranes, the RER membrane. Thus one might expect to find a specific transport peptide-receptor interaction to occur on this membrane, which would result in the membrane-site-specificity observed in the translocation mechanism in eukaryotic cells. A receptor complex has been isolated from dog pancreatic RER membranes and the isolated complex has been shown to interact strongly with the transport peptide [52, 53, 97].

All of the transport peptides for secretory and transmembrane proteins that have been sequenced contain a high proportion of hydrophobic amino acid residues. Moreover, every known transport peptide sequence has a region of 9 to 18 hydrophobic residues essentially devoid of charged residues. Many of the hydrophobic segments also contain clusters of large chain hydrophobic residues. These clusters are found towards the amino terminal end of the hydrophobic segment and range from 2 to 8 residues in length. Clusters of this type are apparently less prominent among prokaryotic sequences since they only appear in lengths of 3 residues or less (table I). For example, while *Escherichia coli* λ-receptor protein has a high proportion of large chain hydrophobic residues within the hydrophobic segment there exist no clusters of these residues.

As the marked hydrophobicity is one characteristic shared by all transport peptides studied, it is likely that this is one of the structural features responsible for binding of the nascent preprotein to the membrane and the consequent translocation across the membrane. The hydrophobic clusters may function in the recognition of the membrane-bound binding complex isolated by *Walter and Blobel* [97] and *Meyer and Dobberstein* 52, 53], or, by virtue of its increased hydrophobicity, may facilitate the initial insertion of the transport peptide into the hydrophobic layer of the membrane. As discussed later, the insertion of the hydrophobic segment into the lipid bilayer may provide sufficient energy to translocate the initial segment of the nascent peptide across the membrane.

Another common feature of the transport peptide can be found from an examination of the primary structure. The cleavage of the peptide by the membrane-bound peptidase usually occurs after a small chain uncharged amino acid. Cleavage most often follows an alanine or glycine residue but also occurs after a serine or cysteine residue. In single cases each, cleavage has been observed after a threonine (preprolactin) and a tryptophan (immunoglobulin light chain MOPC 321) residue.

Secondary Structure in Transport Peptides

In an attempt to find a common structural feature that could be related to the function of the transport peptide, several investigators [3, 10, 13, 63] have now calculated secondary structures for transport peptides based on the primary sequence data. In most cases these predictions are based on the rules of *Chou and Fasman* [15] using the parameters developed from water-soluble globular proteins. The difficulties encountered when applying these rules to hydrophobic peptides are well illustrated by a comparison of the predicted structures of *Chan* et al. [13] with those predicted by *Austin* [3]. The structures of 13 transport peptides were predicted by both groups. Only 4 of these predictions show reasonable agreement between the two groups, one group often predicting an α helix for a region that the other group indicated to be a β-structure. When the predictions were reevaluated, the probabilities were found to be high for both α- and β-structure as indicated in figure 2. *Rosenblatt* et al. [66] likewise, recalculated the probabilities for the possible secondary structures of prepro-parathyroid hormone and found two high probability structures, also shown in figure 2. This suggests that the peptide may assume two conformations depending on the stage of interaction with the membrane. Circular dichro-

---

Fig. 2. The predicted secondary structures of 20 selected transport peptides using the rules of *Chou and Fasman* [15] for determining the probability of formation of α-helix ($<P_\alpha>$) and β-structure ($<P_\beta>$). Regions with a high probability of forming β or hairpin turns are represented by the horizontal rectangular boxes. Charged residues are shown with the appropriate sign enclosed by a circle. In addition, all eukaryotic proteins contain a partial positive charge at their amino-termini. Proteolytic cleavage of the transport peptide occurs between residues −1 and +1 (at 0).

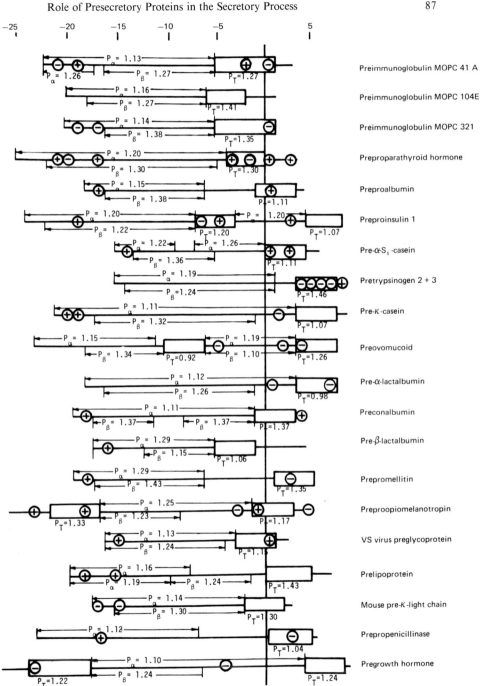

ism measurements by the *Fasman* group have indeed indicated a preference for β-structure formation in aqueous buffer and an α-helix in a nonpolar environment indicating that a structural change might take place
upon interaction of the peptide with the membrane.

A more vigorous analysis of secondary structure for mouse K light
chain was made by *Pincus and Klausner* [63] using the computer method
of *Nemethy and Scheraga* [57] for the calculation of the global minimal
energy of peptides. Their structure, calculated for the transport peptide in
a low dielectric medium, consists of a flattened polar amino-terminal
tetrapeptide (Glu-Thr-Asp-Thr) followed by a tightly packed α-helix for
the next eight residues (Leu-Leu-Leu-Trp-Val-Leu-Leu-Leu-Trp-Val)
followed by a chain reversal at the terminal tetrapeptide sequence (Trp-
Val-Pro-Gly).

A helical structure for the transport peptide within the membrane is
also predicted by the thermodynamic considerations presented in both the
helical hairpin [23] and direct transfer hypotheses [96]. These two theoretical models are based on thermodynamic calculations regarding the
insertion of a hydrophobic peptide into a lipid bilayer without interaction
with a protein receptor complex. In the absence of such a protein-protein
interaction in the membrane, the formation of a β or extended structure by
the transport peptide would result in unpaired hydrogen bond donors and
receptors that would destabilize the transport peptide within the hydrophobic layer of the membrane by approximately 186 kcal/mol of 20 residue (transport) peptides. According to the helical hairpin hypothesis the
transport peptide and the adjacent hydrophilic peptide fold into a helical
hairpin consisting of two antiparallel helices separated by a hairpin turn.
Due to the hydrophobic nature of the transport peptide, the helical hairpin
inserts spontaneously into the hydrophobic regions of the bilayer. The
energy gained by transferring the hydrophobic segment of the transport
peptide from an aqueous to a nonpolar environment was estimated to be
approximately 40 kcal/mol of transport peptide. This is sufficient to partition 20–25 residues from a 'polar' peptide into the membrane, if the
transport peptide enters the membrane as a loop. The insertion of the
transport peptide as an α-helix, in addition to minimizing the unfavorable
energy associated with unpaired hydrogen bonds in a hydrophobic medium, also minimizes the interaction between the transport peptide and
the adjacent segments of the peptide loop structure. An α-structure therefore facilitates the incoming arm of the loop sliding past that already
within the membrane [23]. However, recent evidence indicates that the

transport peptide inserts into the membrane immediately after it emerges from the tunnel in the large ribosomal subunit. In the case of pretrypsinogen 2+3, the maximal extra-ribosomal polypeptide chain length which, during synthesis, allows functional binding of the nascent chain to microsomal membranes is 22 amino acid residues [77]. Since the transport peptide for pretrypsinogen is 15 residues long, binding can be assumed to occur prior to the formation of a 30-residue hairpin loop. In the direct transfer model [96], the transport peptide interacts with the membrane as it emerges from the ribosome. The energy gained by insertion of the hydrophobic residues into the lipid bilayer and by the formation of electrostatic interactions between the ribosome and membrane compensates not only for the five degrees of freedom lost by the ribosome but also for the membrane insertion of polar residues adjacent to the transport peptide. Provided that no regions containing high concentrations of hydrophobic or hydrophilic residues follow in the sequence, little additional energy is required to complete the translocation process. Theoretically, therefore, the hydrophobicity of the transport peptide is sufficient to allow insertion into the membrane of both the transport peptide and an adjacent peptide of equal length from the authentic protein. However, the hydrophobic interaction with the lipid bilayer does not explain the specificity of protein translocation at the level of the RER membrane in eukaryotes. An interaction between the transport peptide and at least one membrane receptor protein must be involved at this level.

Each of the above structural predictions was made on the assumption that the transport peptide does not interact, in the plane of the membrane, with an integral membrane protein. The presence of such a protein, yet to be demonstrated, could conceivably stabilize a β-type structure [10, 73, 85]. Furthermore, the interaction of the transport peptide with a transmembrane protein to form a hybrid β sheet structure would ensure orientation of the peptide such that it spanned the membrane. Based on the average residue repeat distances, 20 residues in an α-helix are required to span the hydrophobic layer of the RER membrane (estimated to be 30 Å in width). Shorter transport peptides, such as those for amylase and αS_1-casein may not have sufficient length to span the membrane as an α-helix, since the hydrophobic peptide associated with each of these proteins contains only 13 amino acid residues.

Despite the difficulty in predicting the occurrence of α- of β-type structure for the transport peptide positioned in the membrane, all sequences of the preproteins for which structural predictions were made have a predicted β-turn within a few residues of the peptide cleavage point. In general, the position of the β-turn is related to the length of the hydrophobic segment such that the length of the peptide between the charged residues at the amino-terminus and the start of the predicted β-turn is between 11 and 18 residues. These β-turns could serve three possible functions. First they could provide a region responsible for chain reversal in the nascent peptide chain that would facilitate insertion of the peptide in the membrane as a loop structure. β-Turns could also be important in providing a region adjacent to the cleavage point that is free of rigid structure (α-helix or β-structure) and therefore allows accessibility for processing by the membrane-bound protease. By analogy, the specific cleavage of proalbumin and several peptide prohormones following paired basic residues has been found to occur at residues close to predicted β-turns. Other paired basic residues not observed to signal cleavage during processing of these proteins are found in region of predicted α-helix [25]. The presence of the small chain, uncharged residues at the cleavage sites for the transport peptides would contribute further to the flexibility of this region. Finally, in cases where the predicted β-turn follows the cleavage point (e.g. pancreatic presecretory proteins) this region may provide the first element of structure in the nascent polypeptide chain during its translocation across the lipid bilayer. The presence of structure, even minimal structure, in the polypeptide chain at the cisternal leaflet of the membrane and the absence of structure (extended chain) at the cytoplasmic leaflet of the membrane will ensure that translocation of the polypeptide chain occurs in a vectorial manner into the vesicular space of the RER. As structure accumulates in the polypeptide chain in the cisternal space, first, secondary and then tertiary structure, the motive force for vectorial transfer will correspondingly increase.

Energy of polypeptide chain folding resulting in the motive force for vectorial transfer of the polypeptide chain may account for the translocation and ultimate release of ovalbumin from the microsomal membrane. This secretory protein is released to the cisternal space of the microsomal vesicle without loss of its corresponding transport peptide. In this case the motive force for vectorial discharge and release to the cisternal space must exceed the interaction between the transport peptide and the microsomal membrane. The energy of chain folding and the absence of a positive charge at the amino-terminus of ovalbumin due to acetylation may allow for translocation of the amino-terminal segment, including the transport peptide, to the cisternal space. Uncleaved preprolactin synthesized in the presence of threonine analogues (see following section) is also apparently released from the RER membrane into the luminal or cisternal space [35].The energy associated with the folded protein must be sufficient to dislodge the transport peptide from the membrane. However, this situation would not normally occur, since in the presence of threonine rather than the analogue, the transport peptide is removed by the membrane peptidase at a relatively early step in the translocation process.

Effect of Amino Acid Substitutions on Transport Peptide Function

Amino acid substitutions, produced by the use of mutagenic techniques or amino acid analogues, have been shown to have marked effects on protein translocation across membranes. An E. coli murine lipoprotein is normally translocated across the inner membrane and inserted into the outer membrane. A mutant form of this protein has been studied in which a glycine residue in position −7 is substituted by an aspartic acid residue.

```
 -20           -15           -10           -5            -1 +1
  |             |             |             |             |  |
  M - K - A - T - K - L - V - L - G - A - V - I - L - G - S - T - L - L - A - G - C -
                                                    ↓
                                                    D
```

Although this substitution does not prevent membrane insertion, it does block cleavage of the nascent protein by the transport peptidase. The efficiency of translocation into the outer membrane is also decreased and significant amounts of the mutant lipoprotein are found associated with the inner membrane [43].

Four mutations in the E. coli λ-receptor, normally an outer membrane receptor protein, prevent the insertion of the nascent protein into either the inner or outer membranes, the mutant proteins being found free in the cytoplasm [22].

```
 -25           -20           -15           -10           -5            -1 +1
  |             |             |             |             |             |  |
  M - M - I - T - L - R - K - L - P - L - A - Y - A - V - A - A - G - V - M - S - A - N - A - M - A - V -
                                      ↓
                                      D  ↓
                                         E  ↓
                                            E
                                                               ↓
                                                               R
```

Similarly, five mutations in the maltose binding protein, normally secreted into the *E. coli* periplasmic space, prevent secretion and the mutant proteins are found in the cytoplasm [4].

```
  -25              -20            -15              -10            -5          -1 +1
   |                |              |                |             |           | |
- M- K- I- K- T- G- A- R- I- L- A- L- S- A- L- T- T- M- M- F- S- A- S- A- L- A- K-
                            |                    |
                            P                    |
                                                 E     |
                                                       K     |
                                                             R   |
                                                                 R
```

With the exception of a leucine to proline substitution, each of these mutations involves the replacement of a hydrophobic residue with a charged residue. No mutations which blocked secretion were found in the basic segment at the amino-termini of these transport peptides. The mutations in the hydrophobic segment do not significantly alter the predicted secondary structure of the transport peptide. Hence for the sites indicated, the maintenance of hydrophobicity appears to be a major requirement for the proper function of the transport peptide.

However, at certain positions charged residues can be tolerated in the hydrophilic sequence of the transport peptide. The case of the mutant *E. coli* lipoprotein has been given above. A second case is that the substitution of the glycine in the *E. coli* Lam B transport peptide with either aspartic acid or arginine does not interfere with the secretion of this protein [*Silhavey,* personal commun.].

Hortin and Boime [36] synthesized rat preprolactin in vitro using the threonine analogue β-hydroxy-norvaline. This analogue has a similar structure to threonine with the exception that the side chain is larger by one carbon. The transport peptide of preprolactin contains two threonine residues, one at position −18 and one at position −1, the cleavage site. Substitution with the analogue, presumably at the cleavage site, resulted in a marked decrease in the efficiency of proteolytic removal of the transport peptide. This binding underscores the need for a small chain residue in this position at the cleavage site. Whether the reduction in cleavage rate is due to steric hindrance in the active site for cleavage or due to the disruption of a possible vicinal β-turn is not known. Uncleaved preprolactin, however, is transported across the membrane of the RER but is not secreted from the cell. Since the precursor containing the threonine analogue can be released from rough microsomes into the medium by sonication, it does not appear to remain bound to the membrane (see p. 95 for an explanation of its release).

Hortin and Boime [34] likewise synthesized bovine and rat prepro-lactin and human placental lactogen in the presence of the leucine analogue, β-hydroxyleucine. Their analogue is considerably less hydrophobic than leucine. Proteins synthesized in the presence of microsomal membranes with this analogue were neither processed nor translocated. Solubilized transport peptidase also failed to process the analogue precursors. With the exceptions described above for *E. coli* Lam B murine lipoprotein, it is clear that alterations in the hydrophobic segment of the transport peptide have profound consequences on the translocation function of this peptide.

Importance of Polypeptide Structure in the Authentic Protein

When the transport peptide sequence from a secretory protein, *E. coli* Lam B, was fused to a cytoplasmic protein, galactosidase, using genetic recombination techniques, the resulting protein did not appear in the periplasmic space but remained partially inserted in the membrane [55]. This fusion product contained approximately half of the authentic lam B sequence. Another fusion product containing only 15 residues of lam B in addition to the transport peptide was neither secreted nor inserted into either the inner or outer membranes [84]. Thus, although the transport peptide is essential for translocation of polypeptide segments across the membrane, it is not sufficient to ensure transport of the remaining portion of the protein. The structure of the entire protein must, therefore, play an essential role in the secretory process [83].

As proposed by *Carne and Scheele* [10] and *Von Heijne and Blomberg* [96], the peptide structure immediately following the transport peptide is important in the translocation process. As already indicated, insertion of the transport peptide into the membrane in a loop configuration will result in the insertion of an adjacent peptide of roughly equal length into the membrane (fig. 1). The presence of polar and charged residues in this adjacent peptide will destabilize the peptide's presence in the hydrophobic layer of the membrane. Such unfavorable energy conditions will result in the transfer of this peptide, which represents the amino-terminus of the authentic protein, to the cisternal space of the RER. This is well illustrated by the structure of pretrypsinogen which contains four aspartic acid residues and a lysine in positions 4 through 8 immediately following the cleavage point. Once inserted, this charged region would not be expected to remain within the lipid bilayer. Following transfer of this segment to the cisternal space, translocation of the remaining polypeptide chain will continue unless transfer is interrupted by a 'stop-transport' sequence. As shown for glycophorin [94] and bacterial rhodopsin [24], stop-transport sequences anchor membrane proteins to the lipid bilayer. Thus, polypeptide structure in the authentic proteins figures importantly not only in the initial translocation event, but also in determining whether a protein will be released to the cisternal space and ultimately secreted or retained by the cell as a membrane protein.

Transfer of the adjacent peptide (amino-terminus of the authentic protein) to the cisternal space, with the consequent and progressive formation of higher levels of structure in this space, would seem to be required before the transport peptide is cleaved by the transport peptidase, believed to be associated with the cisternal leaflet of the membrane. Should cleavage occur prior to the correct insertion of the transport peptide, a condition which leads to the rapid occurrence of the subsequent translocation events mentioned above, transfer of polypeptide segments across the lipid bilayer will not occur.

The region adjacent to the cleavage point is also important for the recognition of the cleavage site by the transport peptidase. A mutant of the major coat protein from the filamentous bacteriophage M13 has a transport peptide identical to that of the wild type, but glutamic acid is replaced by leucine at position 2 in the authentic protein.

```
+1          +5            +10
|           |             |
A - E - G - D - D - P - A - K - A - A - F - N -    M13
    |               |
    ↓               ↓
    L               S                             M13 MUTANT (AMBH1R6)
```

The mutant preprotein is processed more slowly than the wild type, with half-lives of 60 s compared to less than 2 s, respectively. The mutant precoat protein did insert into membranes indicating that the mutation, which occurs outside the transport peptide region, affects the peptidase activity and not the insertion process [7].

Function of Cotranslational Proteolytic Processing in the Secretory Process

Although one might estimate that proteolytic processing of presecretory proteins is necessary for the release of nascent secretory proteins from the RER membrane, ovalbumin is synthesized and secreted by the cell without proteolytic loss of its transport peptide and uncleaved preprolactin synthesized with threonine analogues can be released from microsomal membranes by sonication [36]. These two examples indicate that proteolytic removal of the transport peptide is not an absolute requirement for translocation.

Recent studies by *Scheele and Jacoby* [76] indicate that proper folding of pancreatic exocrine proteins requires the proteolytic removal of transport peptides by the transport peptidase associated with microsomal membranes. Compared to pancreatic secretory proteins, presecretory proteins synthesized with dog pancreas mRNA in the absence of microsomal membranes showed conformational instability as judged by their marked sensitivity to proteolysis, aggregation during isoelectric focusing in the presence of $8M$ urea and 0.1% Triton X-100, aggregation during gel electrophoresis in the presence of 0.1% SDS, and insolubility in 40% ethanol. Conformational instability observed among presecretory proteins, containing amino-terminal transport peptides, was shown to be secondary to the absence or incorrect formation of disulfide bonds and nonspecific protein-protein interactions among nascent polypeptide chains. Proteolytic removal of transport peptides was necessary to allow the formation of stable secretory molecules with the correct sets of disulfide bonds. As shown in a separate study [75], proteolytic removal of transport peptides was not only required to achieve correct conformation in pancreatic secretory proteins, but also for expression of authentic biological activities.

Note Added in Proof

Recent studies on the signal receptor complex have shown it to be composed of two complexes, a signal recognition particle (SRP) [a] composed of 6 polypeptides plus a 7S RNA, and a docking protein [b]. It is proposed that the SRP binds to the ribosome, in the cytoplasm, as the transport peptide emerges. The binding stops translation until the complex interacts with the docking protein on the RER [c], ensuring translocation of the secretory protein.

a   Walter, P.; Blobel, G.: Signal recognition particle contains a 7S RNA essential for protein
     translocation across the endoplasmic reticulum. Nature *299:* 691–698 (1982).
b   Meyer, D. I.; Louvard, D.; Dobberstein, B.: Characterization of molecules involved in
     protein translocation using a specific antibody. J. Cell Biol. *92:* 579–583 (1982).
c   Meyer, D. I.: The signal hypothesis – a working model. TIBS *7:* 320–321 (1982).

Acknowledgements

The authors are indebted to Drs. *S. Emr and T. J. Silhavy,* Drs. *M. Pincus and R. D.
Klausner* and Dr. *R. J. MacDonald* for allowing us to use their sequence and structural data
prior to its publication.

References

1   Air, G. M.: Nucleotide sequence coding for the signal peptide and N-terminus of the
     hemagglutinin from an asian (H2N2) strain of influenza virus. Virology *97:* 468–472
     (1979).
2   Ambler, R. P.; Scott, G. K.: Partial amino acid sequence of penicillinase coded by
     Escherichia coli plasmid R6K. Proc. natn. Acad. Sci. USA *75:* 3732–376 (1978).
3   Austin, B. M.: Predicted secondary structures of amino-terminal extension sequences of
     secreted proteins Febs. Lett. *103:* 308–313 (1979).
4   Bedouelle, H.; Bassford, P. J., Jr.; Fowler, A. V.; Zabin, I.; Beckwith, J.; Hofnung, M.:
     Mutations which alter the function of the signal sequence of the maltose binding protein
     of *Escherichia coli.* Nature, Lond. *285:* 78–81 (1980).
5   Blobel, G.; Dobberstein, B.: Transfer of proteins across membranes. I. Presence
     of proteolytically processed and unprocessed nascent immunoglobin light chains
     on membrane bound ribosomes of murine myeloma. J. Cell Biol. *67:* 835–851
     (1975).
6   Blobel, G.; Walter, P.; Chang, C. N.; Goldman, B. M.; Erickson, A. H.; Linguppa,
     V. R.: Translocation of proteins across membranes. The signal hypothesis and beyond;
     in Hopkins, Duncan, Secretory mechanisms, pp. 9–36 (Cambridge University Press,
     Cambridge 1979).
7   Boeke, J. D.; Russel, M.; Model, P.: Processing of filamentous phage pre-coat protein.
     Effects of sequence variations near the signal peptidase cleavage site. J. molec. Biol. *144:*
     103–116 (1980).
8   Burstein, Y.; Schechter, I.: Amino acid sequence of the NH_2-terminal extra piece
     segments of the precursors of mouse immunoglobulin λ_1 and x-type light chains. Proc.
     natn. Acad. Sci. USA *74:* 716–720 (1977).
9   Burstein, Y.; Schechter, I.: Primary structures of N-terminal extra peptide segments
     linked to the variable and constant regions of immunoglobulin light chain precursors.
     Implications on the organization and controlled expression of immunoglobulin genes.
     Biochemistry *17:* 2392–2400 (1978).
10   Carne, T.; Scheele, G.: Amino acid sequences of transport peptides associated with
     canine exocrine pancreatic proteins. J. Cell Biol. *92:* (1982).

11 Caro, L.G.; Palade, G.E.: Protein synthesis, storage and discharge in the pancreatic exocrine cell. An autoradiographic study. J. Cell Biol. *20:* 473–495 (1964).

12 Chang, S.H.; Majumdar, A.; Dunn, R.; Makabe, O.; RajBhandary, V.L.; Khorana, H.G.; Ohtsuka, E.; Tanaka, T.; Taniyama, Y.; Ikehara, M.: Bacteriorhodopsin. Partial sequence of mRNA provides amino acid sequence in the precursor region. Proc. natn. Acad. Sci. USA *78:* 3398–3402 (1981).

13 Chan, S.J.; Patzelt, C.; Duguid, J.R.; Quinn, P.; Labrecque, A.; Noyes, B.; Keim, P.; Heinrickson, R.L.; Steiner, D.F.: Precursors in the biosynthesis of insulin and other peptide hormones; in Russell et al., From gene to protein. Information transfer in normal and abnormal cells, pp. 361–378 (Academic Press, New York 1979).

14 Case, R.M.: Synthesis, intracellular transport and discharge of exportable proteins in the pancreatic acinar cell and other cells. Biol. Rev. *53:* 211–358 (1978).

15 Chou, P.Y.; Fasman, G.D.: Empirical predictions of protein conformation. A. Rev. Biochem. *47:* 251–276 (1978).

16 Chua, N.-H.; Schmidt, G.W.: Transport of proteins into mitochondria and chloroplasts. J. Cell Biol. *81:* 461–483 (1979).

17 Davis, B.D.; Tai, P.-C.: The mechanism of protein secretion across membranes. Nature, Lond. *283:* 433–438 (1980).

18 Devillers-Thiery, A.; Kindt, T.; Scheele, G.; Blobel, G.: Homology in amino terminal sequence of precursors to pancreatic secretory proteins. Proc. natn. Acad. Sci. USA *72:* 5016–5020 (1975).

19 DiRienzo, J.M.; Nakamura, K.; Inouye, M.: The outer membrane proteins of gram-negative bacteria. Biosynthesis, assembly and functions. A. Rev. Biochem. *47:* 481–532 (1978).

20 Drickamer, K.; Kwoh, J.; Kurtz, D.T.: Amino acid sequence of the precursor of rat liver α_{2u}-globulin. J. biol. Chem. *256:* 3634–3636 (1981).

21 Dugan, E.S.; Bradshaw, R.A.; Simms, E.S.; Eisen, H.N.: Amino acid sequence of the ligth chain of a mouse myeloma protein (MOPC-315). Biochemistry *12:* 5400–5416 (1973).

22 Emr, S.D.; Hedgpeth, J.; Clement, J.M.; Silhavy, T.J.; Hofnung, M.: Sequence analysis of mutations that prevent export of λ-receptor, an *Escherichia coli* outer membrane protein. Nature, Lond. *285:* 82–85 (1980).

23 Engelman, D.M.; Steitz, T.A.: The spontaneous insertion of proteins into and across membranes. The helical hairpin hypothesis. Cell *23:* 411–422 (1981).

24 Engelman, D.M.; Zaccai, G.: Bacteriorhodopsin is an inside-out protein. Proc. natn. Acad. Sci. USA *77:* 5894–5898 (1980).

25 Geisow, M.J.: Polypeptide secondary structure may direct the specificity of prohormone conversion. Febs. Lett. *87:* 111–114 (1978).

26 Gorecki, M.; Zeelon, E.P.: Cell free synthesis of rat parotid preamylase. J. biol. Chem. *254:* 525–529 (1979).

27 Greene, L.J.; Hirs, C.H.W.; Palade, G.E.: On the protein composition of bovine pancreatic zymogen granules. J. biol. Chem. *238:* 2054–2070 (1963).

28 Habener, J.F.; Rosenblat, M.; Kemper, B.; Kronenberg, H.M.; Rich, A.; Potts, J.T., Jr.: Preproparathyroid hormone: amino acid sequence, chemical synthesis and some biological studies of the precursor region. Proc. natn. Acad. Sci. USA *74:* 2616–2620 (1977).

29 Hagenbuchle, O.; Bovey, R.; Young, R.A.: Tissue specific expression of mouse α-

amylase genes. Nucleotide sequence of Isoenzyme mRNAs from pancreas and salivary gland. Cell *21:* 179–187 (1980).

30   Hagenbuchle, O.; Tosi, M.; Schibler, V.; Bovey, R.; Wellaver, P. K.; Young, R. A.: Mouse liver and salivary gland α-amylase mRNAs differ only in 5'non-translated sequences. Nature, Lond. *289:* 643–646 (1981).

31   Haugen, D. A.; Armes, L. G.; Yasunobu, K. T.; Loon, M. J.: Amino terminal sequence of phenobarbitol inducible cytochrome P450 from rabbit liver microsomes. Similarity to hydrophobic amino terminal segments of preproteins. Biochem. biophys. Res. Commun. *77:* 967–973 (1977).

32   Hedgpeth, J.; Clement, J.-M.; Marchal, S.; Perrin, D.; Hofnung, M.: DNA sequence encoding the NH_2-terminal peptide involved in transport of λ-receptor, an *E. coli* secretory protein. Proc. natn. Acad. Sci. USA *77:* 2621–2625 (1980).

33   Heidenhain, R.: Beiträge zur Kenntnis der Pancreas. Pflügers Arch. ges. Physiol. *10:* 557 (1875).

34   Hortin, G.; Boime, I.: Inhibition of preprotein processing in ascites tumor lysates by incorporation of a leucine analog. Proc. natn. Acad. Sci. USA *77:* 1356–1360 (1980).

35   Hortin, G.; Boime, I.: Transport of an uncleaved preprotein into the endoplasmic reticulum of rat pituitary cells. J. biol. Chem. *256:* 1491–1494 (1981).

36   Hortin, G.; Boime, I.: Miscleavage at the presequence of rat preprolactin synthesized in pituitary cells incubated with a threonine analog. Cell *24:* 453–461 (1981).

37   Inouye, I.; Halegoua, S.: Secretion and membrane localization of proteins in *Escherichia coli.* CRC crit. Rev. Biochem. *7:* 339–371 (1980).

38   Inouye, S.; Wang, S.; Sekizawa, J.; Halegoua, S.; Inouye, M.: Amino acid sequence for the peptide extension on the prolipoprotein of the *Escherichia coli* outer membrane. Proc. natn. Acad. Sci. USA *74:* 1004–1008 (1977).

39   Jamieson, J. D.; Palade, G. E.: Intracellular transport of secretory proteins in the pancreatic exocrine cell. I. Role of the peripheral elements of the golgi complex. J. Cell Biol. *34:* 577–596 (1967).

40   Jamieson, J. D.; Palade, G. E.: Intracellular transport of secretory proteins in the pancreatic exocrine cell. II. Transport to condensing vacuoles and zymogen granules. J. Cell Biol. *34:* 597–615 (1967).

41   Kreil, G.: Transport of proteins across membranes. A. Rev. Biochem. *50:* 317–348 (1981).

42   Lane, C. D.; Champion, J.; Haiml, L.; Kreil, G.: The sequestration, processing and retention of honey-bee promelittin made in amphibian oocytes. Eur. J. Biochem. *113:* 273–281 (1981).

43   Lin, J. J. C.; Kanazawa, H.; Ozols, J.; Wu, H. C.: An *Escherichia coli* mutant with an amino acid alteration within the signal sequence of outer membrane prelipoprotein. Proc. natn. Acad. Sci. USA *75:* 4891–4895 (1978).

44   Lingappa, V. R.; Katz, F. N.; Lodish, H. F.; Blobel, G.: A signal sequence for the insertion of a transmembrane glycoprotein. Similarities to the signals of secretory proteins in primary structure and function. J. biol. Chem. *253:* 8667–8670 (1978).

45   Lingappa, V. R.; Lingappa, J. R.; Blobel, G.: Chicken ovalbumin contains an internal signal sequence. Nature, Lond. *281:* 117–121 (1979).

46   Lodish, H. F.; Rothman, J. E.: The assembly of cell membranes. The two sides of a biological membrane differ in structure and function; studies of animal viruses and bacteria have helped to reveal how this asymmetry is preserved as the membrane grows. Scient. Am. *240:* 48–63 (1979).

47   MacDonald, R.J.; Crerar, M.M.; Swain, W.F.; Picket, R.L.; Thomas, G.; Rutter, W.J.: Structure of a family of rat amylase genes. Nature, Lond. *287:* 117–122 (1980).
48   MacDonald, R.J.: Private commun.
49   McKean, D.J.; Maurer, R.A.: Complete amino acid sequence of the precursor region of rat prolactin. Biochemistry *17:* 5215–5219 (1978).
50   Meek, R.L.; Walsh, K.A.; Palmiter, R.: Ovalbumin contains an NH_2-terminal functional equivalent of a signal sequence. Fed. Proc. *39:* Abstr. 1370 (1980).
51   Mercier, J.-C.; Haze, G.; Gaye, P.; Hue, D.: Amino terminal sequence of the precursor of ovine β-lactalbumin. Biochem. biophys. Res. Commun. *82:* 1236–1245 (1978).
52   Meyer, D.I.; Dobberstein, B.: Identification and characterization of a membrane component essential for the translocation of nascent proteins across the membranes of the endoplasmic reticulum. J. Cell Biol. *87:* 503–508 (1980).
53   Meyer, D.I.; Dobberstein, D.: A membrane component essential for vectorial translocation of nascent proteins across the endoplasmic reticulum. Requirements for its extraction and reassociation with the membrane. J. Cell Biol. *87:* 498–502 (1980).
54   Milstein, C.; Brownlee, G.G.; Harrison, T.M.; Mathews, M.B.: A possible precursor of immunoglobulin light chains. Nature New Biol. *239:* 117–120 (1972).
55   Moreno, F.; Fowler, A.V.; Hall, M.; Silhavy, T.J.; Zabin, I.; Schwartz, M.: A signal sequence is not sufficient to lead β-galactosidase out of the cytoplasm. Nature, Lond. *286:* 356–359 (1980).
56   Nakanishi, S.; Inoue, A.; Kita, T.; Nakamura, M.; Chang, A.C.Y.; Cohen, S.N.; Numa, S.: Nucleotide sequence of cloned cDNA for bovine corticotrophin-β-lipotropin precursor. Nature, Lond. *278:* 423–427 (1979).
57   Nemethy, G.; Scheraga, H.A.: Protein folding. Q. Rev. Biophys. *10:* 239–252 (1977).
58   Palade, G.E.; Functional changes in the structure of cell components; in Hayashi, Subcellular particles, pp.64–83 (Ronald Press, New York 1959).
59   Palade, G.E.: Intracellular aspects of the process of protein secretion. Science *189:* 347–358 (1975).
60   Palmiter, R.D.; Gagnon, J.; Ericsson, L.H.; Walsh, K.A.: Precursor of egg white lysozyme. Amino acid sequence of an NH_2-terminal extension. J. biol. Chem. *252:* 6386–6390 (1977).
61   Palmiter, R.D.; Gagnon, J.; Walsh, K.A.: Ovalbumin: A secreted protein without a transient hydrophobic leader sequence. Proc. natn. Acad. Sci. USA *75:* 94–98 (1978).
62   Patzelt, C.; Labrecque, A.D.; Duguid, J.R.; Carroll, R.J.; Keim, P.; Heinrikson, R.L.; Steiner, D.F.: Detection and kinetic behaviour of preproinsulin in pancreatic islets. Proc. natn. Acad. Sci. USA *75:* 1260–1264 (1978).
63   Pincus, M.R.; Klausner, R.D.: Prediction of the three dimensional structure of the leader sequence of pre-kappa light chain, a hexadecapeptide (submitted for publication).
64   Redman, C.M.; Sabatini, D.D.: Vectorial discharge of peptides released by puromycin from attached ribosomes. Proc. natn. Acad. Sci. USA *56:* 608–615 (1966).
65   Redman, C.M.; Siekevitz, P.; Palade, G.E.: Synthesis and transfer of amylase in pigeon pancreatic microsomes. J. biol. Chem. *241:* 1150–1158 (1966).
66   Rosenblatt, M.; Beaudette, N.V.; Fasman, G.D.: Conformational studies on the synthetic precursor-specific region of preproparathyroid hormone. Proc. natn. Acad. Sci. USA *77:* 3983–3987 (1980).

67   Rosenblatt, M.; Majzoub, J.A.; Kronenberg, H.M.; Habener, J.F.; Potts, J.T.: The precursor of the specific region of preproparathyroid hormone: chemical synthesis and preliminary studies of its effect on post-translational modification of hormones; in Gross, Meienhofer, Peptides. Structure and biological function, pp. 535–538 (Pierce Chemical, Rockford 1980).

68   Sabatini, D.D.; Blobel, G.: Controlled proteolysis of nascent polypeptides in rat liver cell fractions. II. Location of the polypeptides in rough microsomes. J. Cell Biol. 45: 146–157 (1970).

69   Sabatini, D.D.; Kreibich, G.; Morimoto, T.; Adesnik, M.: Mechanisms for the incorporation of proteins in membranes and organelles. J. Cell Biol. 92: 1–22 (1982).

70   Schaller, H.; Beck, E.; Takanami, M.: Sequence and regulatory signals of the filamentous phage genome; in Denhardt, Dressler, Ray, The single stranded DNA phages, pp. 139–153 (Cold Spring Harbour Laboratory, Cold Spring Harbor 1978).

71   Schechter, I.; Burstein, Y.: Marked hydrophobicity of the NH_2-terminal extra piece of immunoglobin light chain precursors. Possible physiological functions of the extra piece. Proc. natn. Acad. Sci. USA 73: 3273–3277 (1976).

72   Scheele, G.: The secretory process in the pancreatic exocrine cell. Mayo Clin. Proc. 54: 420–427 (1979).

73   Scheele, G.: Biosynthesis, segregation and secretion of exportable proteins in the exocrine pancreas. Am. J. Physiol. 238: G467–G477 (1980).

74   Scheele, G.A.; Dobberstein, B.; Devillers-Thiery, A.; Blobel, G.: In vitro translation of cell fractions from dog pancreas. J. Cell Biol. 67: 385a (1975).

75   Scheele, G.; Jacoby, R.: Proteolytic processing of presecretory proteins is required for development of biological activities in pancreatic exocrine proteins. J. biol. Chem. (in press, 1983).

76   Scheele, G.; Jacoby, R.: Conformational changes associated with proteolytic processing of presecretory proteins allow glutathione-catalyzed formation of native disulfide bonds. J. biol. Chem. 257: 12277–12282 (1982).

77   Scheele, G.; Jacoby, R.; Carne, T.: Mechanism of compartmentation of secretory proteins. Transport of exocrine pancreatic proteins across the microsomal membrane. J. Cell Biol. 87: 611–628 (1980).

78   Scheele, G.; Palade, G.; Tartakoff, A.: Cell fractionation studies on the guinea pig pancreas. Redistribution of exocrine proteins during tissue homogenization. J. Cell Biol. 78: 110–130 (1978).

79   Seeburg, P.H.; Shine, J.; Martial, J.A.; Baxter, J.D.; Goodman, H.M.: Nucleotide sequence and amplification of structural gene for rat growth hormone. Nature, Lond. 240: 486–494 (1977).

80   Sherwood, L.M.; Burstein, Y.; Schechter, I.: Primary structure of the NH_2-terminal extra piece of the precursor to human placental lactogen. Proc. natn. Acad. Sci. USA 76: 3819–3823 (1979).

81   Shields, D.; Blobel, G.: Cell free synthesis of fish preproinsulin and processing by heterologous mammalian microsomal membranes. Proc. natn. Acad. Sci. USA 74: 2059–2063 (1977).

82   Siekevitz, P.; Palade, G.E.: A cytochemical study on the pancreas of the guinea pig. V. In vivo incorporation of leucine-1-^{14}C into the chymotrypsinogen of various cell fractions. J. biophys. biochem. Cytol. 7: 619–630 (1960).

83   Silhavy, T.J.; Bassford, P.J., Jr.; Beckwith, J.R.: A genetic approach to the study of

protein localization in *E. coli;* in Inouye, Bacterial outer membranes: biogenesis and functions, pp. 203–254 (Wiley, New York 1979).

84   Silhavy, T. J.; Shuman, H. A.; Beckwith, J.; Schwartz, M.: Use of gene fusions to study outer membrane protein localization in *Escherichia coli.* Proc. natn. Acad. Sci. USA *74:* 5411–5415 (1977).

85   Steiner, D. F.; Quinn, P. S.; Chan, S. J.; Marsh, J.; Tager, H. S.: Processing mechanisms in the biosynthesis of proteins. Ann. N.Y. Acad. Sci. *343:* 1–6 (1980).

86   Straus, A. W.; Bennett, C. D.; Donohue, A. M.; Rodkey, J. A.; Alberts, A. W.: Rat liver preproalbumin. Complete amino acid sequence of the prepiece. Analysis of the direct translation product of albumin messenger RNA. J. biol. Chem. *252:* 6846–6855 (1977).

87   Straus, A. W.; Boime, I.: Compartmentation of newly synthesized proteins. CRC crit. Rev. Biochem. *12:* 205–236 (1982).

88   Sugimoto, K.; Sugisaki, H.; Okamoto, T.; Takanami, M.: Studies on bacteriophage f$_d$ DNA. IV. The sequences of messenger RNA for the major coat protein gene. J. molec. Biol. *11:* 487–507 (1977).

89   Sures, I.; Goeddel, D. W.; Gray, A.; Ullrich, A.: Nucleotide sequence of human pre-proinsulin complementary DNA. Science *208:* 57–59 (1980).

90   Sutcliffe, J. G.: Nucleotide sequence of the ampicillin resistance gene of *Escherichia coli* plasmid pBR 322. Proc. natn. Acad. Sci. USA *75:* 3737–3741 (1978).

91   Tanaka, Y.; DeCamilli, P.; Meldolesi, J.: Membrane interactions between secretion granules and plasmalemma in three exocrine glands. J. Cell Biol. *84:* 438–453 (1980).

92   Thibodeau, S. N.; Lee, D. C.; Palmiter, R. D.: Identical precursors for serum transferrin and egg white conalbumin. J. biol. Chem. *253:* 3771–3774 (1978).

93   Thibodeau, S. N.; Palmiter, R. D.; Walsh, K. A.: Precursor of egg white ovomucoid. Amino acid sequence of an NH$_2$-terminal extension. J. biol. Chem. *253:* 9018–9023 (1978).

94   Tomita, M.; Marchesi, V. T.: Amino acid sequence and oligosaccharide attachment sites of human erythrocyte glycophorin. Proc. natn. Acad. Sci. USA *72:* 2964–2968 (1975).

95   Unwin, P. N. T.: Three-dimensional model of membrane-bound ribosomes determined by electron microscopy. Nature, Lond. *269:* 118–122 (1977).

96   Von Heijne, G.; Blomberg, C.: Trans-membrane translocation of proteins. The direct transfer model. Eur. J. Biochem. *97:* 175–181 (1979).

97   Walter, P.; Blobel, G.: Purification of a membrane associated protein complex required for protein translocation across the endoplasmic reticulum. Proc. natn. Acad. Sci. USA *77:* 7112–7116 (1980).

98   Ward, C. W.; Dapherdi, T. A.: Primary structure of the Hong Kong (H3) haemagglutinin. Br. med. Bull. *35:* 51–56 (1979).

99   Waterfield, M. D.; Espelie, K.; Elder, K.; Skehel, J. J.: Structure of the haemagglutinin of influenza virus. Br. med. Bull. *35:* 57–63 (1979).

Thomas Carne, PhD, Department of Chemistry, University of Calgary, Calgary, Alberta T2N 1N4 (Canada)

Cell Biology of the Secretory Process, pp. 102–147 (Karger, Basel 1984)

Role of the Golgi Complex in the Secretory Process

Gary Bennett

Department of Anatomy, McGill University, Montreal, Canada

Historically, the Golgi apparatus has been postulated as having a role in secretion almost from its original discovery. Although originally demonstrated by *Camillo Golgi* in neurons [70], the apparatus was subsequently found to occur in a great variety of other cell types including many secretory cells [33]. In secretory cells, the Golgi apparatus was found to usually lie between the nucleus and the secretion granules, and its state of development often correlated with the degree of secretory activity occurring within the cell [33]. Frequently, the smallest secretion granules occurred within the meshwork of the Golgi apparatus, leading *Nassonov* [119] and *Bowen* [25, 27, 28] to suggest that the apparatus gave rise to these secretion granules. In many cell types, the large number of secretion granules made it difficult to observe the early steps in secretory granule formation. In one cell type, the spermatid, however, the Golgi apparatus had been well described, and could clearly be seen to be involved in the formation of granules leading to the development of the acrosome [24, 26, 65]. Although the spermatid was certainly not thought of as a secretory cell at that time, *Bowen* suggested that the acrosome content was a secretion product which might be released from the sperm to play some functional role in the process of fertilization. Many years later, this was indeed shown to be the case [7], and the spermatid thus serves as an excellent example of the role of the Golgi apparatus in secretion. Because of its historical significance, along with our detailed modern knowledge of the structure of its Golgi apparatus, the spermatid will be used as a model cell, in the present article, to relate some of the development of our knowledge of the Golgi apparatus, and to illustrate what is currently known about its structure and its function in the secretory process. Several other more general reviews have been written over the years [6, 27, 28, 33, 44, 47, 53, 56, 63, 76, 86, 88, 97, 104, 116, 122, 146, 164, 173, 174], wherein details of Golgi apparatus structure and function in other cell types may be obtained.

The Golgi apparatus was first observed in the spermatid by *La Valette St. George* [101, 102], many years before Golgi's discovery, and was described in the same cell type by *Platner* [134]. The structure was described in more detail by *Gatenby and Woodger* [65] (fig. 1), and by *Bowen* [24, 26, 27] who showed it to be originally spherical in young spermatids [steps 1–2 of *Leblond and Clermont*, 105]. Later it became hemispherical in shape with the open face oriented towards the nucleus. Only the periphery of the hemisphere took up stain with the osmium or silver impregnation methods used to identify the Golgi apparatus at the time. The central region was unstained and referred to as the idiosome or archoplasm. Thus, even at that time some heterogeneity and polarity in the makeup of the Golgi apparatus was recognized. *Bowen* suggested that the two regions constituted one functional unit which he called the acroblast or Golgi complex. The idiosome had been previously noted as the original site of appearance of a clear vesicle containing a spherical granule [8, 114], and the granule was later named the acrosome [106]. The formation of the acrosome in guinea pig spermatids was described by *Gatenby and Woodger* [65] as follows: the proacrosomic granules appeared to form from the Golgi elements (also called dictyosomes) which surrounded the archoplasm (idiosome) (fig. 1A). Each proacrosomic granule lay within a clear archoplasmic vacuole. The archoplasmic vacuoles then fused to form a single proacrosomic vacuole containing one large proacrosomic granule (fig. 1B, C). The Golgi complex now moved to the anterior end of the spermatid nucleus and became applied to the nuclear membrane (fig. 1C). At this site, the Golgi elements were pushed aside so that the Golgi complex became hemispherical. Meanwhile, the proacrosomic granule differentiated into a darker staining inner zone and a lighter outer zone and together these constituted the acrosome (fig. 1C). The inner zone flattened itself against the nuclear membrane, and the acrosomic vacuole extended down over the nucleus to form an umbrella-shaped headcap (fig. 1D, E). The Golgi complex now broke away from the acrosome and drifted to the posterior end of the spermatid (fig. 1E, F) where it subsequently degenerated.

Although excellent light microscopic descriptions of the formation of the acrosomic system in various species continued to be published using the traditional silver and osmium impregnation and other methods [i.e.

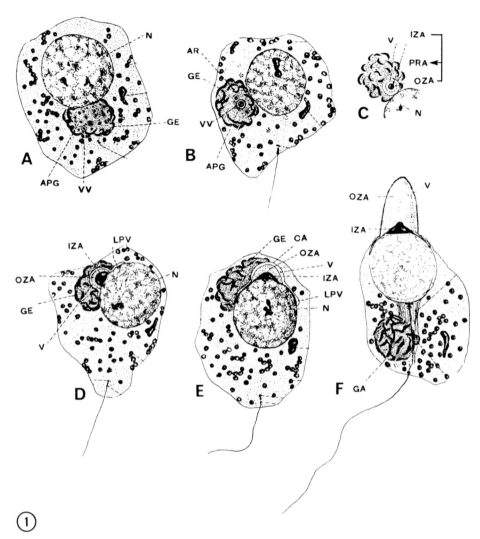

Fig. 1. Development of the acrosome in the guinea pig spermatid. Details described in text. APG = Archoplasmic granule; AR = archoplasm; GA = Golgi apparatus; GE = Golgi element (dictyosome); IZA = inner zone of acrosome; LPV = head cap; N = nucleus; OZA = outer zone of acrosome; PRA = proacrosome; V = acrosomic vacuole; VV = archoplasmic (proacrosomic) vacuoles [from *Gatenby and Woodger,* 65].

Gatenby and Beams, 64], no further great advances were made. Indeed, when *Walker and Allen* [169] and later *Palade and Claude* [131] contended that the fixed impregnated Golgi apparatus of all cells was an artifact based on unspecific staining of various cell components, the state of knowledge seemed even less certain. As stated by *Gatenby* [63] (himself a strong believer in the reality of the structure) at a symposium on the Golgi apparatus held by the Royal Microscopical Society, 'From time to time, sometimes part, sometimes all, of the work on the Golgi apparatus seems to have been disproved.'

Among the proponents of the Golgi apparatus existence during this period were *Leblond and Clermont* who histochemically demonstrated the existence of carbohydrates in this region in spermatids using the periodic acid-Schiff technique, and provided further details on the formation of the acrosome in a variety of species [41, 105]. With this technique the idiosome itself stained only lightly (fig. 2-1). This region elaborated, however, several more heavily stained tiny proacrosomic granules (fig. 2-2) which fused to form an acrosomic granule (fig. 2-3). With some fixatives the granule was uniformly stained, but with others it could be differentiated into a more intensely stained inner zone (corresponding to the acrosomic granule described by traditional methods) and a less intensely stained or vacuolated outer zone (corresponding to the acrosomic vacuole of traditional methods) (fig. 2-3).

The acrosomic granule applied itself to the nucleus, whereupon the outer zone formed an umbrella-like structure, the headcap, which spread to cover a considerable portion of the nuclear surface (fig. 2: 4–7). Meanwhile, the inner zone of the acrosomic granule enlarged and became the acrosome [41]. In some species, i.e., the guinea pig, the inner zone was itself differentiated into an outer and inner zone, thus corresponding to the original description of *Gatenby and Woodger* [65].

A more detailed knowledge of the structure of the spermatid Golgi apparatus awaited the development of the electron microscope, when the first ultrastructural descriptions were published [32, 38, 41, 129]. At this time, a heterogeneity and polarity in structure were confirmed. Thus, the outer portion of the Golgi complex (cortex), which had stained with osmium or silver impregnation methods in the light microscope, corresponded in the EM to stacks of flattened saccules (fig. 5), while the inner portion of the Golgi complex (medulla) contained various vesicles. In the formation of the acrosome, some of the inner vesicles became enlarged (proacrosomic vesicles or vacuoles) and were observed to contain a dense

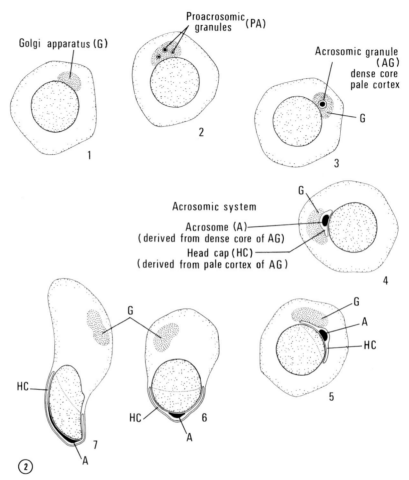

Fig. 2. Development of the acrosome in the rat spermatid, as seen in light microscope sections stained for carbohydrate by the periodic acid-Schiff technique. Details described in text. Courtesy of Dr. *Y. Clermont.*

granule (the proacrosomal granule) surrounded by an almost clear space (fig. 3). The proacrosomic vesicles fused to form the acrosomic vesicle which situated itself against the anterior pole of the nucleus (fig. 4). The dense granule (now called the acrosomic granule) became fixed to the membranous wall of the vesicle adherent to the nuclear membrane, and continued to grow. The Golgi complex then left the acrosomal region and

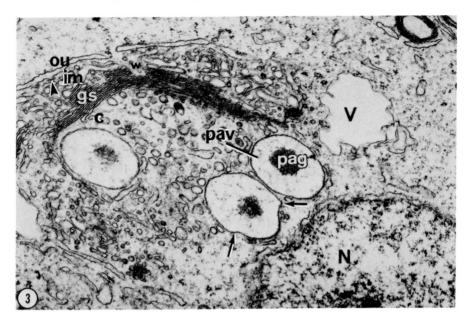

Fig. 3. Golgi region of a step 2 spermatid. The cortex of the Golgi complex consists of the Golgi stack (gs). The convex or immature face of the stack (im) faces the peripheral or outer Golgi region (ou). Bordering on this region is a cisterna of endoplasmic reticulum, which at certain locations forms fuzz-coated buds (arrowhead) projecting into the peripheral Golgi region. The concave face of the Golgi stack (c) faces the medulla region which contains smooth and fuzz-coated vesicles and larger proacrosomic vacuoles (pav). These contain a dense granule, the proacrosomic granule (pag) surrounded by an almost clear space. At some sites along the periphery of these vacuoles, there are bulges suggesting that a small vesicle has recently fused with the vacuole (arrows). Large irregular vacuoles (V) are noted at the periphery of the Golgi stack. N = nucleus; w = well. × 22,000 [162].

moved back along the side of the nucleus to a new position in the caudal portion of the cell. Meanwhile, the acrosomic vesicle collapsed and grew down over the pole of the nucleus, forming the double-layered headcap (fig. 5).

Structure of the Golgi Apparatus in Spermatids

With more modern fixation techniques, a modern interpretation of the structure of the spermatid Golgi apparatus has emerged [78, 79, 112, 148, 162]. Thus, in a young mammalian spermatid of the cap phase (steps

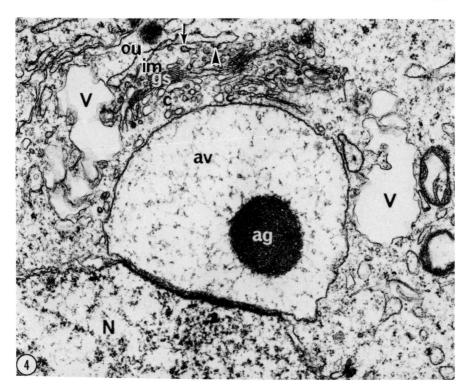

Fig. 4. Step 7 spermatid. One large acrosomic vacuole (av), containing a dense acrosomic granule (ag) is applied to the surface of the nucleus (N). The structure of the Golgi complex
is as described in figure 3. In the peripheral Golgi region, some fuzz-coated vesicles are noted
(arrow). × 30,000 [162].

4–7) or the acrosome phase (steps 8–14) [105], the Golgi apparatus forms a
compact hemispherical mass next to the acrosomic granule (fig. 12) or to
the developing acrosome capping the nuclear surface (fig. 5). The cortex is
made up of several stacks, each composed of 3–9 closely apposed parallel
saccules. Seen in face view, these saccules have an irregular outline (fig. 6).
On the outer or cis-face of the stacks, 3 or 4 of the outer saccules are
frequently interrupted by spaces called wells (fig. 3, 10, 12) which are circular when seen in face view (fig. 6, 7).

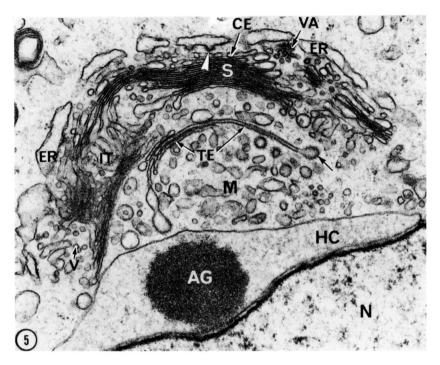

Fig. 5. Golgi region of a step 4 spermatid. In the cortex of the Golgi apparatus, stacks of saccules (S) and intersaccular connecting tubules (IT) are seen. In the medulla (M), the trans element (TE) is present, surrounded by vesicular profiles. The ends of the trans elements are often dilated and fuzz coated (arrow). Cisternae of endoplasmic reticulum (ER) border on the convex face of the Golgi stacks, and form fuzz-coated buds (arrowhead) which protrude into the peripheral Golgi region. This region contains small spherical vesicles (V) and sometimes aggregates of vesicles (VA). In the acrosomic system, the dense acrosomic granule (AG) is located near the nucleus (N), while the acrosomic vacuole (vesicle) grows over the pole of the nucleus to form the headcap (HC). × 34,500 [79].

Between the individual Golgi stacks, the intersaccular regions vary from area to area. In some areas, the stack ends abruptly with all its saccules ending in register (fig. 6, 7, 12). This leaves a gap between adjacent stacks which often contains cisternae of endoplasmic reticulum (fig. 7). More frequently, the saccules of the Golgi stacks become continuous at their edges with membranous tubules with an average diameter of 50 nm (fig. 5, 11, 12). Some of these tubules run toward and become continuous with saccules located at the same level in adjacent stacks (fig. 6, 7). In other

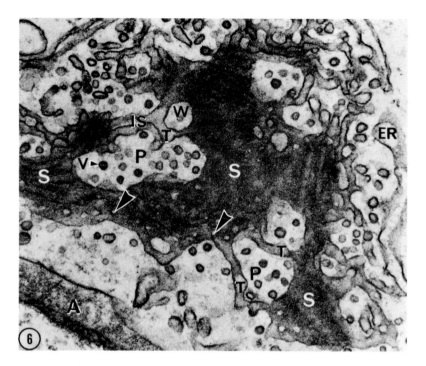

Fig. 6. Face view of the cortex region of the spermatid Golgi apparatus. Saccular (S) and intersaccular connecting regions (IS) are seen. The saccules show minute pores (arrowheads) in their peripheral region. Larger circular areas correspond to wells (W). Numerous vesicles (V) are seen in the gaps delimited by the saccules and the membranous tubules (T) of the intersaccular connecting regions. ER = Endoplasmic reticulum; A = acrosomic system. × 36,000 [79].

instances the tubules loop, intertwine, and anastomose in a complex manner and, in doing so, often protrude up into the peripheral Golgi region (fig. 11). In such regions the membranous tubules sometimes connect to more distal saccules of adjacent stacks, and even sometimes establish connections between saccules of the same stack.

Along the cis face of the Golgi stacks, adjacent to the first saccule on the cis face, is a regular meshwork of anastomotic tubules referred to as the cis element of the Golgi stack (fig. 5, 11, 12). In the intersaccular regions, the cis elements frequently lose their regular meshwork appearance and the tubules connect with tubules arising from the subjacent saccules.

Proximal to the cis element of the Golgi stacks is the peripheral Golgi region (fig. 5, 12), which contains smooth and fuzz-coated vesicles, plus an

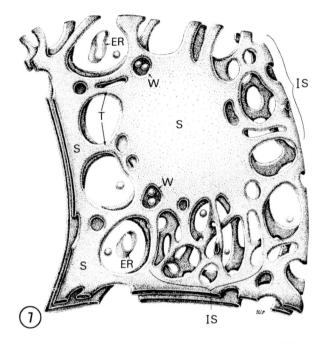

Fig. 7. Diagram showing a face view of two superimposed intermediate saccules of a Golgi stack. Labeling as in figure 6. In the intersaccular connecting regions (IS), tubules frequently connect a saccule of one stack to more distal saccules of other stacks or even the same stack. In the gaps delimited by saccules and intersaccular connecting regions, profiles of endoplasmic reticulum (ER) are sometimes observed [79].

intricate network of membranous tubules 40–50 nm in diameter. Some of these tubules are connected with the cis element. Adjacent to the outer surface of the Golgi apparatus are several cisternae of endoplasmic reticulum. These cisternae are devoid of ribosomes on the side facing the Golgi apparatus, and here they show fuzz-coated buds directed towards the Golgi apparatus.

As one progresses distally (inwardly) across the stack to the inner or trans face, flattened saccules are observed which differ from the overlying saccules in having a wider lumen (30–40 nm) (fig. 5, 9, 11, 12). These saccules are part of the trans element (GERL). In some instances, portions of these flattened saccules (rigid lamellae) are closely applied to the trans face of the Golgi stacks while their extremities show a peeling off configuration (fig. 9, 12). In other instances, the trans elements are completely separated from the stacks, and include both rigid lamellae and tubular networks, which are intermixed with the components of the medulla portion of the Golgi apparatus (fig. 5). The ends of these trans elements are often dilated and fuzz-coated, suggesting that they give rise to the large fuzz-coated vesicles (90–120 nm) in the neighboring medullary cytoplasm (fig. 5, 9, 12). Also present in this region are smooth walled vesicles and tubules 100–150 nm in diameter (fig. 5, 12).

In spermatids subjected to aldehyde-tannic acid fixation, the membranes of Golgi saccules change in appearance from the cis to the trans face of the stack [112]. Those of saccules on the cis face resemble endoplasmic reticulum membranes in exhibiting a symmetrical dark-light-dark pattern. As one progresses across the stack, however, the outer (luminal) dark leaflet becomes less distinct, and the membranes of the GERL element and the acrosome have an asymmetric pattern, as does the plasma membrane.

Histochemistry of the Golgi Apparatus in Spermatids

Staining for Carbohydrate

As noted above, the PAS technique lightly stained the Golgi zone (idiosome) of differentiating spermatocytes, and more intensely stained the proacrosomic granules and the developing acrosomic system (fig. 2) [41, 103, 105]. This revealed the presence of carbohydrates in these structures. Clermont et al. [39] succeeded in extracting the carbohydrates from isolated guinea pig acrosomes and chromatographically identified mannose, galactose, and fucose residues. When spermatids were examined with the periodic acid-chromic acid-methenamine silver (PA-silver) method to detect carbohydrates at the electron-microscope level [138, 162], some stain was observed over the Golgi saccules (fig. 8). The saccules exhibited a staining gradient, the cis element and outer saccules being almost unstained and the inner saccules and trans elements exhibiting an increasingly heavier staining. Thus, those elements (i.e., the outer Golgi saccules or tubules) which had been traditionally identified as the Golgi apparatus using silver or osmium impregnation methods were unstained using the PA-silver technique, while the inner region (idiosome), unstained by traditional methods, was stained with the PA-silver technique. Within the medulla of the Golgi region, various vesicles, especially the large-coated

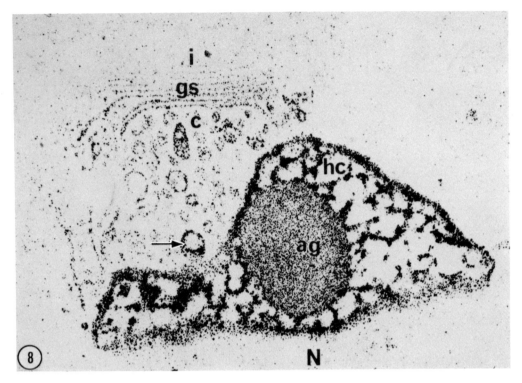

Fig. 8. Thin section of a step 5 spermatid Golgi region stained with the periodic acid-chromic acid-silver methenamine technique to reveal carbohydrates. A gradation of staining from the immature face (i) to the concave face (c) may be observed in the Golgi saccules (gs). Vesicles in the medulla region are stained (arrow). In the acrosomic system, the acrosomic granule is moderately but homogeneously stained, while the wispy material of the head cap (hc) stains intensely. N = Nucleus. × 60,000 [162].

vesicles were positively stained (fig. 8). Lysosomal dense and multivesicular bodies also stained positively for carbohydrates in these cells [138].

During formation of the acrosomic system, the acrosomic vesicle and acrosomic granule were found to be differentially stained by the PA-silver method. The acrosomic granule, dense in electron-micrographic sections stained by the traditional uranyl acetate and lead citrate method (fig. 5), was moderately, but homogeneously stained by PA-silver (fig. 8). The remaining contents of the acrosomic vesicle had exhibited only light, wispy staining with uranyl acetate and lead citrate treatment (fig. 5), but with PA-silver this wispy material stained intensely (fig. 8).

Osmium Impregnation

Although not, strictly speaking, a histochemical method, since the mechanism is unknown, the osmium impregnation technique helps to demonstrate the heterogeneous nature of the Golgi apparatus. In several other cell types the transfer vesicles between the RER and the Golgi apparatus, and a regular network of anastomotic membranous tubules adjacent to the cis face of Golgi stacks have been shown to be osmiophilic [61, 137]. This network is very similar to the cis element observed in spermatids, and probably represents the only portion of the Golgi apparatus stained by the traditional light microscopic silver and osmium impregnation techniques.

Localization of Phosphatase Enzymes

When sections of spermatids are incubated to reveal the presence of cytidine monophosphatase (CMPase) activity [40, 163], the reaction in the Golgi apparatus is primarily restricted to the trans element (GERL) adjacent to the trans face of the Golgi stacks (fig. 9). The flat saccular portions of these trans elements (i.e. rigid lamellae) are highly reactive, and the fuzz-coated buds emanating from these trans elements also show intense reaction. These trans elements thus probably correspond to the CMPase-positive GERL originally described by Novikoff in several other tissues [123, 124]. Reaction product is also found in some vesicles and tubules in the medullary region. During the development of the acrosomic system, reaction is seen in the proacrosomic and acrosomic vesicles (both in the acrosomic granule and in the remainder of the vesicle) (fig. 12). Multivesicular bodies located near the Golgi apparatus at this time exhibit reaction which is concentrated within their small vesicles (fig. 9: inset). At later stages, the acrosome and headcap both show considerable reaction (fig. 9).

When incubated to reveal thiamine pyrophosphatase (TPPase) activity, reaction in the Golgi apparatus is usually limited to one or two saccules on the trans aspect of the Golgi stacks (fig. 10). These trans saccules are seen to exhibit tiny pores or perforations in both their central and peripheral regions. A similar distribution of TPPase activity has been observed in neurons and in other secretory cell types [69, 124]. At the intersaccular regions, the membranous tubules connected to the edge of the TPPase-positive saccules are equally filled with dense reaction product (fig. 10). The acrosomic system reacts unevenly for TPPase activity, sometimes showing little or no reaction (fig. 10, main figure) and at other times

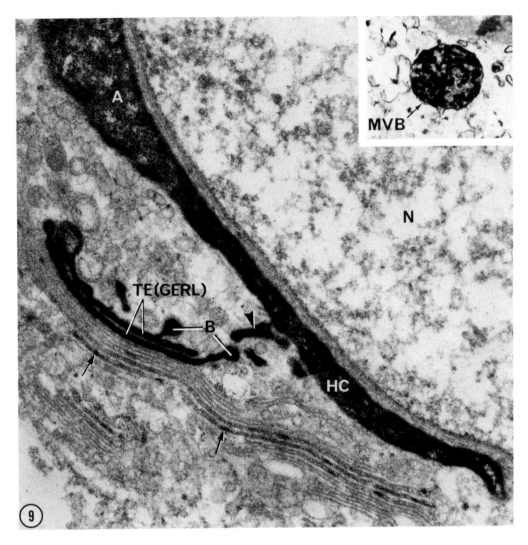

Fig. 9. Section of the Golgi region of a step 7 spermatid showing cytidine monophosphatase (CMPase) reactive elements. The rigid lamellae of the trans elements (TE) (GERL) as well as the buds (B) associated with them are strongly positive. Some reactive profiles (arrowhead) are also seen adjacent to the strongly reactive headcap (HC) and acrosome (A). A weak and spotty reaction can be seen in some intermediate saccules (arrows). N = nucleus. × 40,000. *Inset:* A multivesicular body (MVB) shows a strong reaction in its vesicles. × 38,000. Courtesy of Dr. *Y. Clermont.*

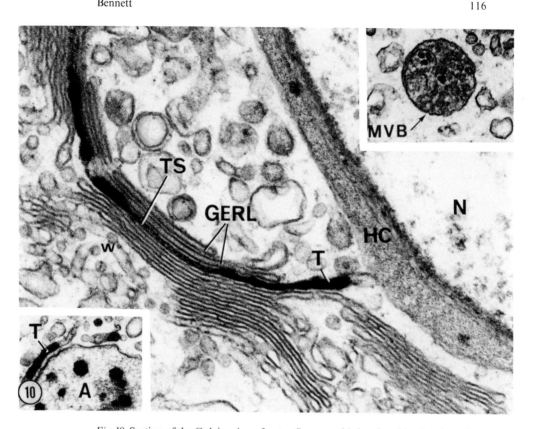

Fig. 10. Section of the Golgi region of a step 7 spermatid showing thiamine pyrophosphatase (TPPase) reactive elements. In the Golgi stack, reaction is limited to the trans saccules (TS) and tubules (T) connected to them. The trans element (GERL) is unreactive. In the acrosomic system the headcap (HC) and acrosome (A) are often unreactive, but sometimes exhibit dispersed lead deposits (lower inset). Multivesicular bodies (MVB in upper inset) were consistently unreactive. w = Well. × 40,000. Courtesy of Dr. *Y. Clermont.*

exhibiting considerable dispersed reaction (lower left inset). Multivesicular bodies do not show TPPase activity (upper right inset).

When incubated for nicotinamide adenine dinucleotide phosphatase (NADPase) activity, reaction in the Golgi apparatus is restricted to the intermediate saccules of the Golgi stack (fig. 11). The cis element (CE) and the first saccule on the cis face (S_1) are unreactive, but the subjacent 3–5 saccules usually contain reaction product. Often a gradient of reaction is observed with a mid-saccule (i.e. S_4) containing more reaction product than those on either side. These saccules are not perforated in their central

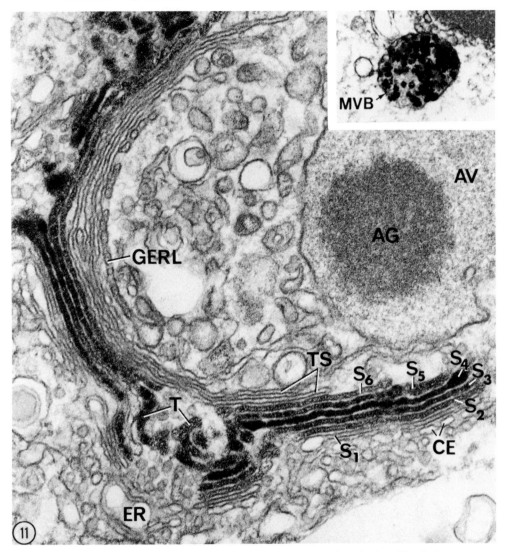

Fig. 11. Section of the Golgi region of a step 3 spermatid showing nicotinamide adenine dinucleotide phosphatase (NADPase) reactive elements. In the Golgi stack, reaction is restricted to the intermediate saccules and their adjacent membranous tubules [T]. These intermediate saccules often exhibit a gradient of reaction, with a mid-saccule (S_4) exhibiting more reaction product than those on the cis side (S_2 and S_3) or on the trans side (S_5 and S_6). The endoplasmic reticulum (ER), cis element (CE), first saccule on the cis face (S_1), trans saccules (TS), and the acrosomic system (AG and AV) are unreactive. There is a very light and spotty reaction in the rigid lamella of the trans element (GERL). *Inset:* A multivesicular body shows strong reaction in its vesicles. × 35,000. Courtesy of Dr. Y. Clermont.

regions, but contain pores in their peripheral regions, as well as the large wells mentioned above (fig. 6). In the intersaccular region, the membranous tubules arising from the edges of the saccules are equally reactive (fig. 10, T), and connect the NADPase-positive saccules of adjacent saccular regions. These tubules sometimes connect to more distal NADPase positive saccules of other stacks or even the same stack (fig. 7). No tubular connections between NADPase-positive intermediate saccules and TPPase-positive trans saccules have been observed. These trans saccules usually show little or no NADPase activity. The trans element (GERL) and vesicles in the medulla are usually negative for TPPase activity (fig. 11), but sometimes show a light spotty reaction. The acrosomic system itself is usually negative for NADPase (fig. 11). On the other hand, the small vesicles within multivesicular bodies are strongly positive (fig. 11, inset).

Thus, the Golgi apparatus of the spermatid is seen to be an extremely complex assembly of membrane-bound structures, which are heterogenous both in their morphology and in their chemical composition (as indicated by differences in histochemical staining). On these criteria, for example, at least 14 different compartments or regions of the spermatid Golgi complex can be visualized, i.e. (1) the fuzz-coated and smooth vesicles and tubules in the peripheral Golgi region; (2) the regular meshwork-like osmiophilic cis element; (3) the NADPase-negative first saccule on the cis face of Golgi stacks, and (4) its associated tubules in the intersaccular regions; (5) the NADPase-positive intermediate saccules, and (6) their associated intersaccular tubules; (7) the well regions occurring within the intermediate saccules containing vesicles; (8) the TPPase-positive trans saccules, and (9) their associated intersaccular tubules; (10) the gap regions between adjacent stacks often containing cisternae of RER; and the CMPase-positive trans element (GERL) consisting of (11) rigid lamellae; (12) tubular networks, and (13) swollen fuzz-coated extensions, and finally (14) the fuzz-coated and smooth tubules and vesicles in the medulla region, some of which give rise to proacrosomic vesicles.

In some other cell types, histochemical or biochemical heterogeneity has been demonstrated within individual saccules. Thus 5'-nucleotidase and adenylate cyclase activity have been shown to be concentrated along the dilated edges of Golgi saccules in liver fractions while the flattened centers of the saccules exhibited little or no activity [36, 52]. Affinity separation techniques have provided evidence suggesting that the edges of Golgi saccules are rich in enzymes usually associated with microsomes, while the central part of the saccules were rich in glycosyltransferase activities [92]. Finally, in the Golgi apparatus of snail multifid gland cells, the individual Golgi saccules exhibit clear zones rich in glycoprotein content and dense zones devoid of glycoproteins [127]. In view of this extreme complexity, it should not be surprising if the Golgi complex is found to be the site of many different activities, and perhaps more than one different route of secretion.

The Role of the Golgi Apparatus in Secretion

Passage of Secretory Material through the Golgi Apparatus

The general pathway of secretion has now been worked out in a great variety of different secretory cell types [130] and it is clear that in virtually all cases the Golgi apparatus is involved. As mentioned at the beginning of this article, the Golgi had been seen to be associated with secretory products almost from its original discovery [27]. When electron-microscopic images of the Golgi apparatus became available, new evidence for the Golgi's role in secretion was provided, since the large Golgi vesicles (at the trans face), and often some of the saccules as well, frequently contained material identical to that seen in nearby secretion granules [43, 55, 60, 89, 178]. Such images indicated that the Golgi apparatus concentrated as well as packaged the secretory product, and although by themselves static, these images suggested that the apparatus was a very dynamic structure.

The final proof for this dynamic activity came with the availability of suitable radioactive biological tracers, coupled with the techniques of cell fractionation and radioautography. Thus, after administration of radioactive amino acids (in vivo or in vitro) to such secretory cells as pancreatic acinar cells [34, 35, 167, 170], chondroblasts [144], and thyroid follicular cells [118], the label (which had been incorporated into protein molecules) was initially taken up in the rough endoplasmic reticulum, but then, within 10–30 min, passed to the Golgi apparatus and thence to secretion products.

The work of *Caro and Palade* [35] suggested that small 'intermediate' or 'transfer vesicles' served to carry the newly synthesized proteins from the special transitional elements of the rough endoplasmic reticulum to the Golgi apparatus. The exact destination of these transfer vesicles varied in different cell types: in guinea pig pancreatic cells they opened by exocytosis into condensing vacuoles at the trans face of the Golgi apparatus, while in many other cell types (and in stimulated pancreatic cells) they

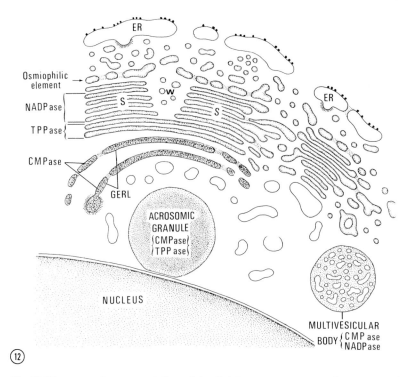

Fig. 12. Diagrammatic representation of the Golgi apparatus, acrosomic system, and associated multivesicular body in a step 3 spermatid. The reactivity of the various elements for the three phosphatases, NADPase, TPPase, and CMPase is indicated. Er = Endoplasmic reticulum; S = saccules; w = well [163].

opened into the cis Golgi saccule or the ends of intermediate Golgi saccules [130]. In spermatids, the cisternae of endoplasmic reticulum adjacent to the Golgi region were seen to have a similar transitional character [115, 148, 162]. These exhibit buds which appear to give rise to vesicles observed in the peripheral Golgi region (fig. 3–5, 12). These buds and vesicles are frequently coated, and at least in some cell types [147] the coat contains clathrin. Some of these vesicles may, in turn, fuse with tubules of the cis element, while others may fuse with more distal Golgi elements, i.e. the first saccule or the intermediate saccules and their adjacent intersaccular tubules.

The driving force which transports the transfer vesicles from the rough endoplasmic reticulum to the Golgi apparatus is not clearly understood. Continued protein synthesis is not necessary [93] but the process is stopped by uncoupling oxidative phosphosylation [93, 165]. Microtubule disruption by colchicine or vinblastine results in inhibition of passage of material and an accumulation of vesicles at the cis face in some cells [110].

In some other cell types, i.e. hepatocytes, continuous tubular connections may exist between the transitional rough endoplasmic reticulum and Golgi saccules [37, 117, 126], and microtubular disrupting drugs do not appear to impede passage of secretory material between the two sites [139].

Passage of Secretion Products from Golgi Apparatus

At the trans face of the Golgi apparatus in spermatids and in many cell types, are special saccular or tubular elements which exhibit acid phosphatase enzymatic activity and are referred to as GERL ('trans element' in spermatids) [123, 124]. *Novikoff* considered GERL as a specialized part of the endoplasmic reticulum situated in close proximity to the Golgi apparatus and involved in the formation of lysosomes. There is evidence that GERL elements give rise to newly formed secretion granules in many secretory cell types [75, 123]. In spermatids the rigid lamellae or tubular elements of GERL form fuzz-coated (clathrin) protuberances which appear to give rise to similar fuzz-coated vesicles in the medulla region [72, 115, 148, 162]. In the early stages of acrosome formation, these lose their coats and fuse to form the proacrosomic granules. Later on, these vesicles fuse directly with and empty their contents into the acrosomic vesicle or developing acrosome.

In many (although not all) secretory cell types, the drugs colchicine and vinblastine inhibit secretion, probably by causing the disruption of microtubules [110, 139, 141]. Recent radioautographic studies on a large variety of cell types from rats injected with these drugs have shown an inhibition of intracellular migration of both secretory glycoproteins and membrane glycoproteins bound for the plasma membrane [17]. When observed in the electron microscope, the ^{3}H-fucose-labeled glycoproteins are seen to pass from the Golgi saccules to the nearby secretory (or other) vesicles, but these vesicles are then inhibited from leaving the Golgi region, suggesting that microtubules are necessary for this process [18, 176].

Modification of Secretion Products in the Golgi Apparatus

Although the above-mentioned early tracer experiments showed that newly synthesized proteins passed through the Golgi apparatus, they did not indicate whether or not this organelle served solely to package and

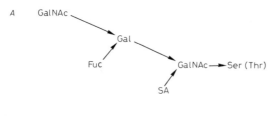

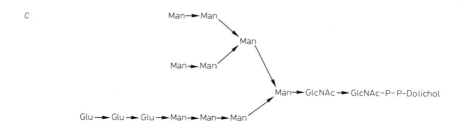

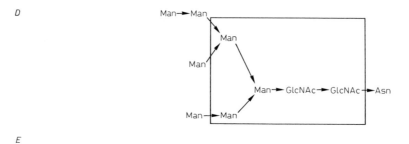

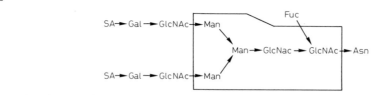

Fig. 13A, B. Serine (threonine)-linked side chains of mucous glycoproteins. *C* Dolichol phosphate glycolipid intermediate involved in the biosynthesis of asparagine-linked glyco-protein side chains. *D* High mannose type of asparagine-linked glycoprotein side chain of ovalbumin. *E* Complex type of asparagine-linked glycoprotein side chain. Core region is enclosed within dotted line. Ser = Serine; Thr = threonine; GalNAc = N-acetyl-galactos-amine; GlcNAc = N-acetyl-glucosamine; Gal = galactose; Fuc = fucose; SA = sialic acid; Man = mannose; Glu = glucose [98].

perhaps concentrate the secretion products or whether it served some additional role in synthesis or modification. Some cells, i.e., goblet cells or chondroblasts, were known to secrete glycoproteins or mucopolysaccharides high in carbohydrate content. It was also becoming increasingly evident that many other secretory products, once thought to be purely protein, were actually glycoproteins [48].

The fact that carbohydrate material (as indicated by positive PAS staining) appeared in the Golgi zone of spermatids and many other cell types [66, 103, 105], suggested that the Golgi apparatus might add carbohydrate to secretions which passed through it. The carbohydrate side chains of glycoproteins are made up of a variety of different sugar residues (fig. 13A, B, D, E) which are sometimes sulfated (fig. 13B). Radioactive sulfate was one of the earliest precursors commercially available, and when administered to chondroblasts, light microscopic radioautographic studies showed that the label was initially localized in the Golgi region, indicating that at least sulfate residues might be added in the Golgi apparatus [95]. Most of the sugar residues in the carbohydrate side chains, on the other hand, are derived (during normal metabolism) from glucose. When ^{3}H-glucose became commercially available, this precursor was injected into young rats by *Neutra and Leblond* [120–122] and the mucus-secreting goblet cells of the large intestine were examined by light and then electron microscope radioautography. The label was found to be incorporated in the Golgi apparatus at short time intervals after injection. With time, the label passed out of the Golgi apparatus and to mucus-secreting granules which then migrated apically in the theca to be ultimately discharged from the cell. It is now known that mucous secretion consists mainly of glycoproteins whose carbohydrate side chains are linked by N-acetylgalactosamine residues to serine or threonine residues of the polypeptide chain (fig. 13A, B) [98]. These side chains also contain galactose, N-acetylglucosamine, fucose, and sialic acid residues; and biochemical studies had shown that all of these residues became labeled after administration of radioactive glucose to rat colonic mucosa [45]. Thus, all of these sugar residues appeared to be added to mucous glycoproteins in the Golgi apparatus.

A more common type of side chain occurring in many other glycoproteins, however, is linked by N-acetylglucosamine to asparagine residues of the polypeptide chain [98, 157]. These may be either of a high mannose type (fig. 13D), consisting of several mannose residues linked to a pair of proximal N-acetylglucosamine residues, or a more complex type

(fig. 13E), consisting of a core of two N-acetylglucosamine and three mannose residues to which are attached more peripheral N-acetylglucosamine,
galactose, fucose, and sialic acid residues [98]. Early biochemical experiments indicated that the core N-acetylglucosamine and mannose residues
were added to glycoproteins very soon after completion or even during the
synthesis of their polypeptide chain on membrane-bound ribosomes [113,
158]. It is now known that these residues are first added one by one (in
reactions catalyzed by specific glycosyltransferases) to a dolichol phosphate glycolipid intermediate which contains the two proximal N-acetylglucosamine residues, several mannose residues as well as some peripheral
glucose residues (fig. 13C). The oligosaccharide is then transferred to an
acceptor glycoprotein molecule [151, 156, 161, 168].

The first radioautographic study on the site of incorporation of one of
the core sugars (i.e. mannose) into glycoproteins was carried out by *Whur*
et al. [175]. ^{3}H-mannose was administered to thyroid follicular cells of rats,
and these were examined by electron-microscopic radioautography. The
label was found to be initially incorporated in the rough endoplasmic
reticulum, and only at later time intervals migrated to the Golgi apparatus
and secretion products. After administration of a peripheral sugar precursor, i.e. ^{3}H-galactose, on the other hand, the label was initially taken up in
the Golgi apparatus, indicating that this residue was added to glycoproteins only after they had migrated from the rough endoplasmic reticulum
to this organelle [175]. A similar initial uptake of ^{3}H-galactose in the Golgi
apparatus was observed in several other cell types, i.e. neurons [46],
duodenal villus columnar cells [9], ameloblasts [171], and spermatids [149].
Relatively pure Golgi fractions have now been isolated from a number of
different cell types and in every case these are highly enriched in galactosyltransferase, the specific glycosyltransferase (or class of glycosyltransferases) which catalyzes the addition of galactose residues from uridine
diphosphate-galactose molecules to glycoproteins [151], possibly also by
way of a glycolipid intermediate [128]. Mannosyltransferase enzymes, on
the other hand, had been found to be enriched in the microsomal fraction
[151]. It is now known that before the peripheral sugar residues are added,
the glucose and some mannose residues of the oligosaccharide side chain
(fig. 13C) are trimmed off by specific glucosidases and mannosidases. This
process is thought to be completed in the cis portion of the Golgi apparatus
[77].

Two sugar residues which always occupy terminal positions on carbohydrate side chains are fucose and sialic acid (fig. 13A, B, D, E). In

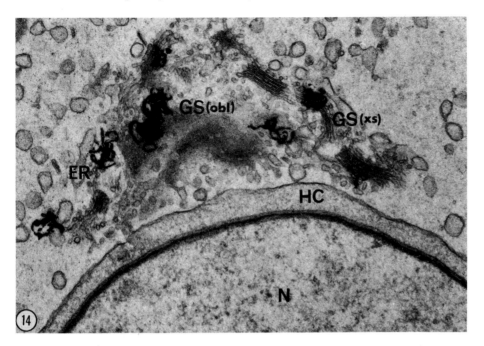

Fig. 14. Electron-microscopic radioautograph of the Golgi region of a step 7 spermatid from a rat sacrificed one hour after an intratesticular injection of ^{3}H-fucose. The silver grains are localized over the Golgi stacks, which in this section are cut both in cross section [GS (xs)] and obliquely [GS (obl)]. The acrosomic system, represented here by the head cap (HC) is unlabeled. × 27,000. Courtesy of Dr. *Y. Clermont.*

asparagine-linked side chains, the fucose is attached directly to one or the other of the paired proximal N-acetylglucosamine residues of the core region, but is still terminal in that no other sugar residue is linked distally to it. After administration of ^{3}H-fucose, the label was again found to be initially incorporated in the Golgi apparatus, as observed in electron microscope radioautographic studies of a number of cell types, i.e., rat intestinal columnar cells [12], thyroid follicular cells [74], hepatocytes [13], kidney tubule cells [15], and thymic epithelial cells and macrophages [10]. Recently, electron microscope radioautographic studies of rat testis after intratesticular injection of ^{3}H-fucose have shown that in spermatids the label is initially taken up in the Golgi apparatus (fig. 14) [163].

The localization of the site of incorporation of sialic acid residues into glycoproteins was not as straightforward as in the case of the previous sugars, since exogenously administered sialic acid does not effectively enter intact cells in amounts sufficient for radioautographic investigation [90]. ^{3}H-N-acetylmannosamine was found to be a quite specific precursor for sialic acid residues of glycoproteins in liver and small intestine, however, [16], and was used for electron microscope radioautographic studies. In both small intestinal columnar cells and in hepatocytes, the label was initially incorporated in the Golgi apparatus (fig. 19) [11, 16]. Light microscope radioautographs of a variety of other cell types also showed initial localization of the ^{3}H-N-acetylmannosamine label to the Golgi region, indicating that the Golgi uptake of sialic acid is a general phenomenon [11]. Within the Golgi apparatus of hepatocytes, the site of initial uptake of ^{3}H-galactose or ^{3}H-fucose labels appeared to be somewhat proximal to that of ^{3}H-N-acetylmannosamine label. Thus, at early time intervals after injection of ^{3}H-galactose or ^{3}H-fucose, most of the silver grains overlay the intermediate region of the Golgi stacks, with the remainder distributed over the cis and trans faces. Shortly after ^{3}H-N-acetylmannosamine injection, however, most of the silver grains overlay the trans face of the Golgi stacks and the adjacent secretory vesicles, suggesting that sialic acid residues may be added to newly formed glycoproteins slightly later than galactose and fucose residues. Assays of glycosyltransferase activity in different subfractions of Golgi elements have indicated that galactosyltransferase and sialyltransferase are evenly distributed in all Golgi subfractions [29] or that they are concentrated in subfractions of trans elements [62, 87, 145]. Such studies measure the enzymatic potential of the transferases using exogenous acceptors (or immunocytochemical localization of the enzyme), however, and may not necessarily be measuring their physiological activity which would also depend on the amounts of endogenous newly formed acceptor glycoproteins available for glycosylation in the different Golgi subfractions. Histochemical visualization of terminal galactose residues of glycoproteins using a peanut lectin-horseradish peroxidase procedure reveals staining only in the intermediate saccules of the Golgi stack [150]. Thus, the newly formed glycoproteins may have most of their potential sites for galactose residues filled while in the intermediate portion of the Golgi stacks, and terminal residues would then be added more distally. The cells may, therefore, not incorporate appreciably more galactose residues in the trans element, even though the necessary glycosyltransferases are present in this compartment [22]. It should be noted, however, that *Kramer and Geuze* [100] have provided radioautographic evidence that in stomach surface mucous cells, ^{3}H-galactose label is incorporated preferentially in the trans cisternae of Golgi stacks. In addition, evidence from genetic reconstitution experiments in cell-free systems (from vesicular stomatitis virus-infected mutant CHO cells), on the other hand, has suggested the existence of two functionally different Golgi compartments, a 'cis' compartment where mannose residues are trimmed from glycoprotein side chains and fatty acyl groups are added, and a 'trans' compartment where terminal glycosylation occurs [146].

Since fucose and sialic acid are terminal sugar residues on the carbohydrate side chains of glycoproteins, their addition in the Golgi apparatus may mark the completion of synthesis of these molecules within this organelle. A variety of glycosyltransferases has been shown to be present at the surface of cells in recent years [133] and could play some role in further glycosylation of newly synthesized glycoproteins. The levels of these surface enzymes are much lower than those occurring within the Golgi apparatus, however, and it has been proposed that these surface ectoglycosyltransferase enzymes may play other roles, such as surface markers for cell recognition or adhesion.

Thus, the terminal glycosylation of glycoproteins appears to occur in the Golgi apparatus. Electron microscope radioautographic studies have confirmed that sulfation of glycoproteins and proteoglycans occurs in the Golgi apparatus of a variety of cell types [68, 177], and sulfotransferase activities have been localized to Golgi fractions [58, 96, 153]. Fatty acids are added to some proteins in the Golgi apparatus [152], and many secretory proteins undergo selective proteolysis in this organelle, i.e., in the conversion of proinsulin to insulin [159], proalbumin to albumin [136, 140], or of proparathormone to parathormone [73]. This process may continue within the Golgi-derived secretory granules. Finally, concentration of many (but not all) secretory products occurs within the trans Golgi saccules and condensing vacuoles [54, 94] either by crystal formation [5], or perhaps interaction with molecules of opposite charge, i.e. sulfated glycosaminoglycans [2, 142]. In some instances, structural assembly of complex secretion products can be visualized within Golgi saccules, i.e., in the formation of collagen in odontoblasts [172] or of scales in certain algal cells [30].

Involvement of the Golgi Apparatus in the
Production of Lysosomes

Secretion products have traditionally been thought of as products which are synthesized by cells, perhaps stored for variable periods of time

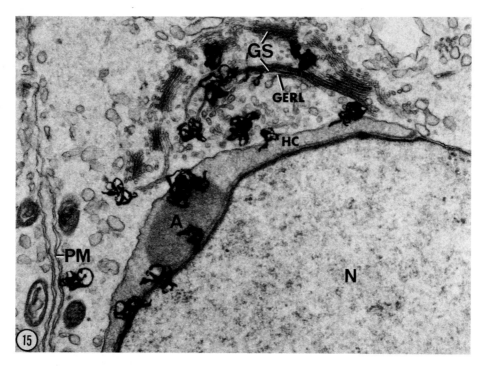

Fig. 15. Electron-microscopic radioautograph of the Golgi region of a step 7 spermatid from a rat sacrificed 4 h after an intratesticular injection of ³H-fucose. In the Golgi region, the silver grains are associated mostly with the trans face of the Golgi stacks and the medullary region (including GERL). Grains are also localized over the acrosome (A) and head cap (HC). One grain lies adjacent to the plasma membrane (PM). × 19,000. Courtesy of Dr. *Y. Clermont.*

within the cell as secretion granules, and then discharged from the cell by exocytosis. *Bowen* [27] was one of the first workers to propose a more expanded view of the secretory process in suggesting that the acrosome be considered a secretion product that would be stored within the spermatid for its lifetime and discharged only at fertilization. It has since been shown that the acrosome contains lysosomal enzymes [1], and these are indeed released from the cell by a process of multiple exocytosis at the time of fertilization [7]. With the advent of electron microscopy and the discovery of lysosomes, it has become apparent that virtually all cells produce and contain lysosomes, and in this sense almost every cell in the body may be considered a secretory cell [130]. In some cell types, i.e. leukocytes, the content of these lysosomes may be released extracellulary at certain occasions, while in many other cell types no such extracellular release may ever occur. To the extent, however, that secondary lysosomes are often formed by fusion of primary lysosomes with phagosomes derived from the extracellular milieu, these secondary lysosomes themselves could be considered as an internalized portion of the extracellular space, and delivery of material to them from the Golgi apparatus has analogies to extracellular secretion.

That the Golgi apparatus produces the lysosomal granules of leukocytes has long been known [3, 5], and *Cohn* et al. [42] have shown the passage of³H-leucine label from rough endoplasmic reticulum through the Golgi apparatus to lysosomes in mononuclear phagocytes. Radioautographic studies after ³H-fucose injection into rats have shown a migration of labeled glycoproteins from the Golgi apparatus to lysosomal dense and multivesicular bodies in a great variety of cell types [10, 13, 14, 67] including spermatids [163] (fig. 14–18).

Routes of Passage of Secretory Material through the
Golgi Apparatus

The general problem of how secretory material moves through the Golgi apparatus remains unresolved. Morpholigical evidence in many cell types, including the spermatid, has suggested that the saccules on the cis face with their contained secretory product physically move across the stack becoming intermediate and then trans face saccules. These latter would be used up in the formation of secretion granules, while at the opposite cis face new saccules would be being continually formed from vesicles derived from the endoplasmic reticulum [71]. The radioautographic studies of *Neutra and Leblond* [121] in goblet cells suggested that a new saccule would be formed every 2 min and a similar turnover rate has been indicated in some other cell types [115]. This model, while very attractive morphologically and quite possibly correct, leaves several questions unanswered. One such question involves the transport of membrane components. The turnover model of Golgi function historically gave rise to the hypothesis of bulk membrane flow [59]. This hypothesis suggested that intracellular membranes turned over at the same rate as secretion products, i.e., the membrane would flow continually, along with the secretory products, from the ER to the Golgi to secretion granules and, after exocytosis, would finally become plasma membrane. The radioautographic studies of *Bennett* et al. [15] in a variety of cell types provided

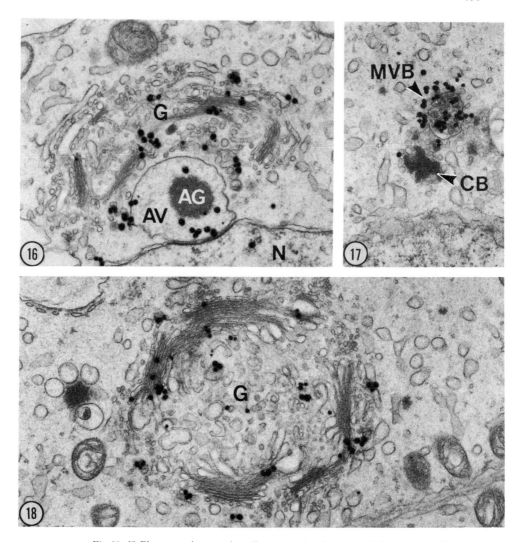

Fig. 16–18. Electron-microscopic radioautographs of spermatids from rats sacrificed 8 h after an intratesticular injection of ³H-fucose. Solution physical development. × 20,000 [163].

Fig. 16. Step 3 spermatid. The silver grains are localized over the Golgi stacks (G), medullary region, and acrosomic system (AG and AV). N = Nucleus.

Fig. 17. Step 3 spermatid. The silver grains are concentrated over the multivesicular body (MVB). A nearly chromatoid body (CB) is unlabeled.

Fig. 18. Step 14 spermatid. The Golgi apparatus, which has now migrated away from the acrosomic system, continues to exhibit a marked radioautographic reaction. The surrounding cytoplasm is unlabeled.

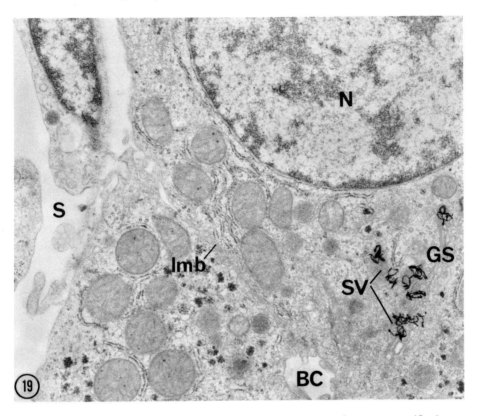

Fig. 19. Electron-microscopic radioautograph of a hepatocyte from a rat sacrificed 10 min after intravenous injection of ^{3}H-N-acetylmannosamine (as a precursor for sialic acid residues of glycoproteins). The silver grains are strongly localized to the Golgi region of the cell. Some grains occur over the central regions and cis face of the Golgi stack (GS), but most are localized over the trans face of the stack and adjacent secretory vesicles (VS). The bile canalicular (BC), lateral membrane (lmb) and the plasma membrane facing the sinusoid (S) are unlabeled. N = Nucleus. × 13,000.

evidence that ^{3}H-fucose and other labeled sugars are incorporated into membrane glycoproteins in the Golgi apparatus, and that these rapidly migrate by way of secretory (or other cytoplasmic) vesicles to the cell surface where they become components of the plasma membrane. Recent observations on the synthesis of viral membrane glycoproteins have indicated that these too are initially manufactured on membrane-bound ribosomes of infected cells and are incorporated into the membrane of the endoplasmic reticulum. Subsequently, they migrate by way of the Golgi apparatus and cytoplasmic vesicles to the plasma membrane [21, 91, 107–109]. Such results would appear to support a membrane flow model. Biochemical studies by *Meldolesi* [111] and others, on the other hand, have shown that many membrane components of intracellular compartments turn over at a much slower rate than secretory components, and the turnover rates of individual components differ from one another. In addition, if simple bulk membrane flow occurred, one would expect a greater similarity in membrane composition between the membranes of different intracellular compartments than actually exists [111]. Finally, the question arises as to what would happen to all of the membrane material arriving at the plasma membrane. In the past it was felt that this was mainly internalized to lysosomes and degraded [115, 125]. Recent evidence, however, indicates that some membrane components are recycled from the plasma membrane to the Golgi apparatus [23, 49, 51, 80, 83–85, 166].

An alternative model to the simple bulk membrane flow hypothesis has been a system whereby vesicles would shuttle back and forth [130]. Thus, transfer vesicles would ferry secretory material from the transitional elements of the ER to the Golgi apparatus and then shuttle back again to the endoplasmic reticulum.

From the trans saccules (and/or GERL where it exists), secretory vesicles would ferry material to the cell surface, and then the membrane would in some fashion be recycled to the Golgi trans saccules or GERL. While this model allows for preservation of the individuality of different intracellular membrane compartments, in a strict form it does not provide for the above-mentioned fact that trans membrane integral proteins of distal compartments (i.e., Golgi apparatus, plasma membrane) are known to be synthesized in the rough endoplasmic reticulum and subsequently travel within membrane to these sites.

A model which allows for most of the existing data has evolved in recent years with the realization that a membrane consists of a viscous but liquid two-dimensional solution of protein molecules mixed with lipid molecules [154]. Within this solution, both proteins and lipids are free to move (subject to certain constraints) within the plane of the membrane. Thus, while many membrane components of transfer or secretory vesicles (perhaps even the bulk of the membrane) may be shuttled back to their original compartment after the vesicle has ferried secretory material distally, a selective flow of those membrane components destined for more distal sites would occur [14, 20, 115, 116]. The transitional rough endoplasmic reticulum, for example, would form transfer vesicles destined to migrate to the Golgi apparatus. Membrane proteins destined for more distal compartments would flow into these vesicles, where many of those destined to remain in the ER would not enter the vesicles. When the vesicle arrived at the cis Golgi element, membrane molecules destined for more distal compartments would enter the cis element membrane, while others would remain in the transfer vesicle membrane to be shuttled back to the ER. The same process would be repeated again with secretory vesicles travelling from the Golgi apparatus to more distal compartments, such as the acrosome in spermatids or to the plasma membrane in most secretory cells. Thus, there would be a flow of individual membrane molecules from compartment to compartment, but no net bulk flow of the entire membrane. In this way, the characteristic membrane proteins would remain in their own compartments long enough to explain the slow turnover times observed by *Meldolesi* and co-workers. In addition, the different individual chemical compositions of different compartments would be explained by only the appropriate components remaining in any one compartment while those of other compartments either would not enter the compartment or would pass on to the next compartment.

The proportion of the total membrane components involved in this selective membrane flow would vary greatly depending on the particular cell type involved. In instances where the plasma membrane is not enlarging, i.e., most nongrowing cells, only a small proportion of the total membrane components might travel to and remain in the plasma membrane. In instances where much new plasma membrane is being formed, on the other hand, i.e. tip growing cells such as neurons [132], the proportion of total membrane components remaining in the plasma membrane might be great enough to approximate a true bulk membrane flow.

The fundamental question remains whether or not turnover in the Golgi saccules exists, although in some instances it seems very probable. The most dramatic example is seen in the formation of scales in the Golgi saccules of certain algae [30, 31]. Here each Golgi saccule in the stack contains a large disc-like, nondeformable scale. As these scales are continuously delivered to the cell surface, their enclosing Golgi saccules become elongated secretion granules and of necessity travel to the cell surface to deliver the scale to the cell exterior by exocytosis. Thus, at least in this cell type, continual turnover of the Golgi saccules must occur. Some system of shuttle vesicles would probably also operate even in these cases, however, to allow certain molecules to enter or leave a saccule during its migration such that it could differentiate en route [146].

In many other cell types, turnover of Golgi saccules is somewhat difficult to visualize. In the spermatid, for example (fig. 12), the heterogeneous structure and chemical composition of the Golgi stack would make this a complex process. In its passage across the Golgi stack of a spermatid, for example, the meshwork osmiophilic cis element would have to suddenly change to a continuous plate-like nonosmiophilic cis saccule. This saccule would then need to acquire NADPase activity, then lose the former, become fenestrated and acquire TPPase activity. Finally, it would have to lose the latter and acquire the CMPase activity and morphology of the GERL element. During all of this, each saccule would need to maintain continuity with similarly transforming tubules in the intersaccular regions. It is not impossible that such transformations could occur, however, especially in the spermatid Golgi which would be less active than that of the goblet cell. Gradients of certain properties from saccule to saccule are also observed, i.e., in intensity of carbohydrate staining (fig. 8), in NADPase activity (fig. 11) and in membrane structure [112], which are suggestive of gradual transformation of a saccule as it would move across the stack.

In cells where the secretory produce is highly distendable, it is conceivable that turnover of the Golgi saccules might not occur and that the secretory product would travel from saccule to saccule via either continuous connections or shuttle vesicles. In cross-sectional images of Golgi stacks in most cell types, connections between adjacent saccules are rarely, if ever, observed. In the spermatid, the tubules of the intersaccular regions sometimes connect a saccule of one stack to a more distal saccule of another stack (or sometimes the same stack). No such tubules have been observed, however, connecting NADPase-positive intermediate saccules to TPPase-positive trans saccules. In other cell types, tubular connections between saccules have not been documented and if present their number and size would not likely be sufficient to transport secretory material at the rate known to occur in actively secreting cells. It is possible, on the other hand, that in some cell types secretion products could travel from saccule to saccule via shuttle vesicles. As discussed above, in such a situation, the chemical heterogeneity of the stack could then be maintained by only certain membrane components entering or leaving these shuttle vesicles.

A second fundamental question in Golgi function is whether or not all secretory products travel through all saccules. As mentioned earlier in this article, there is evidence that in some cells (i.e., pancreatic acinar) secretory products may pass from the transitional endoplasmic reticulum directly to condensing vacuoles on the Golgi's trans face [130]. It has been suggested by *Morré* et al. [115] that the main function of the Golgi stack may often be one of membrane biogenesis and differentiation, while the function of modification and packaging of secretory products would occur mostly in the trans elements.

The relationship between the Golgi stack and the special acid phosphatase positive GERL elements on its trans face has long been a topic of controversy. In several cell types, including the spermatid, secretory granules or vesicles have been observed to arise from GERL. *Novikoff* has considered GERL as a specialized region of the endoplasmic reticulum which maintains a structural continuity with the remainder of the RER system. He has suggested that the Golgi does not directly participate in the formation of these secretory granules, but rather they may arise from material passing directly from the endoplasmic reticulum to GERL [123]. Many studies, however, have shown the presence of secretory product in all of the saccules of the Golgi stacks [4, 75, 82] indicating that the stack itself must be involved in the secretory process. In Golgi fractions of hepatocytes, *Bergeron* et al. [19] have provided evidence for a sequential progression of labeled secretory protein through the cis and trans components of the Golgi apparatus. *Hand and Oliver* [75] have provided evidence that there is a transitional saccule at the trans face of the Golgi which contains the secretion product, peroxidase, and exhibits both TPPase and acid phosphatase activities. These authors suggest that the GERL may arise from the Golgi saccules. In spermatids, the studies on comparative membrane ultrastructure by *Mollenhauer* et al. [112] described above would support such a concept.

Recently, a new aspect of Golgi apparatus function has been demonstrated by the work of *Regoeczi* et al. [143] who have shown that asialotransferrin molecules, taken up from the circulating blood by hepatocytes, are routed to the Golgi apparatus where they are resialylated. Thus, the Golgi apparatus appears to have the capacity to repair previously secreted secretion products.

Segregation of Secretion Products in the Golgi Apparatus

Since many secretory cell types secrete more than one type of secretion product at the same time, and since virtually all secretory cells also produce lysosomes, the question arises as to how the cell organizes the proper intracellular routing of these different products. In some cases, no sorting occurs, and different products are secreted together in the same secretion granules [99]. In snail multifid gland cells, as discussed earlier, the individual Golgi saccules contain areas with dense content and other clear areas. These presumably different secretion products remain separate throughout the secretory process [127]. Similarly, in the formation of the acrosome in spermatids, the dense (acrosome) and clear (headcap) portions of the acrosomic system remain continually distinct from one another (fig. 3–5).

In all cell types, lysosomal enzymes travel to different intracellular compartments from other secretion products, i.e. the lysosomes. In this case, there is evidence that a residue of N-acetylglucosaminylphosphate is added to a mannose residue of the core region of an asparagine-linked side chain of glycoproteins destined to be delivered to lysosomes. The N-acetylglucosamine is then removed, leaving mannose phosphate to act as a signal designating the glycoprotein involved as a lysosomal glycoprotein. Both of these events appear to occur in the cis part of the Golgi apparatus [135]. These lysosomal glycoproteins then bind to specific receptors, and collect into vesicles which bud off of Golgi or GERL to form primary lysosomes [reviewed by *Sly* et al., 155]. The fact that immature secretion granules frequently exhibit some acid phosphatase activity may reflect an inability of the cell to completely separate lysosomal enzymes from other secretory products bound for the cell exterior [76].

It is not known if secretion products bound for different cellular compartments are normally derived from different parts of the Golgi apparatus. In the spermatid, the recent histochemical findings of *Tang* et al. [163] illustrated in the present article, suggest that different subcompartments of the Golgi apparatus may be specialized for the production of different products. Thus, the acrosomic system (CMPase and TPPase positive) probably receives material from GERL (CMPase positive) by way of coated vesicles [57, 72, 112, 148, 162], but may also receive material via an independent route directly from the trans Golgi saccules (TPPase positive). The multivesicular bodies (CMPase and NADPase positive), on the other hand, may receive material from GERL (CMPase positive) as well as independently from the intermediate Golgi saccules (NADPase positive). These findings suggest rather than prove independent pathways, however, since it is possible that other reasons such as inhibitory factors might be causing the lack of specific phosphatase activities in those sites where no reaction is observed.

Conclusion

While modern techniques of radioactive tracer biology combined with radioautography and cell fractionation have fully confirmed the original ideas of *Bowen* and *Nassanov* on the secretory nature of the Golgi apparatus, these techniques along with present day cytochemical and morphological analysis have shown the Golgi apparatus to be an extremely complex structure. Thus, not only does the Golgi apparatus appear to function in numerous different activities, i.e., glycosylation, sulfation, proteolytic digestion, packaging, concentration, but a whole range of secretory and lysosomal products may also simultaneously pass through the apparatus, possibly by a variety of different routes. Finally, apart from its role in secretion, the Golgi apparatus has become increasingly recognized as a vital center in the membrane economy of the cell. In spite of the wealth of data now accumulated, however, some of the most fundamental aspects of Golgi apparatus structure and function remain unclarified, i.e., the mechanism of passage of secretory and membrane components through the Golgi stack, and the relationship of the stack to GERL. It is hoped that future studies using more precise techniques of histochemistry, cell fractionation, and genetic and drug manipulation will provide answers to some of these questions.

Acknowledgements

The author gratefully acknowledges the advice and assistance of Dr. *Yves Clermont* in the preparation of this manuscript. Thanks is especially due for permission to use the many excellent diagrams and photographs of Dr. *Clermont* and his collegues (Drs. *L. Hermo, M. F. Lalli, C. P. Leblond, A. Rambourg, F. R. Susi,* and *X. M. Tang*).

This work was carried out with the assistance of a grant from the Medical Research Council of Canada.

References

1   Allison, A.C.; Hartree, E.F.: Lysosomal enzymes in the acrosome and their possible role in fertilization. J. Reprod. Fertil. *21:* 501 (1970).
2   Avila, J.L.; Convit, J.: Physicochemical characteristics of the glycosaminoglycan-lysosomal enzyme interaction in vitro. Biochem. J. *160:* 129–136 (1976).
3   Bainton, D.F.; Farquhar, M.G.: Origin of granules in polymorphonuclear leukocytes. J. Cell Biol. *28:* 277–302 (1966).

4 Bainton, D.F.; Farquhar, M.G.: Segregation and packaging of granule enzymes in eosinophilic leukocytes. J. Cell Biol. *45:* 54–73 (1970).

5 Bainton, D.F.; Nichols, B.A.; Farquhar, M.G.: Primary lysosomes of blood leukocytes; in Dingle, Fell, Lysosomes in biology and pathology, vol.5, pp.3–32 (North Holland, Amsterdam 1976).

6 Beams, H.W.; Kessel, R.G.: The Golgi apparatus. Structure and function. Int. Rev. Cytol. *23:* 209–276 (1968).

7 Bedford, J.M.: Ultrastructural changes in the sperm head during fertilization in the rabbit. Am. J. Anat. *123:* 329–358 (1968).

8 Benda, C.: Untersuchungen über den Bau des funktionirenden Samenkanalchens einiger Saugethiere und Folgerungen fur die spermatogenese dieser Wirbelthierklasse. Arch. mikrosk. Anat. Entw Mech. *30:* 49–110 (1887).

9 Bennett, G.: Migration of glycoprotein from Golgi apparatus to cell coat in the columnar cells of the duodenal epithelium. J. Cell Biol. *45:* 668–673 (1970).

10 Bennett, G.: Synthesis and migration of glycoproteins in cells of the rat thymus as shown by radioautography after ^{3}H-fucose injection. Am. J. Anat. *152:* 223–256 (1978).

11 Bennett, G.; Kan, F.W.K.; O'Shaughnessy, D.: The site of incorporation of sialic acid residues into glycoproteins and the subsequent fates of these molecules in various rat and mouse cell types as shown by radioautography after injection of [^{3}H]-N-acetylmannosamine. II. Observations in tissues other than liver. J. Cell Biol. *88:* 16–28 (1981).

12 Bennett, G.; Leblond, C.P.: Formation of cell coat material for the whole surface of columnar cells in the rat small intestine, as visualized by radioautography with L-fucose ^{3}H. J. Cell Biol. *46:* 409–416 (1970).

13 Bennett, G.; Leblond, C.P.: Passage of fucose-^{3}H label from the Golgi apparatus into dense and multivesicular bodies in the duodenal columnar cells and hepatocytes of the rat. J. Cell Biol. *51:* 875–881 (1971).

14 Bennett, G.; Leblond, C.P.: Biosynthesis of the glycoproteins present in plasma membrane, lysosomes and secretory materials, as visualized by radioautography. Histochem. J. *9:* 393–417 (1977).

15 Bennett, G.; Leblond, C.P.; Haddad, A.: Migration of glycoprotein from the Golgi apparatus to the surface of various cell types as shown by radioautography after labeled fucose injection into rats. J. Cell Biol. *60:* 258–284 (1974).

16 Bennett, G.; O'Shaughnessy, D.: The site of incorporation of sialic acid residues into glycoproteins and the subsequent fate of these molecules in various rat and mouse cell types as shown by radioautography after injection of [^{3}H]-N-acetylmannosamine. I. Observations in hepatocytes. J. Cell Biol. *88:* 1–15 (1981).

17 Bennett, G.; Parsons, S.; Carlet, E.: Influence of colchicine and vinblastine on the intracellular migration of secretory and membrane glycoproteins. I. Inhibition of glycoprotein migration in various rat cell types as shown by light microscope radioautography after injection of ^{3}H-fucose. Am. J. Anat. (in press, 1982).

18 Bennett, G.; Carlet, E.; Wild, G.; Parsons, S.: Influence of colchicine and vinblastine on the intracellular migration of secretory and membrane glycoproteins. III. Inhibition of intracellular migration of membrane glycoproteins in intestinal columnar cells and hepatocytes as visualized by light and electron microscope radioautography after ^{3}H-fucose injection. Am. J. Anat. (in press, 1982).

19   Bergeron, J.J.M.; Borts, D.; Cruz, J.: The passage of serum destined proteins through
      the Golgi apparatus of rat liver. An examination of heavy and light Golgi fractions. J.
      Cell Biol. *76:* 87–97 (1978).

20   Bergeron, J.J.M.; Evans, W.H.; Geschwind, I.I.: Insulin binding to rat liver Golgi
      fractions. J. Cell Biol. *59:* 771–776 (1973).

21   Bergeron, J.J.M.; Kotwal, G.J.; Levine, G.; Bilan, P.; Rachubinski, R.; Hamilton, M.;
      Shore, G.C.; Ghosh, H.: Intracellular transport of the transmembrane glycoprotein G
      of vesicular stomatitis virus through the Golgi apparatus as visualized by electron
      microscope radioautography. J. Cell Biol. *94:* 36–41 (1982).

22   Bergeron, J.J.M.; Rachubinski, R.; Sikstrom, R.A.; Posner, B.I.; Paiement, J.: Ga-
      lactose transfer to endogenous acceptors within Golgi fractions of rat liver. J. Cell Biol.
      92: 139–146 (1982).

23   Bergeron, J.J.M.; Sikstrom, R.; Hand, A.R.; Posner, B.I.: Binding and uptake of
      ^{125}I-insulin into rat liver hepatocytes and endothelium: an in vivo radioautographic
      study. J. Cell Biol. *80:* 427–443 (1979).

24   Bowen, R.H.: On the idiosome, Golgi apparatus, and acrosome in the male germ cells.
      Anat. Rec. *24:* 159–180 (1922).

25   Bowen, R.H.: On a possible relation between the Golgi apparatus and secretory prod-
      ucts. Am. J. Anat. *33:* 197–217 (1924).

26   Bowen, R.H.: On the acrosome of the animal sperm. Anat. Rec. *28:* 1–13 (1924).

27   Bowen, R.H.: The Golgi apparatus. Its structure and functional significance. Anat.
      Rec. *32:* 151–193 (1926).

28   Bowen, R.H.: The cytology of glandular secretion. Q. Rev. Biol. *4:* 299–324, 484–519
      (1929).

29   Bretz, R.; Bretz, H.; Palade, G.: Distribution of terminal glycosyltransferases in hepatic
      Golgi fractions. J. Cell Biol. *84:* 87–101 (1980).

30   Brown, R.M., Jr.: Observations on the relationship of the Golgi apparatus to wall
      formation in the marine chrysophycean alga *Pleurochrysis Scherffelii Pringsheim.* J.
      Cell Biol. *41:* 109–123 (1969).

31   Brown, R.M., Jr.; Romanovicz, D.K.: Proceedings of the 8th Cellulose Conference.
      Complete-tree utilization and biosynthesis and structure of cellulose. Appl. Polymer.
      Symp. *28:* 537–585 (1976).

32   Burgos, M.H.; Fawcett, D.W.: Studies on the fine structure of the mammalian testis. I.
      Differentiation of the spermatids in the cat *(Felis domestica).* J. biophys. biochem.
      Cytol. *1:* 287–300 (1955).

33   Cajal, S.R.: Algunas variaciones fisiológicas y patológicas del aparato reticular de
      Golgi. Trab. Lab. Invest. biol. *12:* 127–227 (1914).

34   Caro, L.G.: Electron microscopic radioautography of thin sections. The Golgi zone as a
      site of protein concentration in pancreatic acinar cells. J. biophys. biochem. Cytol. *10:*
      37–45 (1961).

35   Caro, L.G.; Palade, G.E.: Protein synthesis, storage and discharge in the pancreatic
      exocrine cell. J. Cell Biol. *20:* 473–495 (1964).

36   Cheng, H.; Farquhar, M.G.: Presence of adenylate cyclase activity in Golgi and other
      fractions from rat liver. J. Cell Biol. *70:* 660–684 (1976).

37   Claude, A.: Growth and differentiation of cytoplasmic membranes in the course of
      lipoprotein granule synthesis in the hepatic cell. I. Elaboration of elements of the Golgi
      complex. J. Cell Biol. *47:* 745–766 (1970).

38 Clermont, Y.: The Golgi zone of the rat spermatid and its role in the formation of cytoplasmic vesicles. J. biophys. biochem. Cytol. *2:* 119–124 (1956).

39 Clermont, Y.; Glegg, R. E.; Leblond, C. P.: Presence of carbohydrates in the acrosome of the guinea pig spermatozoon. Expl Cell Res. *8:* 453–458 (1955).

40 Clermont, Y.; Lalli, M.; Rambourg, A.: Ultrastructural localization of nicotinamide adenine dinucleotide phosphatase (NADP'ase) thiamine pyrophosphatase (TPP'ase) and cytidine monophosphatase (CMP'ase) in the Golgi apparatus of early spermatids of the rat. Anat. Rec. *201:* 613–622 (1981).

41 Clermont, Y.; Leblond, C. P.: Spermiogenesis of man, monkey, ram and other mammals as shown by the 'periodic acid-Schiff' technique. Am. J. Anat. *96:* 229–250 (1955).

42 Cohn, Z.; Fedorko, M.; Hirsch, J.: The in vitro differentiation of mononuclear phagocytes. V. The formation of macrophage lysosomes. J. exp. Med. *123:* 757–766 (1966).

43 Dalton, A. J.: In Brochet, Mirsby, The cell. Biochemistry, physiology, morphology, p. 603 (Academic Press, New York 1961).

44 Dauwalder, M.; Whaley, W. G.; Kephart, J. E.: Functional aspects of the Golgi apparatus. Sub-cell. Biochem. *1:* 225–275 (1972).

45 Draper, P.; Kent, P. W.: Biosynthesis of intestinal mucins. IV. Utilization of [I-^{14}C]-glucose by sheep colonic mucosa in vitro. Biochem. J. *86:* 248–254 (1963).

46 Droz, B.: L'appareil de Golgi comme site d'incorporation du galactose-^{3}H dans les neurones ganglionnaires spinaux chez le rat. J. Microsc. *6:* 419–424 (1967).

47 Duesberg, J.: Trophospongien und Golgischer Binnenapparat. Verh. anat. Ges., Jena *28:* 11–80 (1914).

48 Eylar, E.: On the biological role of glycoproteins. J. theor. Biol. *10:* 89–113 (1965).

49 Farquhar, M. G.: Recovery of surface membrane in anterior pituitary cells. Variations in traffic detected with anionic and cationic ferritin. J. Cell Biol. *77:* R-35–42 (1978).

50 Farquhar, M. G.: Traffic of products and membranes through the Golgi complex; in Silverstein, Transport of macromolecules in cellular systems, pp. 341–362 (Dahlen Konferenzen, Berlin 1978).

51 Farquhar, M. G.: Membrane recycling in secretory cells. Implications for traffic of products and specialized membranes within the Golgi complex; in Hand, Oliver, Basic mechanisms of cellular secretion. Methods in cell biology, vol. 23, pp. 400–428 (Academic Press, New York 1981).

52 Farquhar, M. G.; Bergeron, J. J. M.; Palade, G. E.: Cytochemistry of Golgi fractions prepared from rat liver. J. Cell Biol. *60:* 8–25 (1974).

53 Farquhar, M. G.; Palade, G. E.: The Golgi apparatus (complex) – 1954–1981 – from artifact to center stage. J. Cell Biol. *91:* 77s–103s (1981).

54 Farquhar, M. G.; Reid, J. A.; Daniell, L.: Intracellular transport and packaging of prolactin. A quantitative electron microscope autoradiographic study of mammotrophs dissociated from rat pituitaries. Endocrinology *102:* 296–311 (1978).

55 Farquhar, M. G.; Wellings, S. R.: An electron microscope study of the glomerulus in nephrosis, glomerulonephritis, and lupus erythematosus. J. biophys. biochem. Cytol. *3:* 319 (1957).

56 Favard, P.: The Golgi apparatus; in Lima de Faria, Handbook of molecular cytology, pp. 1130–1155 (North Holland, Amsterdam 1969).

57 Fawcett, D.: Morphogenesis of the mammalian sperm acrosome in new perspective; in

Afzelius, The functional anatomy of the spermatozoon, pp.199–210 (Pergamon Press, New York 1973).

58 Fleischer, B.; Smigel, M.: Solubilization and properties of galactosyltransferase and sulfotransferase activities of Golgi membranes in Triton X-100. J. biol. Chem. *253:* 1632–1638 (1978).

59 Franke, W.W.; Morré, D.J.; Deumling, B.; Cheetham, R.D.; Kartenbeck, J.; Jarasch, E.D.; Zentgraf, H.W.: Synthesis and turnover of membrane proteins in rat liver. An examination of the membrane flow hypothesis. Z. Naturforsch. *26b:* 1031–1039 (1971).

60 Frei, J.V.; Sheldon, H.: A small granular component of the cytoplasm of keratinizing epithelia. J. biophys. biochem. Cytol. *11:* 719–729 (1961).

61 Friend, D.S.; Murray, M.: Osmium impregnation of the Golgi apparatus. Am. J. Anat. *117:* 135–150 (1965).

62 Fries, E.; Rothman, J.E.: Transient activity of Golgy-like membranes as donors of vesicular stomatitis viral glycoproteins in vitro. J. Cell Biol. *90:* 697–704 (1981).

63 Gatenby, J.B.: The Golgi apparatus. J. R. microsc. Soc. *74:* 134–161 (1954).

64 Gatenby, J.B.; Beams, H.W.: The cytoplasmic inclusions in the spermatogenesis of man. Qu. Jl microsc. Sci. *78:* 1–30 (1935).

65 Gatenby, J.B.; Woodger, J.H.: The cytoplasmic inclusions of the germ cells. IX. On the origin of the Golgi apparatus on the middle piece of the ripe sperm of Cavia, and the development of the acrosome. Q. Jl microsc. Sci. *65:* 265–291 (1921).

66 Gersch, I.: A protein component of the Golgi apparatus. Archs Path. *47:* 99–109 (1949).

67 Ginsel, L.A.; Onderwater, J.J.M.; Daems, W.T.: Transport of radiolabeled glycoprotein to cell surface and lysosome-like bodies of adsorptive cells in cultured small-intestinal tissue from normal subjects and patients with a lysosomal storage disease. Virchows Arch. Abt.B Zellpath. *30:* 245–273 (1979).

68 Godman, G.C.; Lane, N.: On the site of sulfation in the chondrocyte. J. Cell Biol. *21:* 353–366 (1964).

69 Goldfischer, S.; Essner, E.; Schiller, B.: Nucleoside diphosphatase and thiamine pyrophosphatase activities in the endoplasmic reticulum and Golgi apparatus. J. Histochem. Cytochem. *19:* 349–360 (1971).

70 Golgi, C.: Sur la structure des cellules nerveuses. Archs ital. Biol. *30:* 60–71 (1898).

71 Grassé, P.P.: Ultrastructure, polarité et reproduction de l'appareil de Golgi. C. r. hébd. Séanc. Acad. Sci., Paris *245:* 1278–1281 (1957).

72 Griffiths, G.; Warren, G.; Stuhlfauth, I.; Jockusch, B.M.: The role of clathrin coated vesicles in acrosome formation. Eur. J. Cell Biol. *26:* 52–60 (1981).

73 Habener, J.J.; Chang, H.T.; Potts, J.T., Jr.: Enzymic processing of proparathyroid hormone by cell-free extracts of parathyroid glands. Biochemistry *16:* 3910–3917 (1977).

74 Haddad, A.; Smith, M.; Herscovics, A.; Nadler, N.; Leblond, C.P.: Radioautographic study of in vivo and in vitro incorporation of fucose-^{3}H into thyroglobulin by rat thyroid follicular cells. J. Cell Biol. *49:* 856–882 (1971).

75 Hand, A.; Oliver, C.: Cytochemical studies of GERL and its role in secretory granule formation in exocrine cells. Histochem. J. *9:* 375–392 (1977).

76 Hand, A.; Oliver, C.: The Golgi apparatus: protein transport and packaging in secretory cells; in Hand, Oliver, Basic mechanisms of cellular secretion. Methods in cell biology, vol.23, pp.137–153 (Academic Press, New York 1981).

77 Hasilik, A.; Klein, U.; Waheed, A.; Strecker, G.; Von Figura, K.: Phosphorylated oligosaccharides in lysosomal enzymes: identification of α-N-acetylglucosamine-(1)-phospho-(6)-mannose diester groups. Proc. natn. Acad. Sci. USA 77: 7074–7078 (1980).

78 Hermo, L.; Clermont, Y.; Rambourg, A.: Endoplasmic reticulum-Golgi apparatus relationships in the rat spermatid. Anat. Rec. 193: 243–256 (1979).

79 Hermo, L.; Rambourg, A.; Clermont, Y.: Three dimensional architecture of the cortical region of the Golgi apparatus in rat spermatids. Am. J. Anat. 157: 357–373 (1980).

80 Herzog, V.: Pathways of endocytosis in secretory cells. Trends biochem. Sci. 6: 319–322 (1981).

81 Herzog, V.; Farquhar, M.: Luminal membrane retrieved after exocytosis reaches most Golgi cisternae in secretory cells. Proc. natn. Acad. Sci. USA 74: 5073–5077 (1977).

82 Herzog, V.; Miller, F.: The localization of endogenous peroxidase in the parotid gland of the rat. Z. Zell. Microsc. 107: 403–420 (1970).

83 Herzog, V.; Miller, F.: Membrane retrieval in secretory cells. Morphological evidence for participation of Golgi cisternae in internalization of plasma membrane following stimulated exocytosis. Symp. Soc. exp. Biol. 33: 101–116 (1979).

84 Herzog, V.; Miller, F.: Membrane retrieval in epithelial cells of isolated thyroid follicles. Eur. J. Cell Biol. 19: 203–215 (1979).

85 Herzog, V.; Reggio, H.: Pathways of endocytosis from luminal plasma membrane in rat exocrine pancreas. Eur. J. Cell Biol. 21: 141–150 (1980).

86 Hibbard, H.: Current status of our knowledge of Golgi apparatus in animal cell. Q. Rev. Biol. 20: 1–19 (1945).

87 Hino, Y.; Asano, A.; Sato, R.: Biochemical studies on rat liver Golgi apparatus. III. Subfractionation of fragmented Golgi apparatus by counter current distribution. J. Biochem. 83: 935–942 (1978).

88 Hirsch, G.C.: Protoplasma-Monographien, vol.18 (Borntrager, Berlin 1939).

89 Hirsch, G.C.: The external secretion of the pancreas as a whole and the communication between the endoplasmic reticulum and the Golgi bodies; in Goodwin, Lindberg, Biological structure and function, vol.1, pp.195–208 (Academic Press, New York 1961).

90 Hirschberg, C.B.; Goodman, S.; Green, C.: Sialic acid uptake by fibroblasts. Biochemistry 15: 3591–3599 (1976).

91 Irving, R.A.; Toneguzzo, F.; Rhee, S.H.; Hofmann, T.; Ghosh, H.P.: Synthesis and assembly of membrane glycoproteins: presence of leader peptide in non-glycosylated precursor of membrane glycoprotein of vesicular stomatitis virus. Proc. natn. Acad. Sci. USA 76: 570–574 (1979).

92 Ito, A.; Palade, G.: Presence of NADPH-cytochrome P-450 reductase in rat liver Golgi membranes. J. Cell Biol. 79: 590–597 (1978).

93 Jamieson, J.D.: Transport and discharge of exportable proteins in pancreatic exocrine cells. In vitro studies; in Bronner, Kleinzeller, Current topics in membranes and transport, vol.3, pp.273–338 (Academic Press, New York 1972).

94 Jamieson, J.D.; Palade, G.E.: Intracellular transport of secretory proteins in the pancreatic exocrine cell. II. Transport to condensing vacuoles and zymogen granules. J. Cell Biol. 34: 597–615 (1967).

95 Jennings, M. A.; Florey, H. W.: Autoradiographic observations on mucous cells of stomach and intestine. Q. Jl exp. Physiol. *41:* 131–152 (1956).

96 Katona, E.: Incorporation of inorganic sulfate in rat-liver Golgi. Eur. J. Biochem. *63:* 583–590 (1976).

97 Kirkman, H.; Severinghaus, A. E.: A review of the Golgi apparatus. Parts I, II and III. Anat. Rec. *70:* 413–431, 557–573, *71:* 79–103 (1938).

98 Kornfeld, R.; Kornfeld, S.: Structure of glycoproteins and their oligosaccharide units; in Lennarz, The biochemistry of glycoproteins and proteoglycans, pp. 1–34 (Plenum Press, New York 1980).

99 Kraehenbuhl, J. P.; Racine, L.; Jamieson, J. D.: Immunocytochemical localization of secretory proteins in bovine pancreatic exocrine cells. J. Cell Biol. *72:* 406–423 (1977).

100 Kramer, M. F.; Geuze, J. J.: Comparison of various methods to localize a source of radioactivity in ultrastructural autoradiographs. J. Histochem. Cytochem. *28:* 381–387 (1980).

101 La Valette St. George: Über die Genese der Samenkörper. Arch. mikrosk. Anat. *1:* 403–414 (1865).

102 La Valette St. George: Über die Genese der Samenkörper. Arch. mikrosk. Anat. *3:* 263–273 (1867).

103 Leblond, C. P.: Distribution of periodic acid-reactive carbohydrates in the adult rat. Am. J. Anat. *86:* 1–49 (1950).

104 Leblond, C. P.; Bennett, G.: Role of the Golgi apparatus in terminal glycosylation; in Brinkley, Porter, International cell biology 1976–1977, pp. 326–336 (Rockefeller University Press, New York 1977).

105 Leblond, C. P.; Clermont, Y.: Spermiogenesis of rat, mouse, hamster and guinea pig as revealed by the 'periodic acid-fuchsin sulfurous acid' technique. Am. J. Anat. *90:* 167–216 (1952).

106 Lenhossek, M. Von: Untersuchungen über Spermatogenese. Arch. mikrosk. Anat. EntwWech. *51:* 215–318 (1898).

107 Lodish, H. F.; Braell, W. A.; Schwartz, A. L.; Strous, G. J.; Zilberstein, A.: Synthesis and assembly of membrane and organelle proteins. Int. Rev. Cytol. *12:* suppl., pp. 247–307 (1981).

108 Lodish, H. F.; Rothman, J. E.: The assembly of cell membranes. Scient. Am. *240:* 48–63 (1979).

109 Lodish, H. F.; Silverstein, A.; Porter, M.: Synthesis and assembly of transmembrane viral and cellular glycoproteins; in Hand, Oliver, Basic mechanisms of cellular secretion. Methods in cell biology, vol. 23, pp. 5–25 (Academic Press, New York 1981).

110 Malaisse-Lagae, F.; Amherdt, M.; Ravazzola, M.; Sener, A.; Hutton, J. C.; Orci, L.; Malaisse, W. J.: Role of microtubules in the synthesis, conversion and release of (pro) insulin. A biochemical and radioautographic study in rat islets. J. clin. Invest. *63:* 1284–1296 (1979).

111 Meldolesi, J.: Membranes and membrane surfaces. Dynamics of cytoplasmic membranes in pancreatic acinar cells. Phil. Trans. R. Soc. Lond. B *268:* 39–53 (1974).

112 Mollenhauer, H. H.; Hass, B. S.; Morré, D. J.: Membrane transformations in Golgi apparatus of rat spermatids. J. Microsc. Biol. cell. *27:* 33–36 (1976).

113 Molnar, J.; Teegarden, D.; Winzler, R.: The biosynthesis of glycoproteins. VI. Pro-

duction of extracellular radioactive macromolecules by Ehrlich ascites carcinoma cells during incubation with glucosamine-^{14}C. Cancer Res. *25:* 1860 (1965).

114 Moore, J.E.S.: Some points in the spermatogenesis of mammalia. Int. Mschr. Anat. Physiol. *11:* 129–166 (1894).

115 Morré, D.J.; Kartenbeck, J.; Franke, W.: Membrane flow and interconversions amongst endomembranes. Biochim. biophys. Acta *559:* 71–152 (1979).

116 Morré, D.J.; Ovtracht, L.: Dynamics of Golgi apparatus. Membrane differentiation and membrane flow. Int. Rev. Cytol., suppl. 5, pp. 61–188 (1977).

117 Morré, D.J.; Ovtracht, L.: Structure of rat liver Golgi apparatus. Relationship to lipoprotein secretion. J. ultrastruct. Res. *74:* 284–295 (1981).

118 Nadler, N.J.; Young, B.A.; Leblond, C.P.; Mitmaker, B.: Incorporation of thyroglobulin in the thyroid follicle. Endocrinology *74:* 333–354 (1964).

119 Nassonov, D.N.: Das Golgische Binnennetz und seine Beziehungen zu der Sekretion. Untersuchungen über einige Amphibiendrüsen. Arch. mikrosk. Anat. EntwMech. *97:* 136–186 (1923).

120 Neutra (Peterson), M.; Leblond, C.P.: Synthesis of complex carbohydrates in the Golgi region as shown by radioautography after injection of labeled glucose. J. Cell Biol. *21:* 143–148 (1964).

121 Neutra, M.; Leblond, C.P.: Synthesis of the carbohydrate of mucus in the Golgi complex as shown by electron microscope radioautography of goblet cells from rats injected with glucose-H^3. J. Cell Biol. *30:* 119–136 (1966).

122 Neutra, M.; Leblond, C.P.: The Golgi apparatus. Scient. Am. *220:* 100–107 (1969).

123 Novikoff, A.B.; Novikoff, P.M.: Cytochemical contributions to differentiating GERL from the Golgi apparatus. Histochem. J. *9:* 525–551 (1977).

124 Novikoff, P.M.; Novikoff, A.B.; Quintana, N.; Hauw, J.: Golgi apparatus, GERL, and lysosomes of neurons in rat dorsal root ganglia studied by thick section and thin section cytochemistry. J. Cell Biol. *50:* 859–886 (1971).

125 Oliver, C.; Hand, A.: Uptake and fate of luminally administered horseradish peroxidase in resting and isoproterenol-stimulated rat parotid acinar cells. J. Cell Biol. *76:* 207–220 (1978).

126 Ovtracht, L.; Morré, D.J.; Cheetham, R.D.; Mollenhauer, H.H.: Subfractionation of Golgi apparatus from rat liver. Method and morphology. J. Microsc. *18:* 87–102 (1973).

127 Ovtracht, L.; Thiéry, J.P.: Mise en évidence par cytochimie ultrastructurale de compartiments physiologiquement différents dans un même saccule Golgien. J. Microsc. *15:* 135–170 (1972).

128 Paiement, J.; Rachubinski, R.A.; Ng Ying Kin, N.M.K.; Sikstrom, R.A.; Bergeron, J.J.M.: Membrane fusion and glycosylation in the rat hepatic Golgi apparatus. J. Cell Biol. *92:* 147–154 (1982).

129 Palade, G.E.: Studies on endoplasmic reticulum simple dispositions in cells in situ. J. biophys. biochem. Cytol. *1:* 567–582 (1955).

130 Palade, G.E.: Intracellular aspects of the process of protein synthesis. Science *189:* 347–358 (1975).

131 Palade, G.E.; Claude, A.: The nature of the Golgi apparatus. J. Morph. *85:* 35–70 (1949).

132 Pfenninger, K.H.; Bunge, R.P.: Freeze-fracturing of nerve growth cones and young fibers. J. Cell Biol. *63:* 180–196 (1974).

133 Pierce, M.; Turley, E. A.; Roth, S.: Cell surface glycosyltransferase activities. Int. Rev. Cytol. *65:* 1–47 (1980).

134 Platner, G.: IV. Die Entstehung und Bedeutung der Nebenkerne im Pankreas, ein Beitrag zur Lehre von der Sekretion. Arch. mikrosk. Anat. EntwMech. *33:* 180–216 (1889).

135 Pohlmann, R.; Waheed, A.; Hasilik, A.; Von Figura, K.: Synthesis of phosphorylated recognition marker in lysosomal enzymes is located in the cis part of the Golgi apparatus. J. biol. Chem. *257:* 5323–5325 (1982).

136 Quinn, P.; Judah, J.: Calcium-dependent Golgi vesicle fusion and cathepsin B in the conversion of proalbumin into albumin in rat liver. Biochem. J. *172:* 301–309 (1978).

137 Rambourg, A.; Clermont, Y.; Marraud, A.: Three dimensional structure of the osmium-impregnated Golgi apparatus as seen in the high voltage electron microscope. Am. J. Anat. *140:* 27–46 (1974).

138 Rambourg, A.; Hernandez, W.; Leblond, C. P.: Detection of complex carbohydrates in the Golgi apparatus of rat cells. J. Cell Biol. *40:* 395–414 (1969).

139 Redman, C. M.; Banerjee, D.; Howell, K.; Palade, G.: Colchicine inhibition of plasma protein release from rat hepatocytes. J. Cell Biol. *66:* 42–59 (1975).

140 Redman, C. M.; Banerjee, D.; Manning, C.; Huang, C. Y.; Green, K.: In vivo effect of cholchicine on hepatic protein synthesis and on the conversion of proalbumin to serum albumin. J. Cell Biol. *77:* 400–416 (1978).

141 Redman, C. M.; Banerjee, D.; Yu, S.: The effect of colchicine on the synthesis and secretion of rat serum albumin; in Hand, Oliver, Methods in cell biology, vol. 23, pp. 231–246 (Academic Press, New York 1981).

142 Reggio, H. A.; Palade, G. E.: Sulfated compounds in the zymogen granules of the guinea pig pancreas. J. Cell Biol. *77:* 288–314 (1978).

143 Regoeczi, E.; Chindemi, P. A.; Debanne, M. T.; Charlwood, P. A.: Partial resialylation of human asialotransferrin type 3 in the rat. Proc. natn. Acad. Sci. USA *79:* 2226–2230 (1982).

144 Revel, J. P.; Hay, S. D.: An autoradiographic and electron microscopic study of collagen synthesis in differentiating cartilage. Z. Zellforsch. *61:* 110–144 (1963).

145 Roth, J.; Berger, E. G.: Immunocytochemical localization of galactosyltransferase in Hela cells. Codistribution with thiamine pyrophosphatase in trans-Golgi cisternae. J. Cell Biol. *92:* 223–229 (1982).

146 Rothman, J. E.: The Golgi apparatus. Two organelles in tandem. Science *213:* 1212–1219 (1981).

147 Rothman, J. E.; Fine, R. E.: Coated vesicles transport newly synthesized membrane glycoproteins from endoplasmic reticulum to plasma membrane in two successive stages. Proc. natn. Acad. Sci. USA *77:* 780–784 (1980).

148 Sandoz, D.: Evolution des ultrastructures au cours de la formation de l'acrosome du spermatozoïde chez la souris. J. Microsc. *9:* 535–558 (1970).

149 Sandoz, D.: Etude autoradiographique de l'incorporation in vitro de galactose ^{3}H dans les spermatides de souris. J. Microsc. *15:* 403–408 (1972).

150 Sato, A.; Spicer, S. S.: Ultrastructural visualization of galactosyl residues in various alimentary epithelial cells with the peanut lectin-horseradish peroxidase procedure. Histochemistry *73:* 607–624 (1982).

151 Schachter, H.; Roseman, S.: Mammalian glycosyltransferases. Their role in the synthesis and function of complex carbohydrates and glycolipids; in Lennarz, The bio-

chemistry of glycoproteins and proteoglycans, pp. 85–160 (Plenum Press, New York 1980).

152 Schmidt, M.; Schlesinger, M.: Relation of fatty acid attachment to the translation and maturation of vesicular stomatitis and sindbis virus membrane glycoproteins. J. biol. Chem. *255:* 3334–3339 (1980).

153 Silbert, J.; Freilich, L. S.: Biosynthesis of chondroitin sulfate by a Golgi apparatus enriched preparation from cultures of mouse mastocytoma cells. Biochem. J. *190:* 307–313 (1980).

154 Singer, S. J.; Nicolson, G. L.: The fluid mosaic model of the structure of cell membranes. Science *175:* 720–731 (1972).

155 Sly, W. S.; Fischer, H. D.; Gonzalez-Noriega, A.; Grubb, J. H.; Natowicz, M.: Role of the 6-phosphomannosyl-enzyme receptor in intracellular transport and adsorptive pinocytosis of lysosomal enzymes; in Hand, Oliver, Basic mechanisms of cellular secretion. Methods of cell biology, vol. 23, pp. 191–214 (Academic Press, New York 1981).

156 Snider, M. D.; Robbins, P. W.: Synthesis and processing of asparagine-linked oligosaccharides of glycoproteins; in Hand, Oliver, Basic mechanisms of cellular secretion. Methods of cell biology, vol. 23, pp. 89–100 (Academic Press, New York 1981).

157 Spiro, R. G.: Glycoproteins. A. Rev. Biochem. *39:* 599–637 (1970).

158 Spiro, R. G.; Spiro, M.: Glycoprotein biosynthesis. Studies in thyroglobulin. Characterization of a particulate precursor and radioisotope incorporation by thyroid slices and particle systems. J. biol. Chem. *241:* 1271–1282 (1966).

159 Steiner, D. F.; Clark, J. L.; Nolan, C.; Rubenstein, A. H.; Margoliash, E.; Melani, F.; Oyer, P. E.: In Cesari, Luft, The pathogenesis of diabetes mellitus, pp. 123–132 (Almqvist & Wiksell, Stockholm 1970).

160 Steiner, D. F.; Quinn, P. S.; Chan, S. J.; Marsh, J.; Tager, H. S.: Processing mechanisms in the biosynthesis of proteins. Ann. N.Y. Acad. Sci. *343:* 1–16 (1980).

161 Struck, D. K.; Lennarz, W. J.: The function of saccharide-lipids in synthesis of glycoproteins; in Lennarz, The biochemistry of glycoproteins and proteoglycans, pp. 35–83 (Plenum Press, New York 1980).

162 Susi, F. R.; Leblond, C. P.; Clermont, Y.: Changes in the Golgi apparatus during spermiogenesis in the rat. Am. J. Anat. *130:* 251–268 (1971).

163 Tang, X. M.; Lalli, M. F.; Clermont, Y.: A cytochemical study of the Golgi apparatus of the spermatid during spermiogenesis in the rat. Am. J. Anat. *163:* 283–294 (1982).

164 Tartakoff, A. M.: The Golgi complex. Crossroads for vesicular traffic. Int. Rev. exp. Path. *22:* 227–251 (1980).

165 Tartakoff, A.; Vassalli, P.: Plasma cell immunoglobulin secretion. Arrest is accompanied by alterations of the Golgi complex. J. exp. Med. *146:* 1332–1345 (1977).

166 Thyberg, J.: Internalization of cationized ferritin into the Golgi complex of cultured mouse peritoneal macrophages. Effects of colchicine and cytochalasin B. Eur. J. Cell Biol. *23:* 95–103 (1980).

167 Van Heyningen, H.: Secretion of protein by the acinar cells of the rat pancreas, as studied by electron microscopic radioautography. Anat. Rec. *148:* 485–498 (1964).

168 Waechter, C. J.; Lennarz, W. J.: The role of polyprenol-linked sugars in glycoprotein synthesis. A. Rev. Biochem. *45:* 95–112 (1976).

169 Walker, C.; Allen, M.: On the nature of 'Golgi bodies' in fixed material. Proc. R. Soc. Lond., Ser. B *101:* 468–483 (1927).

170 Warshawsky, H.; Leblond, C. P.; Droz, B.: Synthesis and migration of proteins in the cells of the exocrine pancreas as revealed by specific activity determination from radioautographs. J. Cell Biol. *16:* 1–24 (1963).

171 Weinstock, A.; Leblond, C. P.: Elaboration of the matrix glycoproteins of enamel by the secretory ameloblasts of the rat incisor as revealed by radioautography after galactose-^{3}H injection. J. Cell Biol. *51:* 26–51 (1971).

172 Weinstock, M.; Leblond, C. P.: Radioautographic visualization of the deposition of a phosphoprotein at the mineralization front in the dentin of the rat incisor. J. Cell Biol. *56:* 838–845 (1973).

173 Whaley, W. G.: The Golgi apparatus. Cell Biol. Monogr., vol. 2 (Springer, New York 1976).

174 Whaley, W. G.; Dauwalder, M.: The Golgi apparatus, the plasma membrane, and functional integration. Int. Rev. Cytol. *58:* 199–245 (1979).

175 Whur, P.; Herscovics, A.; Leblond, C. P.: Radioautographic visualization of the incorporation of galactose-^{3}H and mannose-^{3}H by rat thyroids in vitro in relation to the stages of thyroglobulin synthesis. J. Cell Biol. *43:* 289–311 (1969).

176 Wild, G.; Bennett, G.: Influence of colchicine and vinblastine on the intracellular migration of secretory and membrane glycoproteins. II. Inhibition of secretion of thyroglobulin in thyroid follicular cells as visualized by light and electron microscope radioautography after ^{3}H-fucose injection. Am. J. Anat. (in press, 1982).

177 Young, R. W.: The role of the Golgi apparatus in sulfate metabolism. J. Cell Biol. *57:* 175–189 (1973).

178 Zeigel, R. F.; Dalton, A. J.: Speculations based on the morphology of the Golgi systems in several types of protein-secreting cells. J. Cell Biol. *15:* 45–54 (1962).

Gary Bennett, PhD, Department of Anatomy, McGill University,
3640 University Street, Montreal, Que. H3A 2B2 (Canada)

Cell Biology of the Secretory Process, pp. 148–170 (Karger, Basel 1984)

The Role of GERL in the Secretory Process

Arthur R. Hand, Constance Oliver

Laboratory of Biological Structure, National Institute of Dental Research,
National Institutes of Health, Bethesda, Md, USA

Introduction

The concept of GERL originated in 1964 with studies on the localization of phosphatases in the Golgi region of dorsal root ganglion neurons by *Novikoff* [70], and has been expanded upon in several subsequent publications [71, 76, 80]. *Novikoff* observed that, in addition to its localization in lysosomes, the cytochemical reaction product of acid phosphatase activity was present in a smooth membrane system at the inner (trans) face of the Golgi apparatus. This membrane system appeared to give rise to several types of lysosomes, and to have direct membrane continuities with the endoplasmic reticulum in the Golgi region. The lack of nucleoside diphosphatase or thiamine pyrophosphatase activity (enzymatic markers of the trans-Golgi saccules) and the frequent separation from the stack of saccules, further differentiated these smooth-surfaced membranes from the Golgi apparatus. The acronym GERL was chosen to emphasize its location in the Golgi region, its derivation from the Endoplasmic Reticulum, and its role in Lysosome formation. Subsequent studies by *Novikoff* and others revealed the presence of GERL, or GERL-like structures, in a number of other cell types [6, 8, 12, 19, 26, 27, 32, 40, 42, 43, 49, 50, 72, 74, 75, 77, 78, 82, 89, 90, 92, 93, 97].

Early cytochemical studies of several exocrine and endocrine secretory cells demonstrated acid phosphatase activity in immature secretory granules (condensing vacuoles) [68] and what appeared to be the trans-Golgi saccule [87, 102, 104]. Subsequent studies revealed that the acid phosphatase reactive saccule had many of the characteristics of GERL, as described in neurons. Moreover, the trans-Golgi saccules were shown to have nucleoside diphosphatase/thiamine pyrophosphatase activity, but usually not acid phosphatase activity, indicating that the two enzymes were present in different membrane systems. These observations suggested that GERL, as well as the Golgi apparatus, plays an important role in the transport and packaging of secretory proteins [42, 43, 71, 75, 78].

The involvement of the Golgi apparatus in the secretory process has long been recognized. Light microscopic studies of secretory cells stained by metallic impregnation techniques revealed a close association of the Golgi apparatus with secretory granules, and changes in the size and distribution of the Golgi apparatus during the secretory cycle were clearly evident [10, 65, 66]. Early electron microscopic studies defined the basic structure of the Golgi apparatus, confirmed its close association with forming secretory granules, and suggested its role in the transport and condensation of secretory product [20, 30, 39, 96, 99]. Radioautographic studies provided additional evidence for the function of the Golgi apparatus in intracellular transport [16, 53, 54, 109] and demonstrated its role in the chemical modification of newly synthesized secretory and membrane proteins [3, 51, 60, 67, 111]. Isolation of the Golgi apparatus or Golgi subfractions from secretory cells [25, 31, 64] has allowed biochemical correlations with many of the morphological observations. These and other aspects of Golgi apparatus structure and function are discussed thoroughly in several reviews [2, 17, 29, 36, 45, 91]. In this chapter we discuss the structure and cytochemistry of GERL, its relationship to the Golgi apparatus, and its functional implications in secretory cells.

Structure and Cytochemistry of GERL

In exocrine secretory cells GERL appears as a cisternal structure of variable length at the trans-face of the Golgi apparatus, located adjacent to or frequently separated from the trans-Golgi saccule (fig. 1, 4). The membranes of GERL are slightly thicker than those of the Golgi saccules and often delineate a very uniform cisternal space, giving it the appearance of a 'rigid lamella' [18]. Direct membranous continuity of GERL with the forming secretory granules is observed occasionally. Unequivocal continuities between GERL and smooth or rough endoplasmic reticulum have not been demonstrated. However, GERL, and immature granules forming from GERL, are closely related to modified cisternae of rough endoplasmic reticulum. These cisternae of endoplasmic reticulum closely parallel

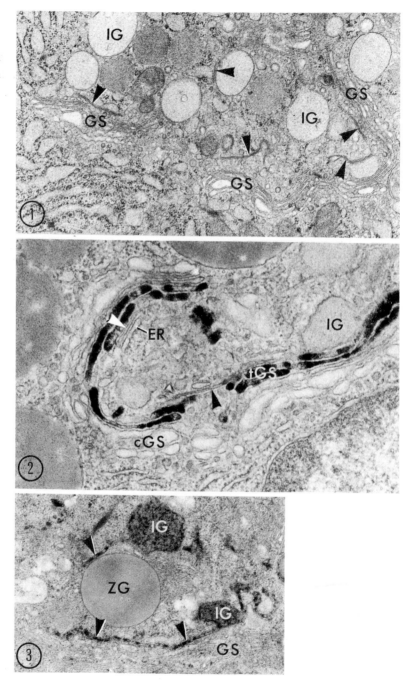

the membrane of GERL or the immature granules and lack ribosomes on the surface adjacent to GERL (fig. 2, 4–6).

Positive identification of GERL using morphological criteria alone is often difficult due to the abundance of cisternal, tubular and vesicular structures in the Golgi region. Phosphatase cytochemistry has proved invaluable for differentiating GERL from these other membranous structures [41] in most cells. In preparations incubated for nucleoside diphosphatase or thiamine pyrophosphatase activity, reaction product is typically present in the trans-Golgi saccules, but not in GERL (fig. 2, 5). Likewise, immature secretory granules are unreactive for nucleoside diphosphatase or thiamine pyrophosphatase. In contrast, incubation for acid phosphatase results in deposits of reaction product in GERL and immature granules (fig. 3, 6). The trans-Golgi saccule occasionally may exhibit slight reactivity (fig. 6), but the remainder of the Golgi saccules and the mature secretory granules are unreactive. (Although this pattern of enzyme distribution is typical, different localizations of nucleoside diphosphatase and acid phosphatase have been observed in various cells [9, 21, 33, 52, 62, 89, 90; *Hand,* unpublished observations]. Factors such as cell type, physiological state, tissue fixation and preparation, and cyto-

---

Fig. 1–19. All electron micrographs are of rat or guinea pig tissues fixed in a glutaraldehyde-formaldehyde fixative [57] and prepared for transmission electron microscopy by standard methods. Some tissues were incubated for the cytochemical localization of acid phosphatase activity at pH 5.0 [69], using cytidine monophosphate or sodium β-glycerophosphate as substrate, nucleoside diphosphatase activity at pH 7.2 [73], using inosine diphosphate or thiamine pyrophosphate as substrate, or endogenous peroxidase activity at pH 7.0 [43], using an H_2O_2-diaminobenzidine medium. Figure 11 is of tissue embedded in glycol methacrylate, sectioned and stained for glycoprotein with phosphotungstic acid [94].

Fig. 1–3. Serous secretory cells.

Fig. 1. Golgi apparatus of a secretory cell of the rat exorbital lacrimal gland. Several stacks of saccules (GS) are present; the cis-saccules are usually somewhat dilated. Several cisternae of GERL (arrowheads) are present in the trans Golgi region. Immature secretory granules (IG). ×16,700.

Fig. 2. Nucleoside diphosphatase reaction product is present in 1–2 trans-Golgi saccules (tGS) in a cell of the rat lingual serous gland. The cis-Golgi saccules (cGS), GERL (arrowheads), and immature secretory granules (IG) are unreactive. Short cisternae of endoplasmic reticulum (ER) lie adjacent to GERL. ×33,700.

Fig. 3. Acid phosphatase reaction product is present in GERL (arrowheads) and immature secretory granules (IG) of an exocrine cell of the guinea pig pancreas. The Golgi saccules (GS) and zymogen granule (ZG) are unreactive. ×30,500.

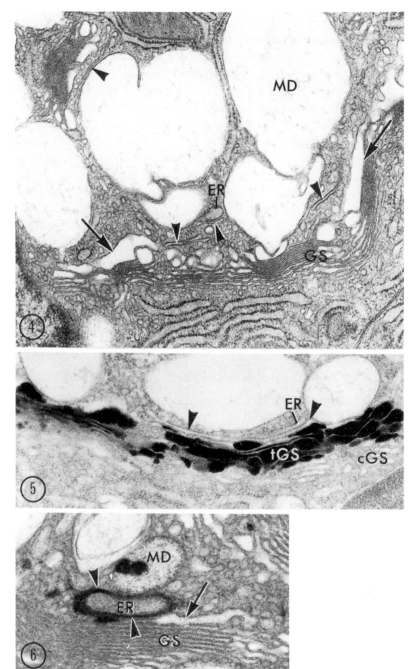

chemical substrate may influence the localization of enzyme reaction product.) Thick sections (up to 0.5 µm) of tissues incubated for acid phosphatase activity reveal an anastomosing tubular network at the periphery of the cisternal regions of GERL (fig. 7). These tubules connect the forming secretory granules to the cisternal portion and connect adjacent cisternal regions. The structural details revealed by the thick sections emphasize that recognition of GERL in thin sections of unincubated preparations is dependent on nearly perpendicular sections through the cisternal region. The anastomosing tubular network will usually appear as vesicles in the trans-Golgi region, and the numerous continuities with forming secretory granules will frequently be undetectable in routine thin sections.

Localization of Secretory Proteins

Studies on the localization of secretory proteins in the Golgi apparatus and GERL have provided interesting but somewhat perplexing results. In general, enzyme or immunocytochemical localization of secretory proteins has confirmed their expected distribution in the intracellular compartments involved in transport, packaging and storage [14, 35, 42, 43]. For example, reaction product for endogenous peroxidase, a secretory enzyme of the acinar cells of the rat exorbital lacrimal gland, is present in the rough endoplasmic reticulum, vesicles between the endoplasmic reticulum and Golgi apparatus, the Golgi saccules, and in immature and

---

Fig. 4–6. Mucous secretory cells of the rat sublingual gland.

Fig. 4. The large Golgi apparatus consists of 9 saccules (GS); numerous vesicles are present at both the cis- and trans-faces. The trans-saccules appear to dilate (arrows) as they accumulate secretory product. GERL can be recognized as narrow cisternae with thick membranes (arrowheads) slightly separated from the trans-face. Short cisternae of endoplasmic reticulum (ER) are often present adjacent to GERL. Mucous secretory droplets (MD). ×24,300.

Fig. 5. Reaction product of nucleoside diphosphatase is present in 5 saccules (tGS) on the trans-side of the Golgi apparatus. The cis-saccules (cGS) are unreactive. One cisterna (arrowheads) at the trans-face, presumably GERL, is also unreactive. A short cisterna of endoplasmic reticulum (ER) lies adjacent to part of GERL. ×27,800.

Fig. 6. Acid phosphatase reaction product is present in GERL (arrowheads) and a small immature mucous droplet (MD). The partially dilated trans Golgi saccule is weakly reactive (arrow); the remainder of the Golgi saccules (GS) are unreactive. A cisterna of endoplasmic reticulum (ER) appears surrounded by GERL. ×51,000.

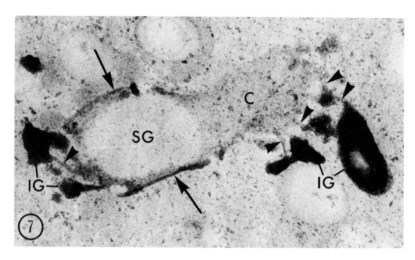

Fig. 7. Rat exorbital lacrimal gland, incubated for acid phosphatase activity; 0.2-µm thick section observed at 100 kV. Reaction product is present in GERL and immature secretory granules (IG). The cisternal portion of GERL is seen in face view (C) and on edge (arrows) as it curves around a secretory granule (SG). Anastomosing tubules (arrowheads) arising from the periphery of the cisternal portion are in continuity with the immature granules. ×59,000.

mature secretory granules (fig. 8–10). Surprisingly, reaction product is usually (although not always) absent from GERL (fig. 8–10), even though direct continuities between GERL and strongly reactive immature granules are evident [42, 43]. In contrast, the presence of glycoprotein can clearly be demonstrated in GERL as well as the trans-Golgi saccules, immature and mature secretory granules (fig. 11). The nature of the stained material present in GERL has not yet been determined. Assuming the validity of the cytochemical results (i.e., that they are not due to unrecognized preparative artifacts), these observations complicate our understanding of the function of GERL in secretory cells. These aspects will be discussed more fully in the final section.

Fig. 8–10. Peroxidase localization in rat exorbital lacrimal gland. Reaction product for this secretory enzyme is present in the endoplasmic reticulum, vesicles (V) between the endoplasmic reticulum and the Golgi apparatus, Golgi saccules (GS), immature (IG) and mature (SG) secretory granules. Little or no reaction product is present in GERL (arrow-

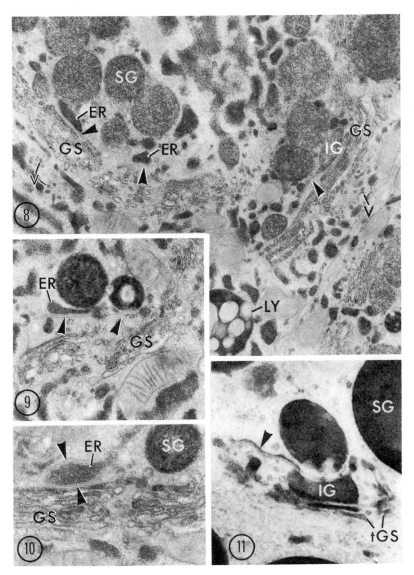

heads), but the adjacent cisternae of endoplasmic reticulum (ER) are strongly reactive. Lysosomes (LY) are also frequently reactive. *8* ×18,300. *9* ×21,300. *10* ×32,800.

Fig. 11. Rat parotid gland, stained for glycoprotein. The trans-Golgi saccules (tGS), GERL (arrowhead), immature (IG) and mature (SG) secretory granules are strongly stained. The endoplasmic reticulum and cis-Golgi saccules are unstained. ×33,400.

Functional Modulations of GERL Structure and Cytochemistry

Alterations in the physiologic state of many cells are reflected in the structure and biochemical properties of their constituent organelles. In exocrine cells, the stimulation of secretory activity results in the exocytotic discharge of stored secretory granules, recovery of granule membrane from the cell surface, and an increase in the size of the Golgi apparatus as newly synthesized secretory proteins are packaged into granules [1, 8, 44, 47, 61, 98]. Similar changes occur in endocrine cells and other cells which produce large amounts of exportable protein.

The Golgi apparatus and GERL of parotid and lacrimal acinar cells undergo significant structural and cytochemical changes in response to secretagogue stimulation [44]. At early times after stimulation (up to 5 h), GERL appears more extensive than in resting cells, and acid phosphatase activity is markedly enhanced in GERL and forming secretory granules (fig. 12–14). Reaction product is also seen more frequently in the trans-Golgi saccule (fig. 14). During the period of maximal secretory granule formation (approximately 6–12 h after stimulation), the distribution of thiamine pyrophosphatase activity is altered. Reaction product is frequently present in immature granules and in cisternae at the trans-face of the Golgi apparatus which resemble GERL (fig. 15–17). The trans-Golgi saccules, as in the unstimulated state, also contain thiamine pyrophosphatase reaction product. The GERL-like cisternae have many of the structural characteristics of GERL, such as a narrow luminal space, separation from the stack of saccules, and a close relationship to modified cisternae of endoplasmic reticulum. Similar modulations of enzyme distribution in the Golgi apparatus and GERL occur during postnatal differentiation of parotid acinar cells [22] and during recovery from ethionine intoxication [86].

Alterations in the structure and relationship of the Golgi apparatus and GERL have also been reported in other secretory cells following stimulation. Hypertrophy of GERL-like tubules and vesicles at the trans-face of the Golgi apparatus occurs after hyperstimulation of guinea pig pancreatic exocrine cells in vivo [58] and in vitro [55]. Estrogen-stimulated pituitary mammotrophic cells exhibit an increase in the number of nucleoside diphosphatase-reactive Golgi saccules, and reaction product is present in forming secretory granules and in what appears to be GERL [103]. Hypothalamic neurosecretory neurons from hyperosmotically stressed mice also exhibit an increase in thiamine pyrophosphatase activ-

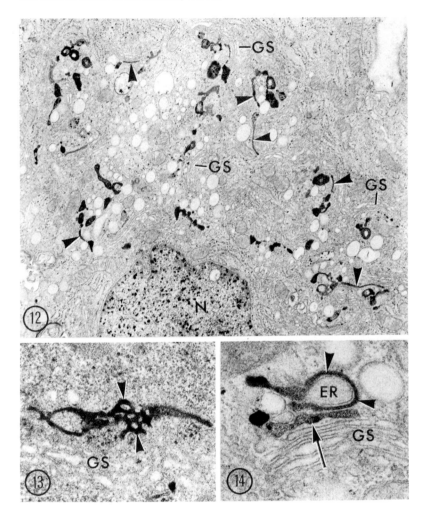

Fig. 12–14. Acid phosphatase localization in stimulated serous cells.

Fig. 12. Rat exorbital lacrimal gland, 1 h after pilocarpine (50 mg/kg). GERL (arrowheads) and the numerous irregularly shaped immature granules are strongly reactive. The Golgi saccules (GS) are unreactive. The nucleus (N) contains non specific lead phosphate deposits. ×9,700.

Fig. 13. Rat parotid gland, 1 h after isoproterenol (30 mg/kg). The tubular portion of GERL (arrowheads) is seen in face view. The Golgi saccules (GS) are unreactive. ×29,100.

Fig. 14. Rat exorbital lacrimal gland, 1 h after pilocarpine (50 mg/kg). Reaction product is present in GERL (arrowheads), and in an irregular saccule (arrow) between GERL and the unreactive Golgi saccules (GS). Endoplasmic reticulum (ER). ×44,800.

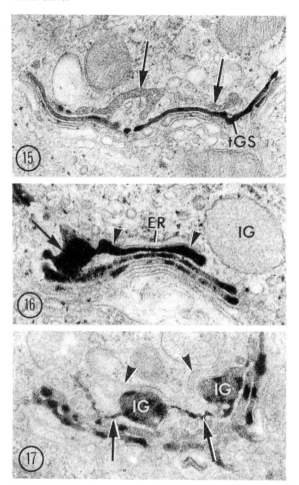

Fig. 15–17. Thiamine pyrophosphatase localization in stimulated serous cells.

Fig. 15. Rat exorbital lacrimal gland, 1 h after pilocarpine (50 mg/kg). Reaction product is present in the trans-Golgi saccule (tGS); weaker activity is also present in the adjacent saccule. A slightly dilated irregular saccule (arrows) containing reaction product is partially separated from the trans-face of the Golgi apparatus. ×34,000.

Fig. 16. Rat parotid gland, 8 h after isoproterenol (30 mg/kg). An intensely reactive cisterna (arrowheads) of variable width is present at the trans-face. This GERL-like cisterna is in continuity with a small reactive immature granule (arrow), and is paralleled by a portion of endoplasmic reticulum (ER). Two trans-Golgi saccules also contain reaction product. Unreactive immature granule (IG). ×44,500.

Fig. 17. Rat parotid gland, 5 h after isoproterenol (30 mg/kg). Two reactive immature granules (IG) are in continuity with narrow, reactive GERL-like cisternae (arrows) partially separated from the trans-face. Unreactive cisternae of GERL (arrowheads) are also present. ×35,700.

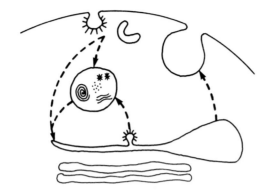

Fig. 18. Diagram summarizing the probable functions of GERL in secretory cells. GERL may participate in posttranslational modification of secretory and membrane proteins; package secretory proteins into granules for discharge by exocytosis (right); segregate lysosomal hydrolases into primary lysosomes (center) for delivery to endocytic or autophagic vacuoles; and receive membrane and membrane-bound ligands endocytosed from the cell surface (left) for recycling and/or degradation.

ity, and reaction product is present in GERL and forming secretory granules [12]. In contrast to the observations made in exocrine cells, acid phosphatase activity in these stimulated neurons appears reduced. An additional interesting feature of the stimulated neurosecretory cells is that neurosecretory granules form from all Golgi saccules as well as GERL. In all of the examples cited above, the structural and cytochemical modulations are temporally related to periods of increased secretory granule formation.

Functions of GERL in Secretory Cells

The structure and cytochemistry of GERL in secretory cells suggest that it has a number of functions. Its participation in secretory granule formation is presently the best documented, albeit incompletely understood, function. The various roles that GERL may play in secretory cells are discussed below and summarized in figure 18.

Protein Transport and Packaging

The location of GERL at the trans-face of the Golgi apparatus, its continuity with immature secretory granules, and the presence of acid phosphatase activity in both GERL and the immature granules, indicate

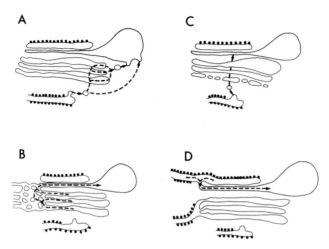

Fig. 19. Diagram illustrating the possible routes for secretory protein transport in the Golgi region. *A* Vesicular transport between the endoplasmic reticulum, Golgi saccules, and immature granules. Some proteins may bypass the Golgi saccules and be delivered directly to the immature granules. *B* Proteins may move between different saccules, adjacent Golgi stacks, GERL and immature granules through an intersaccular network of anastomosing tubules. *C* Protein transport from endoplasmic reticulum to immature granules by movement of whole saccules from cis- to trans-face. The membrane of GERL and the immature granules may be derived from the trans-saccule. *D* Direct transfer between the endoplasmic reticulum and immature granules via GERL. Transfer from the endoplasmic reticulum to Golgi saccules may also occur through tubular continuities.

that GERL is intimately involved in the formation of secretory granules. Despite these close structural and cytochemical relationships, the role of GERL in secretory protein transport between the Golgi saccules and forming granules is not well understood. Although the absence of enzymatically or immunologically detectable secretory protein in GERL may be due to technical considerations, the consistency of this observation suggests that the concentration of secretory proteins may be very low in GERL, or that GERL does not represent the main route of protein transfer between the Golgi apparatus and the forming granules. Figure 19 illustrates several possible routes by which secretory proteins may traverse the Golgi apparatus to reach the secretory granules. Figure 19 A: *Vesicular transport* – vesicles shuttling between the Golgi saccules and the forming granules, or directly between the transitional endoplasmic reticulum and the forming granules, may constitute the primary transport mechanism

[53, 54]. If this is the case, then GERL would have little or no role in protein transport, and the occasional presence of secretory protein in GERL may represent 'backflow' from the attached immature granules. Figure 19 B: *Intersaccular network* – an anastomosing network of tubules located between adjacent stacks of saccules would allow movement of secretory protein between adjacent stacks and between saccules at different levels within the same stack [95]. It also allows for protein transfer between the trans-saccule and GERL. Figure 19 C: *Saccule-GERL conversion* – movement of saccules and content from the cis- to the trans-face with conversion of the trans-saccule to GERL concomitant with the segregation of the secretory proteins in the immature granules [42, 43, 52, 86]. The movement of Golgi saccules from the cis- to the trans-face is known to occur in certain organisms [15] and has been postulated for a number of different vertebrate cell types [67, 110]. This mechanism could also account for the usual absence of secretory proteins in GERL, as well as the modulations of enzyme activity observed following secretory stimulation. The available data is insufficient to discriminate among these possibilities (fig. 19 A–C) or even some combination of the three. Figure 19 D: *Direct ER-transfer* – transfer of secretory protein from the endoplasmic reticulum to GERL through direct continuities [71]. This alternative is considered less likely due to the lack of confirmed continuities between the endoplasmic reticulum and GERL in secretory cells.

Modification of Secretory and Membrane Proteins

Following their synthesis, secretory and membrane proteins may undergo one or more posttranslational modifications, including glycosylation, sulfation, hydroxylation, crosslink formation, and specific proteolysis. Several observations suggest that GERL may be involved in some of these processing steps. *Bennett and O'Shaughnessy* [4] and *Bennett* et al. [5] have recently shown that in many cells sialic acid is incorporated into glycoproteins at the trans-aspect of the Golgi apparatus and suggest that GERL may be one of the organelles involved in the reaction. Acid phosphatase in GERL could conceivably participate in the glycosylation reaction through the hydrolysis of cytidine monophosphate liberated during the transfer of sialic acid from cytidine monophosphosialic acid to the endogenous acceptor glycoprotein. A similar function has been attributed to the nucleoside diphosphatase present in the Golgi saccules, which hydrolyzes uridine diphosphate after transfer of galactose from uridine diphosphogalactose to the acceptor glycoprotein [11, 59].

Certain secretory proteins, notably hormones such as insulin, para-thyroid hormone, vasopressin and corticotropin-endorphin, are synthe-sized as larger precursors (prohormones) which subsequently are con-verted to active hormones at an intracellular site [34, 38, 46, 105–107]. The proteolysis, at least for proinsulin and proparathormone, is accomplished through tryptic and carboxypeptidase B-like reactions which apparently begin during transport through the Golgi region and probably continue within the secretory granules [38, 105–107]. If GERL is involved in lysos-omal hydrolase segregation and transport (see below), it is tempting to speculate that lysosomal proteases present in GERL may also play a role in the proteolytic processing of secretory proteins.

Segregation of Secretory Proteins and Lysosomal Hydrolases

Both secretory proteins and lysosomal hydrolases are synthesized on polysomes bound to the endoplasmic reticulum and are transported through the same intracellular compartments. Posttranslational modifi-cation of lysosomal hydrolases provides them with a recognition marker, mannose-6-phosphate, which binds to specific receptors on intracellular membranes [100, 101]. One function of GERL in secretory cells, as well as in other cell types, thus may be to collect and direct receptor-bound hydrolases to lysosomes. The membrane of GERL occasionally exhibits coated areas, which conceivably could generate coated vesicles containing bound hydrolases. Residual hydrolase activity in immature granules may represent enzyme which has escaped the segregation mechanism. Sup-porting this possible function in secretory cells is the well-documented role of GERL in lysosome formation in neurons and hepatocytes [27, 70, 77, 80], as well as in other cells such as regressing corpus luteum [90], mega-karyocytes [6], macrophages [26] and Kupffer cells [93]. However, the presence of other lysosomal hydrolases, such as arylsulfatase, nonspecific esterase, and trimetaphosphatase, has not been demonstrated in GERL of secretory cells, even though these enzymes are present in typical secondary lysosomes and basal lysosomes in exocrine cells [23, 84, 85].

Intracellular Degradation of Secretory Products

In contrast to the potential specific proteolytic modification of secre-tory proteins described earlier, GERL may be involved in the intracellular degradation of various secretory products. For example, in hepatocytes of rats fed sucrose-enriched diets containing orotic acid with or without clofibrate, lipoprotein particles accumulate in GERL and appear to be undergoing degradation [79, 81, 83]. Cultured fibroblasts have been shown to degrade as much as 40% of their newly synthesized collagen prior to its secretion [7]. Recent cytochemical and cell fractionation studies have characterized GERL and lysosomes in similar cells [97], and procollagen has been demonstrated in lysosomes of odontoblasts by immunocyto-chemical methods [56]. While the possible mechanisms for GERL in-volvement in these degradative processes have not been elucidated, they could include failure to adequately segregate lysosomal and secretory proteins, and alteration of secretory granule membranes allowing fusion with lysosomes, such as occurs during crinophagy [24, 102]. An additional possibility is incomplete release of granule content during exocytosis and/or endocytic uptake of a portion of the secreted product with subse-quent delivery to GERL or lysosomes. This potential mechanism is dis-cussed more fully in the following paragraph.

Ligand Sequestration and Membrane Recycling

Work by *Gonatas* et al. [37] has demonstrated that cell surface-bound lectins are internalized and sequestered in GERL and lysosomes of cul-tured neurons. Similar results were later obtained for lectins in hepatoma ascites cells [63], horseradish peroxidase in mouse anterior pituitary so-matotrophs [13] and peritoneal macrophages [108], and colloidal silver in alveolar macrophages of beige mice [26]. Other than the eventual incor-poration of the endocytosed material into the lysosomal system for deg-radation, the specific function(s) served by segregation in GERL is unclear. It may reflect the more general process of membrane recycling believed to be occurring between the cell surface and the Golgi region [29]. In cells in which secretory granules form from GERL, reutilization of membrane during the formation of new secretory granules would clearly involve GERL. Surface membranes have been shown to reach the Golgi apparatus and forming granules of secretory cells under certain conditions [28, 47, 48, 88]. The specific route followed by the membrane has not yet been elucidated nor have the biochemical modifications which may be required for reutilization of the membrane.

Conclusion

In the preceding sections we have attempted to define the structure, characteristics and functions of GERL in secretory cells. GERL possesses distinct structural and cytochemical properties which permit its differentiation from both the endoplasmic reticulum and the Golgi apparatus. It may be argued that the description of GERL as an entity separate from the Golgi saccules is artificial, and that its structure and properties represent only one extreme of the heterogeneity typical of the Golgi apparatus. Clearly, GERL is closely related spatially and functionally to the Golgi apparatus, especially in regard to its participation in secretory protein transport and packaging. The modulation in enzyme activities observed in various physiological states further strengthens the relationship between these two structures. Although inclusion of GERL as an integral component of the Golgi apparatus might serve to simplify our concepts of Golgi apparatus function, we believe that the continued identification of GERL in secretory cells focuses attention to the key position it occupies and the important events in the secretory process which occur in this region of the cell.

References

1   Amsterdam, A.; Ohad, I.; Schramm, M.: Dynamic changes in the ultrastructure of the acinar cell of the rat parotid gland during the secretory cycle. J. Cell Biol. *41:* 753–773 (1969).

2   Beams, H. W.; Kessel, R. G.: The Golgi apparatus: structure and function. Int. Rev. Cytol. *23:* 209–276 (1968).

3   Bennett, G.; Leblond, C. P.; Haddad, A.: Migration of glycoprotein from the Golgi apparatus to the surface of various cell types as shown by radioautography after labeled fucose injection into rats. J. Cell Biol. *60:* 258–284 (1974).

4   Bennett, G.; O'Shaughnessy, D.: The site of incorporation of sialic acid residues into glycoproteins and the subsequent fates of these molecules in various rat and mouse cell types as shown by radioautography after injection of [^{3}H]N-acetylmannosamine. I. Observations in hepatocytes. J. Cell Biol. *88:* 1–15 (1981).

5   Bennett, G.; Kan, F. W. K.; O'Shaughnessy, D.: The site of incorporation of sialic acid residues into glycoproteins and the subsequent fates of these molecules in various rat and mouse cell types as shown by radioautography after injection of [^{3}H]N-acetylmannosamine. II. Observations in tissues other than liver. J. Cell Biol. *88:* 16–28 (1981).

6   Bentfield, M. E.; Bainton, D. F.: Cytochemical localization of lysosomal enzymes in rat megakaryocytes and platelets. J. clin. Invest. *56:* 1635–1649 (1975).

7   Bienkowski, R. S.; Cowan, M. J.; McDonald, J. A.; Crystal, R. G.: Degradation of newly synthesized collagen. J. biol. Chem. *253:* 4356–4363 (1978).

8   Bogart, B. I.: Secretory dynamics of the rat submandibular gland. An ultrastructural and cytochemical study of the isoproterenol-induced secretory cycle. J. Ultrastruct. Res. *52:* 139–155 (1975).

9 Boutry, J.-M.; Novikoff, A. B.: Cytochemical studies on Golgi apparatus, GERL, and lysosomes in neurons of dorsal root ganglia in mice. Proc. natn. Acad. Sci. USA 72: 508–512 (1975).

10 Bowen, R. H.: The cytology of glandular secretion. Rev. Biol. 4: 484–519 (1929).

11 Brandan, E.; Fleischer, B.: Localization and role of nucleoside diphosphatase and 5′-nucleotidase in Golgi vesicles of rat liver. J. Cell Biol. 87: 200a (1980).

12 Broadwell, R. D.; Oliver, C.: Golgi apparatus, GERL, and secretory granule formation within neurons of the hypothalamo-neurohypophysial system of control and hyperosmotically stressed mice. J. Cell Biol. 90: 474–484 (1981).

13 Broadwell, R. D.; Oliver, C.: An enzyme cytochemical study of the endocytic pathways in anterior pituitary cells of the mouse in vivo. J. Histochem. Cytochem. (in press, 1982).

14 Broadwell, R. D.; Oliver, C.; Brightman, M. W.: Localization of neurophysin within organelles associated with protein synthesis and packaging in the hypothalamo-neurohypophysial system: an immunocytochemical study. Proc. natn. Acad. Sci. USA 76: 5999-6003 (1979).

15 Brown, R. M., Jr.; Willison, J. H. M.: Golgi apparatus and plasma membrane involvement in secretion and cell surface deposition with special emphasis on cellulose biogenesis; in Brinkley, Porter, International cell biology, 1976–1977, pp.267–283 (Rockefeller University Press, New York 1977).

16 Caro, L. G.; Palade, G. E.: Protein synthesis, storage, and discharge in the pancreatic exocrine cell. An autoradiographic study. J. Cell Biol. 30: 473–495 (1964).

17 Case, R. M.: Synthesis, intracellular transport and discharge of exportable proteins in the pancreatic acinar cell and other cells. Biol. Rev. 53: 211–354 (1978).

18 Claude, A.: Growth and differentiation of cytoplasmic membranes in the course of lipoprotein granule synthesis in the hepatic cell. I. Elaboration of elements of the Golgi complex. J. Cell Biol. 47: 745–766 (1970).

19 Clermont, Y.; Lalli, M.; Rambourg, A.: Ultrastructural localization of nicotinamide adenine dinucleotide phosphatase (NADPase), thiamine pyrophosphatase (TPPase), and cytidine monophosphatase (CMPase) in the Golgi apparatus of early spermatids of the rat. Anat. Rec. 201: 613–622 (1981).

20 Dalton, A.J.; Felix, M. D.: Cytologic and cytochemical characteristics of the Golgi substance of epithelial cells of the epididymis – in situ, in homogenates and after isolation. Am. J. Anat. 94: 171–207 (1954).

21 Decker, R. S.: Lysosomal packaging in differentiating and degenerating anuran lateral motor column neurons. J. Cell Biol. 61: 599–612 (1974).

22 Doine, A. I.; Oliver, C.; Hand, A. R.: Cytochemical studies of the Golgi apparatus and GERL in parotid acinar cells of developing rats. J. Histochem. Cytochem. 28: 601–602 (1980).

23 Doty, S. B.; Smith, C. E.; Hand, A. R.; Oliver, C.: Inorganic trimetaphosphatase as a histochemical marker for lysosomes in light and electron microscopy. J. Histochem. Cytochem. 25: 1381–1384 (1977).

24 de Duve, C.: The lysosome in retrospect; in Dingle, Fell, Lysosomes in biology and pathology, vol.1, pp.3–40 (North Holland, Amsterdam 1969).

25 Ehrenreich, J. H.; Bergeron, J.J.M.; Siekevitz, P.; Palade, G. E.: Golgi fractions prepared from rat liver homogenates. I. Isolation procedure and morphological characterization. J. Cell Biol. 59: 45–72 (1973).

26   Essner, E.; Haimes, H.: Ultrastructural study of GERL in beige mouse alveolar macro-
     phages. J. Cell Biol. *75:* 381–387 (1977).
27   Essner, E.; Oliver, C.: Lysosome formation in hepatocytes of mice with Chediak-
     Higashi syndrome. Lab. Invest. *30:* 596–607 (1974).
28   Farquhar, M.G.: Recovery of surface membrane in anterior pituitary cells. Variations in
     traffic detected with anionic and cationic ferritin. J. Cell Biol. *77:* R35–R42 (1978).
29   Farquhar, M.G.; Palade, G.E.: The Golgi apparatus (complex) – (1954–1981) – from
     artifact to center stage. J. Cell Biol. *91:* 77s–103s (1981).
30   Farquhar, M.G.; Rinehart, J.F.: Cytologic alterations in the anterior pituitary gland
     following thyroidectomy: an electron microscope study. Endocrinology *55:* 857–876
     (1954).
31   Fleischer, B.; Fleischer, S.; Ozawa, H.: Isolation and characterization of Golgi mem-
     branes from bovine liver. J. Cell Biol. *43:* 59–79 (1969).
32   Friend, D.S.; Farquhar, M.G.: Functions of coated vesicles during protein absorption
     in the rat vas deferens. J. Cell Biol. *35:* 357–376 (1967).
33   Fujita, H.; Okamoto, H.: Fine structural localization of thiamine pyrophosphatase and
     acid phosphatase activities in the mouse pancreatic acinar cell. Histochemistry *64:*
     287–295 (1979).
34   Gainer, H.; Sarne, Y.; Brownstein, M.J.: Biosynthesis and axonal transport of rat
     neurohypophyseal proteins and peptides. J. Cell Biol. *73:* 366–381 (1977).
35   Geuze, J.J.; Slot, J.W.; Tokuyasu, K.T.: Immunocytochemical localization of amy-
     lase and chymotrypsinogen in the exocrine pancreatic cells with special attention to the
     Golgi complex. J. Cell Biol. *82:* 697–707 (1979).
36   Goldfischer, S.: The internal reticular apparatus of Camillo Golgi: a complex, heter-
     ogeneous organelle, enriched in acid, neutral, and alkaline phosphatases, and involved
     in glycosylation secretion, membrane flow, lysosome formation, and intracellular
     digestion. J. Histochem. Cytochem. *30:* 717–733 (1982).
37   Gonatas, N.K.; Kim, S.U.; Stieber, A.; Avrameas, S.: Internalization of lectins in
     neuronal GERL. J. Cell Biol. *73:* 1–13 (1977).
38   Habener, J.F.; Amherdt, M.; Ravazzola, M.; Orci, L.: Parathyroid hormone biosyn-
     thesis. Correlation of conversion of biosynthetic precursors with intracellular protein
     migration as determined by electron microscope autoradiography. J. Cell Biol. *80:*
     715–731 (1979).
39   Haguenau, F.; Bernhard, E.: L'appareil de Golgi dans les cellules normales et cancé-
     reuses de vertébrés. Rappel historique et étude au microscope électronique. Arch.
     Anat. microsc. Morphol. exp. *44:* 27–55 (1955).
40   Hand, A.R.: Morphology and cytochemistry of the Golgi apparatus of rat salivary
     gland acinar cells. Am. J. Anat. *130:* 141–158 (1971).
41   Hand, A.R.: Cytochemical differentiation of the Golgi apparatus from GERL. J. His-
     tochem. Cytochem. *28:* 82–86 (1980).
42   Hand, A.R.; Oliver, C.: Cytochemical studies of GERL and its role in secretory granule
     formation in exocrine cells. Histochem. J. *9:* 375–392 (1977).
43   Hand, A.R.; Oliver, C.: Relationship between the Golgi apparatus, GERL, and secre-
     tory granules in acinar cells of the rat exorbital lacrimal gland. J. Cell Biol. *74:* 399–413
     (1977).
44   Hand, A.R.; Oliver, C.: Effects of secretory stimulation on the Golgi apparatus and
     GERL of rat parotid acinar cells. J. Cell Biol. *87:* 304a (1980).

45   Hand, A.R.; Oliver, C.: The Golgi apparatus: protein transport and packaging in
      secretory cells; in Hand, Oliver, Basic mechanisms of cellular secretion. Methods in cell
      biology, vol. 23, pp. 137–153 (Academic Press, New York 1981).

46   Herbert, E.; Phillips, M.; Budarf, M.: Glycosylation steps involved in processing of
      pro-corticotropin-endorphin in mouse pituitary tumor cells; in Hand, Oliver, Basic
      mechanisms of cellular secretion. Methods in cell biology, vol. 23, pp. 101–118 (Aca-
      demic Press, New York 1981).

47   Herzog, V.; Farquhar, M.G.: Luminal membrane retrieved after exocytosis reaches
      most Golgi cisternae in secretory cells. Proc. natn. Acad. Sci. USA 74: 5073–5077
      (1977).

48   Herzog, V.; Miller, F.: Membrane retrieval in secretory cells. Morphological evidence
      for participation of Golgi cisternae in internalization of plasma membrane following
      stimulated exocytosis; in Hopkins, Duncan, Secretory mechanisms. Symp. Soc. exp.
      Biol., vol. 33, pp. 101–116 (Cambridge University Press, Cambridge 1979).

49   Holtzman, E.; Dominitz, R.: Cytochemical studies of lysosomes, Golgi apparatus and
      endoplasmic reticulum in secretion and protein uptake by adrenal medulla cells of the
      rat. J. Histochem. Cytochem. 16: 320–336 (1968).

50   Inoué, K.; Kurosumi, K.: Cytochemical and three-dimensional studies of Golgi appa-
      ratus and GERL of rat anterior pituitary cells by transmission electron microscopy. Cell
      Struct. Funct. 2: 171–186 (1977).

51   Ito, S.: Structure and function of the glycocalyx. Fed. Proc. 28: 12–25 (1969).

52   Jaeken, L.; Thines-Sempoux, D.; Verheyen, F.: A three-dimensional study of organelle
      interrelationships in regenerating rat liver. I. The GERL-system. Cell Biol. int. Rep. 2:
      501–513 (1978).

53   Jamieson, J.D.; Palade, G.E.: Intracellular transport of secretory proteins in the pan-
      creatic exocrine cell. I. Role of the peripheral elements of the Golgi complex. J. Cell
      Biol. 34: 577–596 (1967).

54   Jamieson, J.D.; Palade, G.E.: Intracellular transport of secretory proteins in the pan-
      creatic exocrine cell. II. Transport to condensing vacuoles and zymogen granules. J.
      Cell Biol. 34: 597–615 (1967).

55   Jamieson, J.D.; Palade, G.E.: Synthesis, intracellular transport, and discharge of
      secretory proteins in stimulated pancreatic exocrine cells. J. Cell Biol. 50: 135–158
      (1971).

56   Karim, A.; Cournil, I.; Leblond, C.P.: Immunohistochemical localization of procol-
      lagens. II. Electron microscopic distribution of procollagen I antigenicity in the odon-
      toblasts and predentin of rat incisor teeth by a direct method using peroxidase linked
      antibodies. J. Histochem. Cytochem. 27: 1070–1083 (1979).

57   Karnovsky, M.J.: A formaldehyde-glutaraldehyde fixative of high osmolality for use in
      electron microscopy. J. Cell Biol. 27: 137a (1965).

58   Kern, H.F.; Kern, D.: Elektronenmikroskopische Untersuchungen über die Wirkung
      von Kobaltchlorid auf das exokrine Pankreasgewebe des Meerschweinchens. Virchows
      Arch. Abt. B Zellpathol. 4: 54–70 (1969).

59   Kuhn, N.J.; White, A.: The role of nucleoside diphosphatase in a uridine nucleotide
      cycle associated with lactose synthesis in rat mammary-gland Golgi apparatus. Bio-
      chem. J. 168: 423–433 (1977).

60   Lane, N.; Caro, L.; Otero-Vilardebó, L.R.; Godman, G.C.: On the site of sulfation in
      colonic goblet cells. J. Cell Biol. 21: 339–351 (1964).

61 Lillie, J. H.; Han, S. S.: Secretory protein synthesis in the stimulated rat parotid gland. Temporal dissociation of the maximal response from secretion. J. Cell Biol. *59:* 708–721 (1973).

62 Mayahara, H.; Ishikawa, T.; Ogawa, K.; Chang, J. P.: The three dimensional structure of the Golgi complex in cultured fibroblasts. An ultracytochemical study with thin and thick sections. Acta histochem. cytochem. *11:* 239–251 (1978).

63 Moller, P. C.; Wang, J. J.; Yokoyama, M.; Chang, J. P.: Cytochemical localization of lectin labeled vesicles in GERL region of hepatoma ascites cells. Histochemistry *62:* 289–297 (1979).

64 Morré, D. J.; Hamilton, R. L.; Mollenhauer, H. H.; Mahley, R. W.; Cunningham, W. P.; Cheetham, R. D.; Lequire, V. S.: Isolation of a Golgi apparatus-rich fraction from rat liver. I. Method and morphology. J. Cell Biol. *44:* 484–491 (1970).

65 Nassonov, D.: Das Golgische Binnennetz und seine Beziehungen zu der Sekretion. Untersuchungen über einige Amphibiendrüsen. Arch. mikroskop. Anat. EntwMech. *97:* 136–186 (1923).

66 Nassonov, D.: Das Golgische Binnennetz und seine Beziehungen zu der Sekretion. Morphologische und experimentelle Untersuchungen an einigen Säugetierdrüsen. Arch. mikroskop. Anat. EntwMech. *100:* 433–472 (1924).

67 Neutra, M.; Leblond, C. P.: Synthesis of the carbohydrate of mucus in the Golgi complex as shown by electron microscope radioautography of goblet cells from rats injected with glucose-H^3. J. Cell Biol. *30:* 119–136 (1966).

68 Novikoff, A. B.: Cytochemical staining methods for enzyme activities: their application to the rat parotid gland. Jewish meml Hosp. Bull. *7:* 70–93 (1962).

69 Novikoff, A. B.: Lysosomes in the physiology and pathology of cells: contributions of staining methods; in de Reuck, Cameron, Ciba Foundation Symposium on Lysosomes, pp. 36–73 (Little, Brown, Boston 1963).

70 Novikoff, A. B.: GERL, its form and function in neurons of rat spinal ganglia. Biol. Bull. *127:* 358 (1964).

71 Novikoff, A. B.: The endoplasmic reticulum: A cytochemist's view (a review). Proc. natn. Acad. Sci. USA *73:* 2781–2787 (1976).

72 Novikoff, A. B.; Albala, A.; Biempica, L.: Ultrastructural and cytochemical observations on B-16 and Harding-Passey mouse melanomas. The origin of premelanosomes and compound melanosomes. J. Histochem. Cytochem. *16:* 299–319 (1968).

73 Novikoff, A. B.; Goldfischer, S.: Nucleoside diphosphatase activity in the Golgi apparatus and its usefulness for cytological studies. Proc. natn. Acad. Sci. USA *47:* 802–810 (1961).

74 Novikoff, A. B.; Leuenberger, P. M.; Novikoff, P. M.; Quintana, N.: Retinal pigment epithelium. Interrelations of endoplasmic reticulum and melanolysosomes in the black mouse and its beige mutant. Lab. Invest. *40:* 155–165 (1979).

75 Novikoff, A. B.; Mori, M.; Quintana, N.; Yam, A.: Studies of the secretory process in the mammalian exocrine pancreas. I. The condensing vacuoles. J. Cell Biol. *75:* 148–165 (1977).

76 Novikoff, A. B.; Novikoff, P. M.: Cytochemical contributions to differentiating GERL from the Golgi apparatus. Histochem. J. *9:* 525–551 (1977).

77 Novikoff, A. B.; Roheim, P. S.; Quintana, N.: Changes in rat liver cells induced by orotic acid feeding. Lab. Invest. *15:* 27–49 (1966).

78 Novikoff, A. B.; Yam, A.; Novikoff, P. M.: Cytochemical study of secretory process in

transplantable insulinoma of Syrian golden hamster. Proc. natn. Acad. Sci. USA *72:* 4501–4505 (1975).

79 Novikoff, P.M.; Edelstein, D.: Reversal of orotic acid-induced fatty liver in rat by clofibrate. Lab. Invest. *36:* 215–231 (1977).

80 Novikoff, P.M.; Novikoff, A.B.; Quintana, N.; Hauw, J.-J.: Golgi apparatus, GERL, and lysosomes of neurons in rat dorsal root ganglia, studied by thick section and thin section cytochemistry. J. Cell Biol. *50:* 859–886 (1971).

81 Novikoff, P.M.; Roheim, P.S.; Novikoff, A.B.; Edelstein, D.: Production and prevention of fatty liver in rats fed clofibrate and orotic acid diets containing sucrose. Lab. Invest. *30:* 732–750 (1974).

82 Novikoff, P.M.; Yam, A.: Sites of lipoprotein particles in normal rat hepatocytes. J. Cell Biol. *76:* 1–11 (1978).

83 Novikoff, P.M.; Yam, A.: The cytochemical demonstration of GERL in rat hepatocytes during lipoprotein mobilization. J. Histochem. Cytochem. *26:* 1–13 (1978).

84 Oliver, C.: Cytochemical localization of acid phosphatase and trimetaphosphatase activities in exocrine acinar cells. J. Histochem. Cytochem. *28:* 78–81 (1980).

85 Oliver, C.: Enzyme cytochemical studies of basal lysosomes in exocrine acinar cells. J. Histochem. Cytochem. *29:* 898 (1981).

86 Oliver, C.; Auth, R.E.; Hand, A.R.: Morphological and cytochemical alterations of the Golgi apparatus and GERL in rat parotid acinar cells during ethionine intoxication and recovery. Am. J. Anat. *158:* 275–284 (1980).

87 Osinchak, J.: Electron microscopic localization of acid phosphatase and thiamine pyrophosphatase activity in hypothalamic neurosecretory cells of the rat. J. Cell Biol. *21:* 35–47 (1964).

88 Ottosen, P.D.; Courtoy, P.J.; Farquhar, M.G.: Pathways followed by membrane recovered from the surface of plasma cells and myeloma cells. J. exp. Med. *152:* 1–19 (1980).

89 Paavola, L.G.: The corpus luteum of the guinea pig. II. Cytochemical studies on the Golgi complex, GERL, and lysosomes in luteal cells during maximal progesterone secretion. J. Cell Biol. *79:* 45–58 (1978).

90 Paavola, L.G.: The corpus luteum of the guinea pig. III. Cytochemical studies on the Golgi complex and GERL during normal postpartum regression of luteal cells, emphasizing the origin of lysosomes and autophagic vacuoles. J. Cell Biol. *79:* 59–73 (1978).

91 Palade, G.E.: Intracellular aspects of the process of protein secretion. Science *189:* 347–358 (1975).

92 Pelletier, G.; Novikoff, A.B.: Localization of phosphatase activities in the rat anterior pituitary gland. J. Histochem. Cytochem. *20:* 1–12 (1972).

93 Pino, R.M.; Pino, L.C.; Bankston, P.W.: The relationships between the Golgi apparatus, GERL, and lysosomes of fetal rat liver Kupffer cells examined by ultrastructural phosphatase cytochemistry. J. Histochem. Cytochem. *29:* 1061–1070 (1981).

94 Rambourg, A.: Morphological and histochemical aspects of glycoproteins at the surface of animal cells. Int. Rev. Cytol. *31:* 57–114 (1971).

95 Rambourg, A.; Clermont, Y.; Hermo, L.: Three-dimensional structure of the Golgi apparatus; in Hand, Oliver, Basic mechanisms of cellular secretion. Methods in cell biology, vol. 23, pp. 155–166 (Academic Press, New York 1981).

96 Rhodin, J.: Correlation of ultrastructural organization and function in normal and

experimentally changed proximal convoluted tubule cells of the mouse kidney; PhD thesis Karolinska Institutet, Stockholm (1954).

97 Rome, L.H.; Garvin, A.H.; Alietta, M.M.; Neufeld, E.: Two species of lysosomal organelles in cultured human fibroblasts. Cell *17:* 143–153 (1979).

98 Simson, J.V.; Discharge and restitution of secretory material in the rat parotid gland in response to isoproterenol. Z. Zellforsch. mikrosk. Anat. *101:* 175–191 (1969).

99 Sjöstrand, F.S.; Hanzon, V.: Ultrastructure of Golgi apparatus of exocrine cells of mouse pancreas. Expl Cell Res. *7:* 415–429 (1954).

100 Sly, W.S.; Fischer, H.D.: The phosphomannosyl recognition system for intracellular and intercellular transport of lysosomal enzymes. J. Cell Biochem. *18:* 67–85 (1982).

101 Sly, W.S.; Fischer, H.D.; Gonzalez-Noriega, A.; Grubb, J.H.; Natowicz, M.: Role of the 6-phosphomannosyl-enzyme receptor in intracellular transport and adsorptive pinocytosis of lysosomal enzymes; in Hand, Oliver, Basic mechanisms of cellular secretion. Methods in cell biology, vol.23, pp.191–214 (Academic Press, New York 1981).

102 Smith, R.E.; Farquhar, M.G.: Lysosome function in the regulation of the secretory process in cells of the anterior pituitary gland. J. Cell Biol. *31:* 319–347 (1966).

103 Smith, R.E.; Farquhar, M.G.: Modulation in nucleoside diphosphatase activity of mammotrophic cells of the rat adenohypophysis during secretion. J. Histochem. Cytochem. *18:* 237–250 (1970).

104 Sobel, H.J.; Arvin, E.: Localization of acid phosphatase activity in rat pancreatic acinar cells: a light and electron microscopic study. J. Histochem. Cytochem. *13:* 301–303 (1965).

105 Steiner, D.F.; Kemmler, W.; Tager, H.S.; Rubenstein, A.H.: Molecular events taking place during the intracellular transport of exportable proteins. The conversion of peptide hormone precursors; in Ceccarelli, Clementi, Meldolesi, Advances in cytopharmacology, vol.2, pp.195–205 (Raven Press, New York 1974).

106 Steiner, D.F.; Quinn, P.S.; Patzelt, C.; Chan, S.J.; Marsh, J.; Tager, H.S.: Proteolytic cleavage in the posttranslational processing of proteins; in Prescott, Goldstein, Cell biology: a comprehensive treatise, vol.4, pp.175–202 (Academic Press, New York 1980).

107 Tager, H.S.; Steiner, D.F.; Patzelt, C.: Biosynthesis of insulin and glucagon; in Hand, Oliver, Basic mechanisms of cellular secretion. Methods in cell biology, vol.23, pp.73–88 (Academic Press, New York 1981).

108 Thyberg, J.; Stenseth, K.: Endocytosis of native and cationized horseradish peroxidase by cultured mouse peritoneal macrophages. Variations in cell surface binding and intracellular traffic and effects of colchicine. Eur. J. Cell Biol. *25:* 308–318 (1981).

109 van Heyningen, H.E.: Secretion of protein by the acinar cells of the rat pancreas, as studied by electron microscopic radioautography. Anat. Rec. *148:* 485–497 (1964).

110 Wienstock, M.; Leblond, C.P.: Synthesis, migration, and release of precursor collagen by odontoblasts as visualized by radioautography after [³H]proline administration. J.Cell Biol. *60:* 92–127 (1974).

111 Young, R.W.: The role of the Golgi complex in sulfate metabolism. J. Cell Biol. *57:* 175–189 (1973).

Dr. A.R. Hand, Laboratory of Biological Structure, National Institute of Dental Research, National Institutes of Health, Bethesda, MD 20205 (USA)

Cell Biology of the Secretory Process, pp.171–195 (Karger, Basel 1984)

The Cell Surface during the Secretory Process

Tony Antakly

Laboratory of Pathobiology, Clinical Research Institute of Montreal, Que., Canada

Introduction

The anterior pituitary secretes six major types of hormones: prolactin, GH, FSH, LH, TSH, and ACTH. The secretory activities of pituitary cells are regulated both by neurohormones (neuropeptides, biogenic amines) of hypothalamic origin and by peripheral hormones (sex steroids, thyroid hormones, glucocorticoids and others). Each of these regulatory hormones (or factors) exerts either a stimulatory or an inhibitory action on one or more types of pituitary cells [41]. Because most of these regulatory hormones are circulating in the blood, the study of the specificity of their action at the pituitary level has been extremely complicated in the in vivo approach. However, such studies have been greatly facilitated by the development of the rat pituitary cell culture system [67]. In fact, adenohypophyseal cells in culture have been extremely useful, not only as a precise biological assay for analogs of the neurohormones: TRH, LHRH, and somatostatin [12, 41, 67] but also for the determination of the characteristics of interaction between hypothalamic and peripheral hormones [8, 30, 31]. Using primary cultures of pituitary cells, we initiated [9] a series of studies to investigate the effects of these regulatory factors at the cellular level. These studies [1, 3–10, 42, 59] were performed using electron microscopy, immunocytochemistry and radioimmunoassays. Because monolayer cultures offer an attractive system to study the cells by scanning electron microscopy (SEM), we examined the cell surface topography of these cells during various states of secretory activity. We observed dramatic changes in the surface morphology of these cells which correlated with changes in pituitary hormone secretion [1–10, 42, 59]. This dynamic change in the cell surface during altered secretory activity is not unique to pituitary cells but also occurs in other cell types as well. In order to extend our studies we examined rat hepatocytes maintained in short-term pituitary culture. Indeed, rat hepatocytes maintained in short-term culture [28, 51, 63] secrete serum proteins such as albumin [28, 51] and α2u globulin [2, 21]. The rate of albumin and α2u globulin secretion gradually decreases with time in culture. However, when these cells were grown in the presence of dexamethasone, their secretory activity was not only increased as compared to control cells, but they were found to secrete plasma proteins over a longer period of time extending to several days [14, 16, 35, 43, 51, 54]. We found that this hormonally induced increase in secretory activities was accompanied by dramatic changes in their surface morphology. No changes in surface morphology were induced by glucocorticoids on hepatoma cells that do not respond to glucocorticoids (see below). The secretion of plasma protein and pituitary hormones is also inhibited by colchicine, a drug that disrupts microtubules. Concomitant to its inhibitory effect on protein secretion, colchicine induced characteristic changes in the surface morphology of both hepatocytes and pituitary cells. This article attempts to review the studies mentioned above in the light of our recent unpublished observations and accumulating reports emphasizing the role of the cell surface in the secretory process.

Studies on Prolactin Cells in Primary Culture

Among the six major hormones secreted by the pituitary, prolactin appears of particular interest to us since prolactin-secreting cells represent over 70% of the total secretory cells observed between days 3 and 12 in culture (fig. 1). Moreover, prolactin secretion can be stimulated by estrogens [3–5, 22, 71], TRH [3–5, 15], cyclic AMP derivatives [60] or specifically inhibited by dopamine or one of its agonists CB-154 [3–5, 26, 58, 62]. Prolactin secretion is also inhibited by colchicine [10, 40].

Effects of Different Agents on Prolactin Secretion

Already a 4-hour incubation of the cells with the potent dopamine agonist, CB-154 (1 μM), reduced the basal prolactin release by 93% as compared to non-treated controls (fig. 2) and this inhibitory effect persisted at that level for up to 24 h. Incubation of the cells with colchicine for 4 h reduced basal prolactin release by 42% ($p < 0.01$) [3–5]; this inhibition

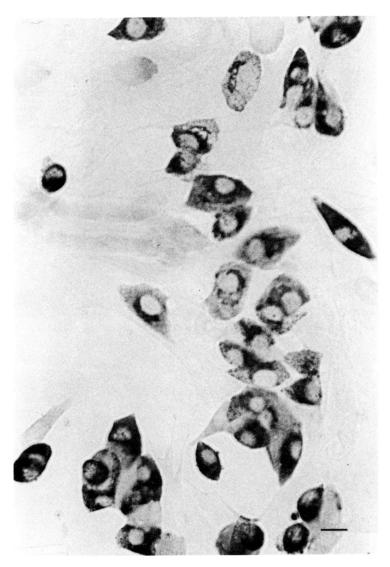

Fig. 1. Immunocytochemical localization of prolactin in pituitary cells maintained for 6 days in primary culture. As determined in our previous studies [3–10] prolactin cells constitute almost 70% of the total secretory cells in these cultures. Bar = 10 μ*M*.

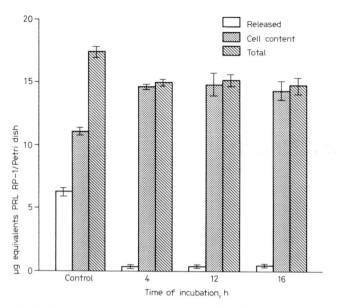

Fig. 2. Effect of short-term incubation with CB-154 on prolactin (PRL) release, hormone cell content and total hormone (release and cell content) in rat anterior pituitary cells in culture. 3 days after plating, cells were incubated for the indicated time intervals in the presence or absence of 1 μ*M* CB-154 and prolactin content measured at the end of corresponding incubations [from ref. 8, with permission].

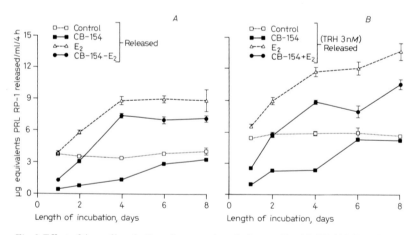

Fig. 3. Effect of time of incubation of rat anterior pituitary cells with CB-154, E₂, CB-154 with E₂ or the vehicle alone on spontaneous *(A)* or TRH-induced *(B)* prolactin (PRL) release measured at the end of the 4-hour incubation in the absence or presence of 3n*M* TRH. Data are presented as means ± SEM [from ref. 8, with permission].

was proportional to time, reaching a value of 84% after 16 h [7], (fig. 1, 2). Estradiol (E_2) increased both the release (260% over control at day 4) and cell content of prolactin (100% over control at day 8, fig. 3). The amount of prolactin released following a 4-hour final stimulation with TRH was significantly greater than that of basal and E_2-pretreated cells (fig. 3).

Changes of the Cell Surface Topography during Various Secretory States

Immediately following cell dissociation, the cells appear rounded (fig. 4a) and generally smooth. After plating, the cells gradually flatten and attach to the dishes. Within 3 days after plating and thereafter two types of cells were observed: the rounded-up ones (7–12 μm) and the fibroblasts (fig. 4b). Numerous protrusions were seen at the surface of the former consisting of microvilli (~0.1 μm in diameter) and blebs (0.4–1 μm). Sometimes, filamentous extensions (filopodia) were also present; they seem to anchor the cells to the dishes and to each other. The rounded-up cells are secretory because they stain for one of the pituitary hormones [1, 3–10, 42]. Prolactin cells represent 70% of these cells as demonstrated by immunocytochemistry [1, 3–10, 42] (fig. 1). LH cells are ~2%, FSH 4–6%, ACTH 4–8%, TSH 1–3%, GH 10–18% [3–5, 8]. Following 4 h of treatment with colchicine, noticeable changes of secretory cell shape could be observed: microvilli became rare [fig. 4–7 of ref. 10]. In some cells, the number of blebs as well as their size was increased, whereas in others, the surface became smoother than in controls. At stages later than 16 h, blebs of large size (1–3 μm) seemed to be shed off from the cells. Following 8–16 h of CB-154 treatment, many cells showed folds, but the general feature was a marked decrease in the number of microvilli [3–6, 8, 10]. When the cells were incubated in the presence of 17β-estradiol (E_2), a prominent feature was an increase in the number of cell surface protrusions (microvilli and blebs, fig. 5). Similar effects of E_2, but of lesser magnitude, were observed when the cells were incubated for 4 h in the presence of TRH alone. When the cells were pretreated (1–8 days) with E_2 before incubation for 4 h with TRH, the number of surface protrusions was further increased indicating a potentiation by estrogen of the neurohormone-induced changes in cell surface activity (fig. 5). Cyclic AMP and its derivatives which are known to affect hormonal secretion of most pituitary cell types [41, 60] as well as the phosphodiesterase inhibitor theophylline (which essentially prevents degradation of cyclic AMP), had an effect on the pituitary cell surface morphology. 1 h after the addition of 8-Br-cyclic AMP or theophylline, cyto-

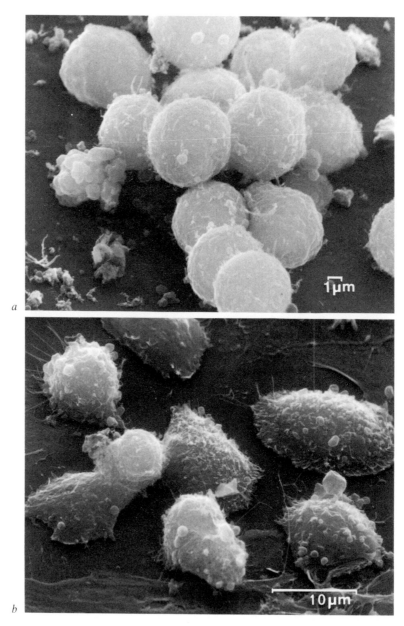

Fig. 4. Surface morphology of freshly dissociated pituitary cells *(a)* and those maintained in primary culture for 6 days *(b)*. In *B* note two types of cells: the rounded-up ones, which are secretory, and the very flattened cells which are the fibroblasts [3–10].

plasmic extensions resembling pseudopods were observed on some cells. After 4 h most cells (presumably secretory) displayed these extensions which displayed blebs, microvilli and membrane folds (fig. 6).

Studies on Hepatocytes in Primary Culture

Recent studies have focussed on the changes induced by glucocorti-coids [48–50] and colchicine [Antakly et al., unpublished] on the surface topography as well as on the intracellular cytoskeleton of rat hepatocytes maintained in short-term primary culture according to established in vivo perfusion technique [63]. Cell fractionation analysis by velocity sedimen-tation at 1 g [27, 29] revealed that the cell preparations are at least 90% hepatic parenchymal cells. Hepatic epithelial cells, which were isolated by Dispase I treatment [27, 29, 47, 52, 65], and two Morris hepatoma cell lines, designated 7777 and 7795, were also studied [48].

Modulation of Surface Morphology of Hepatocytes, Hepatic Epithelial Cells and Hepatoma Cells in Culture

Effect of Dexamethasone

As shown in figures 7a and 9a, hepatocytes isolated from normal adult liver are covered with microvilli and blebs. Similar observations were reported by Williams et. al. [70] and Penasse et al. [61]. With increasing time in culture, hepatocytes have a tendency to spread and the degree of spreading is greatly influenced by the addition of dexamethasone to the medium [51]. We have previously found that this dexamethasone-influenced flattening is accompanied by rearrangements of cytoskeletal fibrillar networks [47, 49]. In view of the link that is known to exist between cell shape and extracellular matrix [37, 55], we have examined by SEM the changes in surface morphology following treatment of the cells with dexamethasone. Between 4 and 48 h in culture in the absence of dexamethasone, a great number of hepatocytes spread onto the substratum and gradually lost their microvilli. After 48 h a great number of cells formed a 'carpet' of flattened cells, some of them overlaid by more rounded-up cells (fig. 7a); the cells progressively detached from the culture substratum. In the presence of dexamethasone, the most prominent obser-vation was the induction of an extracellular matrix consisting of an elabo-

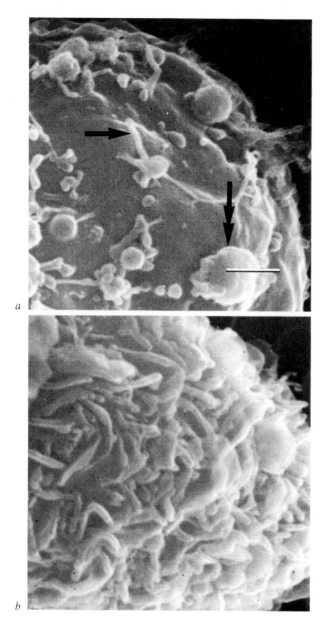

Fig. 5. SEM of pituitary cells cultured for 8 days. *a* Control (untreated), note the presence of occasional microvilli (arrow) and blebs (double-head arrow) on the cell surface. *b* SEM of E₂ stimulated cell: the cell surface is covered with microvilli and blebs. Bar = 1 μm.

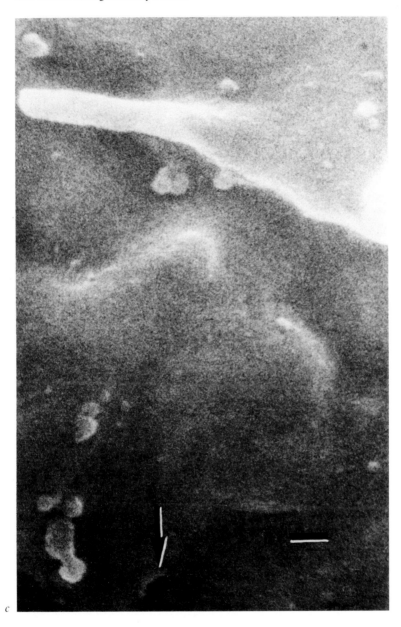

c Ultrahigh resolution SEM of the surface of a control cell demonstrating the presence of a microvillus, few particles on the cell surface as well as a pit (arrow). Bar = 0.1μm.

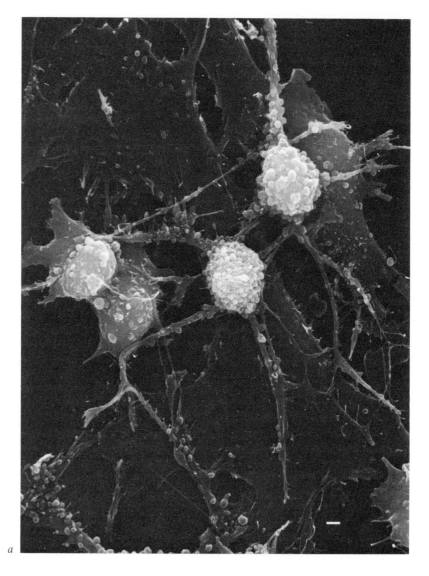

Fig. 6. a SEM of pituitary cells cultured for 6 days and preincubated with dibutyryl cyclic AMP (1 m*M*) for 4 h. Note the extensive branching of the rounded-up cells. A higher

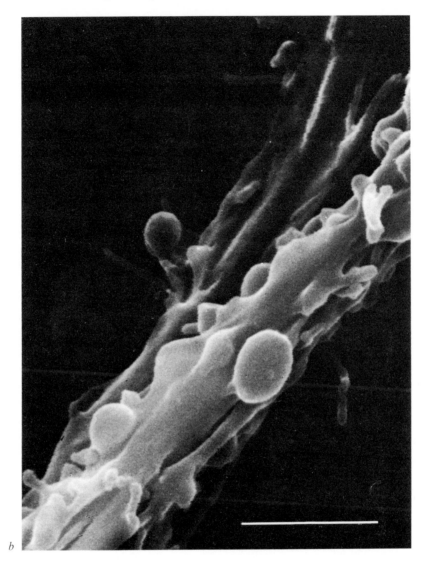

b

magnification of one cytoplasmic branch is shown in figure 6b. Note the presence of several cell surface blebs. Bar = 1μm *(a)* and 0.1μm *(b)*.

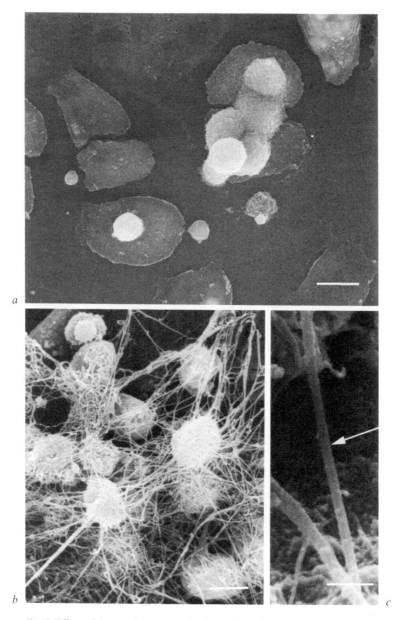

Fig. 7. Effect of dexamethasone on the formation of extracellular biomatrix in hepatocytes. *a, b* SEM of hepatocytes maintained in primary culture for 48 h as described [47–52] in the absence *(a)* or presence *(b, c)* of dexamethasone. Note in figure *7b* the presence of a network of extracellular fibers (termed extracellular biomatrix). The chemical composition

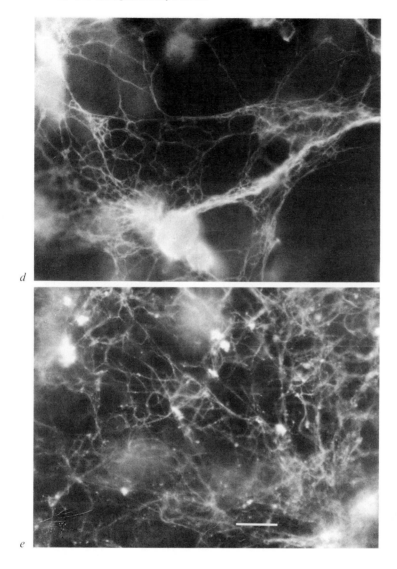

of this extracellular biomatrix was determined by immunofluorescence *(d)* localization of fibronectin and collagen type I *(e)* in cells treated for 48 h with 1 μ*M* dexamethasone. Bars = 10 μm *(a, b, d, e)*, bar = 1 μm *(c)* [from ref. 48, with permission].

rate network of fibers overlaying the cells (fig. 7b). The chemical composition of these biomatrices was analyzed by immunofluorescence using specific antibodies [47, 48] to fibronectin and collagen type I and type IV. The fibers reacted with the anti-fibronectin and collagen type I but not with anti-collagen type IV (fig. 7d, e). Dexamethasone was found not to induce formation of extracellular biomatrix in either hepatic epithelial cells or hepatoma cells 7777 and 7795 (SEM/81). These observations on the inability of dexamethasone to induce biomatrix formation in hepatic epithelial and hepatoma cells is similar to that reported for transformed fibroblasts in culture [37, 55]. This may either be due to the absence of glucocorticoid receptor and/or molecular defect of this receptor in the hepatic epithelial and hepatoma cells or to other factors interfering with the gene expression of fibronectin or collagens. Another explanation could be that the absence of biomatrix formation is due to a breakdown of the newly synthesized proteins (fibronectin and collagen) by proteases. In this context it is known that surface-exposed fibronectin is indeed very sensitive to degradation by exogenous trypsin or plasmin [37] and transformed cells including hepatoma have elevated plasminogen activator levels [70].

Effect of Colchicine on the Hepatocyte Cell Surface and on the Secretion of Albumin and α2u Globulin

As can be seen in figure 8 colchicine significantly inhibited albumin release by about 60% and α2u globulin by 56% after 4 or 24 h of incubation. Concomitant to the inhibitory effect on protein secretion, colchicine induced dramatic changes in the hepatocyte surface morphology (fig. 9). These changes include loss of most of the microvilli and formation of large pocket-like bulgings from the cell surface and of numerous blebs of various sizes which seem to be released from the cell surface. Also colchicine-treated cells display a typical deformed shape presumably due to the loss of microtubules as structural elements of the cells [7]. Structural elements include not only microtubules but also microfilaments which also seem to play an important role in secretion since drugs that disrupt microfilaments are known to inhibit protein secretion from secretory cells [46]. The mechanisms responsible for the inhibitory action of colchicine on secretory processes are still not completely elucidated. It is possible that this drug might act directly on the cell menbrane by inducing structural changes in the cell membrane as suggested by the present studies. In this context it is known that colchicine affects cell membrane fluidity [39].

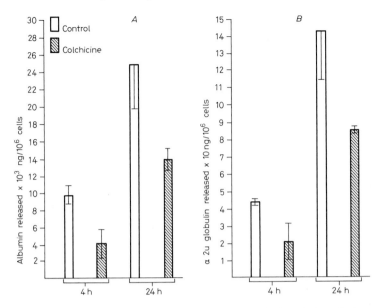

Fig. 8. Effects of colchicine on albumin *(A)* and α2u globulin *(B)* release by hepatocytes maintained in primary culture for the indicated period of time, either in the absence or presence of 1 μ*M* colchicine. Note a significant decrease in protein release (p <0.05 in all cases) in cells incubated with colchicine [*Antakly* et al., unpublished].

Hormonal Changes Induced by Drugs and Hormones on Various Types of Secretory Cells: General Discussion

The secretagogue-induced modulations of the cell morphology described above do not only occur in pituitary and hepatocytes in culture but also occur in a variety of other cell types as well both in vivo and in vitro as indicated below. In addition, such modulations are not an exclusive feature of protein-secreting cells, but also occur in steroid and biogenic amine-secreting cells. A well-studied system is that of granulosa cells [18, 25] (fig. 10). *Chang* et al. [19] first called attention to surface changes of rat granulosa cells in situ following pregnant mare's serum gonadotropin (PMSG) stimulation. Such changes include the induction of the formation of microvilli on the surface of large follicle granulosa cells and their absence from smaller sized follicles [20]. Such findings were also reported in the case of human follicular granulosa cells [64]. Other investigators extended these studies to the modulation of lutein [69] and rat granulosa cell surface exposed to FSH. The dramatic effect of FSH on the formation

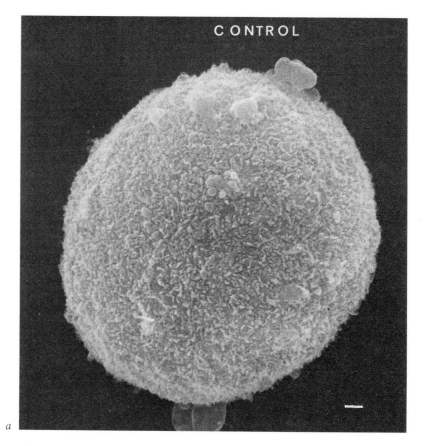

a

Fig. 9. Effect of colchicine on surface morphology of hepatocytes. 1 h after dissociation, hepatocytes were incubated in the absence *(a)* or presence *(b)* of 1 μ*M* colchicine for 4 additional h before fixation and subsequent processing for SEM as described [3–10]. Note the

of cytoplasmic extensions in rat granulosa cell surface is preceded by a dose-dependent increase in intracellular cyclic AMP (which have comparable effects to those of FSH), is potentiated by cyclic nucleotide phosphodiesterase inhibitors, and is mimicked by cyclic AMP derivatives. This phenomenon involves rearrangement of cytoplasmic microfilament bundles [44]. *Centola* [18] showed a correlation between prolactin-stimulated progesterone secretion by rat granulosa cell cultures and the increase in the number of microvilli on the cell surface. *Crisp and Alexander* [25] have elegantly investigated the surface morphology of granulosa cells in various states of culture conditions. In adrenocortical tumor cells, ACTH, which

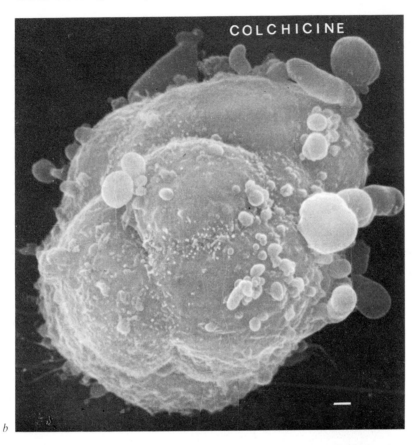

COLCHICINE

b

dramatic effect of colchicine on the surface morphology. Note in figure 9a that the cell surface is covered with microvilli and some blebs. Most of the microvilli have disappeared following colchicine treatment. Bars = 1 μm.

stimulated steroidogenic activity, was found to increase the number of microvilli and filopodia on the cell surface [53]. The modulation of cell surface topography, namely the formation of microvilli, by hormones and secretagogues also occurs in vivo. Practically all cell surfaces of the body which are exposed either to a lumen or to a space are known to display surface structures such as microvilli and blebs [56]. The best example is perhaps that following in vivo injection of LHRH into the rabbit third ventricle, the ependymal cells responded by numerous irregular microvillous eruptions of the apical membrane [17]. Also upon in vivo stimulation of canine thyroid follicular cells by thyrotropin and 5-HT, the

a

Fig. 10. SEM of porcine granulosa cells cultured essentially as described for the rat cells [25] from immature *(a)* and mature follicles *(b)*. With maturation of the follicle by TSH and LH *(b)*, cell surface features undergo differentiation (i. e. increase in blebs, microvilli, some filopodia and beaded filaments). These morphological changes correlate well with increases

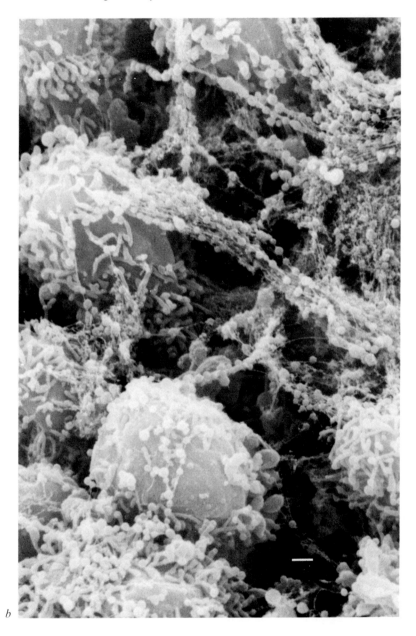

b

in cell surface gonadotropins [25]. Bar = 10 μm. Courtesy of Dr. *Thomas M. Crisp,*
Georgetown University, Washington, DC. See references by *Crisp and Alexander* [25] and
Centola [18] for additional SEM description on granulosa cells.

apical surface of these cells responded by the formation of pseudopods and other surface structures [57]. Another in vivo example is the response of human, rabbit and mouse uterine lining epithelium to hormonal stimulation [11, 33]. *Ferenczy* [33] found that estradiol stimulates the formation of a great number of cilia and microvilli. In the absence of estrogen or in hormonally unstimulated postmenopausal endometrial surface, microvilli and cilia as well as apocrine secretion by these cells is decreased or absent. The increase in the density and length of microvilli in the pituitary and in other cell types as a response to stimulatory agents, in particular estrogen, might be of great biological importance. Recently, *Vic* et al. [68] observed that estradiol (but not testosterone) specifically and progressively increased the number and the length of microvilli at the cell surface of NCF_7 human breast cancer cells in culture. In absence of estradiol, these cancer cells showed no evidence of secretory activity as evidenced by the rare presence of Golgi complexes, RER or secretory granules. Estradiol progressively transformed these breast cancer cells into secretory cells containing rough endoplasmic reticulum, Golgi complexes and secretory granules. Microvilli were also shown to play a role in the response of the toad bladder granular cells to vasopressin [45]. Nerve growth factor (NGF) was found to have profound effects on the morphology of cultured sympathetic neurons [24] as well as PC12 clonal NGF-responsive pheochromocytoma cells [23]. In these studies, *Connolly* et al. [23, 24] showed the formation, within minutes, of dorsal ruffles as a response to NGF, which were followed 10 h later by the appearance of microvilli. Epidermal growth factor (EGF) was also found to induce changes in the cell surface and membrane properties of other cell types [13, 34, 36] as well.

Conclusions

The surface where the cell faces its immediate environment is a specialized part which includes the plasma membrane and adjacent cytoplasm. During the last few years there has been a considerable interest in studying the cell surface during alterations in cellular activities such as secretion and transformation. Scanning electron microscopy, a relatively new technique, has been extremely useful in visualizing alterations in the cell surface. The present article has stressed the notion of the morphodynamic aspects of the cell surface and showed that hormones which affect protein and steroid secretion by target cells induce dramatic changes in the cell surface morphology. These changes involve alterations in structures such as microvilli, blebs, filopodia, membrane folds and pits. These changes presumably indicate a restructuring of the cell surface: (1) to accommodate new cell-surface components such as secretory granule membranes that fuse with the plasma membrane during exocytosis; (2) to increase the cell surface without significantly increasing cell volume during transport of ions, endocytosis and pinocytosis, as well as in order to expose more membrane receptors, and (3) some cell types such as hepatocytes were shown to secrete extracellular biomatrix in response to hormonal stimulation: this biomatrix is important for cell adhesion and cell survival. More investigation is needed to elucidate the specific role of these cell surface phenomena that accompany or precede the secretory process.

References

1   Antakly, T.: Hormonally induced changes in cultured pituitary cell (including gonadotrophs); in Hafez, Kenemans, Human reproduction in three dimensions, p.69 (University of Nijmegen, Nijmegen 1981).
2   Antakly, T.; Feigelson, P.: Light and electron microscopic localization of α2u globulin in cultured hepatocytes. Proc. 6th Int. Congr. Histochem., Cytochem. Brighton 1980.
3   Antakly, T.; Pelletier, G.: An approach to the study of cultured secretory cells in various states of activity, integrating data from SEM, TEM, immunocytochemistry, autoradiography, and radioimmunoassay; in Scanning electron microscopy/1980, vol.II, pp. 213–222 (SEM Inc., AMF O'Hare, Chicago 1980).
4   Antakly, T.; Pelletier, G.; Lagace, L.; Labrie, F.: Cell surface changes of aging pituitary cells in primary culture. J.Cell Biol. 87: 296a (1980).
5   Antakly, T.; Pelletier, G.; Zeytinoglu, F.: Morphological studies on the prolactin cell stimulatory and inhibitory secretory response. 61st Ann. Meet. Endocr. Soc., Los Angeles 1979, p.288.
6   Antakly, T.; Pelletier, G.; Zeytinoglu, F.; Labrie, F.: Scanning electron microscopy of rat pituitary cells in monolayer culture; in Scanning electron microscopy/1979, vol.III, pp. 387–396 (SEM Inc., AMF O'Hare, Chicago 1979).
7   Antakly, T.; Pelletier, G.; Zeytinoglu, F.; Labrie, F.: Effect of colchicine on the morphology and prolactin secretion on anterior pituitary cells in primary culture. Am. J. Anat. 156: 353 (1979).
8   Antakly, T.; Pelletier, G.; Zeytinoglu, F.; Labrie, F.: Changes of cell morphology and prolactin secretion induced by 2-Br-α-ergocryptine, estradiol and TRH in anterior pituitary cells in culture. J.Cell Biol. 86: 377 (1980).
9   Antakly, T.; Zeytinoglu, F.; Pelletier, G.; Labrie, F.: Surface morphology and secretory activity of rat anterior pituitary cells in primary culture. 9th Int. Congr. Electron Microscopy, 1978, vol.II, p.584.

10 Antakly, T.; Zeytinoglu, F.; Pelletier, G.; Labrie, F.: Autohistoradiographic localization of TRH (thyrotropin-releasing hormone) binding sites in cultured rat pituitary cells. Expl Cell Res. *122:* 394 (1979).

11 Barberini, F.; Sartori, S.; Motta, P.: Changes in the surface morphology of the rabbit endometrium related to the estrous and progestational stages of the reproductive cycle. Cell Tiss. Res. *190:* 207 (1978).

12 Bélanger, A.; Labrie, F.; Borgeat, P.; Savary, M.; Coté, J.; Drouin, J.; Schally, A. V.; Coy, D. H.; Immer, H.; Sestanj, K.; Nelson, V.; Gotz, M.: Inhibition of growth hormone and thyrotropin-release by growth hormone-release-inhibiting hormone. Mol. cell. Endocrinol. *1:* 329 (1974).

13 Berliner, J. A.; Aharanov, A.; Pruss, R.: Cell surface changes associated with EGF and insulin-induced cell adhesion. Expl Cell Res. *133:* 227 (1981).

14 Bissel, D. M.; Hammeker, L.; Meyer, U. A.: Parenchymal cells from adult rat liver in non-proliferating monolayer culture. J. Cell Biol. *59:* 722 (1973).

15 Blackwell, R. E.; Guillemin, R.: Hypothalamic control of adenohypophyseal secretions. A. Rev. Physiol. *35:* 357 (1973).

16 Bonney, R. J.; Becker, J. E.; Walker, P. R.: Primary monolayer cultures of adult rat liver parenchymal cells suitable for study of the regulation of enzyme synthesis. In Vitro *9:* 399 (1974).

17 Bruni, J. E.; Montemurro, D. G.; Clattenburg, R. E.: Morphology of the ependymal lining of the rabbit third ventricle following intraventricular administration of synthetic luteinizing hormone-releasing hormone (LHRH): a scanning electron microscopic investigation. Am. J. Anat. *150:* 411 (1977).

18 Centola, G. M.: Correlation of cell surface features with progesterone secretion in rat granulosa cell cultures. Biol. Reprod. *20:* 1067 (1979).

19 Chang, S. C. C.; Anderson, W.; Lewis, J. C.; Ryan, R. J.; Kang, Y. H.: The porcine ovarian follicle. II. Electron microscopic study of surface features of granulosa cells at different stages of development. Biol. Reprod. *16:* 349 (1977).

20 Chang, S. C. C.; Ryan, K. J.: A time study of the effect of PMSG on surface characteristics of the granulosa cells in rat ovaries. Mayo Clinic Proc. *51:* 621 (1976).

21 Chen, C. L.; Feigelson, P.: Glucocorticoid induction of α2u globulin protein synthesis and its mRNA in rat hepatocytes in vitro. J. biol. Chem. *253:* 7880 (1978).

22 Chen, C. L.; Meites, J.: Effects of estrogen and progesterone on serum and pituitary prolactin levels in ovariectomized rats. Endocrinology *86:* 503 (1970).

23 Connolly, J. L.; Green, J. L.; Viscarello, R. R.; Riley, W. D.: Rapid, sequential changes in surface morphology of PC12 pheochromocytoma cells in response to nerve growth factor. J. Cell Biol. *82:* 820 (1980).

24 Connolly, J. L.; Green, S. A.; Green, L. A.: Pit formation and rapid changes in surface morphology of sympathetic neurons in response to nerve growth factor. J. Cell Biol. *90:* 176 (1981).

25 Crisp, T. M.; Alexander, J. S.: Optimal conditions for the study of surface topography of granulosa cell cultures. 11th Int. Congr. Anat. Advances in the Morphology of Cells and Tissues, p. 49 (Liss, New York 1981).

26 Del Pozo, E.; Brun Del Re, R.; Varga, L.; Friesen, H.: The inhibition of prolactin secretion in man by CB-154 (2-Br-α-ergocryptine). J. clin. Endocr. Metab. *35:* 768 (1972).

27 Deschênes, J.; Lafleur, L.; Marceau, N.: Sedimentation velocity distribution of

cells from adult and fetal rat liver and ascites hepatoma. Expl Cell Res. *103:* 183 (1976).

28 Deschênes, J.; Valet, J. P.; Marceau, N.: Hepatocytes from newborn and weanling rats in monolayer culture: isolation by perfusion, fibronectin-mediated adhesion, spreading and functional activities. In Vitro *16:* 722 (1980).

29 Deschênes, J.; Valet, J. P.; Marceau, N.: The relationship between cell volume, ploidy and functional activity in differentiating hepatocytes. Cell Biophys. *3:* 321 (1981).

30 Drouin, J.; De Léan, A.; Rainville, D.; Lachance, R.; Labrie, F.: Characteristics of the interaction between TRH and somatostatin for thyrotropin and prolactin release. Endocrinology *98:* 514 (1976).

31 Drouin, J.; Lavoie, M.; Labrie, F.: Effect of gonadal steroids on the LH and FSH response to 8-Br-cyclic AMP in anterior pituitary cells in culture. Endocrinology *102:* 358 (1978).

32 Eroschenko, V. P.: Surface changes in oviduct, uterus and vaginal cells of neonatal mice after estradiol-17 beta and the insecticide chlordecone treatment: a scanning electron microscopy study. Biol. Reprod. *26:* 707 (1982).

33 Ferenczy, A.: Surface ultrastructural response to the human uterine lining epithelium to hormonal environment. A scanning electron microscopic study. Acta cytol. *21:* 566 (1977).

34 Gonzalez, A.; Ganido, J.; Vial, J. D.: Epidermal growth factor inhibits cytoskeleton-related changes in the surface of parietal cells. J. Cell Biol. *88:* 108 (1981).

35 Guguen, C.; Guillouzo, A.; Boisnard, M.; Le Cam, A.; Bourel, M.: Etude ultrastructurale de monocouches d'hépatocytes de rat adulte cultivés en présence d'hémisuccinate d'hydrocortisone. Biol. Gastroentérol. *8:* 223 (1975).

36 Heine, U. I.; Keski-Oja, J.; Wetzel, B.: Rapid membrane changes in mouse epithelial cells after exposure to epidermal growth factor. J. Ultrastruct. Res. *77:* 335 (1981).

37 Hynes, R. O.: Cell surface proteins and malignant transformation. Biochim. biophys. Acta *258:* 73 (1976).

38 Hynes, R. O.; Wike, J. A.; Bye, J. M.: Are proteins involved in altering cell surface protein during viral transformation?; in Reich, Shaw, Rifkin, Proteases and biological control, p. 931 (Cold Spring Harbor Laboratory, Cold Spring Harbor 1980).

39 Knutton, S.; Summer, M. C.; Pasternak, C. A.: Role of microvilli in surface changes of synchronized P8154 mastocytoma cells. J. Cell Biol. *66:* 568 (1975).

40 Labrie, F.; Gauthier, M.; Pelletier, G.; Borgeat, P.; Lemay, A.; Gouge, J. J.: Role of microtubules in basal and stimulated release of growth hormone and prolactin in the rat adenohypophysis in vitro. Endocrinology *93:* 903 (1973).

41 Labrie, F.; Lagace, L.; Beaulieu, M.; Ferland, L.; De Léan, A.; Drouin, J.; Borgeat, P.; Kelly, P. A.; Cusan, L.; Dupont, A.; Lemay, A.; Antakly, T.; Pelletier, G.; Barden, N.: Mechanisms of action of hypothalamic and peripheral hormones in anterior pituitary gland; in Li, Hormonal proteins and peptides, vol. VIII, p. 205 (Academic Press, New York 1979).

42 Lagace, L.; Labrie, F.; Antakly, T.; Pelletier, G.: Changes of sensitivity of anterior pituitary cells to estradiol and hypothalamic hormones during long-term primary culture. Am. J. Physiol. *240:* E602 (1981).

43 Laishes, B. A.; Williams, G. M.: Conditions affecting primary cell cultures of functional adult rat hepatocytes. I. The effect of insulin. In Vitro *12:* 521 (1976).

44   Lawrence, T.S.; Ginsberg, R.D.; Gilula, N.B.; Beers, W.H.: Hormonally induced cell shape change in cultured rat ovarian granulosa cells. J.Cell Biol. *80:* 21 (1979).

45   Le Furgey, A.; Tisher, C.C.: Time course of vasopressin-induced formation of microvilli in granular cells of toad urinary bladder. J.Membrane Biol. *61:* 13 (1981).

46   Malaisse, W.J.; Leclerq-Meyer, V.; Van Obberghen, E.; Sommers, G.; Devis, G.; Ravazzola, M.; Malaisse-Lagae, F.; Orci, L.: The role of microtubular-microfilamentous system in insulin and glucagon release by the endocrine pancreas; in Borgers, Brafander, Microtubule and microtubule inhibitors, p.143 (North-Holland, Amsterdam 1973).

47   Marceau, N.; Goyette, R.; Deschênes, J.: Morphological differences between epithelial and fibroblast cells in rat liver cultures and the roles of cell surface fibronectin and cytoskeletal element organization in cell shape. Ann. N.Y. Acad. Sci. *349:* 138 (1980).

48   Marceau, N.; Goyette, R.; Guidoin, R.; Antakly, T.: Hormonally induced formation of extracellular biomatrix in cultured normal and neoplastic liver cells. Effect of dexamethasone; in Scanning Electron Microscopy/1982 (in press).

49   Marceau, N.; Goyette, R.; Pelletier, G.; Antakly, T.: Hormonally induced changes in the cytoskeleton organization of adult and newborn rat hepatocytes cultured on fibronectin precoated substratum. Effect of dexamethasone and insulin. Cell. mol. Biol. incl. Cyto-Enzymol. *29* (in press).

50   Marceau, N.; Goyette, R.; Valet, J.P.: The effect of dexamethasone on formation of a fibronectin extracellular matrix by rat hepatocytes. Expl Cell Res. *125:* 497 (1979).

51   Marceau, N.; Noel, M.; Deschênes, J.: Growth and functional activities of neonatal and adult rat hepatocytes cultured on fibronectin-coated substratum in serum-free medium. In Vitro *18:* 1 (1982).

52   Marceau, N.; Robert, A.; Mailhot, D.: The major surface protein of epithelial cells from newborn and adult rat livers in primary cultures. Biochem. biophys. Res. Commun. *75:* 1092 (1977).

53   Mattson, P.; Kowal, J.: Effects of cytochalasin B on unstimulated and adrenocorticotropin-stimulated adrenocortical tumor cells in vitro. Endocrinology *111:* 1632 (1982).

54   Michalopoulos, G.; Pitot, H.C.: Primary culture of parenchymal liver cells on collagen membranes. Expl Cell Res. *94:* 70 (1975).

55   Mosher, D.F.; Saksela, O.; Keski-Oja, J.: Distribution of a major surface-associated glycoprotein, fibronectin, in cultures of adherent cells. J. supramol. Struct. *6:* 551 (1977).

56   Motta, P.; Andrews, P.; Porter, K.R.: Microanatomy of cell and tissue surfaces (Lea & Febiger, Philadelphia 1977).

57   Nunez, E.A.; Gershon, M.D.: Formation of apical pseudopods by canine thyroid follicular cells: induction by thyrotropin and 5-hydroxytryptamine: antagonism by reserpine. Anat. Rec. *192:* 215 (1978).

58   Pasteels, J.L.; Danguy, A.; Frerotte, M.; Ectors, F.: Inhibition de la sécrétion de prolactine par l'ergocornine et la 2-Br-α-ergocryptine: action directe sur l'hypophyse en culture. Annls Endocr. *32:* 188 (1971).

59   Pelletier, G.; Antakly, T.; Labrie, F.: The prolactin cells: structure and function; in Mena, Valverdre, Frontiers and perspectives of prolactin secretion: a multidisciplinary approach, Mexico 1982 (Academic Press, in press 1983).

60   Pelletier, G.; Lemay, A.; Beraud, G.; Labrie, F.: Ultrastructural changes accompanying

the stimulatory effect of monobutyryl cAMP on the release of growth hormone. Endocrinology 91: 1355 (1972).

61 Penasse, W.; Bernaert, D.; Mosselmans, R.: Scanning electron microscopy of adult rat
hepatocytes in situ, after isolation in pure fractions by elutriation and after culture. Biol.
cell. 34: 175 (1979).

62 Ravault, J.P.; Courot, M.; Garnier, D.; Pelletier, J.; Terqui, M.: Effect of 2-bromo-
α-ergocryptine (CB-154) on plasma prolactine, LH and testosterone levels, accessory
reproductive glands and spermatogenesis in lambs during puberty. Biol. Reprod. 17: 192
(1977).

63 Seglen, P.O.: Preparation of rat liver cells. II. Effects of ions and chelators on tissue
dispersion. Expl Cell Res. 76: 25 (1973).

64 Steger, R.W.; Peluso, J.J.; Isamail, M.; Hafez, E.S.E.; Hallen, L.J.: Surface ultrastructure of granulosa cells of human and rat ovarian follicles; in Johari, Scanning
electron microscopy, vol. II, p. 349 (IIT Research Institute, Chicago 1977).

65 Takaoka, T.; Yasumoto, S.; Katsuta, H.: A simple method for the cultivation of rat
liver cells. J. exp. Med. 45: 317 (1975).

66 Thom, M.H.; Davies, K.J.; Senkers, R.J.; Allen, J.M.; Studd, J.W.: Scanning electron
microscopy of the endometrial cell surface in postmenopausal women receiving
estrogen therapy. Br. J. Obstet. Gynaec. 88: 904 (1981).

67 Vale, W.; Grant, G.; Amoss, M.; Blackwell, R.; Guillemin, R.: Culture of enzymatically dispersed anterior pituitary cells: functional validation of a method. Endocrinology 91: 562 (1972).

68 Vic, P.; Vignon, F.; Derocq, D.; Rochefort, H.: Effect of estradiol on the ultrastructure
of the MCF7 human breast cancer cells in culture. Cancer Res. 42: 667 (1982).

69 Wilkinson, R.F.; Anderson, E.; Aalberg, J.: Cytological observations of dissociated rat
corpus luteum. J. Ultrastruct. Res. 57: 168 (1976).

70 Williams, G.M.; Bermudez, E.; San, R.H.C.: Rat hepatocyte primary cultures. IV.
Maintenance in defined medium and the role of production of plasminogen activator
and other proteases. In Vitro 14: 824 (1978).

71 Yen, S.S.C.; Ehra, Y.; Siler, T.M.: Augmentation of prolactin secretion by estrogen in
hypogonadal women. J. clin. Invest. 53: 652 (1974).

T. Antakly, MD, Laboratory of Pathobiology, Clinical Research Institute of Montreal,
110 Pine Avenue West, Montreal, Quebec H2W 1R7 (Canada)

Cell Biology of the Secretory Process, pp. 196–213 (Karger, Basel 1984)

The Secretory Process in the Anterior Hypophysis

Georges Pelletier

MRC Group in Molecular Endocrinology, Le Centre Hospitalier de l'Université Laval, Que., Canada

Introduction

The anterior pituitary gland, also named 'pars distalis of the pituitary gland', is known to produce six different hormones with a well-documented activity. These are adrenocorticotropin (ACTH), thyrotropin (TSH), prolactin (PRL), growth hormone (GH), luteinizing hormone (LH) and follicle-stimulating hormone (FSH). With the exception of FSH and LH which are produced by the same cells, all the other hormones are produced by different cell types.

This heterogeneity in the cell population has made very difficult the study of the secretory process in the anterior pituitary with biochemical techniques. For example, cell fractionation which is currently used to analyze secretory processes in the exocrine glands has not been a fruitful approach to study anterior pituitary secretion. In fact, much of what we know about the cell biology of the anterior pituitary has been learned from application of morphological techniques, i.e. light microscopy, electron microscopy, autoradiography, enzyme histochemistry and immunocytochemistry, which can be applied to intact tissues. Moreover, during the last decade, the discovery of three hypothalamic hormones (capable of stimulating or inhibiting specifically the secretion of pituitary hormones) has greatly contributed to increase our knowledge about the mechanisms involved in the regulation of anterior pituitary secretion [for a review, see 13].

In this chapter, we will summarize what is known about the morphology of the anterior pituitary and the intracellular processing of secretory products. Since the great majority of the studies on the processing of anterior pituitary hormones have been performed in the rat, we will only discuss the results obtained in this species.

Cell Types in the Rat Anterior Pituitary

During the last years, the extensive use of immunocytochemical techniques at both light and electron microscope levels [1, 2, 4, 8, 16–18, 29–33] has contributed to confirm and extend previous observations [5, 10, 21] on the morphological characteristics of the secretory cells of the anterior pituitary.

Somatotrophs
The GH-secreting cells are the most abundant cells in the anterior pituitary. They are characterized by the presence of ovoid or round secretory granules of diameter of about 300–400 nm. The rough endoplasmic reticulum is abundant and consists of parallel narrow elongated cisternae containing dense material (fig. 1).

Mammotrophs
The mammotroph or PRL-secreting cell is more abundant in females than in males (especially during lactation). This cell type contains the largest secretory granules of any anterior pituitary cell. The mature granules are often ovoid and elliptical with a diameter ranging from 600 to 900 nm (fig. 2). The Golgi apparatus is generally extensive.

Opiocorticotrophs
The cell type first described as responsible for the secretion of ACTH [16, 21] is now known to also produce β-lipotropic hormone (β-LPH) and cleavage products of β-LPH, especially β-endorphin and γ-endorphin [30, 39]. The opiocorticotrophic cell is characterized by its angular shape and the presence of a single row of secretory granules (≈ 200 nm in diameter) along the plasma membrane.

Thyrotrophs
The TSH-secreting cell has an angular shape similar to that of the opiocorticotroph. Its secretory granules (≈ 140 nm in diameter) are the smallest ones of any cell type and are either distributed throughout the cytoplasm or

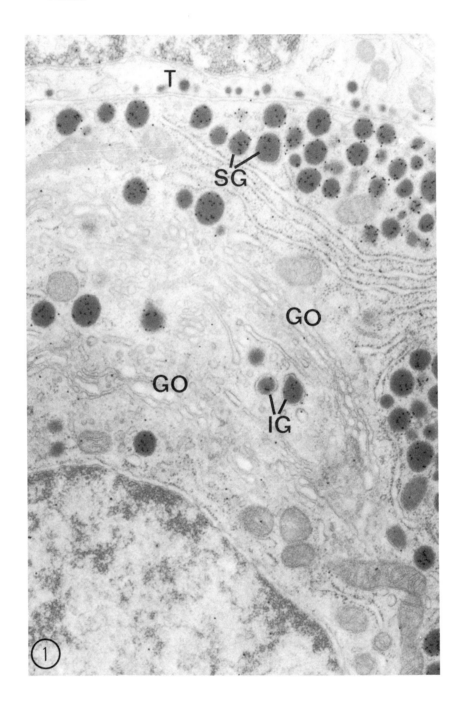

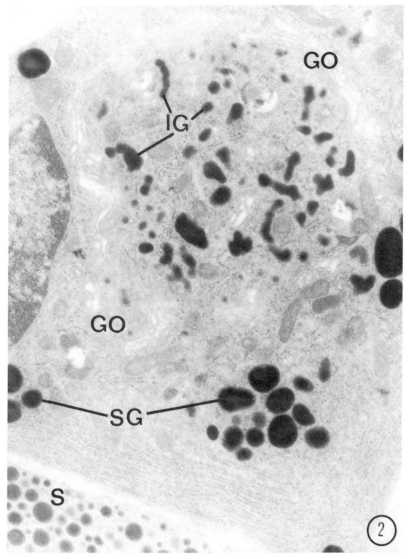

Fig. 2. Mammotroph immunostained for PRL. Staining is observed in the secretory granules (SG). Immature secretory granules (IG) found in the Golgi area (GO) are small and irregular and lightly stained. An adjacent somatotroph (S) is unstained. Protein A-peroxidase techniques. ×6,200.

Fig. 1. Somatotroph immunostained for GH. Accumulation of molecules of the protein-gold complex indicates that GH is mostly concentrated in the secretory granules (SG). Note the presence of immature granules (IG) in the Golgi area (GO). An adjacent thyrotropic cell (T) is unstained. Protein A-gold techniques. ×23,000.

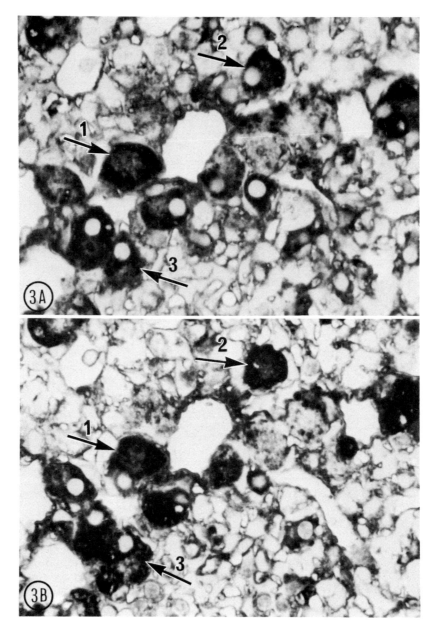

Fig. 3. Consecutive semithin (1μm sections through the rat pituitary immunostained for β-LH (*A*) and β-FSH (*B*). The same cells (arrows) contain both LH and FSH. Peroxidase-anti-peroxidase technique. ×750.

located in the periphery of the cell. The rough endoplasmic reticulum (RER) consists of flattened cisternae which become dilated after thyroidectomy.

Gonadotrophs

On the basis of recent immunocytochemical results, it seems now very likely that both LH and FSH are produced by the same cells [1, 8, 29, 38] (fig. 3). Nevertheless, there exists still some controversy about this question since *Childs and Ellison* [2] have recently found some gonadotrophic cells containing only one of the two gonadotrophins. Two types of gonadotrophs (both containing FSH and LH) can be distinguished: type I containing two types of secretory granules (200 nm and 300–700 nm in diameter) and type II containing one type of granules of a uniform size (200–300 nm in diameter) [2, 10]. In gonadotrophs, the RER generally consists of dilated cisternae.

Biosynthesis of Pituitary Hormones

The available information about the biosynthesis of pituitary hormones is mostly coming from studies on GH, PRL and, more recently, on the ACTH-β-endorphin system. Using incorporation of radioactive amino acids and isolation of GH and PRL on polyacrylamide gel electrophoresis, *Labrie* et al. [11] and *Farquhar* et al. [6] have shown that amino acids are rapidly (within 30 min) incorporated into GH and PRL. With the help of radioautography, it has been shown that this rapid incorporation occurred exclusively in the areas of the RER [6, 11, 34]. These data suggest that like other proteins, PRL and GH are synthesized only by bound ribosomes which are attached to the outer surface of the RER [20]. The presence of specific receptors in these membranes [9] permits the transmembranal growth and transport of polypeptide chains, which lead eventually to the segregation of the finished hormone molecules within the RER cisternae. Following synthesis, these hormones are rapidly transported to the Golgi apparatus, probably via small vesicles.

Following our initial observations [14, 30] that ACTH and β-LPH can be localized not only in the same pituitary cells but also in the same secretory granules, it has been recently shown that both ACTH and β-LPH (and endorphins) were coming from a large glycoprotidic precursor named proopiocortin [15, 35]. In this case, there is first biosynthesis of a large precursor (molecular weight of about 31K) which is subsequently cleaved into active hormones (ACTH, α-MSH and endorphins) and a fragment

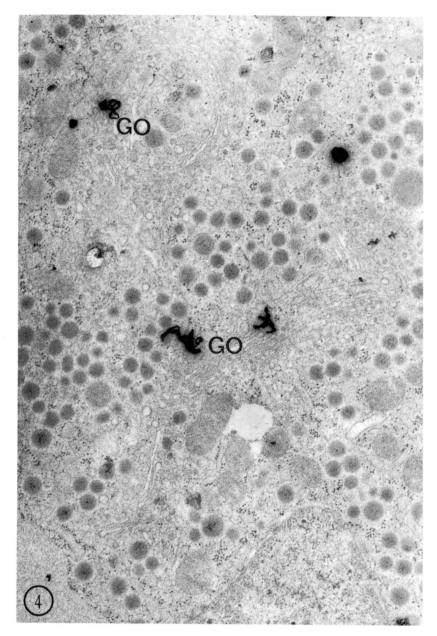

Fig. 4. Radioautograph of a gonadotrophic cell 10 min after intravenous injection of [³H]-fucose. Silver grains are detected only in the Golgi apparatus (GO), indicating that terminal sugars are added to the glycoproteins in the Golgi apparatus. ×15,300.

(16K) of unknown biological significance. The exact intracellular site of cleavage is still unknown. As in β-cells, it is possible that cleavage of the precursor occurs in the Golgi apparatus [37].

The biosynthesis of glycoprotidic hormones (FSH, LH and TSH) has not been extensively studied so far. As for most glycoprotein hormones, it has been demonstrated that the terminal sugars are added to the polypeptidic backbones in the Golgi apparatus [26, 27] (fig. 4).

In both gonadectomy and thyroidectomy cells where the dilated cisternae of the RER contain granules lacking membrane, there is a strong indication that TSH and gonadotrophins can be packaged by the RER. The first evidence has been obtained from incorporation of ^{3}H-labeled fucose which is one of the last sugars to be added to a glycoprotein. In both gonadectomy and thyroidectomy cells, much of the [^{3}H]-fucose-labeling has been detected over the RER cisternae [26, 27], thus suggesting that final stages of TSH and gonadotrophin synthesis can take place in this organelle. This hypothesis has been supported by recent immunocytochemical studies [17] demonstrating staining for the β-chains of TSH and the two gonadotrophins in the dilated cisternae of RER of thyroidectomy and gonadectomy cells, respectively. Staining was usually stronger over intracisternal granules, thus indicating some condensation of immunoreactive material in these structures. The fate of these hormones synthesized and packaged in the RER is still unknown, although it is tempting to speculate that they might be released directly from the cisternae to the extracellular spaces without going through the usual processing of immature and mature granules and exocytosis.

Formation and Processing of Secretory Granules

From observations obtained with electron microscope radioautography [6, 11, 34], it appears that soon after synthesis the newly synthesized hormones are transported to the Golgi apparatus. Nothing is known about this transport, except that it is energy-dependent [11], probably involving fission and fusion of small vesicular elements. Then, the concentration of secretory products occurs in the stacked cisternae of the Golgi apparatus while formation of immature secretory granules takes place in the forming (or maturing) face of this organelle.

The exact way by which the hormones are packaged in secretory granules is still undefined. On the basis of experiments involving solubi-

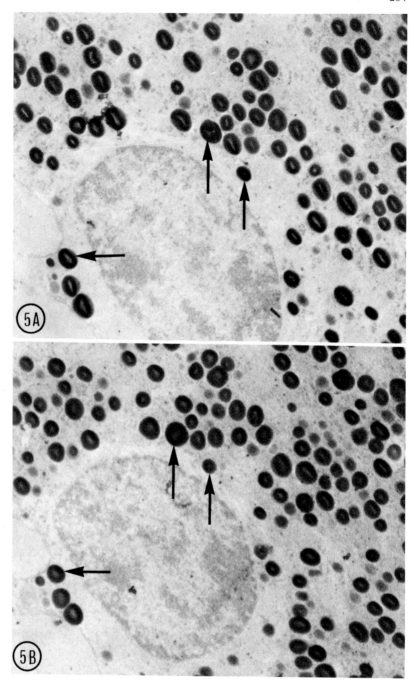

lization of PRL-containing granules, *Zanini* et al. [40] concluded that in secretion granules, PRL was part of a solid supramolecular organization. At the electron microscopic level, the immature secretion granules found in the Golgi apparatus are generally characterized by the presence of wider space between the core and limiting membrane (fig. 1, 2). When immunohistochemical staining of hormones is performed, the immature granules are usually less reactive than the mature ones, indicating that they contain smaller amounts of active hormones. These granules seem to become mature and get their definite form during migration from the Golgi area to the plasma membrane.

In the case of PRL-secreting cells, the formation of immature granules seems to be different from other pituitary cells, since the mature granules come from the aggregation of small immature granules [5] (fig. 2).

As mentioned above, the Golgi apparatus is also the organelle where terminal sugars are added to the glycoprotein backbone. All the secretory cell types of the pituitary gland, including those which do not secrete glycoprotidic hormones, can incorporate labeled sugars in their granules [26, 27]. From kinetic analysis performed after [³H]-fucose injection, it appears that a large proportion of newly synthesized glycoproteins migrates subsequently to the secretory granules. In the case of glycoprotein-secreting cells, the migration of newly synthesized glycoproteins to the granules likely represents the packaging of hormones (fig. 9). In opiocorticotrophs, since a glycoprotidic fragment (16K fragment) of the common precursor for ACTH and β-LPH is present within the secretory granules [7, 32] (fig. 5), the transfer of sugar-labeled proteins to granules is probably related not to the hormones themselves but rather to the precursor. It is also possible that some glycoproteins might serve as a constituent of the matrix and/or the membrane of the granules [3]. Glycoproteins which become incorporated into the plasma membrane can reach directly this organelle without being transported through the secretory granules (fig. 9).

Fig. 5. Serial sections through a opiocorticotroph from a human pituitary gland. *A* Staining for ACTH. *B* Staining from a fragment (16K) of the common precursor for ACTH and β-LPH (*B*) (antibodies supplied by Dr. *Mains* and Dr. *Eipper*). The same secretory granules (arrows) are labeled. ×16,000.

Fate of Secretory Granules

At some distance from the Golgi apparatus, the secretory granules generally lose their features of immaturity (fig. 1, 2). Although the exact time-course of granule maturation is not known, it certainly occurs rapidly. In fact, 30 min after the beginning of incorporation of labeled material into PRL cells, about 25% of the total amount of silver grains (representing radioactivity) are seen over mature secretory granules [11].

In order to release their content outside the cell, the secretory granules must ultimately migrate to the plasma membrane. Nothing certain is known about the factors or organelles susceptible to influence the movement of granules. Although the microtubule-microfilament system seems to be involved in the oriented movement of some organelles, the role of such a system in the migration of anterior pituitary granules has not been clearly defined. Inhibitors of microtubules such as colchicine and vinblastine have been found to decrease the secretion of anterior pituitary hormones [12, 23]. The exact mechanism of action of these inhibitors on pituitary secretion is still obscure.

It is generally admitted that hormonal release occurs as a consequence of granular discharge outside the secretory cell. The fusion of the granule membrane with the plasma membrane, creating a stoma which allows passage of the granule content to the extracellular space, has been named exocytosis (or emiocytosis) (fig. 6). Although it is not known if all the secretory granules will be ultimately involved in exocytosis, it appears that under conditions of intense stimulation, some pituitary cells, especially PRL and GH cells, show extensive exocytosis and become rapidly depleted of most of their granules [22, 24]. In the case of strong stimulation by cyclic AMP (cAMP), the proportion of immature granules due to the increase of biosynthetic activity is greatly enhanced after 1 h of stimulation [24].

In cell types other than PRL and GH cells, even after induction of intense hormonal release, exocytosis is not a commonly observed phenomenon [5, 22, 24, 25], probably due to a more rapid dissolution of the granular content or to a lower frequency of exocytosis per cell. In this respect, it should be recalled that in ultrathin sections, frequent exocytosis cannot be properly identified because of the infrequency with which the point of fusion of granule membrane and plasma membrane coincides with the plane of the section. The use of horseradish peroxidase (HRP) which rapidly impregnates granules in contact with the extracellular space

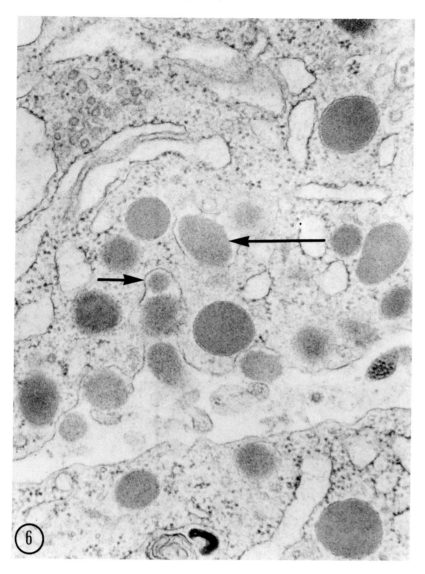

Fig. 6. Portion of PRL cell in a hemipituitary incubated 'in vitro'. 5 min after the addition of cAMP, extensive exocytosis of granules can be observed. Note the presence of three granules (short arrow) extruded in the same channel connecting with the extracellular space. The long arrow points out two granules which are probably in the process of exocytosis, but since the plane of section is not adequate, exocytosis cannot be undoubtedly identified. ×44,000.

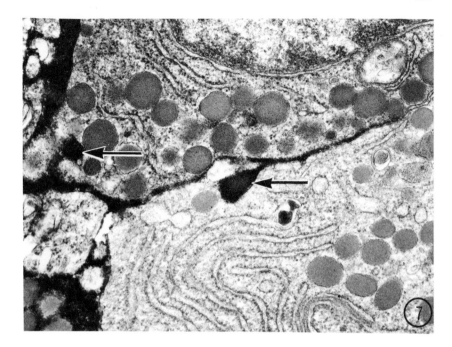

Fig. 7. Portions of a GH and a PRL cell after 5 min of incubation with HRP and cAMP. Impregnation by HRP of two granules (arrows) has permitted the identification of exocytosis. Note the invagination of the plasma membrane which is forming a channel. ×27,000.

has allowed the observation of a much larger number of exocytotic processes [25]. Again, during stimulation of all the cell types by cAMP, exocytosis detected by this procedure was much more frequent in PRL and GH cells [21, 36]. This procedure has also allowed to observe earlier processes of exocytosis such as the invagination of the plasma membrane prior to the expulsion of the granule content (fig. 7).

In a reverse situation, when there is acute suppression of hormonal release, an accumulation of secretory granules in the cytoplasm occurs secondarily. *Smith and Farquhar* [36] have demonstrated that in PRL cells, lysosomes can dispose of undischarged granules. This action of lysosomes could explain the rapid lowering of PRL pituitary content after cessation of lactation.

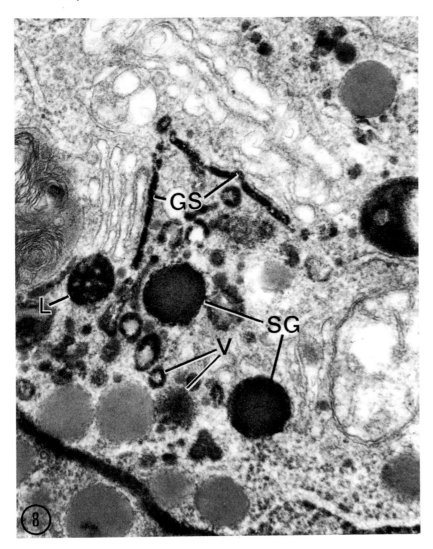

Fig. 8. Portion of a GH cell 4 h after the beginning of incubation with HRP and cAMP. A number of vesicles (V) containing HRP have reached the Golgi area. Some Golgi saccules (GS), usually the innermost ones, as well as secretory granules (SG) and lysosomes (L) are also HRP-positive. ×50,000.

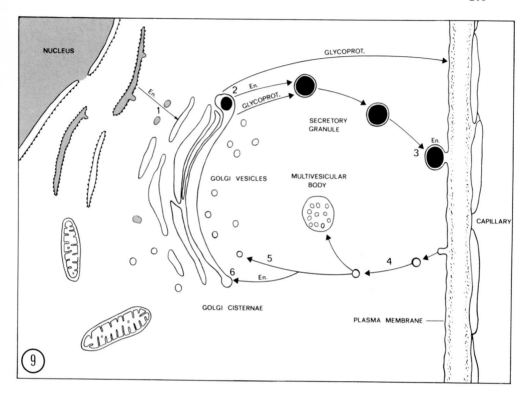

Fig. 9. Diagram representing the formation of secretory granules and circulation of membranes in an anterior pituitary cell. Newly synthesized proteins migrate to the Golgi area (1) probably via small vesicles to be packaged into the secretory granules (2). The secretory granules then migrate to the plasma where exocytosis takes place (3). After exocytosis, fragments of intact membranes migrate back to the Golgi area (4) where they either increase the number of Golgi vesicles (5) or are destroyed by lysosomes (multivesicular bodies). Some of these vesicles are probably fusing with innermost Golgi cisternae (6), permitting reutilization of membranes. Glycoproteins, the synthesis of which is completed in the Golgi area, can either be incorporated in the secretory granules or directly migrate to the plasma membrane. Steps 1, 2, 3 and 6 are energy-dependent (En.).

Retrieval of the Granule Membrane

Since, during stimulation of secretion, there is a marked increase in the number of vesicles located close to the plasma membrane and of vesicles in the Golgi area [24] and since the increase of vesicles is not prevented by inhibition of protein synthesis [25], we hypothesized that these vesicles could originate from retrieval of granule membranes. To test this hypothesis, we have used HRP as a tracer to investigate the fate and circulation of membranes during secretion and define possible sites of relocation of recirculated membranes [25]. It was observed that in soma-totrophs and mammotrophs stimulated by cAMP, there was a rapid incorporation of HRP in peripheral vesicles. Within 30 min, the labeled vesicles had already migrated into the Golgi area. Lysosomes were also seen to contain HRP. After 1 h of incubation, HRP could also be found in the stacked Golgi cisternae, especially those found along the concave surface of the Golgi apparatus (fig. 8). At the longest time intervals studied (4 h), some somatotrophic and mammotrophic granules were HRP-posi-tive. These findings, which have been fully confirmed by *Farquhar* et al. [6], strongly suggest that the granule membrane relocated to the cell sur-face during exocytosis is recaptured and recirculated back to the Golgi where it can be reutilized for the formation of new granules (fig. 9). Very similar findings have been recently found in β-cells of the endocrine pan-creas [19].

References

1   Bugnon, C.; Fellmann, D.; Levys, D.; Bloch, B.: Etude cytoimmunologique des cellules gonadotropes et de cellules thyréotrophes de l'adénohypophyse du rat. C.r. Séanc. Soc. Biol. *4:* 907–912 (1975).
2   Childs, G. V.; Ellison, D. G.: Quantitative immunocytochemical studies of serially sectioned gonadotrophs in the adult male rat. Endocrine Society Meeting, Washington, Abstr. 625 (1980).
3   Costoff, A.; McShan, W. H.: Isolation and biological properties of secretory granules from rat anterior pituitary glands. J. Cell Biol. *43:* 564–574 (1969).
4   Duello, T. M.; Halmi, N. S. Ultrastructural immunocytochemical localization of growth hormone and prolactin in human pituitaries. J. clin. Endocr. Metab. *49:* 189–196 (1979).
5   Farquhar, M. G.: Processing of secretory products by cells of the anterior pituitary gland. Mem. Soc. Endocr. *19:* 79–122 (1971).
6   Farquhar, M. G.; Skutelsky, E. H.; Hopkins, C. R.: Structure and function of the ante-rior pituitary and dispersed pituitary cells. In *in vitro* studies; in Tixier-Vidal, Farquhar, The anterior pituitary, pp. 83–135 (Academic Press, New York 1975).
7   Guy, J.; Leclerc, R.; Pelletier, G.: Localization of a fragment (16K fragment) of the common precursor for adrenocorticotropin (ACTH) and β-lipotropin (β-LPH) in the rat and human pituitary gland. J. Cell Biol. *86:* 825–830 (1980).
8   Herbert, D. C.: Localization of antisera to LHB and FSH in the rat pituitary glandj. Am. J. Anat. *144:* 379–383 (1975).

9 Krulich, G.; Frerenstein, C. N.; Pereyra, B. N.; Ulrich, B. L.; Sabatini, D. D.: Proteins of rough microsomal membranes related to specific membrane proteins exposed at the binding sites. J. Cell Biol. *77:* 488–496 (1978).

10 Kurosumi, K.: Functional classification of cell types of the anterior pituitary gland accomplished by electron microscopy. Histol. Jap. *29:* 329–362 (1968).

11 Labrie, F.; Pelletier, G.; Lemay, A.; Borgeat, P.; Barden, N.; Dupont, A.; Savary, M.; Côté, J.; Boucher, R.: Control of protein synthesis in anterior pituitary gland; in Diczfalusy, Karolinska Symp. on Research Methods in Reproductive Endocrinology, Geneva, p. 301 (1973).

12 Labrie, F.; Gauthier, M.; Pelletier, G.; Borgeat, P.; Lemay, A.; Gouge, J. J.: Role of microtubules in basal and stimulated release of growth hormone and prolactin in rat adenohypophysis *in vitro.* Endocrinology *93:* 903–914 (1973).

13 Labrie, F.; Lagacé, L.; Beaulieu, M.; Ferland, L.; De Léan, A.; Drouin, J.; Borgeat, P.; Kelly, P. A.; Cusan, L.; Dupont, A.; Lemay, A.; Antakly, T.; Pelletier, G.; Barden, N.: Mechanisms of action of hypothalamic and peripheral hormones in the anterior pituitary gland; in Li, Hormonal proteins and peptides, pp. 206–278 (Academic Press, New York 1979).

14 Leclerc, R.; Pelletier, G.; Chrétien, M.: Immunohistochemical localization of ACTH and β-LPH. Proc. microsc. Soc. Canada *2:* 92–93 (1975).

15 Mains, R. E.; Eipper, B. A.: Common precursor to corticotropins and endorphins. Proc. natn. Acad. Sci. USA *74:* 3014–3018 (1977).

16 Moriarty, G. C.; Halmi, N. S.: Electron microscopic study of the adrenocorticotrophin-producing cell with the use of unlabeled antibody and the soluble peroxidase–antiperoxidase complex. J. Histochem. Cytochem. *20:* 590–603 (1972).

17 Moriarty, G. C.: Immunocytochemistry of the pituitary glycoprotein hormones. J. Histochem. Cytochem. *24:* 846–849 (1976).

18 Nakane, P. K.: Classifications of anterior pituitary cell types with immunoenzyme histochemistry. J. Histochem. Cytochem. *18:* 9–20 (1970).

19 Orci, L.; Penelet, A.; Gorden, P.: Less-understood aspects of the morphology of insulin secretion and binding. Recent Prog. Horm. Res. *34:* 95–121 (1978).

20 Palade, G. E.: Intracellular aspects of the process of protein secretion. Science *189:* 347–355 (1975).

21 Pelletier, G.; Racadot, J.: Identification des cellules hypophysaires sécrétant l'ACTH chez le rat. Z. Zellforsch. mikrosk. Anat. *116:* 228–239 (1971).

22 Pelletier, G.; Peillon, F.; Vila-Porcile, E.: An ultrastructural study of sites of granule extrusion in the anterior pituitary of the rat. Z. Zellforsch. mikrosk. Anat. *115:* 501–507 (1971).

23 Pelletier, G.; Bornstein, M. B.: Effect of colchicine on the rat anterior pituitary gland in tissue culture. Expl Cell Biol. *70:* 221–223 (1972).

24 Pelletier, G.; Lemay, A.; Béraud, G.; Labrie, F.: Ultrastructural changes accompanying the stimulatory effect of N^6-monobutyryl adenosine 3′,5′-monophosphate on the release of growth hormone (GH), prolactin (PRL) and adrenocorticotropin hormone (ACTH) in rat anterior pituitary gland *in vitro.* Endocrinology *91:* 1355–1371 (1972).

25 Pelletier, G.: Secretion and uptake of peroxidase by rat adenohypophyseal cells. J. Ultrastruct. Res. *43:* 445–459 (1973).

26 Pelletier, G.; Puviani, R.: Detection of glycoproteins and autoradiographic localization

of [^{3}H] fucose in the thyroidectomy cells of rat anterior pituitary gland. J. Cell Biol. *56:* 600–605 (1973).

27 Pelletier, G.: Autoradiographic studies of synthesis and intracellular migration of glycoproteins in the rat anterior pituitary gland. J. Cell Biol. *62:* 185–197 (1974).

28 Pelletier, G.; Puviani, R.: Permeability of capillaries to different tracers and uptake of horseradish peroxidase by the secretory cells in rat anterior pituitary gland. Z. Zellforsch. *147:* 361–372 (1974).

29 Pelletier, G.; Leclerc, R.; Labrie, F.: Identification of gonadotropic cells in the human pituitary by immunoperoxidase technique. J. mol. cell. Endocr. *6:* 123–128 (1976).

30 Pelletier, G.; Leclerc, R.; Labrie, F.; Côté, J.; Chrétien, M.; Lis, M.: Immunohistochemical localization of β-lipotropic hormone in the pituitary gland. Endocrinology *100:* 770–776 (1977).

31 Pelletier, G.; Robert, F.; Hardy, J.: Identification of human anterior pituitary cells by immunoelectron microscopy. J. clin. Endocr. Metab. *46:* 534–542 (1978).

32 Pelletier, G.; Puviani, R.; Leclerc, R.: Immunoelectron microscopic localization of ACTH, β-LPH and a fragment (16K) of their common precursor in rat brain. Proc. Endocrine Soc. Meet., Abstr. No. 227 (1979).

33 Pelletier, G.; Puviani, R.; Bosler, O.; Descarries, L.: Immunocytochemical detection of peptides in osmicated and plastic-embedded tissue. An electron microscopic study. J. Histochem. Cytochem. *29:* 759–764 (1981).

34 Racadot, J.; Olivier, L.; Porcile, E.; Droz, B.: Appareil de Golgi et origine des grains de sécrétion dans les cellules adénohypophysaires chez le rat. Etude radioautoradiographique en microscopie électronique après injection de leucine tri-tiée. C.r. hebd. Séanc. Acad. Sci., Paris *261:* 2972–2974 (1965).

35 Roberts, J.L.; Herbert, E.: Characterization of a common precursor to corticotropin and β-lipotropin: identification of β-lipotropin peptides and their arrangements relative to corticotropin in the precursor synthesized in a cell free system. Proc. natn. Acad. Sci. USA *74:* 5300–5304 (1977).

36 Smith, R.E.; Farquhar, M.G.: Lysosome function in the regulation of secretory process in cells of the anterior pituitary gland. J. Cell Biol. *31:* 319–347 (1966).

37 Steiner, D.F.; Kemmler, W.; Tazer, H.S.; Rubenstein, A.H.: Molecular events taking place during intracellular transport of exportable proteins. The conversion of peptide hormone precursors; in Ceccarelli, Clementi, Meldolesi, Advances in cytopharmacology, vol. 2, pp. 195–205 (Raven Press, New York 1974).

38 Tougard, C.; Picart, R.; Tixier-Vidal, A.: Immunocytochemical localization of glycoprotein hormones in the rat anterior pituitary. J. Histochem. Cytochem. *28:* 101–114 (1980).

39 Vaudry, H.; Pelletier, G.; Guy, J.; Leclerc, R.; Jegou, S.: Immunohistochemical localization of γ-endorphin in the rat pituitary gland and hypothalamus. Endocrinology *106:* 1515–1520 (1980).

40 Zanini, A.; Giannathasio, G.; Meldolesi, J.: Intracellular events in prolactin secretion; in Jutisz, McKerns, Synthesis and release of adenohypophyseal hormones, pp. 105–124 (Plenum Press, New York 1980).

Georges Pelletier, MRC Group in Molecular Endocrinology, Le Centre Hospitalier de l'Université Laval, Laval, Quebec G1V 4G2 (Canada)

Cell Biology of the Secretory Process, pp. 214–246 (Karger, Basel 1984)

Processing of Peptide Hormone and Neuropeptide Precursors[1]

M. Chrétien, G. Boileau, C. Lazure, N. G. Seidah

Protein and Pituitary Hormone Laboratory, Clinical Research Institute of Montreal, Montreal, Que., Canada

General Introduction

The biosynthesis of ACTH, β-endorphin, and their related peptides included in pro-opiomelanocortin (POMC), involves the synthesis of a larger precursor as described for most polypeptides and proteins destined for secretion from cells. After subsequent maturation of the precursor, the active substances are released by proteolysis. The mechanisms involved in their secretion are common to all eukaryotic cells, since the packaging of lysosomal enzymes and the insertion of protein in the cell membranes proceed via the same route. The study of the secretory mechanisms in pancreatic exocrine cells leads *Palade* [79] to postulate the existence of six well-defined steps in the process: (1) synthesis of the protein on membrane-bound ribosomes; (2) segregation of the newly synthesized protein in the cisternal space of the rough endoplasmic reticulum (RER); (3) intracellular transport of the protein to the Golgi complex; (4) concentration of the dilute protein solution in condensing vacuoles; (5) intracellular storage in secretion granules, and (6) discharge in the extracellular liquid of the content of the secretory granules.

During the past few years, much attention has been focused on the study of the molecular interactions involved in the biosynthesis and segregation (steps 1 and 2) of secretory proteins. The development of techniques to isolate specific messenger ribonucleic acids (mRNA) and the availabil-

[1] Work supported by the Medical Research Council of Canada (PG-2), the National Cancer Institute of Canada, and the National Institutes of Health (#16315-03).

ity of cell-free protein synthesis systems have stimulated the study of the primary translation products and their processing. Analysis of the primary translation product of mRNAs coding for secretory proteins revealed the presence of an hydrophobic peptide located at the N-terminal end of the proteins. When purified microsomal membranes were added to the cell-free protein synthesis systems, the peptide, called signal peptide, was essential for the binding of the ribosomes to the membranes and segregation of the nascent polypeptide chain. The results of the studies suggest the following sequence of events for the biosynthesis and segregation of secretory proteins [10]: (1) initiation of protein synthesis on free ribosomes; (2) attachment of the ribosomes to the membrane of the endoplasmic reticulum, mediated by the presence of the signal peptide, and (3) vectorial discharge of the peptide chain in the cisternal space of the RER with the concomitant enzymatic cleavage of the signal peptide and glycosylation [7, 50, 91, 114]. The maturation of the proprotein into active molecules is completed during the transport through the Golgi complex and in the secretory granules.

We will present in the next pages the molecular interactions involved in the secretory process, and we will summarize the present knowledge about the maturation of human POMC into β-endorphin, ACTH, MSH [22, 23] and a new N-terminal fragment which is a major secretory product [99, 106]. We will then apply this type of maturation to other models of pro-hormones and pro-peptides in order to demonstrate what secretory products are or shall be expected. The conclusion will be that this type of maturation was foreseen when we presented in 1967 the hypothesis of pro-hormone using β-lipotropin as model [16].

Biosynthesis and Segregation of the Prohormones

The biosynthesis of a secretory protein, as for any protein of the cell, starts in the cytosol by the formation of a complex between the ribosome, the mRNA, the initiator transmitter RNA (tRNA), and a group of protein factors [116]. The synthesis of the protein proceeds until the chain is about 60 or 70 amino acids long. Then the signal peptide that has emerged from the big subunit of the ribosome promotes the formation of a junction between the 60S subunit of the ribosome and the membrane of the endoplasmic reticulum. The exact mechanism by which the signal peptide

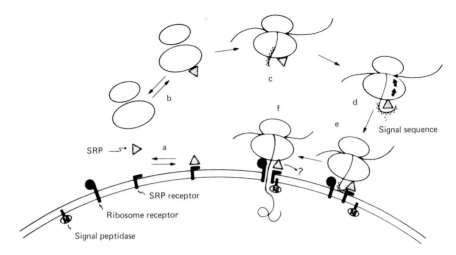

Fig. 1. Schemes of the proposed models for the synthesis and segregation of secretory proteins [reproduced with the permission of *Blobel*, ref.114]. A signal recognition protein (SRP) is in equilibrium between a membrane-bound and a free soluble form of SRP (a). The free form of SRP is in equilibrium with monomeric ribosomes (b). The nascent secretory protein's signal sequence causes an enhancement in the apparent affinity of SRP for polysomes (c) which arrests the synthesis of the nascent chain (d) unless appropriate membranes containing an SRP receptor are added. This assembled complex constitutes a functional translocation machinery which permits the translocation of the secretory protein.

initiates the binding of the ribosome to the membrane and the translocation of the nascent polypeptide chain through the membrane is not known. However, *Blobel and Dobberstein* [10] first proposed that the interaction of the signal peptide with the membranes recruits some membrane protein receptors and causes their association to form a tunnel across the membrane. The tunnel is further stabilized by the binding of the big ribosomal subunit to the membrane. The newly formed peptide chain is then inserted in the tunnel and driven across the membrane by the sole force of the translational machinery. Once the translocation of the nascent chain is initiated, the signal peptide is cleaved by an endopeptidase called signal peptidase. More recently, *Walter and Blobel* [114] described one additional intermediary (fig. 1) which is a signal recognition protein (SRP) which is attached to the nascent polysomes and has high affinity for the signal peptide. The SRP-signal peptide complex is then attached to a putative

SRP-receptor on the membrane of the endoplasmic reticulum, the translocation occurs.

The examination of the primary structure of the signal peptides of different pre-pro-proteins listed in figure 2 reveals some common characteristics: the signal peptides are 15–30 amino acids in length and are characterized by a region rich in hydrophobic amino acids (boxed in figure 2) followed by a region containing small neutral amino acids. Based on these physical properties of all signal peptides, *Steiner* et al. [109] proposed an alternative hypothesis called the β-transportation model. This model proposes that the recognition of the ribosome-binding site on the membrane of the endoplasmic reticulum, the proper orientation of the nascent chain across the membrane, and the motive force to translocate it result from the interaction of the highly hydrophobic central region of the signal peptide with protein components of the endoplasmic reticulum membrane. This interaction would also stabilize and orient the prepeptide with respect to the signal peptidase. At this moment, the data available are insufficient to prove or disapprove either hypothesis. However, the role of the signal peptide is clear: to promote the ribosome binding to the endoplasmic reticulum membrane and to initiate the nascent polypeptide chain translocation across the membrane via the SRP receptor and the complex SRP-signal peptide.

The interaction between the membrane of the endoplasmic reticulum and the signal peptide is not the only factor involved in the binding of the ribosome to the membrane. Two proteins specific to the RER have recently been shown by *Sabatini* et al. [see review in ref. 52] to be at the junction between the membrane and the ribosome in rat hepatocyte cells. These proteins, called ribophorins, are transmembrane glycoproteins and are found associated with the polysomes when the RER is treated with neutral detergents. The recovery of polysome-associated ribophorins in neutral detergents suggests that the interaction between the ribosome and the membrane is of ionic nature. The presence of ribophorins was found to be a regular feature of RER prepared from different vertebrate species tissues. Two forces are then involved in the binding of the ribosome to the membrane during the course of protein biosynthesis: (1) the interaction of the complex SRP-signal peptide → with the SRP receptor with some hydrophobic components of the membrane and (2) the ionic interaction of the ribosome 60S subunit with the ribophorins. However, these forces probably do not act simultaneously: the complex SRP-signal peptide assumes the initial binding of the ribsome to its site on the membrane and

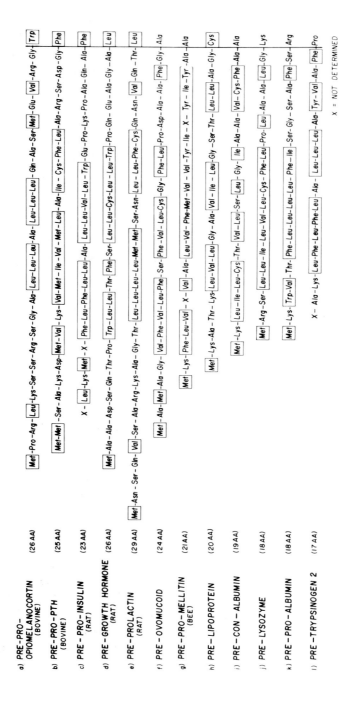

Fig. 2. Amino acid sequence of the 'signal peptides' of some secretory proteins.

the interaction between the ribophorins, and the 60S subunit anchors the ribosome during most of the translation since the signal peptide is cleaved co-translationally [91].

The cleavable signal peptide is a common feature to most of the secretory proteins of eukaryotic cells and to bacterial membrane glyco-proteins [44]. The only exception found so far is ovalbumin, a protein secreted in the chicken oviduct [80]. However, the ovalbumin molecule is segregated and glycosylated as are the proteins with a cleavable signal peptide.

The cotranslational cleavage of the signal peptide is done by an endopeptidase located at the luminal side of the endoplasmic reticulum membrane. An enzyme from the bacterium *Escherichia coli* has been isolated [118] and found to be a protein with a molecular weight of 39,000 daltons. The bacterial enzyme does not need a soluble factor to be active. The characterization of the eukaryotic enzyme is not yet done, since only recently has it been solubilized in detergent and its activity and specificity conserved [48]. The specificity of the enzyme is not very high. Examina-tion of the primary structure of the pre-proteins around the cleavage site indicates no common sequence of amino acids. However, neutral or hydrophobic amino acids with small side chains are always found at the carboxyl side of the site (fig. 2). One can imagine a signal peptidase specific to each pre-protein or to a tissue. However, cleavage of the signal peptide and segregation of the pro-protein from different sources by cell-free pro-tein synthesis systems supplemented with dog pancreatic microsomal or ascite tumor cell membranes ruled out that possibility [9, 10, 58–60].

Posttranslational Modification of Prohormones

During its biosynthesis and subsequent transport through the Golgi complex, the secretory protein is subjected to different modifications including N- or O-glycosylation, proteolysis, and chemical modifications of amino acid residues such as acetylation, amidation, and hydroxylation. In the past few years, much attention was focused on the mechanisms involved in the glycosylation of secretory proteins (human chorionic go-nadotropin) [8, 46] and of virus protein (vesicular stomatitis virus G pro-tein) [51, 57, 91, 111]. From the results of these studies emerges a probable scheme of events for the glycosylation of secretory proteins. It can be divided into three main steps. The first step occurs during the translation

Table I. Sequences at main cleavage sites in precursor proteins

Pro-opiomelanocortin (MW 29,000)

NH_2 (76 AA) —— RR .. (12 AA) ..... KR —— (31 AA) .. KR —— (39 AA) ..... KR ..... (56 AA) —— KR ..... (31 AA) ..... COOH
N-terminus   Joining peptide   ACTH   γ-LPH   β-Endorphin

Pre-pro-enkephalin (MW 55,000)

NH_2 ..... (97 AA) ...... KR ▮ KR ▮ KK ..... (20 AA) ..... KR ▮ KK ..... (41 AA) ..... KR ▮ RGLKR .....
... (12 AA) ... KR ▮ RR ...(11 AA) ... KR ▯▯▯ KR ..... (22 AA) ... KR ▮ RF-COOH

Pre-pro-somatostatin (MW 12,500)

NH_2 .... (92 AA) ...... R ... (12 AA) ... RK .... (14 AA) ... COOH
Somatostatin 1–14

Pro-Insulin (MW 9,000)

NH_2 ..... (30 AA) .... RR ..... (31 AA) ..... KR .. (21 AA) .. COOH
β-Chain   C-peptide   α-chain

Pre-pro-renin (MW 38,000)

NH_2 ..... (61 AA) .... KR .... (287 AA) .... RR ..... (46 AA) .... COOH
α-Chain   β-Chain

Pro-parathyroid hormone (MW 10,000)

NH_2 ...... (5 AA) ..... KR .... (84 AA) ....COOH
PTH

Pro-glucagon (MW 11,000)

NH₂ ... (14 AA) ... KR .... (29 AA) .... KR .. (5 AA) . KR ..... (31 AA) ... RRE
                                 Glucagon   Glucagon related

Pro-pressophysin (MW 17,000)

NH₂ .... (10 AA) .... KR .... (92 AA) .... RRVR ..... (39 AA) ..... COOH
            Vasopressin   Neurophysin   CPP

Pro-calcitonin (MW 12,500)

NH₂ .... (59 AA) .... KR ...... (32 AA) ..... GKKR ... (16 AA) ... COOH
                                    Calcitonin

E = Glutamic acid; F = phenylalanine; G = glycine; K = lysine; L = leucine; R = arginine; V = valine; ■ = Met-enkephalin; ▢▢▢ = Leu-enkephalin; ——— = MSHs'.

of the protein and consists in the attachment of a core of high-mannose oligosaccharide to asparagine residues of the nascent polypeptide chain via a N-glycosidic linkage. The oligosaccharides are carried in the membrane of the endoplasmic reticulum by a polyisoprenoid (dolichol). The second step consists in the trimming of glucose and mannose residues from the high-mannose core by glucosidases and mannosidases. This step is not carried by the RER, since cell-free synthesis systems supplemented with microsomal membranes accumulate high-mannose intermediates. Addition of a postribosomal supernatant fraction to a cell-free system results in the trimming of the high-mannose core. In a third step, peripheral sugars (galactose, N-acetylglucosamine, and N-acetylneuraminic acid) are linked by glycosyl transferase to the mannose residues of the core. The intracellular location of the mannosidases and glucosidases and of the peripheral glycosyl transferase has not been determined, but it is likely to be in the Golgi and the plasma membrane.

Oligosaccharides are also linked to serine or threonine residues of protein by an O-glycosylic linkage. Unlike the glycosylation of asparagine, which is done by addition of a core of oligosaccharides, the glycosylation of serine and threonine residues proceeds by the sequential addition of galactose, N-acetylgalactosamine, and N-acetylneuraminic acid by separate glycosyl transferases. The intracellular location of the O-glycosyl transferases is not known, but the total absence of glycosylation in the presence of microsomal membranes treated with the antibiotic tunycamycin, which inhibits specifically the N-glycosylation of asparagine residues, suggests a different location than the endoplasmic reticulum membrane. However, the possibility cannot be ruled out that a conformational change induced by N-glycosylation is necessary for the O-glycosylation to proceed.

The maturation of secretory proteins involves in many cases the proteolytic cleavage of the precursor into active substances. Examination of the primary structure of the pro-hormones and pro-proteins listed in table I indicates that a general mechanism for their processing into active substances involves proteolytic cleavage at sequences that contain paired basic residues.

This structural requirement became first apparent in the β-LPH model proposed by *Chrétien and Li* [16] and reinforced when the structure of pro-insulin became known in 1968 [13] (fig. 3). This mechanism seems to require the action of two membrane-bound proteases: (1) a trypsinlike endopeptidase cleaves specifically the polypeptide chain at the carboxyl end of the paired dibasic residues, and (2) the basic amino acids are sub-

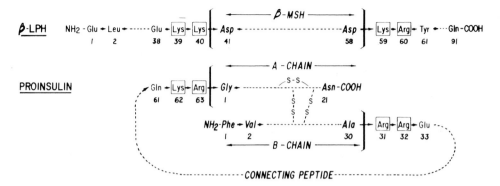

Fig. 3. Amino acid sequences of β-LPH and pro-insulin at their site of enzymatic cleavage.

sequently removed by a carboxypeptidase B-like exopeptidase. However, not all paired basic amino acids are cleaved, and it is suggested that the secondary structure of the molecule determines the accessibility of the residues to the proteases [41]. A protease with high specificity for paired basic residues has been described in the pituitary gland [2]. The enzyme is found associated with subcellular fractions containing secretory granules indicating that proteolytic cleavage occurs late in the secretory process. *Seidah* et al. [102] recently found that tonin, a submaxillary enzyme, is able to release endorphin peptides from β-LPH, and ACTH from POMC. Studies done on the maturation of pro-insulin [108] and pro-parathyroid hormone [65] showed that the cleavage of these pro-hormones is initiated in the Golgi complex (see the review by *B. Kemper,* this book).

Many other enzymes have been described with trypsinlike or chymotrypsinlike activity that process pro-proteins into fragments displaying some biological activities. However, structural studies of the fragments were never done, and the possibility exists that an unspecific fragment contains in part or in total the biologically active peptide. More work needs to be done on the pro-hormone-maturating enzymes. We feel that we may be faced with a new set of proteolytic activities which will be different and more specific than trypsinlike and carboxypeptidase B-like described by most authors.

During biosynthesis and the subsequent intracellular transport, polypeptide chains can be subject to modifications such as disulfide bridge formation, acetylation, amidation, hydroxylation of proline and lysine residues, and formation of an N-terminal pyroglutamic residue. Some of these modifications occur early during the biosynthesis of the protein (e.g., hydroxylation of collagen). Others occur only later in the Golgi complex or in the secretory granules after processing of the pro-protein by the maturation enzymes (e.g., acetylation of α-MSH). These modifications help the molecule to assume a given secondary structure and may be involved in the interaction of the molecule with its receptor.

Pro-Opiomelanocortin

First Biosynthesis of β-LPH, γ-LPH, and β-Endorphin in Porcine Pituitary Gland

In 1974–1976 we demonstrated unequivocally that β-LPH and γ-LPH were biosynthesized in bovine pituitary glands. Soon after the isolation and the characterization of sheep and human β-endorphin [17, 19], we used the highly purified sheep β-endorphin as a marker to identify its radioactive biosynthetic counterpart. Bovine pituitary slices and isolated cells of the pars intermedia incorporate radioactive methionine, leucine, and lysine into substances with properties identical with those of β-endorphin [19–23, 27, 102].

The material extracted from the bovine pituitary tissue was purified by two CMC chromatography columns. An aliquot of the resultant solution was analyzed by polyacrylamide gel electrophoresis at pH 4.5 and was shown to contain a peptide with the same mobility as fragment 61–91 of ovine β-lipotropin. When rechromatographed on a CMC column with a less steep gradient, this material was further resolved into two components. One contained material migrating as a single band on polyacrylamide gel at pH 4.5 with the same mobility as ovine β-endorphin. Sequencing of this material showed radioactivity only in methionine at position 5. No radioactivity remained in the cup of the sequencer. These results proved that this radioactive material is pure β-endorphin. This type of characterization was later used for the identification of rat β-LPH, γ-LPH, ACTH, α-MSH and β-endorphin [100] (see next section).

The possibility that β-endorphin is produced by degradation of β-lipotropin during the isolation process can be ruled out because *Chrétien and Gilardeau* [15] have shown that β-lipotropin is not broken down chemically or enzymatically during the isolation process. Moreover, circular dichroism studies by *St-Pierre* et al. [95] show that β-lipotropin is very stable over a wide range of pH and temperature. One can also exclude the possibility that the appearance of β-endorphin could be due to hydrolytic trypsinlike activity released from broken cells during the incubation, for three reasons: (1) protein labeling is linear with time, suggesting little lysis of pituitary cells; (2) a trypsin inhibitor was included in the incubation medium, and (3) a crude pituitary extract was used as carrier and served as substrate for any enzymatic activity which might still have been present.

Biosynthesis of β-LPH, γ-LPH, and β-Endorphin in Rat Pituitary Gland

Although bovine pars intermedia cells incorporated enough radioactivity to prove the biosynthesis of β- and γ-lipotropin as published in the mid 1970s [5, 18] and β-endorphin in 1977 [27], these cells were not stable enough for pulse-chase experiments. We then started using rat pars intermedia which is a more suitable tissue for incubation experiments.

Cells isolated from rat pars intermedia and incubated for 3 h incorporated radioactive amino acids into four main fractions, which were repurified from each protein extract and corresponded to labeled peptides eluting with or slightly before ovine β-LPH, γ-LPH, β-endorphin, and ACTH on CMC [42, 100].

The rat γ-LPH peptide showed an amino terminal sequence with Leu at 2, 10 and 14, Lys at position 20, and Val at 13 and 27. On CMC, it eluted slightly before its sheep homologue and was slightly more acidic on PAGE, but had the same molecular weight on PAGE/SDS/urea. However, no Met-labeled γ-LPH could be seen, from which one can deduce that its β-MSH (β-LPH 41–48) fragment lacked methionine [100]. This has also been found to be true for murine β-MSH by *Roberts* et al. [90] through their sequencing of AtT-20 cloned cDNA and has been confirmed by the DNA sequencing of rat POMC by *Drouin and Goodman* [34].

The rat β-LPH showed the same amino terminal sequence as γ-LPH, but eluted slightly before ovine β-LPH on CMC. However, its molecular weight and migration on PAGE were identical with those of ovine β-LPH. The final proof of its identity was obtained when we blocked the ε-NH$_2$ group of lysines by citraconylation and brought about selective tryptic cleavage at arginine residues, whereby the C-terminal segment, β-endorphin (β-LPH 61–91), was released [100, 101], a property not observed with rat γ-LPH. Rat β-LPH and γ-LPH both possess different N-terminal sequences from their ovine, bovine, porcine, and human counterparts. *Rubinstein* et al. [92] have isolated rat β-LPH in sufficient quantity to be able to determine its amino acid composition. They found slight differences from ovine β-LPH which agrees with our sequence analysis except that we find only one methionine residue instead of two. We have carried out further studies on the partial sequence of rat β- and γ-LPH, and they reveal many differences from the other species [42]. This has been confirmed and completed by the elegant DNA sequencing of *Drouin and Goodman* [34].

The β-endorphin peptide behaved in all respects like its sheep homologue and showed the presence of Phe 4, Met 5, Lys 9, Leu 14, Val 15, and Leu 17. A computer data bank search showed that the probability of finding other molecules with identically positioned key amino acids is less than one in a hundred million (the authors thank Dr. *Margaret Dayhoff* for her collaboration). Moreover, no trace of Leu$_5$-β-endorphin was seen on sequencing of the ^{3}H-Leu labeled peptide, so that Leu$_5$-β-endorphin seems not to exist in rat pituitaries which is also confirmed by *Drouin and Goodman* [34]. Our results agree with those of *Rubinstein* et al. [93] who purified rat β-endorphin and found it to have the same amino acid composition and an identical tryptic digest pattern.

Biosynthesis of ACTH and Related Peptides

Although the structures of ACTH and of α-MSH had been known for almost 15 years, the structural relationship between the two molecules in relation to their biosynthesis did not attract much attention until two additional facts were published: (1) that ACTH was present in high molecular weight forms [26, 63, 77, 78, 117] and (2) that CLIP was present in the pars intermedia [97, 98]. The most important in vitro experiments were carried out in the mid 1970s using simultaneously two experimental approaches. The first one, carried out by *Eipper and Mains* [35–38], *Eipper* et al. [39], and *Mains and Eipper* [66, 67], used a double-immunoprecipitation technique to isolate labeled proteins from cells incubated with radioactive amino acids. The second, carried out by *Roberts and Herbert* [87, 88] and *Roberts* et al. [89, 90], used a cell-free translation system from a poly(A)-containing mRNA fraction obtained from AtT-20 cells. Both studies confirmed the hypothesis that ACTH is biosynthesized as a large precursor form of molecular weight 28,000–31,000 which is later transformed into 13- and 4.5-kilodalton (K) molecules. *Nakanishi* et al.

[72–75] and *Jones* et al. [49] confirmed the presence of a large precursor using mRNA from bovine pituitary glands.

Pulse-chase experiments were carried out recently showing the maturation from one molecule to the other in the rat pars intermedia [28] and in mouse AtT-20 tumors [68].

β-LPH and ACTH: Part of the Same Precursor

It had been known for more than a decade that β-MSH and ACTH were under similar physiological control and were also increased simultaneously in pathological conditions [4]. This raised the logical suspicion that both molecules came from the same cell.

Phifer et al. [84], using immunoperoxidase staining with anti-ACTH[17–39], anti-α-MSH, and anti-β-MSH, demonstrated that all these molecules were present in the so-called corticotrophs (present in both pars distalis and pars intermedia), a feature which was suspected by *Dessy* et al. [32] and *Furth* et al. [40] who described the β-LPH-containing cells. More recently, *Pelletier* et al. [82] have successfully identified β-LPH in the ACTH cells of human pituitary glands as well as in pituitaries of several other species. Their antiserum [31] was directed at the N-terminal portion of β-LPH and did not crossreact with β-MSH. Finally, other groups have confirmed that β-endorphin is located in the same cells [71, 115].

Although these facts had been known for some years, no attempt was made before 1976 to study the cross-reactivity of the large forms of ACTH with either a β-MSH, a β-LPH or a β-endorphin-directed antiserum. Since our approach to elucidating the β-LPH maturation process was mainly to characterize chemically the secretory products, we could not do it directly, since the amount of common precursor is too low. However, we were able to show that ACTH-producing tumors (AtT-20 and MtT-F4) also contain β-LPH and β-endorphin [11, 96], thus completing the analogy of their secretory material with the normal pituitary gland.

Lowry et al. [62] were the first to describe the presence of ACTH and LPH in the same molecule. *Mains* et al. [70] and *Roberts and Herbert* [87, 88] almost simultaneously confirmed by radioimmunoassay the evidence of *Lowry* et al. [62] that ACTH and LPH could be part of the same precursor. *Rubinstein* et al. [94] were able to purify a small quantity of the common precursor from rat pituitaries. *Crine* et al. [28] were the first to show the precursor's maturation during pulse-chase experiments, while *Nakanishi* et al. [72–75] and *Roberts* et al. [90] were able to obtain the final proof with their cloned cDNA sequence analysis. Such a precursor has also been found in a human ACTH ectopic tumor [6], in frog [83], and toad pituitaries [61].

This common precursor was first called '31K-precursor' [35–38, 67] or 'precursor to ACTH, LPH, and β-endorphin' [70, 87, 88] and 'pro-opiocortin' [94]. We felt that this last piece of nomenclature was incomplete, and we renamed the precursor: pro-opiomelanocortin, a name which takes into account the three main biological activities of its known secretory components [22].

While continuing to work on the biosynthesis of β-LPH and its related peptides, we added the obviously important end product ACTH. Those working on ACTH biosynthesis added β-LPH to their model. There is a great deal of literature on the measurement and release of these peptides indicating indirectly the type of maturation that occurs in the anterior and intermediate lobes as well as in pituitary tumors. We will concentrate this section on the in vitro labeling experiments which show the maturation of the precursor during pulse-chase experiments.

A number of criteria need to be met before the existence of a biosynthetic precursor for a peptide hormone or secreted protein can be established with certainty [for similar criteria see ref. 112].

(1) Pulse-labeling and pulse-chase experiments are essential to establish the precursor-product relationship between high and low molecular weight forms of the proteins.

(2) Immunoprecipitation experiments with well-characterized antibodies raised against the hormone must precipitate the higher as well as the lower molecular weight forms of the hormones. It could happen, however, that the high molecular weight precursor might not be recognized by the antibody to the low molecular weight molecule.

(3) Peptide mapping of the precursor by means of a high-resolution two-dimensional separation or by high-pressure liquid chromatography of tryptic fragments must demonstrate the existence, within the larger molecular form, of peptides characteristic of the active hormone, together with additional fragments.

(4) Sequence analysis of the putative precursor either directly or through its nucleotide sequence must reveal the existence of an additional peptide covalently linked to the hormone.

Any of these findings taken alone constitutes suggestive evidence for the existence of a precursor for a given hormone, but conclusive proof can be obtained only through careful structural analysis of the putative precursor.

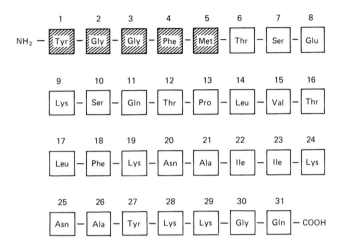

Fig. 4. Sequence of human β-endorphin.

Gene Expression of POMC in the Human

In the normal rat and mouse pituitary and the mouse pituitary tumor cells (AtT-20) pulse-chase experiments have shown that POMC gives rise to β-endorphin, β-LPH, α-MSH, and ACTH depending on the tissue (anterior lobe or pars intermedia). Such studies cannot be made in the human pituitaries, because fresh tissue is rarely available.

One alternate way to look at the expression of POMC gene in human is to isolate and characterize the end products. We had already characterized the human β-endorphin [33] and its sequence is shown in figure 4. The results obtained by *Li* et al. [56] are in agreement with ours. A few articles have been published on the chemical structure of human γ-LPH, and there was some disagreements between groups. We have recently reviewed this structure, and we proposed that it has the sequence shown in figure 5. We still have some differences with the last published results of *Li and Chung* [55] (fig. 5).

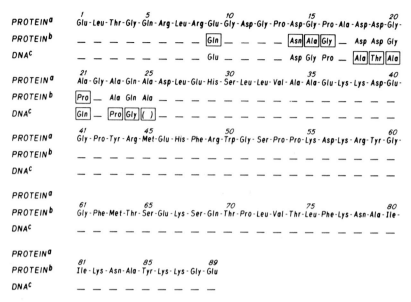

Fig. 5. Comparison of the N-terminal sequences of human β-LPH with that reported by Li and Chung [55] and Chang et al. [14]. [a] This work. [b] Sequence by Li and Chung [55]. [c] DNA sequence [14]. ( ) = Absence of residue.

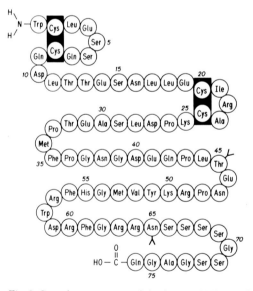

Fig. 6. Complete sequence of the human NH$_2$-terminal glycopeptide, including the disulfide bridge arrangement. Y = Glycosylation site.

The structure of human ACTH has been known for many years, and all workers in the field agree on its sequence.

Up to 1980, more than 50% of the human POMC molecules were not known. From pulse-chase experiments and the cDNA work of different groups it was anticipated that the N-terminal prosthetic group would contain more than 100 residues. We have recently isolated the major secretory products of this N-terminal region.

The anterior lobe of the human pituitary releases a large quantity of a 12-kilodalton glycoprotein which corresponds to the first 76 amino acids of the molecule. It has two sites of glycosylation, threonine No. 45 and asparagine No. 65. It contains two disulfide bridges, Cys 2–8 and Cys 20–24. Its sequence is shown in figure 6.

While this work was in progress, *Chang* et al. [14] published the partial sequence of the genomic DNA of human POMC. These results lead to the conclusion that there has to be a peptide between the 1–76 N-terminal and the ACTH molecule.

The genomic DNA sequence predicted a peptide of 109 amino acid residues preceeding the ACTH and β-lipotropin sequence [14]. It was, therefore, proposed [99, 106] that the pair of basic residues Lys_{77}-Arg_{78}, predicted from the DNA sequence [14], are cleaved during the maturation of POMC in vivo. Therefore, a 31 amino acid 'missing fragment' representing residues 79–109 must also have been generated [14, 99, 106].

The isolation and sequence characterization of such a peptide from human pituitary extracts have been carried out by our group [105]. It was found that the predicted DNA sequence is faithfully translated into its peptide form, but that the isolated peptide (denoted as HJP for human joining peptide) lacks the C-terminal Gly residue [14]. Evidence has been presented showing that the isolated 30-residue (HJP) peptide is amidated at its COOH-terminal glutamic acid residue (fig. 7).

The amino acid composition and sequence of the 30 residues is identical to that expected from the nucleotide sequence [14]. In contrast, the complete amino acid sequence of the human N-terminal glycosylated 1–76 fragment (denoted HNT) of human POMC, an Arg [99, 104, 106], was found to replace the nucleotide sequence predicted Gly [14] at residue 22. *Takahashi* et al. [113] and *Cochet* et al. [24] in a revised structure have now published that the DNA sequence agrees entirely with our protein sequence.

The isolation of only small amounts (0.5 mg compared to 15 mg per 250 glands for HNT) of this HJP fragment in the human pituitary extracts

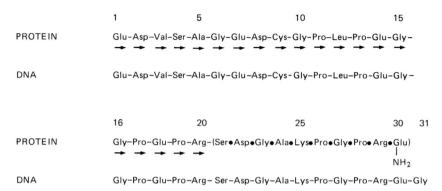

Fig. 7. Sequence of the isolated 30-residue HJP peptide. The arrows indicate direct sequence assignment of the amino acid residues. Beyond residue 20, the sequence is aligned by homology of the amino acid composition and the genomic DNA sequence [14]. A carboxy-terminal Glu-NH_2 is shown at residue 30.

could either be due to rapid protease destruction during the isolation procedure or to the presence of such a peptide in other molecular forms not yet isolated. Since previous work [99, 106] has established that the segment 1–76 (fig. 6) was one of the important pituitary forms of this part of the POMC molecule, this could mean that the HJP fragment isolated could be found attached either C-terminally to the HNT segment [prosegment (16K)] or N-terminally to the ACTH sequence [HJP (4K) + ACTH]. Therefore, a general maturation model is presented in figure 8, which takes into account our results and previously isolated peptides [3, 22, 54, 99, 104, 106] and the results of pulse-chase experiments [22, 29, 30, 69, 85]. Although the prosegment (16K + ACTH) peptide was found to be an intermediate in the production of ACTH [22], no report has yet appeared on the presence of ACTH attached to the HJP segment which could well account for some of the large molecular weight forms of ACTH found in pituitary extracts [76].

Also it is not yet known if the prosegment (16K) made of HNT plus HJP (fig. 8) already contains an amidated C-terminal Glu residue, or that amidation occurs posteriorly to the cleavage of Lys_{77}-Arg_{78} bonds. A carboxy-terminal glycine represents a recognition site for amidation of the penultimate residue [1, 45, 110], and its prediction at residue 109 from the DNA sequence [14] agrees with the finding of a Glu amide in the HJP segment isolated.

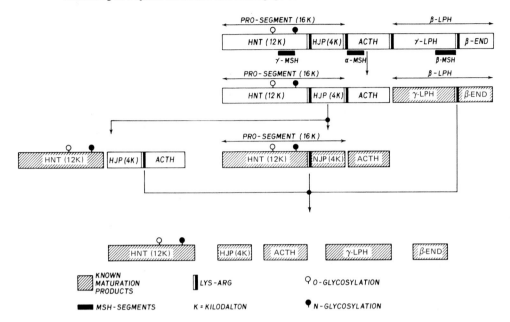

Fig. 8. General model for maturation of human POMC molecule. The hatched areas represent isolated and completely characterized peptides from human pituitaries. The molecular weights of the HNT 76-residue glycosylated peptide and the HJP isolated are given in parentheses as 12 and 4 kilodaltons (K), respectively. The glycosylation sites of HNT are at Thr_{45} and Asn_{65}. Exclusive cleavage at the pair of basic residues Lys-Arg is emphasized, together with the position of the three MSH segments within the POMC precursor. The possibility of generating an HJP peptide attached to HNT or to ACTH is proposed, based on the low yields of the HJP peptide isolated.

Furthermore, from the model in figure 8, it is seen that in the human the maturation enzyme(s) responsible for the generation of all POMC peptides is very selective, since it cleaves preferentially at the pair of basic residues Lys-Arg [99, 106]. The isolation and characterization of the HJP segment add more weight to the validity of this observation, since for its production it would involve cleavage at the Lys_{77}-Arg_{78} and Lys_{110}-Arg_{111} bonds [14], followed by a so-called carboxypeptidase-B-like cleavage of the C-terminal Lys-Arg pairs [22, 109].

Comparison of the length expected for the 'joining peptides' in human [14], bovine [75], and rat [34] homologues shows that it varies between 30, 24, and 19 residues, respectively. The identity of sequence of the C-terminal Pro-Arg-Glu-Gly sequences of the human [14] and rat [34] 'join-ing peptides' would indicate that, analogous to the human situation, the 19-residue rat 'joining peptide', if present, could also be amidated at the penultimate residue, thereby generating an 18-residue peptide amidated at its C-terminal glutamic acid. The 'joining peptide' segment of the POMC precursor molecule [14, 24, 34, 75, 113] represents a sequence-variable region, bearing little homology between species. By analogy to the sequence-variable 'connecting peptide' segment of pro-insulin [107], this possibly indicates a structural rather than a biological function for this region of the POMC molecule.

Therefore, much work needs to be done before we can accurately define the precise maturation pathway and biological functions of the various segments of this pluripotent POMC precursor molecule.

Processing of Other Precursor Molecules: the Use of POMC Model to Predict Their Secretion Products

With the development of DNA structure analysis, the complete sequence of the precursor can be determined either from a cDNA tran-script of the mRNA or from the genomic DNA. Examples have recently appeared which demonstrate the analytical power of this approach. It can be observed from such studies that the pairs of basic amino acid residues are present at most sites of cleavage (table I). This 'magic' rule became first noticed in the β/γ-LPH model proposed in 1967 [16] and reinforced with pro-insulin in 1968 (fig. 3) [13].

Pro-Enkephalin

If one applies this rule to the recently published structure of pro-enkephalin (fig. 9) [25], one could predict that this multipotent gene can give rise to a number of enkephalins. As seen in figure 9, by assuming that the splitting is at the pair Lys-Arg as for POMC maturation, this gene gives rise to one molecule each of Met-enkephalin and Leu-enkephalin and five other longer enkephalins containing 8–48 amino acids. If other pairs are cleaved, it could give rise to numerous other enkephalins including four molecules of Met-enkephalin, one molecule of Leu-enkephalin and two other Met-enkephalins of 7 and 8 amino acids each. Only pulse-chase experiments and/or purification studies will give the answer. However, all the peptides described in figure 9 are candidates to be secreted as such.

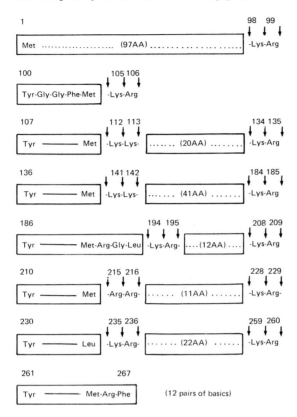

Fig. 9. The complete sequence of pre-pro-enkephalin as determined from the genomic structure by *Comb* et al. [25]. The sites of possible cleavages with the description of the end products: if the Lys-Arg pair is the only one cleavage, the pro-molecule will give rise to one molecule each of Met- and Leu-enkephalin, one Met-enkephalin of 27 residues (amino acid 107–133), one Met-enkephalin of 48 residues (amino acid 136–183), one Met-enkephalin of 8 residues (amino acid 186–193), one Met-enkephalin of 18 residues (amino acid 210–228), one Met-enkephalin of 7 residues (amino acid 261–267), three non-enkephalin related peptides, one of 97 residues (minus the signal peptide), one of 12 residues (amino acid 196–207), and one of 22 residues (amino acid 237–258). If the other pairs are cleaved, the secretory pattern becomes more simple. One gets one molecule of Leu-enkephalin, three types of Met-enkephalin, four molecules of regular Met-enkephalin, one each of 7 and 8 amino acids.

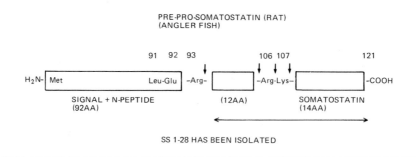

PRE-PRO-SOMATOSTATIN (RAT)
(ANGLER FISH)

Fig. 10. The complete sequence of pre-pro-somatostatin [43]. There is no Lys-Arg, and cleavage at other pairs of basic amino acid residues could give up to three somatostatins of 14, 31 and 39 residues. One has to note that a somatostatin (ss) 1–28 has been isolated which does not imply a cleavage at a pair of basic amino acids.

Prosomatostatin

Lets look now at the other models in table I. The first somatostatin isolated and characterized by *Brazeau* et al. [12] has 14 amino acids. A 28 amino acids long only has been found so far by *Pradayrol* et al. [86]. The secretion of this new larger somatostatin indicates a cleavage at one arginyl. This does not correspond to the rule of pairs of basic residues. Only pulse-chase experiment will be able to confirm that somatostatin 1–28 is a real secretory product (fig. 10).

Pro-Insulin

In pro-insulin (table I, fig. 3), there is a release of the C-peptide which has yet to find a biological role.

Pro-Renin

The recently published pre-pro-renin is interesting [81] (table I, fig. 11). Like pro-insulin, renin has two chains, but is biosynthesized as a single chain and during maturation becomes a two-chain protein with two intra- and one interchain disulfide bridges.

Proparathormone

The PTH model (table I) [65] (see also *B. Kemper's* review, this book) is the simplest of all, since the active component is at the C-terminus while it has a very short prosegment (5 amino acid residues).

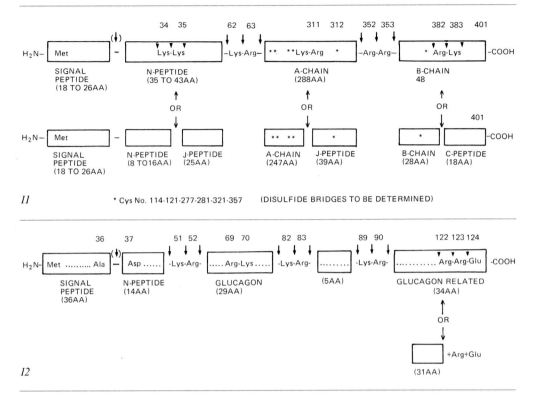

Fig. 11. The complete sequence of pre-pro-renin (rat, possible secretory products). The different possibilities of cleavage open a great deal of different solutions. Those proposed by *Panthier* et al. [81] are well founded except for the signal peptide.

Fig. 12. The complete sequence of pre-pro-glucagon (angler fish). The interesting aspect is the presence of two regions coding for glucagon and a glucagon-related peptide [from ref. 64].

Pro-Glucagon

The pro-glucagon DNA published recently by *Lund* et al. [64] (table I, fig. 12) reveals the interesting feature of the regular glucagon in the center of the molecule and a C-terminus glucagon-related peptide whose biological activity has yet to be determined. All cleavage sites are Lys-Arg.

Pro-Vasopressin

The vasopressin precursor, also called pro-pressophysin, has also been recently sequenced [53]. It is interesting to note that a glycopeptide

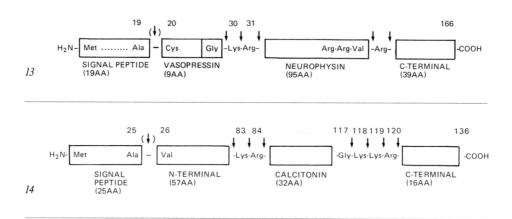

Fig. 13. The complete sequence of pre-pro-pressophysin. It indicates that vasopressin is at the N-terminus, while neurophysin is at the center and new C-terminal fragment has been isolated [from ref. 53].

Fig. 14. The complete sequence of pre-pro-calcitonin (rat). Calcitonin is at the center of the molecule, while two other non-calcitonin-related peptides can be released [from ref. 47].

we have recently isolated [103] and proposed to be part of the vasopressin precursor is the C-terminal fragment of the precursor with vasopressin at the C-terminus and neurophysin in the center (table I, fig. 13).

Pro-Calcitonin

Finally, the calcitonin model (table I, fig. 14) is more standard with calcitonin in the middle between pairs of Lys-Arg and a C-terminus fragment of unknown biological significance.

Conclusion

Through the knowledge of the chemistry of pro-hormones, one can envisage their maturation processes during their transport from the endoplasmic reticulum to the secretion granules. More important, it is possible to foresee the biosynthesis of new peptides whose existence is not suspected by the regular isolation methods which had been the only accessible method so far.

It is important that these concepts are discussed in a book in which morphology is so thoroughly discussed. It provides ways and means to interpret the localization of these peptides in different secretory tissues but most importantly in the brain.

References

1 Amara, S.G.; David, D.N.; Rosenfeld, M.G.; Roos, B.A.; Evans, R.M.: Characterization of rat calcitonin mRNA. Proc. natn. Acad. Sci. USA *77:* 4444–4448 (1980).

2 Austen, B.M.: Isolation from pituitary granules and properties of a protease specific for consecutive basic residues; in Rich, Gross, Peptides, synthesis – structure – function, pp. 493–496 (Pierce Chemical Co., Rockford, 1981).

3 Benjannet, S.; Seidah, N.G.; Routhier, R.; Chrétien, M.: A novel human pituitary peptide containing the γ-MSH sequence. Nature *285:* 415–416 (1980).

4 Berson, S.A.; Yalow, R.S.: Peptide hormones in plasma. Harvey Lect. *62:* 107–163 (1968).

5 Bertagna, X.; Lis, M.; Gilardeau, C.; Chrétien, M.: In vitro biosynthesis of bovine β-lipotropic hormone. Can. J. Biochem. *52:* 349–358 (1974).

6 Bertagna, X.; Nicholson, W.E.; Sorenson, G.D.; Pettengill, O.S.; Mount, C.D.; Orth, D.: Corticotropin, lipotropin, and β-endorphin production by a human non-pituitary tumor in culture: evidence for a common precursor. Proc. natn. Acad. Sci. USA *75:* 5160–5164 (1978).

7 Bielinska, M.; Boime, I.: mRNA dependent synthesis of a glycosylated subunit of human chorionic gonadotropin in cell-free extracts derived from ascites tumor cells. Proc. natn. Acad. Sci. USA *75:* 1768–1772 (1978).

8 Bielinska, M.; Boime, I.: Glycosylation of human chorionic gonadotropin in mRNA-dependent cell-free extracts: post-translational processing of an asparagine-linked mannose-rich oligosaccharide. Proc. natn. Acad. Sci. USA *76:* 1208–1212 (1979).

9 Bielinska, M.; Grant, G.; Boime, I.: Processing of placental peptide hormones synthesized in lysates containing membranes derived from tunicamycin-treated ascites tumor cells. J. biol. Chem. *253:* 7117–7119 (1978).

10 Blobel, G.; Dobberstein, B.: Transfer of proteins across membranes. I. Presence of proteolytically processed and unprocessed nascent immunoglobulin light chains on membrane-bound ribosomes of murine myeloma. J. Cell Biol. *67:* 835–851 (1975).

11 Bourassa, M.; Scherrer, H.; Pezalla, P.D.; Lis, M.; Chrétien, M.: Immunoreactive β-endorphin in the rat mammotropic transplantable tumor (MtT-F4). Cancer Res. *38:* 1568–1571 (1978).

12 Brazeau, P.; Vale, W.; Burgus, R.; Ling, N.; Butcher, M.; Rivier, J.; Guillemin, R.: Hypothalamic polypeptide that inhibits the secretion of immunoreactive pituitary growth hormone. Science *179:* 77–79 (1973).

13 Chance, R.E.; Ellis, R.M.; Bromer, W.W.: Porcine pro-insulin: characterization and amino acid sequence. Science *161:* 165–167 (1968).

14 Chang, A.C.Y.; Cochet, M.; Cohen, S.N.: Structural organization of human genomic DNA encoding the pro-opiomelanocortin peptide. Proc. natn. Acad. Sci. USA *77:* 4890–4894 (1980).

15   Chrétien, M.; Gilardeau, C.: Effects of glacial acetic acid on the extraction, chemical stability, and biological activities of β-LPH. Can. J. Biochem. *48:* 511–516 (1970).

16   Chrétien, M.; Li, C.H.: Isolation, purification and characterization of γ-lipotropic hormone from sheep pituitary glands. Can. J. Biochem. *45:* 1163–1174 (1967).

17   Chrétien, M.; Benjannet, S.; Dragon, N.; Seidah, N.G.; Lis, M.: Isolation of peptides with opiate activity from sheep and human pituitaries: relationship to β-lipotropin. Biochem. biophys. Res. Commun. *72:* 472–478 (1976).

18   Chrétien, M.; Lis, M.; Gilardeau, C.; Benjannet, S.: In vitro biosynthesis of γ-lipotropic hormone. Can. J. Biochem. *54:* 566–570 (1976).

19   Chrétien, M.; Seidah, N.G.; Benjannet, S.; Dragon, N.; Routhier, R.; Motomatsu, T.; Crine, P.; Lis, M.: A β-LPH precursor model: recent developments concerning morphine-like substances. Ann. N.Y. Acad. Sci. *297:* 84–107 (1977).

20   Chrétien, M.; Crine, P.; Lis, M.; Seidah, N.G.: Biosynthesis of β-endorphin. Bull. schweiz. Akad. med. Wiss. *34:* 155–169 (1978).

21   Chrétien, M.; Crine, P.; Lis, M.; Gianoulakis, C.; Gossard, F.; Benjannet, S.; Seidah, N.G.: Biosynthesis of β-endorphin from β-lipotropin and a large precursor molecule; in Van Ree, Terenius, Characteristics and functions of opioids. Dev. Neurosci., vol. IV, pp. 345–357 (Elsevier-North Holland, New York 1978).

22   Chrétien, M.; Benjannet, S.; Gossard, F.; Gianoulakis, C.; Crine, P.; Lis, M.; Seidah, N.G.: From β-lipotropin to β-endorphin and pro-opiomelanocortin. Can. J. Biochem. *57:* 1111–1121 (1979).

23   Chrétien, M.; Crine, P.; Lis, M.; Benjannet, S.; Seidah, N.G.: β-Lipotropin: proendorphin; in Central nervous system effects of hypothalamic hormones and other peptides, pp. 237–251 (Raven Press, New York 1979).

24   Cochet, M.; Chang, A.C.Y.; Cohen, S.N.: Characterization of the structural gene and putative 5′-regulatory sequences for human pro-opiomelanocortin. Nature *297:* 335–338 (1982).

25   Comb, M.; Seeburg, P.H.; Adelman, J.; Eiden, L.; Herbert, E.: Primary structure of the human Met- and Leu-enkephalin precursor and its mRNA. Nature *295:* 663–666 (1982).

26   Coslovsky, R.; Yalow, R.S.: Influence of the hormonal forms of ACTH on the pattern of corticosteroid secretion. Biochem. biophys. Res. Commun. *60:* 1351–1356 (1974).

27   Crine, P.; Benjannet, S.; Seidah, N.G.; Lis, M.; Chrétien, M.: In vitro biosynthesis of β-endorphin in pituitary glands. Proc. natn. Acad. Sci. USA *74:* 1403–1406 (1977).

28   Crine, P.; Gianoulakis, C.; Seidah, N.G.; Gossard, F.; Pezalla, P.D.; Lis, M.; Chrétien, M.: Biosynthesis of β-endorphin from β-lipotropin and a larger molecular weight precursor in rat pars intermedia. Proc. natn. Acad. Sci. USA *75:* 4719–4723 (1978).

29   Crine, P.; Seidah, N.G.; Routhier, R.; Gossard, F.; Chrétien, M.: Processing of two forms of the common precursor to α-melanotropin and β-endorphin in the rat pars intermedia. Eur. J. Biochem. *110:* 387–396 (1980).

30   Crine, P.; Seidah, N.G.; Jeannotte, L.; Chrétien, M.: Two large glycoprotein fragments related to the NH₂-terminal part of the adrenocorticotropin-β-lipotropin precursor are the end-products of the maturation process in the rat pars intermedia. Can. J. Biochem. *58:* 1318–1322 (1980).

31   Desranleau, R.; Gilardeau, C.; Chrétien, M.: Radioimmunoassay of ovine β-lipotropic hormone. Endocrinology *91:* 1004–1010 (1972).

32   Dessy, C.; Herlant, M.; Chrétien, M.: Détection par immunofluorescence des cel-

lules synthétisant la lipotropin. C. r. hebd. Séanc. Acad. Sci., Paris *276:* 335–338 (1973).

33 Dragon, N.; Seidah, N.G.; Lis, M.; Routhier, R.; Chrétien, M.: Primary structure and morphine-like activity of human β-endorphin. Can. J. Biochem. *55:* 666–670 (1977).

34 Drouin, J.; Goodman, H. M.: Most of the coding region of rat ACTH/β-LPH precursor gene lacks intervening sequences. Nature *288:* 610–613 (1980).

35 Eipper, B. A.; Mains, R. E.: High molecular weight forms of adrenocorticotropic hormone in the mouse pituitary and in a mouse pituitary tumor cell line. Biochemistry *14:* 3836–3844 (1975).

36 Eipper, B. A.; Mains, R. E.: Peptide analysis of a glycoprotein form of adrenocorticotropic hormone. J. biol. Chem. *252:* 8821–8832 (1977).

37 Eipper, B. A.; Mains, R. E.: Analysis of the common precursor to corticotropin and endorphin. J. biol. Chem. *253:* 5732–5744 (1978).

38 Eipper, B. A.; Mains, R. E.: Existence of a common precursor to ACTH and endorphin in the anterior and intermediate lobes of the rat pituitary. J. supramol. Struct. *8:* 247–262 (1978).

39 Eipper, B. A.; Mains, R. E.; Guenzi, D.: High molecular weight forms of adrenocorticotropic hormone are glycoproteins. J. biol. Chem. *251:* 4121–4126 (1976).

40 Furth, J.; Chrétien, M.; Lis, M.; Bélanger, A.; Moy, P.; Grauman, J.: Multipotent lipotropic hormones. In search of a pituitary cell producing multipotent LPH. Archs Path. *99:* 572–581 (1975).

41 Geisow, M. J.: Polypeptide secondary structure may direct the specificity of prohormone conversion. FEBS Lett. *87:* 111–114 (1978).

42 Gianoulakis, C.; Seidah, N. G.; Routhier, R.; Chrétien, M.: Further characterization of rat β-LPH, γ-LPH and β-endorphin biosynthesized by isolated cells of pars intermedia and pars distalis. Int. J. Peptide Protein Res. *16:* 97–105 (1980).

43 Goodman, R. H.; Jacobs, J. W.; Chin, W. W.; Lund, P. K.; Dee, P. C.; Habener, J. F.: Nucleotide sequence of a cloned structural gene coding for a precursor of pancreatic somatostatin. Proc. natn. Acad. Sci. *77:* 5869–5873 (1980).

44 Halegoua, S.; Sekizawa, J.; Inouye, M.: A new form of structural lipoprotein of outer membrane of *Escherichia coli.* J. biol. Chem. *252:* 2324–2330 (1977).

45 Harris, J. I.; Lerner, A. B.: Amino acid sequence of the α-melanocyte stimulating hormone. Nature *179:* 1346–1347 (1957).

46 Hussa, R. O.: Biosynthesis of human chorionic gonadotropin. Endocr. Rev. *1:* 268–294 (1980).

47 Jacobs, J. W.; Goodman, R. H.; Chin, W. W.; Dee, P. C.; Habener, J. F.; Bell, N. H.; Potts, J. T., Jr.: Calcitonin messenger RNA encodes multiple polypeptides in a single precursor. Science *213:* 457–459 (1981).

48 Jackson, R. C.; Blobel, G.: Post-translational processing of full-length presecretory proteins with canine pancreatic signal peptidase. Ann. N.Y. Acad. Sci. *343:* 391–403 (1980).

49 Jones, R. E.; Pulkrabek, P.; Grunberger, D.: Mouse pituitary tumor mRNA directed cell-free synthesis of polypeptides that are cross-reactive with adrenocorticotropic hormone antiserum. Biochem. biophys. Res. Commun. *74:* 1490–1495 (1977).

50 Kiely, M. L.; McKnight, G. S.; Schimke, R. T.: Studies on the attachment of carbohydrate to ovalbumin nascent chains in hen oviduct. J. biol. Chem. *251:* 5490–5495 (1976).

51   Kornfeld, S.; Li, E.; Tabas, I.: The synthesis of complex-type oligosaccharides. II. Characterization of the processing intermediates in the synthesis of the complex oligosaccharide units of the vesicular stomatitis virus G protein. J. biol. Chem. *253:* 7771–7778 (1978).

52   Kreibich, G.; Czako-Graham, M.; Grebenau, R.; Sabatini, D.D.: Functional and structural characteristics of endoplasmic reticulum proteins associated with ribosome binding sites. Ann. N.Y. Acad. Sci. *343:* 17–33 (1980).

53   Land, H.; Schütz, G.; Schmale, H.; Richter, D.: Nucleotide sequence of cloned cDNA encoding bovine arginine vasopressin-neurophysin II precursor. Nature *295:* 299–303 (1982).

54   Larivière, N.; Seidah, N.G.; Chrétien, M.: Complete sequence of the glycosylated amino-terminal segment of porcine pro-opiomelanocortin. Int. J. Peptide Protein Res. *18:* 487–491 (1981).

55   Li, C.H.; Chung, D.: Isolation, characterization and amino acid sequence of β-lipotropin from human pituitary glands. Int. J. Peptide Protein Res. *17:* 131–142 (1981).

56   Li, C.H.; Yamashiro, D.; Chung, D.; Doneen, B.A.: Isolation, structure, synthesis and morphine-like activity of β-endorphin from human pituitary glands. Ann. N.Y. Acad. Sci. *297:* 158–166 (1977).

57   Li, E.; Tabas, I.; Kornfeld, S.: The synthesis of complex-type oligosaccharides. I. Structure of the lipid-linked oligosaccharide precursor of the complex-type oligosaccharides of the vesicular stomatitis virus G protein. J. biol. Chem. *253:* 7762–7770 (1978).

58   Lingappa, V.R.; Devillers-Thiery, A.; Blobel, D.: Nascent prehormones are intermediates in the biosynthesis of authentic bovine pituitary growth hormone and prolactin. Proc. natn. Acad. Sci. USA *74:* 2432–2436 (1977).

59   Lingappa, V.R.; Katz, F.N.; Loddish, H.F.; Blobel, G.: A signal sequence for the insertion of a transmembrane glycoprotein. Similarities to the signals of secretory proteins in primary structure and function. J. biol. Chem. *253:* 8867–8870 (1978).

60   Lingappa, V.R.; Prusad, R.; Ebner, K.; Blobel, G.: Coupled cell-free synthesis, segregation, and core glycosylation of a secretory protein. Proc. natn. Acad. Sci. USA *75:* 2338–2342 (1978).

61   Loh, Y.P.: Immunological evidence for two common precursors to corticotropins, endorphins, and melanotropin in the neurointermediate lobe of the toad pituitary. Proc. natn. Acad. Sci. USA *76:* 796–800 (1979).

62   Lowry, P.J.; Hope, J.; Silman, R.E.: The evolution of corticotrophin-melanotrophin and lipotrophins. Excerpta Med. Int. Congr. Ser., No. 402, pp. 71–76 (1976).

63   Lowry, P.J.; Silman, R.E.; Hope, J.; Scott, A.P.: Structure and biosynthesis of peptides related to corticotropins and β-melanotropins. Ann. N.Y. Acad. Sci. *297:* 49–60 (1977).

64   Lund, P.K.; Goodman, R.H.; Dee, P.C.; Habener, J.F.: Pancreatic preproglucagon cDNA contains two glucagon-related coding sequences arranged in tandem. Proc. natn. Acad. Sci. USA *79:* 345–349 (1982).

65   MacGregor, R.R.; Chu, L.L.H.; Lohn, D.V.: Conversion of proparathyroid hormone to parathyroid hormone by a particulate enzyme of the parathyroid gland. J. biol. Chem. *251:* 6711–6716 (1976).

66   Mains, R.E.; Eipper, B.A.: Molecular weights of adrenocorticotropic hormone in

extracts of anterior and intermediate-posterior lobes of mouse pituitary. Proc. natn. Acad. Sci. USA *72:* 3565–3569 (1975).

67 Mains, R.E.; Eipper, B.A.: Biosynthesis of adrenocorticotropic hormone in mouse pituitary tumor cells. J. biol. Chem. *251:* 4115–4120 (1976).

68 Mains, R.E.; Eipper, B.A.: Coordinate synthesis of corticotropins and endorphins by mouse pituitary tumor cells. J. biol. Chem. *253:* 651–655 (1978).

69 Mains, R.E.; Eipper, B.A.: Synthesis and secretion of corticotropins, melanotropins, and endorphins by rat intermediate pituitary cells. J. biol. Chem. *254:* 7885–7894 (1979).

70 Mains, R.E.; Eipper, B.A.; Ling, N.: Common precursor to corticotropins and endorphins. Proc. natn. Acad. Sci. USA *74:* 3014–3018 (1977).

71 Martin, R.; Weber, E.; Voigt, K.H.: Localization of corticotropin- and endorphin-related peptides in the intermediate lobe of the rat pituitary. Cell. Tiss. Res. *196:* 307–319 (1979).

72 Nakanishi, S.; Taii, S.; Hirata, Y.; Matsukura, S.; Imura, H.; Numa, S.: A large product of cell-free translation of messenger RNA coding for corticotropin. Proc. natn. Acad. Sci. USA *73:* 4319–4323 (1976).

73 Nakanishi, S.; Inoue, A.; Taii, S.; Numa, S.: Cell-free translation product containing corticotropin and β-endorphin encoded by messenger RNA from anterior lobe and intermediate lobe of bovine pituitary. FEBS Lett. *84:* 105–109 (1977).

74 Nakanishi, S.; Kita, T.; Taii, S.; Imura, H.; Numa, S.: Glucocorticoid effect on the level of corticotropin messenger RNA activity in rat pituitary. Proc. natn. Acad. Sci. USA *74:* 3283–3286 (1977).

75 Nakanishi, S.; Inoue, A.; Kita, T.; Nakamura, M.; Chang, A.C.Y.; Cohen, S.N.; Numa, S.: Nucleotide sequence of cloned cDNA for bovine corticotropin-β-lipotropin precursor. Nature *278:* 423–427 (1979).

76 Orth, D.N.; Nicholson, W.E.: Different molecular forms of ACTH. Ann. N.Y. Acad. Sci. *297:* 27–45 (1977).

77 Orth, D.N.; Nicholson, W.E.; Shapiro, M.; Byyny, R.: Adrenocorticotropic hormone (ACTH) and melanocyte-stimulating hormone (MSH) production by a single cell (Abstract No.140). 52nd Meet. Endocr. Soc., St. Louis 1970.

78 Orth, D.N.; Nicholson, W.E.; Mitchell, W.M.; Island, D.P.; Shapiro, M.; Byyny, R.L.: ACTH and MSH production by a single cloned mouse pituitary tumor cell line. Endocrinology *92:* 385–393 (1973).

79 Palade, G.E.: Intracellular aspects of the process of protein synthesis. Science *189:* 347–358 (1975).

80 Palmiter, R.D.; Gagnon, J.; Walsh, K.A.: Ovalbumin: a secreted protein without a transient hydrophobic leader sequence. Proc. natn. Acad. Sci. USA *75:* 94–98 (1978).

81 Panthier, J.J.; Foote, S.; Chambraud, B.; Strosberg, A.D.; Corvol, P.; Rougeon, R.: Complete amino acid sequence and maturation of the mouse submaxillary gland renin precursor. Nature *298:* 90–92 (1982).

82 Pelletier, G.; Leclerc, R.; Labrie, F.; Côté, J.; Chrétien, M.; Lis, M.: Immunohistochemical localization of β-lipotropic hormone in the pituitary gland. Endocrinology *100:* 770–776 (1977).

83 Pezalla, P.D.; Seidah, N.G.; Benjannet, S.; Crine, P.; Lis, M.; Chrétien, M.: Biosynthesis of β-endorphin, β-lipotropin and the putative ACTH-LPH precursor in the frog pars intermedia. Life Sci. *23:* 2281–2291 (1978).

84 Phifer, R.F.; Orth, D.N.; Spicer, S.S.: Specific demonstration of the human hypophyseal adrenocortico-melanotropic (ACTH/MSH) cell. J. clin. Endocr. Metab. *39:* 684–692 (1974).

85 Phillips, M.A.; Budarf, M.L.; Herbert, E.: Glycosylation events in the processing and secretion of pro ACTH-endorphin in mouse pituitary tumor cells. Biochemistry *20:* 1666–1675 (1981).

86 Pradayrol, L.; Jornvall, H.; Mutt, V.; Ribet, A.: N-terminally extended somatostatin: the primary structure of somatostatin-28. FEBS Lett. *109:* 55–58 (1980).

87 Roberts, J.L.; Herbert, E.: Characterization of a common precursor to corticotropin and β-lipotropin: cell-free synthesis of the precursor and identification of corticotropin peptides in the molecule. Proc. natn. Acad. Sci. USA *74:* 4826–4830 (1977).

88 Roberts, J.L.; Herbert, E.: Characterization of a common precursor to corticotropin and β-lipotropin: identification of β-lipotropin peptides and their arrangement relative to corticotropin in the precursor synthesized in a cell-free system. Proc. natn. Acad. Sci. USA *74:* 5300–5304 (1977).

89 Roberts, J.L.; Phillips, M.; Rosa, P.A.; Herbert, E.: Steps involved in the processing of common precursor forms of adrenocorticotropin and endorphin in cultures of mouse pituitary cells. Biochemistry *17:* 3609–3618 (1978).

90 Roberts, J.L.; Seeburg, P.H.; Shine, J.; Herbert, E.; Baxter, J.D.; Goodman, H.M.: Corticotropin and β-endorphin: construction and analysis of recombinant DNA complementary to mRNA for the common precursor. Proc. natn. Acad. Sci. USA *76:* 2153–2157 (1979).

91 Rothman, J.E.; Lodish, H.F.: Synchronised transmembrane insertion and glycosylation of a nascent membrane protein. Nature *269:* 775–779 (1977).

92 Rubinstein, M.; Stein, S.; Gerber, L.D.; Udenfriend, S.: Isolation and characterization of the opioid peptides from rat pituitary: β-lipotropin. Proc. natn. Acad. Sci. USA *74:* 3052–3055 (1977).

93 Rubinstein, M.; Stein, S.; Udenfriend, S.: Isolation and characterization of the opioid peptides from rat pituitary: β-endorphin. Proc. natn. Acad. Sci. USA *74:* 4969–4972 (1977).

94 Rubinstein, M.; Stein, S.; Udenfriend, S.: Characterization of pro-opiocortin, a precursor to opioid peptides and corticotropin. Proc. natn. Acad. Sci. USA *75:* 669–671 (1978).

95 St-Pierre, S.; Gilardeau, C.; Chrétien, M.: Circular dichroism studies of sheep β-lipotropic hormone. Can. J. Biochem. *54:* 992–998 (1976).

96 Scherrer, H.; Seidah, N.G.; Benjannet, S.; Lis, M.; Pezalla, P.D.; Bourassa, M.; Chrétien, M.: β-Endorphin and β-lipotropin secretion by an ACTH-secreting mouse pituitary tumor. FEBS Lett. *90:* 353–356 (1978).

97 Scott, A.P.; Bennett, H.P.J.; Lowry, P.J.; McMartin, C.; Ratcliffe, J.G.: Corticotrophin-like intermediate lobe peptide (CLIP) – a new pituitary and tumor peptide. J.Endocr. *55:* 36–37 (1972).

98 Scott, A.P.; Lowry, P.J..; Bennett, H.P.J.; McMartin, C.; Ratcliffe, J.G.: Purification and characterization of porcine corticotrophin-like intermediate lobe peptide. J.Endocr. *61:* 369–380 (1974).

99 Seidah, N.G.; Chrétien, M.: Complete amino acid sequence of human pituitary glycopeptide: an important maturation product of pro-opiomelanocortin. Proc. natn. Acad. Sci. USA *78:* 4236–4240 (1981).

100 Seidah, N.G.; Gianoulakis, C.; Crine, P.; Lis, M.; Benjannet, S.; Routhier, R.; Chrétien, M.: In vitro biosynthesis and chemical characterization of β-lipotropin, γ-lipotropin, and β-endorphin in rat pars intermedia. Proc. natn. Acad. Sci. USA 75: 3153–3157 (1978).

101 Seidah, N.G.; Gianoulakis, C.; Crine, P.; Benjannet, S.; Routhier, R.; Lis, M.; Chrétien, M.: in Characteristics and functions of opioids. Dev. Neurosci. vol. IV, pp.291–292 (Elsevier-North Holland, New York 1978).

102 Seidah, N.G.; Chan, J.; Mardini, G.; Benjannet, S.; Chrétien, M.; Boucher, R.; Genest, J.: Specific cleavage of β-LPH and ACTH by tonin: release of an opiate-like peptide, β-LPH (61-78). Biochem. biophys. Res. Commun. 86: 1002–1013 (1979).

103 Seidah, N.G.; Benjannet, S.; Chrétien, M.: The complete sequence of a novel human pituitary glycopeptide homologous to pig posterior pituitary glycopeptide. Biochem. biophys. Res. Commun. 100: 901–907 (1981).

104 Seidah, N.G.; Benjannet, S.; Routhier, R.; De Serres, G.; Rochemont, J.; Lis, M.; Chrétien, M.: Purification and characterization of the N-terminal fragment of pro-opiomelanocortin from human pituitaries: homology to the bovine sequence. Biochem. biophys. Res. Commun. 95: 1417–1424 (1980).

105 Seidah, N.G.; Rochemont, J.; Hamelin, J.; Benjannet, S.; Chrétien, M.: The missing fragment of the pro-sequence of human pro-opiomelanocortin: sequence and evidence for C-terminal amidation. Biochem. biophys. Res. Commun. 102: 710–716 (1981).

106 Seidah, N.G.; Rochemont, J.; Hamelin, J.; Lis, M.; Chrétien, M.: Primary structure of the major human pituitary glycopeptide pro-opiomelanocortin NH_2-terminal glycopeptide: evidence for an aldosterone-stimulating activity. J. biol. Chem. 256: 7977–7984 (1981).

107 Steiner, D.F.: Peptide hormone precursors: biosynthesis, processing and significance; in Parsons, Peptide hormone, pp.49–64 (MacMillan, London 1976).

108 Steiner, D.F.; Kemmler, W.; Tager, H.S.; Peterson, J.D.: Proteolytic processing in the biosynthesis of insulin and other proteins. Fed. Proc. 33: 2105–2115 (1974).

109 Steiner, D.F.; Quinn, P.S.; Chan, S.J.; Marsh, J.; Tager, H.S.: Processing mechanisms in the biosynthesis of proteins. Ann. N.Y. Acad. Sci. 343: 1–16 (1980).

110 Suchanek, G.; Kreil, G.: Translation of melittin messenger RNA in vitro yields a product terminating with glutaminyl-glycine rather than with glutaminamide. Proc. natn. Acad. Sci. USA 74: 975–978 (1977).

111 Tabas, I.; Kornfeld, S.: The synthesis of complex-type oligosaccharides. III. Identification of an α-d-mannosidase activity involved in a late stage of processing of complex-type oligosaccharides. J. biol. Chem. 253: 7779–7786 (1978).

112 Tager, H.S.; Rubenstein, A.H.; Steiner, D.F.: Methods for the assessment of peptide precursors. Studies on insulin biosynthesis. Meth. Enzym. 37: 326–345 (1975).

113 Takahashi, H.; Teranishi, Y.; Nakanishi, S.; Numa, S.: Isolation and structural organization of the human corticotropin-β-lipotropin precursor gene. FEBS Lett. 135: 97–102 (1981).

114 Walter, P.; Blobel, G.: Translocation of proteins across the endoplasmic reticulum. III. Signal recognition protein (SRP) causes signal sequence-dependent and site-specific arrest of chain elongation that is released by microsomal membranes. J. Cell Biol. 91: 557–561 (1981).

115 Weber, E.; Voigt, K.H.; Martin, R.: Concomitant storage of ACTH- and endorphin-

like immunoreactivity in the secretory granules of anterior pituitary corticotrophs. Brain Res. *157:* 385–390 (1978).

116 Weissbach, H.: in Chambliss, Craven, Davies, J., Davis, K., Kahan, Nomura, Ribosomes: structure, function and genetics, pp. 377–411 (University Park Press, Baltimore 1979).

117 Yalow, R.D.; Berson, S.A.: Size heterogeneity of immunoreactive human ACTH in plasma and in extracts of pituitary glands and ACTH-producing thymoma. Biochem. biophys. Res. Commun. *44:* 439–445 (1971).

118 Zwizinski, C.; Wickner, W.: Purification and characterization of leader (signal) peptidase from *Escherichia coli.* J. biol. Chem. *255:* 7973–7977 (1980).

M. Chrétien, MD, Protein and Pituitary Hormone Laboratory, Clinical Research Institute of Montreal, 110 Pine Avenue West, Montreal H2W 1R7 (Canada)

Cell Biology of the Secretory Process, pp. 247–275 (Karger, Basel 1984)

Secretion in the Posterior Pituitary Gland

B. T. Pickering, R. W. Swann

Department of Anatomy, University of Bristol, The Medical School, Bristol, England

Introduction

The neurohypophysis was first shown to contain biologically active substances when *Howell* [1898] demonstrated that the vasopressor activity of pituitary extracts was localized in the posterior lobe of the gland. Since that time, a number of activities have been described for extracts of mammalian neurohypophyses and these have been ascribed to two peptide hormones, oxytocin, which contracts the smooth muscle of the uterus (uterotonic) and causes milk-ejection during suckling, and vasopressin, which increases blood pressure and is the antidiuretic hormone in mammals. To date, ten biologically active vertebrate neurohypophysial peptides have been identified, all of which have been sequenced and most synthesized [*du Vigneaud*, 1960; *Acher*, 1980; *Chauvet* et al., 1980]. It is now understood (fig. 1) that the magnocellular hypothalamic neurons of the paraventricular (PVN) and supraoptic (SON) nuclei in mammals elaborate the hormones oxytocin and vasopressin, which are then transported to the neural lobe of the pituitary gland whence they are released into the blood stream [*Bargmann and Scharrer*, 1951]. Moreover, these endocrine neurons [*Cross* et al., 1975] show all the characteristics of both nerve cells and secretory cells and, in fact, provide a good model system for the study of secretion [*Pickering*, 1978; *Gainer*, 1981].

Structural Components of the System

The general ultrastructural features of the cells of the PVN and SON are similar, the neuronal perikarya containing extensive endoplasmic

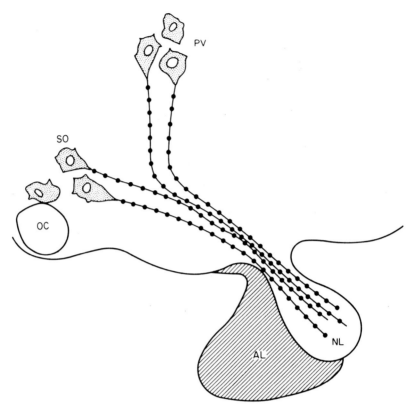

Fig. 1. The hypothalamo-neurohypophysial system. SO=Supraoptic nucleus; PV=pa-raventricular nucleus; AL=anterior lobe; NL=neural lobe of the pituitary gland; OC=optic chiasma [from *Pickering,* 1978].

reticulum and a prominent Golgi complex [*Duncan and Alexander,* 1961; *Sloper and Bateson,* 1965; *Zambrano and de Robertis,* 1966, 1967, 1968; *Flament-Durand,* 1971; *Morris,* 1971; *Cannata and Morris,* 1973]. Neurosecretory granules (NSGs), of about 180 nm diameter, are seen in the perikarya, the non-myelinated nerve fibre and in the neurohypophysis (fig. 2), and it is within these granules that the hormones are packaged by the Golgi and transported to the neural lobe [*Palay,* 1957; *Bargmann* et al., 1957; *Barer* et al., 1963; *Morris,* 1978]. The hormonal stores in the neurohypophysis represent axonal dilatations packed with NSGs (fig. 2) which are then available for release.

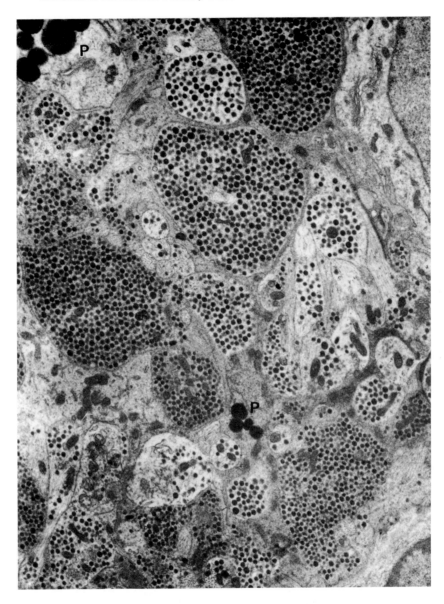

Fig. 2. Electron micrograph of a section of the rat neural lobe. Note the axonal dilatations packed with neurosecretory granules. Attention is drawn to two profiles of pituicytes (P) containing characteristic lipid droplets. ×9,200.

Both SON and PVN contain cells which make oxytocin and those which make vasopressin although it appears that there is regional organization within the nuclei with, for example, oxytocin-producing cells being concentrated in the more rostral parts of PVN [*Swaab* et al., 1975; *Fisher* et al., 1979; *Dierickx,* 1980]. There have been claims that some neurons may make both hormones but the weight of the evidence is against this and in favour of one cell for vasopressin and another for oxytocin [e.g. see *Morris* et al., 1977; *Dierickx* et al., 1978].

It is now completely accepted that the posterior pituitary hormones are neurosecretory products released from the axon terminals in the neural lobe, but this has not always been the case. Although most of the extra-vascular tissue in the neurohypophysis is represented by neuronal components, some 20–30% of it is occupied [*Morris* et al., 1978] by glial type cells known as pituicytes. Before the concept of neurosecretion had been established, the pituicytes had been recognized as the probable source of the neurohypophysial hormones [*Bucy,* 1930; *Gersch,* 1939] and, even now, we must not dismiss them as purely supporting cells with no rôle to play in hormone secretion. *Duchen* [1962] pointed out that stimuli for hormone release led to increased mitotic activity in the pituicytes. There have been many reports about the relationship of pituicyte activity and hormone release [see *Morris* et al., 1978, for recent review] with much significance attached to the appearance and disappearance of the characteristic lipid droplets in these cells. It seems very likely that pituicytes do play a part in the control of neurohypophysial activity, but whether this is biochemical or morphological [*Tweedle and Hatton,* 1980] remains to be shown.

Chemical Components of the System

The Hormones

Figure 3 shows the amino acid sequences of the ten neurohypophysial peptides which have been isolated from vertebrate pituitary glands. They are all structural analogues of oxytocin, the first such peptide to be isolated and synthesized [*du Vigneaud* et al., 1953]. It is not appropriate to consider here either structure-activity relationships [*Berde and Boissonnas,* 1966; *Manning and Sawyer,* 1977; *Pickering,* 1970] or the phylogeny of these peptides [*Acher,* 1974, 1980; *Sawyer,* 1968, 1971; *Heller and Pickering,*

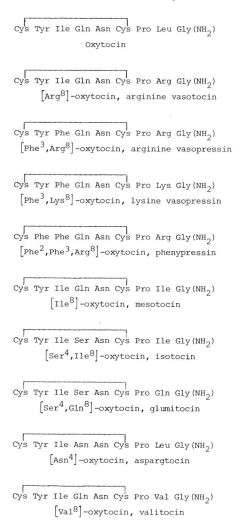

Cys Tyr Ile Gln Asn Cys Pro Leu Gly (NH$_2$)
                    Oxytocin

Cys Tyr Ile Gln Asn Cys Pro Arg Gly (NH$_2$)
    [Arg^8]-oxytocin, arginine vasotocin

Cys Tyr Phe Gln Asn Cys Pro Arg Gly (NH$_2$)
    [Phe^3,Arg^8]-oxytocin, arginine vasopressin

Cys Tyr Phe Gln Asn Cys Pro Lys Gly (NH$_2$)
    [Phe^3,Lys^8]-oxytocin, lysine vasopressin

Cys Phe Phe Gln Asn Cys Pro Arg Gly (NH$_2$)
    [Phe^2,Phe^3,Arg^8]-oxytocin, phenypressin

Cys Tyr Ile Gln Asn Cys Pro Ile Gly (NH$_2$)
        [Ile^8]-oxytocin, mesotocin

Cys Tyr Ile Ser Asn Cys Pro Ile Gly (NH$_2$)
        [Ser^4,Ile^8]-oxytocin, isotocin

Cys Tyr Ile Ser Asn Cys Pro Gln Gly (NH$_2$)
        [Ser^4,Gln^8]-oxytocin, glumitocin

Cys Tyr Ile Asn Asn Cys Pro Leu Gly (NH$_2$)
        [Asn^4]-oxytocin, aspargtocin

Cys Tyr Ile Gln Asn Cys Pro Val Gly (NH$_2$)
        [Val^8]-oxytocin, valitocin

Fig. 3. Amino acid sequences of the ten vertebrate neurohypophysial hormones.

1970] because these have been reviewed extensively; instead we have identified selected reviews. It is, however, worth drawing attention to a development which has occurred since the latest of these reviews, namely the discovery of phenypressin in the neurohypophysis of a marsupial [*Chauvet* et al., 1980].

The Neurophysins

This volume contains a chapter devoted to the chemistry and biology of the neurophysins (p. 276) so that no more will be said than that they are present in NSG in equimolar amounts to the hormones.

Neurohypophysial Glycopeptide

Intracisternal injection of radioactive sugars leads to the incorporation of radioactivity into a molecule of similar molecular size to the neurophysin which has been referred to as neurohypophysial glycopeptide [*Jones and Swann,* 1974; *Gainer and Brownstein,* 1978]. Similarly, a glycopeptide was isolated from posterior pituitary extracts of several species [*Holwerda,* 1972; *Smyth and Massey,* 1979; *Seidah* et al., 1981]. This component, which is probably responsible for the histochemical localization of glycopeptide in NSG [*Tasso,* 1973; *Tasso and Rua,* 1975], is now known to be related to the precursor for vasopressin (p. 253).

The biological rôle of the glycopeptide is still a matter for debate and experiment. In the ACTH system, *Loh and Gainer* [1979] have evidence that the carbohydrate moieties direct the proteolytic cleavage during processing, although this view is challenged by *Herbert* et al. [1980]. Our own findings [*Gonzàles* et al., 1981] in the neurohypophysial system suggest that glycosylated components are concerned with packaging, since the injection of tunicamycin, an inhibitor of N-glycosylation, leads to an accumulation of material in the cisternae of the rough endoplasmic reticulum and an absence of secretory granules. In a recent review *Olden* et al. [1982] have examined evidence from a number of systems and reach the view that carbohydrate 'tags' are attached to molecules to direct them to specific cellular organelles: In other words that glycosylation is concerned with intracellular trafficking. In the particular system with which we are concerned, the neurohypophysial hormones, it is too early to reach a firm conclusion but any such conclusion must embrace the oxytocin precursor for which, at present, there is no evidence for a carbohydrate component.

Biosynthesis and Intraneuronal Transport

Common Precursors for Hormones and Neurophysins

First attempts to study the biosynthetic pathway for vasopressin were made by *H. Sachs* and his colleagues in the 1960s. They were able to show that radioactive cysteine could be incorporated into vasopressin both by the hypothalamus of the dog in vivo and by slices from guinea pig hypothalamus in vitro [*Sachs,* 1960, 1963; *Sachs and Takabatake,* 1964; *Takabatake and Sachs,* 1964]. Moreover, in both systems, the pattern of incorporation was the same: an initial lag phase when no labelled hormone was formed, followed by a steady increase of tracer into vasopressin. The addition of inhibitors of protein synthesis prevented incorporation if made during the lag phase but were ineffective at later times. At that time considerable interest was being shown in the nature of neurophysin(s) and its rôle as a carrier protein(s). *Sachs* and his colleagues showed that the pattern of incorporation of [^{35}S]-cysteine into neurophysin was very similar to that into vasopressin and that it was susceptible to the same inhibitors which acted on neurophysin and hormone in parallel.

In his early experiments, *Sachs* [1963] had shown that the specific radioactivity of vasopressin isolated from the hypothalamus of dogs which had received intraventricular infusions of [^{35}S]-cysteine was not very different from that of vasopressin isolated from the posterior pituitary glands of the same animals. Since it was believed that hormone was synthesized in the hypothalamus and transported to the neurohypophysis, and since the pool size of hormone in the latter was 200 times or more greater than the former, it had been expected that the specific radioactivity of hypothalamic hormone would have been many times that of glandular hormone. *Sachs* interpreted this observation as suggesting that the neuronal perikarya in the hypothalamus synthesized a precursor which subsequently gave rise to vasopressin during transport toward the pituitary. These suggestions, taken together with his later data, led *Sachs* to propose that vasopressin and neurophysin shared a common precursor protein which was cleaved during transport [*Sachs,* 1969; *Sachs* et al., 1969]. This hypothesis is depicted in figure 4 which also includes the subsequent cleavage of the neurophysin in the oxytocin neuron, [*Burford and Pickering,* 1973].

The common precursor hypothesis received great support when it became clear that there is indeed a neurophysin which is related to vasopressin and one which is related to oxytocin. Such evidence came from a number of sources. Firstly, *Dean* et al. [1968] observed that, during the density-gradient centrifugation of bovine NSGs, there was a tendency for oxytocin-containing granules to equilibrate at a slightly lower density than the vasopressin ones and that bovine neurophysin I (bNpI), characterized by its lack of methionine and content of histidine tended to go along with oxytocin and bovine neurophysin II (bNpII) with vasopressin. Moreover,

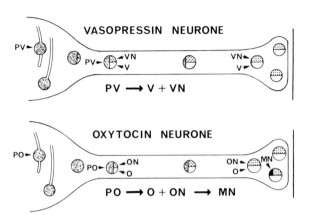

Fig. 4. Representation of the maturation of precursors in vasopressin- and oxytocin-
producing neurons. PV=Provasopressin (propressophysin); VN=vasopressin-neurophysin;
V=vasopressin; PO=pro-oxytocin (pro-oxyphysin); ON=oxytocin-neurophysin; MN=mi-
nor neurophysin; O=oxytocin.

when radioimmunoassays became available for the separate bovine neurophysins, it was found that conditions, or stimuli, associated with the
release of oxytocin (e.g. parturition, milking) led to a rise in the blood level
of bNpI whereas those which are associated with vasopressin release (e.g.
dehydration, haemorrhage) increased the level of bNpII in the circulation
[*Robinson* et al., 1971; *Legros and Louis,* 1974; *Legros,* 1975].

Further evidence for a biosynthetic association of one neurophysin
with oxytocin and another with vasopressin came from the rat when it was
shown that the Brattleboro rat which is unable to synthesize vasopressin is
also lacking one of the rat neurophysins [*Burford* et al., 1971]. There is now
good reason to believe that similar specific associations of one hormone
and one neurophysin occur in other species, e.g. human [*Robinson,* 1975]
and pig [*Pickup* et al., 1973; *Dax and Johnston,* 1978].

Although the hypothesis that vasopressin and neurophysin share a
common precursor was discussed by *Sachs* in 1969, the evidence in favour
was for a long time only circumstantial [*Pickering,* 1976, 1978]. More
definitive evidence began to appear towards the end of the 1970s and
beginning of the eighties. Much of this stemmed from the work of
H. Gainer and his colleagues, who showed [*Gainer* et al., 1977a, b] that
[^{35}S]-cysteine injected into SON was incorporated into polypeptides with

molecular weights of about 20,000. These 20K components, which were subsequently shown [*Brownstein* et al., 1977] to possess neurophysin immunodeterminants, were the first species to be labelled in SON extracts after the injection. As the time between isotope injection and preparation of extract increased, there was a diminution in the labelling of the 20K components and appearance of label in components with the characteristics of the rat neurophysins. Moreover, closer inspection of the 20K putative precursors showed that they were in fact composed of two pairs, one with molecular weight of 20K and one of 17K, so that the *Gainer* group considered them to be a pair of 20K precursors and a pair of 17K intermediates. When *Gainer* and his colleagues applied their techniques to Brattleboro rats they found that one 20K and one 17K component were missing [*Russell* et al., 1980a] so that it was possible to ascribe components to either the vasopressin or the oxytocin pathway. More precise analysis showed that there was a size difference between the precursor for vasopressin and that for oxytocin. The former referred to by *Russell* et al. [1980b] as propressophysin was estimated by these workers to have a molecular weight of 20.5K while they found the oxytocin precursor, prooxyphysin, to behave as a 18.7K species. In our hands (fig. 5), sodium dodecyl sulphate electrophoresis distinguishes two components (19K and 21K), each with immunodeterminants for both vasopressin and its neurophysin, from a component precipitable by antisera to oxytocin and its neurophysin which behaves as 15K.

As soon as estimates of the size of the precursors began to become available it was apparent that the precursor was larger than the simplest possibility: hormone linked to neurophysin. A possible candidate for the additional fragment arose from the discovery of a posterior pituitary component of molecular weight about 10,000 which became labelled after injection of [^{3}H]-fucose into the cerebrospinal fluid or SON of rats [*Jones and Swann,* 1974; *Gainer and Brownstein,* 1978] and which was present within the NSGs [*Jones and Swann,* 1975]. Such a glycopeptide consisting of 39 amino acid residues has been characterized from posterior pituitary extracts of several species [*Holwerda,* 1972; *Smyth and Massey,* 1979; *Seidah* et al., 1981] and, as we shall see later, is almost certainly part of the vasopressin precursor.

Until now, we have been discussing evidence which has accrued from experiments involving isotope injection in vivo. Several laboratories have succeeded in directing cell-free synthesis of neurophysin precursors with mRNA isolated from the supraoptic region of hypothalami of mice [*Lin* et

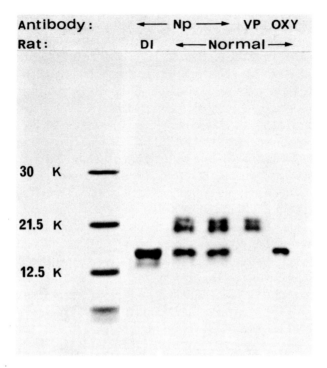

Fig. 5. Autoradiogram of sodium dodecyl sulphate/polyacrylamide gel electrophoresis run with immunoprecipitates prepared from SON extracts taken from normal or Brattleboro (DI) rats 1 h after an intranuclear injection of [^{35} S]-cysteine (10 µCi/side). Antisera used would precipitate total rat neurophysins (Np) or vasopressin (VP) or oxytocin (OXY) [unpubl. data of *C.B. González*].

al., 1979], rats [*Lin* et al., 1979; *Schmale and Richter,* 1981a] or cattle [*Giudice and Chaiken,* 1979; *Schmale and Richter,* 1980; *Richter* et al., 1980]. The results can be summarized in terms of the findings of *Richter* and his colleagues. In the vasopressin-related series, using bovine mRNA, common precursors to vasopressin and bovine vasopressin-neurophysin (bNpII) can be immunoprecipitated with antisera to either of these polypeptides. Such techniques have led *Richter* and his associates to conclude that bovine supraoptic mRNA directs the synthesis of an initial precursor (pre-proform) of apparent molecular weight 21,000. In the presence of microsomal membranes this 21K component is replaced by a 23K one which is glycosylated. The microsomal membranes, besides providing the mechanism for N-glycosylation, would be expected to cleave off the signal peptide (or lead sequence) and this is supported by the observation that

BOVINE PRE-PROVASOPRESSIN

MET-PRO-ASP-ALA-THR-LEU-PRO-ALA-CYS-PHE-LEU-SER-LEU-LEU-ALA-PHE-THR-SER-ALA-

CYS-TYR-PHE-GLN-ASN-CYS-PRO-ARG-GLY-GLY-LYS-ARG-ALA-MET-SER-ASP-LEU-GLU-LEU-

ARG-GLN-CYS-LEU-PRO-CYS-GLY-PRO-GLY-GLY-LYS-GLY-ARG-CYS-PHE-GLY-PRO-SER-ILE-

CYS-CYS-GLY-ASP-GLU-LEU-GLY-CYS-PHE-VAL-GLY-THR-ALA-GLU-ALA-LEU-ARG-CYS-GLN-

GLU-GLU-ASN-TYR-LEU-PRO-SER-PRO-CYS-GLN-SER-GLY-GLN-LYS-PRO-CYS-GLY-SER-GLY-

GLY-ARG-CYS-ALA-ALA-ALA-GLY-ILE-CYS-CYS-ASN-ASP-GLU-SER-CYS-VAL-THR-GLU-PRO-

GLU-CYS-ARG-GLU-GLY-VAL-GLY-PHE-PRO-ARG-ARG-VAL-ARG-*Ala-Asn-Asp-Arg-Ser-Asn*-

Ala-Thr-Leu-Leu-Asp-Gly-Pro-Ser-Gly-Ala-Leu-Leu-Leu-Arg-Leu-Val-Gln-Leu-Ala-

Gly-Ala-Pro-Glu-Pro-Ala-Glu-Pro-Ala-Gln-Pro-Gly-Val-Tyr

VASOPRESSIN
VASOPRESSIN-NEUROPHYSIN
Glycopeptide

Fig. 6. The sequence of pre-provasopressin as deduced by *Land* et al. [1982].

during synthesis in the presence of membranes which had been treated with tunicamycin, an inhibitor of N-glycosylation, the predominant species was a 19K component [*Schmale and Richter,* 1981b], probably representing the unglycosylated pro-form.

The isolation of the appropriate mRNA described above paved the way for the complete characterization of the common vasopressin/vasopressin-neurophysin precursor. *Land* et al., [1982] constructed the complementary DNA for bovine hypothalamic mRNA, inserted it into a plasmid which was amplified in cultured *Escherichia coli.* After selecting clones in which the plasmid was replicated they were able to isolate the appropriate complementary DNA, sequence it and hence predict the amino acid sequence of the precursor as shown in figure 6. This sequence corresponds to the 21K pre-propressophysin detected by *Schmale and*

Richter [1981b], even though its molecular weight is 17,310: Sodium dodecyl sulphate gel electrophoresis seems to give falsely high estimations of molecular weight for neurophysin-like proteins. The precursor has a 19 residue signal peptide, rich in hydrophobic amino acids, followed by the sequence of vasopressin and then that of bovine vasopressin-neurophysin (bNpII). The 39 residue sequence which follows neurophysin is identical to that of the glycopeptide isolated from bovine pituitaries by *Smyth and Massey* [1979].

The putative oxytocin precursor seems to be a smaller molecule, both in cattle [*Schmale and Richter*, 1980] and in rats [*Russell* et al., 1980b] (fig. 5), and is not adsorbed to immobilized concanavalin A [*Schmale and Richter*, 1980; *Russell* et al., 1980b]. Thus there is no evidence to suggest that the oxytocin/oxytocin-neurophysin precursor is glycosylated and, indeed, the Brattleboro rat, which is incapable of making vasopressin, does not have a neural lobe glycopeptide which can be labelled by injection of [^{3}H]-fucose into the SON [*Russell* et al., 1980b].

Processing of Hormone and Neurophysin Precursors

When considering the processing of the precursors there are two questions to be addressed. Firstly, what is the chain of events and, secondly, where do the conversions take place?

It is convenient to take the second question first, since there have been suggestions about this for many years. The experiments which prompted *Sachs* [1963] to suggest that vasopressin is synthesized by way of a precursor protein also gave the first indication that processing of this precursor occurred within the secretory granule as it passed from the cell body in the hypothalamus towards the axon terminals in the neural lobe. Further evidence for 'maturation' in transit came from the observation that radioactive, fully formed hormone [*Jones and Pickering*, 1970, 1972] and neurophysin [*Norström and Sjöstrand*, 1971; *Burford and Pickering*, 1973] arrived in the neural lobe of rats within 1–1.5 h of the presentation of labelled amino acid to the hypothalamus. Since *Takabatake and Sachs* [1964] had shown that the maturation time for the vasopressin precursor was itself 1–1.5 h, the fast transport velocities deduced for the neural lobe components became astronomical if conversion of the precursor was not occurring during transport of the secretory granule along the axon. Confirmation that the precursors do become processed during transport was provided by *Gainer* et al. [1977a,b], who showed that there was a progressive shift of the proportion of precursors and products in favour of the

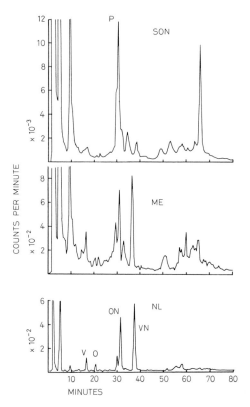

Fig. 7. High performance liquid chromatograms [*Swann* et al., 1982] prepared with extracts of supraoptic nucleus (SON), median eminence (ME) and neural lobe (NL) prepared from rats 2 h after an injection of [^{35}S]-cysteine into SON. P = Precursors; V = vasopressin; O = oxytocin; ON = oxytocin-neurophysin; VN = vasopresin-neurophysin.

latter in extracts made from SON, median eminence and neural lobe. A similar demonstration is made in figure 7.

With the establishment of the complete sequence of pre-propressophysin, the processing of the precursor becomes easy to appreciate. As mentioned earlier, three types of component of the neurosecretory granules have been characterized: hormone, neurophysin and glycopeptide. These three components, together with the expected signal peptide, account for the whole of the initial precursor (fig. 6). The Ala-Cys bond separating the putative signal peptide from the sequence of vasopressin falls within the group of bonds found to link such peptides in pre-proforms

	Vasopressin	α-MSH [*Mains* et al., 1977]	Honeybee mellitin [*Suchanek* et al., 1978]
Precursor sequence	...Pro.Arg.Gly. Gly.Lys.Arg...	...Lys.Pro.Val.Gly. Lys.Lys...	...Arg.Gln.Gln. Gly.OH
Peptide carboxy- terminal sequence	...Pro.Arg.Gly.NH$_2$	...Lys.Pro.Val.NH$_2$	...Arg.Gln.Gln.NH$_2$

Fig. 8. Precursor sequences giving rise to amidated peptides.

of other secretory polypeptides [*Steiner* et al., 1980]. The other two cleavage sites between hormone and neurophysin on the one hand, and neurophysin and glycopeptide on the other, have trypsin-sensitive bonds. Indeed the former consists of the pair of basic residues -Lys-Arg- which is the most common bonding at cleavage sites in propeptides [*Steiner* et al., 1980]. In contrast, neurophysin is linked to the glycopeptide by a single arginyl residue and the sequence -Arg-Ala- gives no indication of why its enzyme sensitivity should be greater than any other arginyl bond in the precursor so that it may be that the intragranular conformation of the precursor contributes to this.

The fully formed neurohypophysial hormones are amidated at their carboxytermini but, so far, no enzyme has been characterized which will perform the required amidation. There have been suggestions that the mechanism is one of transamidation and it has been suggested that such a reaction would be favoured at a bond involving glycine at its carboxyterminal side [*Bradbury* et al., 1976]. Certainly such a sequence is seen in the precursors of a number of amidated polypeptides (fig. 8) and it may be that rather than solely acting as the recognition site for a transamidase the glycine donates its nitrogen by a mechanism which is yet to be determined. (Note added in proof: Evidence for this has been given recently by *Bradbury, A. F., Finnie, M. D. A., and Smyth, D. G.* [Nature, Lond. *298:* 686–688 (1982)].)

One can only guess at possible intermediates in the biosynthetic pathway. *Gainer* and his colleagues have given evidence for intermediates from their studies with the rat in vivo (*Brownstein* et al., 1977) and our own evidence (fig. 5) is also compatible with more than one polyprotein containing sequences of both hormone and neurophysin. It is likely, however, that these components represent different levels of glycosylation and, for this reason, we should not conclude that the smaller component of a pair necessarily represents the product rather than the precursor.

On the other hand, *Schmale and Richter* [1980, 1981] have described an 18K component, produced in vitro under the direction of bovine hypothalamic RNA, which has the immunodeterminants of bNpII but not of vasopressin. Such a component would not be unexpected if it were to represent cleavage of hormone from the precursor, involving the sensitive -Lys-Arg- sequence, leaving a polypeptide consisting of neurophysin linked to the glycopeptide. None of the in vivo experiments has indicated the presence of a corresponding component and it may be, as suggested above, that the conformation imposed upon the precursor within the secretory granule confers an increased sensitivity on the single arginyl bond linking neurophysin and glycopeptide.

As mentioned above, the conversion of prohormone into its products occurs while the secretory granule is passing along the axons and is largely complete by the time it arrives in the neural lobe. The question arises as to whether the in-transit processing is solely a time-dependent phenomenon or requires actual granule movement. In other words, is there any control of the process provided by the extragranular environment through which the granule passes as it moves along? The answer seems to be no, since processing appears to continue within granules which have been arrested in the perikarya by treating with small doses of colchicine [*Parish* et al., 1981]. This treatment has little effect on synthesis of the primary precursor or on the nature of processing [*Birkett* et al., 1982].

Mention must be made of the very large ($>70K$) molecules reported recently [*Béguin* et al., 1981; *Rosenior* et al., 1981] which cross-react with antisera both to the hormones and to the neurophysins. It is tempting to consider these as large precursors, perhaps directed by a different genetic message from that discussed above and, indeed, *Lauber* et al. [1981] have made such a suggestion to explain their findings that one of these large molecules contains components of the ACTH system in addition to neurophysin and vasopressin. However, just because they are larger than hormone and neurophysin does not necessarily mean that they are precursors! It could equally well be that these large molecules arise at later stages in the secretory pathway and represent a membrane-associated product. Whatever their origin, these very large molecules of apparent pluripotency do present an interesting problem for future study.

Mechanism of Transport

NSGs are transported along the axons at velocities of 50–100 mm/day. This puts them into the class of cellular constituents transported at rates which *Willard* and his colleagues have called group II moving at a somewhat lower velocity than that (about 400 mm/day) of the fastest moving components [*Lorenz and Willard*, 1978; *Baitinger* et al., 1982]. It is believed that transport of the vesicles involves interaction with microtubules [*Schmitt*, 1969] and the ability of colchicine to arrest movement supports this view, since this drug inhibits the polymerization of tubulin.

Neurohypophysial Hormone Release

Relationship of Electrical Activity and Release
Various physiological stimuli such as suckling, increased osmotic pressure or reduction in the volume of extracellular fluid, will induce hormone release. These stimuli, however, do not act directly upon the store of hormone in the neural lobe but through the neurosecretory cell bodies, which propagate action potentials along the hypothalamo-neurohypophysial tract to the neurohypophysis [*Koizumi* et al., 1964; *Dreifuss* et al., 1971; *Wakerley and Lincoln*, 1973]. In the case of oxytocin a slow irregular pattern of action potentials is recorded from the antidromically identified oxytocin cells between milk-ejections in the anaesthetized lactating rat. However, just prior to an increase in intramammary pressure and milk-ejection there is a fast continuous burst of firing [*Wakerley and Lincoln*, 1973], which is related to the release of oxytocin (fig. 9). Vasopressin cells have a pattern of firing which enables them to be distinguished from oxytocin cells. Using the vasopressin-specific releasing stimulus of haemorrhage, *Poulain* et al. [1977] showed that vasopressin cells changed from a slow irregular pattern of firing in the unstimulated state to a pattern of high frequency 'bursts' interspersed with periods of quiescence, that is, the cells became phasic. Similar recordings have been obtained from isolated hypothalamic cells in tissue slices [*Haller* et al., 1978; *Gähwiler and Dreifuss*, 1980]. Why the two cell types should show different electrical properties upon stimulation may well reflect their respective patterns of secretion. Oxytocin, during lactation, is required in pulses, each pulse separated by several minutes. Stimulation of vasopressin release, on the other hand, during dehydration or haemorrhage for example, may be an intense and protracted affair, and it has been suggested that a phasic pattern of firing may be optimal for vasopressin release [*Dutton and Dyball*, 1979]. Why phasic firing should be more effective in releasing hormone

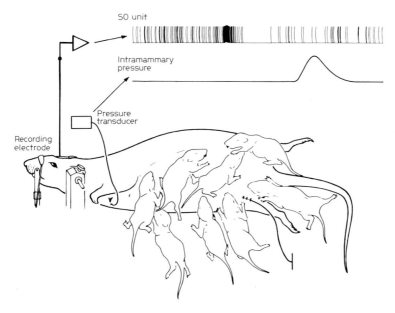

SO unit

Intramammary
pressure

Pressure
transducer

Recording
electrode

Fig. 9. Scheme for measuring the electrical activity of hypothalamic neurosecretory neurons associated with reflex release of oxytocin during suckling in the rat. The upper trace (supraoptic unit: SO unit) shows the action potentials recorded from a cell in the SON while the lower trace is recording of intramammary pressure. Note the burst of potentials which occurs 12 s before the increase in pressure [by courtesy of *D. W. Lincoln*].

may be related to the refractory period which follows electrical stimulation of the neurons [*Bicknell* et al., 1982]. Whether such refractory periods are a consequence of calcium ion distribution remains to be investigated.

On arrival of the electrical impulses at the neurohypophysis there is an influx of calcium ions into the nerve endings and an efflux of hormone out of them [*Douglas*, 1974]. Two ions appear to be essential for release: sodium, as an essential ionic component of the action potential, and calcium which has been found to be intrinsic to the phenomenon of stimulus-secretion coupling [*Rubin*, 1970, and this volume; *Douglas*, 1974]. It is almost certain that the influx of calcium ions into the nerve endings is the initiator of a process which results in the release of hormone by exocytosis [*Nagasawa* et al., 1970]. During this exocytotic event the neurosecretory granule moves towards and fuses with the plasma membrane, an opening to the extracellular space appears and the contents are evacuated [*de Robertis and Vaz Ferreira*, 1957].

Membrane Retrieval after Exocytosis

While it has been generally agreed that membrane retrieval after exocytosis is essential for the maintenance of cell integrity, the mechanism whereby this is brought about has been the source of some controversy. Initially it was suggested that membrane is recaptured by micropinocytosis as microvesicles [*Nagasawa* et al., 1970, 1971].

Although micropinocytosis may account for the recapture of some plasma membrane there is now good evidence that most occurs by 'macropinocytosis'in the form of large vacuoles. Firstly, the volume of uptake of extracellular radioactive tracers after hormone release is larger than would be expected if re-uptake occurs by microvesicles [*Nordmann* et al., 1974]. From our own work [*Swann and Pickering,* 1976] it is clear that labelled granule membrane remains in the gland after stimulation of hormone release and that this membrane has sedimentation properties similar to neurosecretory granules rather than to microvesicles. A population of large vacuoles, about the same size as neurosecretory granules, has also been shown to increase in nerve endings in the neural lobe after stimulation [*Morris and Nordmann,* 1980] and to take up horseradish peroxidase [*Castel,* 1974; *Nordmann* et al., 1974; *Theodosis* et al., 1976]. In spite of this evidence there are some who still support the idea of retrieval by micropinocytosis [*Haddad* et al., 1980].

Rôle of Microvesicles

What then is the rôle of microvesicles if not as a means for membrane recapture? An intriguing suggestion has recently been advanced that they may play a part in the homeostasis of calcium [*Morris* et al. 1981]. Microvesicles are rich in Ca^{2+} [*Shaw and Morris,* 1980], which can be demonstrated by its precipitation using an oxalate-pyroantimonate technique, and are capable of ATP-dependent accumulation of this ion [*Nordmann and Chevallier,* 1980]. These workers have suggested that the function of microvesicles might well be to buffer calcium concentration in the cytosol, acting, as it were, as a calcium sink. There remains, of course, the possibility that 'the microvesicles' represent a heterogeneous collection of a number of organelles with different functions, some of which may be concerned with membrane recycling.

The Problem of the Readily Releasable Pool

Prolonged stimulation of the neural lobe results in an increase followed by a decline in hormone release [*Sachs* et al., 1967; *Sachs* et al.,

1969; *Thorn*, 1966]. To obtain a further rise in hormone release the gland had to be left for a time before further stimulation. This observation led to the concept of the 'readily releasable' pool of hormone and *Sachs* et al. [1967] proposed that this reflected a spatial relationship between the granules and the plasma membrane. Granules closest to the plasma membrane would be released immediately while those further away would be released more slowly as they approached the plasma membrane. This view was seriously challenged when it was shown that the decline in the initial hormone release could be explained by limiting calcium concentrations [*Thorn* et al., 1975; *Nordmann*, 1976]. Firm evidence for this alternative has come from experiments where veratridine, which increases intracellular calcium concentrations, has been used to stimulate hormone release from neurohypophyses in vitro [*Nordmann and Dyball*, 1978; *Lescure and Nordmann*, 1980]. After 60 min incubation in a high K$^+$ solution, when the rate of hormone release had fallen to a low level, a further increase in hormone release could be evoked by veratridine.

Although the idea of separating the granules into two populations based on their spatial orientation does not seem to explain the phenomenon of the 'readily releasable' pool, it does have relevance to the 'last in-first out' phenomenon. Isotope incorporation studies have consistently shown that relatively more radioactively labelled hormone and neurophysin is released from glands when stimulated at short times after isotope injection compared to longer times [*Sachs*, 1971; *Norström*, 1974; *Wong and Pickering*, 1976]. Newly synthesized material is clearly released preferentially. How is this accomplished? To test the spatial hypothesis, *Heap* et al. [1975] prepared electron microscopic autoradiograms of rat neural lobes taken at various times after intracisternal [^{35}S]-cysteine injection. Distinguishing axonal dilatations containing microvesicles which were in close apposition to basement membrane as 'nerve endings' and larger dilatations which lacked these features as 'nerve swellings', they found that the radioactive granules arrived first in the axons and then moved into the nerve endings. With time they then moved into larger and larger nerve swellings. Thus, compatible with the 'last in-first out' principle for isotopically labelled hormone and neurophysin, the neurosecretory granules appear to first enter nerve endings, a release site [*Nordmann and Morris*, 1976; *Morris and Nordmann*, 1980] in the neural lobe before being shunted into even deeper storage pools (fig. 10). A modification of this scheme has been proposed recently by *Chapman* et al. [1982], who envisage a random distribution of granules between 'endings' and 'swellings'

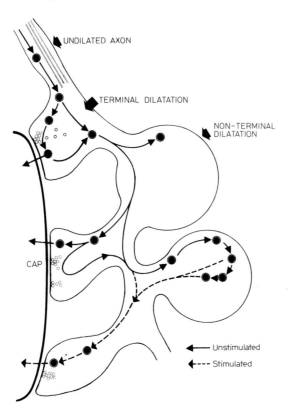

Fig. 10. Scheme for the movement of neurosecretory granules through the compartments of the hypothalamo-neurohypophysial neuron within the neural lobe of the pituitary gland. CAP = Blood capillary [from *Heap* et al., 1975].

according to the space available. This would still result in a preferential movement into 'endings' because this is where release takes place.

In what way granules in the nerve endings differ from those in nerve swellings is not clearly understood. Certainly enzymic activity is continuing within the granules in the neural lobe as can be seen from the degradation of the major rat oxytocin-neurophysin to its breakdown product, the minor rat oxytocin-neurophysin (previously called rNpC) [*Burford and Pickering,* 1973; *North* et al., 1977]. More recently *Nordmann* et al., [1979] have separated two populations of NSGs from rat neural lobe homogenates after fractionation on metrizamide gradients. One population is osmotically stable and takes up [^{35}S]-cysteine which then passes to the second population which is osmotically labile [*Nordmann and La-bouesse,* 1981]. This, together with the observation from dual labelling experiments [*Morris* et al., 1981] that osmotically stable NSGs are more easily released from the neural lobe than are osmotically labile NSGs suggests the former represent young granules while the latter are older granules. It will be interesting to discover exactly what the differences between these two granule types are and at what spatial and/or biochemical stage, if any, a neurosecretory granule is rendered unavailable for release.

This is necessarily a very sketchy discussion of hormone release. There is space to consider neither the afferent pathways leading to the excitation of the neurosecretory neurons [*Poulain and Wakerley,* 1982; *Lincoln and Paisley,* 1982] nor the evidence which is accumulating that opioid peptides play a part in the modulation of their activity [*Clarke* et al., 1979; *Iversen* et al., 1980].

We have confined ourselves to the release of hormones from the posterior pituitary gland. It must be remembered, however, that some of the oxytocin- and vasopressin-containing fibres which leave the hypothalamic nuclei terminate in other parts of the brain [*Buijs,* 1978; *Sofroniew and Weindl,* 1981]. This almost certainly reflects the anatomical correlate of the growing awareness of the importance of peptides like the neurohypophysial hormones on behaviour [*de Wied and Gispen,* 1977].

Other Neurohypophysial Peptides

For the sake of completeness we must record the demonstration of other peptides in neuron terminals of the posterior pituitary: somatostatin [*Hökfelt* et al., 1975], cholecystokinin octapeptide [*Beinfeld* et al., 1980] and enkephalin [*Rossier* et al., 1977]. Indeed, there is some evidence that some of them may coexist with the neurohypophysial hormones in the same neurons [*Vanderhaeghen* et al., 1981; *Martin and Voigt,* 1981] but this needs to be substantiated. It is too early even to speculate on the rôle, if any, that these peptides play in the release of oxytocin and vasopressin.

Likewise, our recent finding of the hormones and neurophysin-like peptides in the ovary and testis [*Wathes and Swann,* 1982; *Wathes* et al., 1982] further widens the biological rôle of these peptides in the control of reproductive processes. It will be interesting to compare their synthesis and release in these peripheral tissues with the processes in the neurosecretory neuron.

References

Acher, R.: Chemistry of the neurohypophysial hormones: an example of molecular evolution; in Handbook of physiology-endocrinology IV, part 1, pp. 119–130 (Am. Physiol. Soc., Washington, D.C. 1974).

Acher, R.: Molecular evolution of biologically active polypeptides; in Neuroactive peptides. Proc. R. Soc. B *210:* 21–43 (1980).

Baitinger, C.; Levine, J.; Lorenz, T.; Simon, C.; Skene, P.; Willard, M.: Characteristics of axonally transported porteins; in Weiss, Axoplasmic transport, pp. 110–120 (Springer, Berlin 1982).

Barer, R.; Heller, H.; Lederis, K.: The isolation, identification and properties of the hormonal granules of the neurohypophysis. Proc. R. Soc. B *158:* 388–416 (1963).

Bargmann, W.; Kroop, A.; Thiel, A.: Electronenmikroskopische Studie an der Neurohypophyse von *Tropidonotus natrix* (mit Berücksichtigung der Pars intermedia). Z. Zellforsch. *47:* 114–126 (1957).

Bargmann, W.; Scharrer, E.: The site of origin of the hormones of the posterior pituitary. Am. Scient. *39:* 255–259 (1951).

Béguin, P.; Nicolas, P.; Boussetta, H.; Fahy, C.; Cohen, P.: Characterization of the 80,000 molecular weight form of neurophysin isolated from bovine neurohypophysis. J. biol. Chem. *256:* 9289–9294 (1981).

Beinfeld, M.C.; Meyer, D.K.; Brownstein, M.J.: Cholecystokinin octapeptide in the rat hypothalamo-neurohypophysial system. Nature, Lond. *288:* 376–378 (1980).

Berde, B.; Boissonnas, R.A.: Synthetic analogues and homologues of the posterior pituitary hormones; in Harris Donovan, The pituitary gland, vol. 3, pp. 624–661 (Butterworths, London 1966).

Bicknell, R.J.; Brown, D.; Ingram, C.D.; Leng, G.: Rapid fatigue of vasopressin secretion from the isolated rat neurohypophysis during sustained electrical stimulation. J. Physiol., Lond. *327:* 38–39 (1982).

Birkett, S.D.; Swann, R.W.; Gonzáles, C.B.; Pickering, B.T.: Analysis of the neurohypophysial components accumulating in the supraoptic nucleus of the rat after injection of colchicine (in preparation, 1982).

Bradbury, A.F.; Smyth, D.G.; Snell, C.R.: Prohormones of β-melanotropin (β-melanocyte-stimulating hormone, β-MSH) and corticotropin (adrenocorticotropic hormone, ACTH): structure and activation. Ciba Fnd Symp. *41:* 61–75 (1976).

Brownstein, M.J.; Robinson, A.G.; Gainer, H.: Immunological identification of rat neurophysin precursors. Nature, Lond. *269:* 259–261 (1977).

Bucy, P.C.: The pars nervosa of the bovine hypophysis. J. comp. Neurol. *50:* 505–519 (1930).

Buijs, R.M.: Intra- and extrahypothalamic vasopressin and oxytocin pathways in the rat. Cell Tiss. Res. *192:* 423–435 (1978).

Burford, G.D.; Jones, C.W.; Pickering, B.T.: Tentative identification of a vasopressin-neurophysin and an oxytocin-neurophysin in the rat. Biochem. J. *124:* 809–813 (1971).

Burford, G.D.; Pickering, B.T.: Intra-axonal transport and turnover of neurophysins in the rat. A proposal for a possible origin of the minor neurophysin component. Biochem. J. *136:* 1047–1052 (1973).

Cannata, M. A.; Morris, J. F.: Changes in the appearance of hypothalamo-neurohypophysial
    neurosecretory granules associated with their maturation. J. Endocr. *57:* 531–538
    (1973).

Castel, M.: In vitro uptake of tracers by neurosecretory axon terminals in normal and
    dehydrated mice. Gen. compar. Endocr. *22:* 336–337 (1974).

Chapman, D. B.; Morris, J. F.; Valtin, H.: How do granules distribute between nerve endings
    and nerve swellings in the neural lobe? Evidence from Brattleboro rats; in Dreifuss,
    Baertschi, Vasopressin, corticoliberin and ACTH related peptides (in press, 1982).

Chauvet, M. T.; Hurpet, D.; Chauvet, J.; Acher, R.: Phenypressin (Phe^2-Arg^8-vasopressin),
    a new neurohypophysial peptide found in marsupials. Nature, Lond. *287:* 640–642
    (1980).

Clarke, G.; Wood, P.; Merrick, L.; Lincoln, D. W.: Opiate inhibition of peptide release
    from the neurohumoral terminals of hypothalamic neurones. Nature, Lond. *282:* 748
    (1979).

Cross, B. A.; Dyball, R. E. J.; Dyer, R. G.; Jones, C. W.; Lincoln, D. W.; Morris, J. F.; Pick-
    ering, B. T.: Endocrine neurones. Recent Prog. Horm. Res. *31:* 243–294 (1975).

Dax, E. M.; Johnston, C. I.: The distribution of neurophysins and posterior pituitary
    hormones in the porcine neurohypophyseal system. J. Neurochem. *31:* 853–860
    (1978).

Dean, C. R.; Hope, D. B.; Kazic, T.: Evidence for the storage of oxytocin with neurophysin-I
    and of vasopressin with neurophysin-II in separate neurosecretory granules. Br. J.
    Pharmacol. *34:* 192P–193P (1968).

Dierickx, K.: Immunocytochemical localization of the vertebrate cyclic nonapeptide neu-
    rohypophyseal hormones and neurophysins. Int. Rev. Cytol. *62:* 119–185 (1980).

Dierickx, K.; Vandesande, F.; Goossens, N.: The one neuron-one hormone hypothesis and
    the hypothalamic magnocellular neurosecretory system of the vertebrates; in Vincent,
    Kordon, Colloques Internationaux du CNRS, No. 280, Biologie cellulaire des pro-
    cessus neurosecretoires hypothalamiques, pp. 391–398 (Centre National de la Re-
    cherche Scientifique, Paris 1978).

Douglas, W. W.: Exocytosis and the exocytosis-vesiculation sequence with special reference
    to neurohypophysis, chromaffin and mast cells, calcium and calcium ionophores; in
    Thorn, Petersen, Secretory mechanisms of exocrine glands, pp. 801–814 (Munksgaard,
    Copenhagen 1974).

Dreifuss, J. J.; Kalnins, I.; Kelly, J. S.; Ruf, K. B.: Action potentials and release of neuro-
    hypophysial hormones in vitro. J. Physiol., Lond. *215:* 805–817 (1971).

Duchen, L. W.: The effects of ingestion of hypertonic saline on the pituitary gland in the rat: a
    morphological study of the pars intermedia and the posterior lobe. J. Endocr. *25:*
    161–168 (1962).

Duncan, D.; Alexander, R.: An electron microscope study of the supraoptic nucleus of the
    rat. Anat. Rec. *139:* 223 (1961).

Dutton, A. I.; Dyball, R. E. J.: Phasic firing enhances vasopressin release from the rat neu-
    rohypophysis. J. Physiol., Lond. *290:* 433–440 (1979).

Fisher, A. W. F.; Price, P. G.; Burford, G. D.; Lederis, K.: A 3-dimensional reconstruction of
    the hypothalamo-neurohypophysial system of the rat. Cell Tiss. Res. *204:* 343–354
    (1979).

Flament-Durand, J.: Ultrastructural aspects of the paraventricular nuclei in the rat. Z.
    Zellforsch. mikrosk. Anat. *116:* 61–69 (1971).

Gähwiler, B. H.; Dreifuss, J. J.: Transition from random to phasic firing induced in neurons cultured from the hypothalamic supraoptic area. Brain Res. *193:* 415–425 (1980).

Gainer, H.: The biology of neurosecretory neurons; in Martin, Reichlin, Bick, Neurosecretion and brain peptides, pp. 5–20 (Raven Press, New York 1981).

Gainer, H.; Brownstein, M. J.: Electrophoretic analyses of proteins transported to the posterior pituitary. J. Neurochem. *30:* 1509–1512 (1978).

Gainer, H.; Sarne, Y.; Brownstein, M. J.: Biosynthesis and axonal transport of rat neurohypophysial proteins and peptides. J. Cell Biol. *73:* 366–381 (1977a).

Gainer, H.; Sarne, Y.; Brownstein, M. J.: Neurophysin biosynthesis: conversion of a putative precursor during axonal transport. Science *195:* 1354–1356 (1977b).

Gersch, I.: The structure and function of the parenchymatous glandular cells in the neurohypophysis of the rat. Am. J. Anat. *64:* 407–443 (1939).

Giudice, L. C.; Chaiken, I. M.: Cell-free biosynthesis of different high molecular weight forms of bovine neurophysins I and II coded by hypothalamic mRNA. J. biol. Chem. *254:* 11767–11770 (1979).

González, C. B.; Swann, R. W.; Pickering, B. T.: Effects of tunicamycin on the hypothalamo-neurohypophysial system of the rat. Cell Tiss. Res. *217:* 199–210 (1981).

Haddad, A.; Guaraldo, S. P.; Pelletier, G.; Brasileiro, I. L. G.; Marchi, F.: Glycoprotein secretion in the hypothalamo-neurohypophyseal system of the rat. Cell Tiss. Res. *209:* 399–422 (1980).

Haller, E. W.; Brimble, M. J.; Wakerley, J. B.: Phasic discharge in supraoptic neurones recorded from hypothalamic slices. Exp. Brain Res. *33:* 131–134 (1978).

Heap, P. F.; Jones, C. W.; Morris, J. F.; Pickering, B. T.: Movement of neurosecretory product through the anatomical compartments of the neural lobe of the pituitary gland. Cell Tiss. Res. *156:* 483–497 (1975).

Heller, H.; Pickering, B. T.: The distribution of vertebrate neurohypophysial hormones and its relation to possible pathways for their evolution; in Heller, Pickering, Pharmacology of the endocrine system and related drugs: the neurohypophysis, pp. 59–79 (Pergamon, Oxford 1970).

Herbert, E.; Budarf, M.; Phillips, M.; Rosa, P.; Policastro, P.; Oates, E.; Roberts, J. L.; Seidah, N. G.; Chrétien, M.: Presence of a pre-sequence (signal sequence) in the common precursor to ACTH and endorphin and the rôle of glycosylation in processing of the precursor and secretion of ACTH and endorphin. Ann. N.Y. Acad. Sci. *343:* 79–93 (1980).

Hökfelt, T.; Efendić, S.; Hellerström, C.; Johansson, O.; Luft, R.; Arimura, A.: Cellular localisation of somatostatin in endocrine-like cells and neurones of the rat with special references to the A$_1$-cells of the pancreatic islets and to the hypothalamus. Acta endocr., Copenh. *200:* suppl., pp. 5–41 (1975).

Holwerda, D. A.: A glycopeptide from the posterior lobe of pig pituitaries. I. Isolation and characterization. Eur. J. Biochem. *28:* 334–339 (1972).

Howell, W. H.: The physiological effects of extracts of the hypophysis cerebri and infundibular body. J. exp. Med. *3:* 245 (1898).

Iversen, L. L.; Iversen, S. D.; Bloom, F. E.: Opiate receptors influence vasopressin release from nerve terminals in rat neurohypophysis. Nature, Lond. *284:* 350–351 (1980).

Jones, C. W.; Pickering, B. T.: Rapid transport of neurohypophysial hormones in the hypothalamoneurohypophysial tract. J. Physiol. Lond. *208:* 73–74 (1970).

Jones, C.W.; Pickering, B.T.: Intra-axonal transport and turnover of neurohypophysial hormones in the rat. J. Physiol., Lond. *227:* 553–564 (1972).

Jones, C.W.; Swann, R.W.: Incorporation of tritiated sugars into rat neural lobe components. J. Endocr. *63:* 53–54 (1974).

Jones, C.W.; Swann, R.W.: A glycoprotein in the neurosecretory granules of the neurohypophysis. J. Physiol., Lond. *245:* 45 (1975).

Koizumi, K.; Ishikawa, T.; Brooks, C. McC.: Control of activity of neurons in the supraoptic nucleus. J. Neurophysiol. *27:* 878–892 (1964).

Land, H.; Schutz, G.; Schmale, H.; Richter, D.: Nucleotide sequence of cloned cDNA encoding bovine arginine vasopressin-neurophysin II precursor. Nature, Lond. *295:* 299–303 (1982).

Lauber, M.; Nicolas, P.; Boussetta, H.; Fahy, C.; Béguin, P.; Camier, M.; Vaudry, H.; Cohen, P.: The M_r 80,000 common forms of neurophysin and vasopressin from bovine neurohypophysis have corticotropin- and β-endorphin-like sequences and liberate by proteolysis biologically active corticotropin. Proc. natn. Acad. Sci. USA *78:* 6086–6090 (1981).

Legros, J.J.: The radioimmunoassay of human neurophysins: contribution to the understanding of the physiopathology of neurohypophyseal function. Ann. N.Y. Acad. Sci. *248:* 281–303 (1975).

Legros, J.J.; Louis, F.: Identification of a vasopressin-neurophysin and of an oxytocin-neurophysin in man. Neuroendocrinology *13:* 371–375 (1974).

Lescure, H.; Nordmann, J.J.: Neurosecretory granule release and endocytosis during prolonged stimulation of the rat neurohypophysis in vitro. Neuroscience *5:* 651–659 (1980).

Lin, C.; Joseph-Bravo, P.; Sherman, T.; Chan, L.; McKelvy, J.F.: Cell-free synthesis of putative neurophysin precursors from rat and mouse hypothalamic poly(A)-RNA. Biochem. biophys. Res. Commun. *89:* 943–950 (1979).

Lincoln, D.W.; Paisley, A.C.: Neuroendocrine control of milk ejection. J. Reprod. Fertil. *65:* 571–586 (1982).

Loh, Y.P.; Gainer, H.: The rôle of the carbohydrate in the stabilization, processing and packaging of the glycosylated adrenocorticotropin-endorphin common precursor in toad pituitaries. Endocrinology *105:* 474–487 (1979).

Lorenz, T.; Willard, M.: Subcellular fractionation of intraaxonally transported polypeptides in the rabbit visual system. Proc. natn. Acad. Sci. USA *75:* 505–509 (1978).

Mains, R.E.; Eipper, B.A.; Ling, N.: Common precursor to corticotropins and endorphins. Proc. natn. Acad. Sci. USA *74:* 3014–3018 (1977).

Manning, M.; Sawyer, W.H.: Structure-activity studies on oxytocin and vasopressin 1954–1976. From empiricism to design; in Moses, Share, Neurohypophysis, pp. 9–21 (Karger, Basel 1977).

Martin, R.; Voigt, K.H.: Enkephalins co-exist with oxytocin and vasopressin in nerve terminals of rat neurohypophysis. Nature, Lond. *289:* 502–504 (1981).

Morris, J.F.: Electron microscopical study of the paraventricular nucleus of the rat hypothalamus. J. Anat. *108:* 592–593 (1971).

Morris, J.F.: Structural aspects of hormone production and storage; in Vincent, Kordon, Cell biology of hypothalamic neurosecretion, pp. 601–618 (Centre National de la Recherche Scientifique, Paris 1978).

Morris, J.F.; Nordmann, J.J.: Membrane recapture after hormone release from nerve endings in the neural lobe of the rat pituitary gland. Neuroscience 5: 639–649 (1980).

Morris, J.F.; Nordmann, J.J.; Dyball, R.E.J.: Structure-function correlation in mammalian neurosecretion. Int. Rev. exp. Path. 18: 1–95 (1978).

Morris, J.F.; Nordmann, J.J.; Shaw, F.D.: Granules, microvesicles and vacuoles. Their rôles in the functional compartments of the neural lobe; in Farner, Lederis, Neurosecretion: molecules, cells, systems, pp.187–196 (Plenum Publishing, New York 1981).

Morris, J.F.; Sokol, H.W.; Valtin, H.: One neuron-one hormone? Recent evidence from Brattleboro rats; in Moses, Share, Neurohypophysis pp.58–66 (Karger, Basel 1977).

Nagasawa, J.; Douglas, W.W.; Schulz, R.A.: Ultrastructural evidence of secretion by exocytosis and of 'synaptic vesicle' formation in posterior pituitary glands. Nature, Lond. 227: 407–409 (1970).

Nagasawa, J.; Douglas, W.W.; Schulz, R.A.: Micropinocytotic origin of coated and smooth microvesicles ('synaptic vesicles') in neurosecretory terminals of posterior pituitary glands demonstrated by incorporation of horseradish peroxidase. Nature, Lond. 232: 341–342 (1971).

Nordmann, J.J.; Evidence for calcium inactivation during hormone release in the rat neurohypophysis. J. exp. Biol. 65: 669–683 (1976).

Nordmann, J.J.; Chevallier, J.: The rôle of microvesicles in buffering $[Ca^{2+}]_i$ in the neurohypophysis. Nature, Lond. 287: 54–56 (1980).

Nordmann, J.J.; Dreifuss, J.J.; Baker, P.F.; Ravazzola, M.; Malaisse-Lagae, F.; Orci, L.: Secretion-dependent uptake of extracellular fluid by the rat neurohypophysis. Nature, Lond. 250: 155–157 (1974).

Nordmann, J.J.; Dyball, R.E.J.: Effects of veratridine on Ca fluxes and the release of oxytocin and vasopressin from the isolated rat neurohypophysis. J. gen. Physiol. 72: 297–304 (1978).

Nordmann, J.J.; Labouesse, J.: Neurosecretory granules: evidence for an aging process within the neurohypophysis. Science 211: 595–597 (1981).

Nordmann, J.J.; Louis, F.; Morris, S.J.: Purification of two structurally and morphologically distinct populations of rat neurohypophysial secretory granules. Neuroscience 4: 1367 (1979).

Nordmann, J.J.; Morris, J.F.: Membrane retrieval at neurosecretory axon endings. Nature, Lond. 261: 723–725 (1976).

Norström, A.: The heterogeneity of the neurohypophysial pool of neurophysin; in Neurosecretion – the final neuroendocrine pathway. VIth Int. Symp. on Neurosecretion, London 1973 (Springer, Berlin 1974).

Norström, A.; Sjöstrand, J.: Axonal transport of proteins in the hypothalamo-neurohypophysial system of the rat. J. Neurochem. 18: 29–39 (1971).

North, W.G.; Valtin, H.; Morris, J.F.; La Rochelle, F.T., Jr.: Evidence for metabolic conversions of rat neurophysins within neurosecretory granules of the hypothalamoneurohypophysial system. Endocrinology 101: 110–118 (1977).

Olden, K.; Parent, J.B.; White, S.L.: Carbohydrate moieties of glycoproteins. A re-evaluation of their function. Biochim. biophys. Acta 650: 209–232 (1982).

Palay, S.L.: The fine structure of the neurohypophysis; in Waelsch, Ultrastructure and cellular chemistry of neural tissue, pp.31–59 (Hoeber, New York 1957).

Parish, D.C.; Rodriguez, E.M.; Birkett, S.D.; Pickering, B.T.; Effects of small doses of colchicine on the components of the hypothalamo-neurohypophysial system of the rat. Cell Tiss. Res. *220:* 809–827 (1981).

Pickering, B. T.: Aspects of the relationships between the chemical structure and biological activity of the neurohypophysial hormones and their synthetic structural analogues; in Heller, Pickering, Pharmacology of the endocrine system and related drugs: the neurohypophysis, pp. 81–110 (Pergamon, Oxford 1970).

Pickering, B. T.: The molecules of neurosecretion: their formation, transport and release. Prog. Brain Res. *45:* 161–179 (1976).

Pickering, B. T.: The neurohypophysial neurone: a model for the study of secretion. Essays Biochem. *14:* 45–81 (1978).

Pickup, J. C.; Johnston, C. I.; Nakamura, S.; Uttenthal, L. O.; Hope, D. B.: Subcellular organization of neurophysins, oxytocin, [8-lysine]-vasopressin and adenosine triphosphatase in porcine posterior pituitary lobes. Biochem. J. *132:* 361–371 (1973).

Poulain, D. A.; Wakerley, J. B.: Electrophysiology of hypothalamic magnocellular neurones secreting oxytocin and vasopressin. Neuroscience *7:* 773–808 (1982).

Poulain, D. A.; Wakerley, J. B.; Dyball, R. E. J.: Electrophysiological differentiation of oxytocin- and vasopressin-secreting neurones. Proc. R. Soc. B *196:* 367–384 (1977).

Richter, D.; Schmale, H.; Ivell, R.; Schmidt, C.: Hypothalamic mRNA-directed synthesis of neuropolypeptides: immunological identification of precursors to neurophysin II/arginine vasopressin and to neurophysin I/oxytocin; in Koch, Richter, Biosynthesis, modification and processing of cellular and viral polyproteins, pp. 43–66 (Academic Press, New York 1980).

Robertis, E. de; Vaz Ferreira, A.: Electron microscope study of the excretion of catechol-containing droplets in the adrenal medulla. Expl Cell Res. *12:* 568–574 (1957).

Robinson, A. G.: Isolation, assay, and secretion of individual human neurophysins. J. clin. Invest. *55:* 360–367 (1975).

Robinson, A. G.; Zimmerman, E. A.; Frantz, A. G.: Physiologic investigation of posterior pituitary binding proteins neurophysin I and neurophysin II. Metabolism *20:* 1148–1155 (1971).

Rosenior, J. C.; North, W. G.; Moore, G. J.: Putative precursors of vasopressin, oxytocin and neurophysins in the rat hypothalamus. Endocrinology *109:* 1067–1072 (1981).

Rossier, J.; Vargo, T. M.; Minick, S.; Ling, N.; Bloom, F. E.; Guillemin, R.: Regional dissociation of β-endorphin and enkephalin contents in rat brain and pituitary. Proc. natn. Acad. Sci. USA *74:* 5162–5165 (1977).

Rubin, R. P.: The rôle of calcium in the release of neurotransmitter substances and hormones. Pharmac. Rev. *22:* 389–428 (1970).

Russell, J. T.; Brownstein, M. J.; Gainer, H.: [^{35}S]-cysteine labelled peptides transported to the neurohypophyses of adrenalectomized, lactating, and Brattleboro rats. Brain Res. *201:* 227–234 (1980a).

Russell, J. T.; Brownstein, M. J.; Gainer, H.: Biosynthesis of vasopressin, oxytocin, and neurophysins: isolation and characterization of two common precursors (propressophysin and prooxyphysin). Endocrinology *107:* 1880–1891 (1980b).

Sachs, H.: Vasopressin biosynthesis. I. In vivo studies. J. Neurochem. *5:* 297–303 (1960).

Sachs, H.: Vasopressin biosynthesis. II. Incorporation of [^{35}S]-cysteine into vasopressin and protein associated with cell fractions. J. Neurochem. *10:* 299–311 (1963).

Sachs, H.: Neurosecretion. Adv. Enzymol. *32:* 327–368 (1969).

Sachs, H.: Secretion of neurohypophysial hormones. Mem. Soc. Endocr. *19:* 965–973 (1971).

Sachs, H.; Fawcett, P.; Takabatake, Y.; Portanova, R.: Biosynthesis and release of vaso-
    pressin and neurophysin. Recent Prog. Horm. Res. 25: 447–491 (1969).
Sachs, H.; Share, L.; Osinchak, J.; Carpi, A.: Capacity of the neurohypophysis to release
    vasopressin. Endocrinology 81: 755–770 (1967).
Sachs, H.; Takabatake, Y.: Evidence for a precursor in vasopressin biosynthesis. Endocri-
    nology 75: 943–948 (1964).
Sawyer, W. H.: Phylogenetic aspects of the neurohypophysial hormones; in Berde, Hand-
    book of experimental pharmacology, vol. 23, pp. 717–747 (Springer, Berlin 1968).
Sawyer, W. H.: Evolution of neurohypophysial peptides among the non-mammalian verte-
    brates; in Wolstenholme, Birch, Neurohypophysial hormones, pp. 5–18 (Churchill
    Livingstone, London 1971).
Schmale, H.; Richter, D.: In vitro biosynthesis and processing of composite common pre-
    cursors containing amino acid sequences identified immunologically as neurophysin
    I/oxytocin and as neurophysin II/arginine vasopressin. FEBS Lett. 121: 358–362
    (1980).
Schmale, H.; Richter, D.: A direct comparison of the rat and bovine arginine vasopres-
    sin/neurophysin II common precursor. Neuropeptides 2: 151–156 (1981a).
Schmale, H.; Richter, D.: Immunological identification of a common precursor to arginine
    vasopressin and neurophysin II synthesized by in vitro translation of bovine hypothal-
    amic mRNA. Proc. natn. Acad. Sci. USA 78: 766–769 (1981b).
Schmitt, F. O.: Fibrous proteins and neuronal dynamics; in Barondes, Cellular dynamics of
    the neuron, pp. 95–111 (Academic Press, New York 1969).
Seidah, N. G.; Benjannet, S.; Chrétien, M.: The complete sequence of a novel human pitu-
    itary glycopeptide homologous to pig posterior pituitary glycopeptide. Biochem. bio-
    phys. Res. Commun. 100: 901–907 (1981).
Shaw, F. D.; Morris, J. F.: Calcium localization in the rat neurohypophysis. Nature, Lond.
    287: 56–58 (1980).
Sloper, J. C.; Bateson, R. G.: Ultrastructure of neurosecretory cells in the supraoptic nucleus
    of the dog and rat. J. Endocr. 31: 139–150 (1965).
Smyth, D. G.; Massey, D. E.: A new glycopeptide in pig, ox and sheep pituitary. Biochem.
    biophys. Res. Commun. 87: 1006–1010 (1979).
Sofroniew, M. V.; Weindl, A.: Central nervous system distribution of vasopressin, oxytocin
    and neurophysin; in Martinez, Jr., Jensen, Messing, Rigter, McGaugh, Endogenous
    peptides and learning and memory processes, pp. 327–369 (Academic Press, New York
    1981).
Steiner, D. F.; Quinn, P. S.; Chan, S. J.; Marsh, J.; Tager, H. S.: Processing mechanisms in the
    biosynthesis of proteins. Ann. N.Y. Acad. Sci. 343: 1–16 (1980).
Suchanek, G.; Kreil, G.; Hemondson, M. A.: Amino acid sequence of honeybee prepro-
    mellitin synthesized in vitro. Proc. natn. Acad. Sci. USA 75: 701–704 (1978).
Swaab, D. F.; Pool, C. W.; Nijveldt, F.: Immunofluorescence of vasopressin and oxytocin in
    rat hypothalamo-neurohypophyseal system. J. neural Transm. 36: 195–215 (1975).
Swann, R. W.; Gonzalez, C. B.; Birkett, S. D.; Pickering, B. T.: Precursors in the biosynthesis
    of vasopressin and oxytocin in the rat. Characteristics of all of the components in
    high-performance liquid chromatography. Biochem. J. 208: 339–349 (1982).
Swann, R. W.; Pickering, B. T.: Incorporation of radioactive precursors into the membrane
    and contents of the neurosecretory granules of the rat neurohypophysis as a method of
    studying their fate. J. Endocr. 68: 95–108 (1976).

Takabatake, Y.; Sachs, H.: Evidence for a precursor in vasopressin biosynthesis. Endocrinology 75: 934–942 (1964).

Tasso, F.: Localisation cytochimique ultrastructurale des glycoprotéines dans les granules neurosécrétoires de la posthypophyse du rat. J. Microscopie 18: 115–118 (1973).

Tasso, F.; Rua, S.: Etude cytochimique ultrastructurale des glycoprotéines dans le complex hypothalamo-hypophysaire du rat. Archs Anat. microsc. Morph. exp. 64: 247–260 (1975).

Theodosis, D. T.; Dreifuss, J. J.; Harris, M. C.; Orci, L.: Secretion related uptake of horseradish peroxidase in neurohypophysial axons. J. Cell Biol. 70: 294–303 (1976).

Thorn, N. A.: In vitro studies of the release mechanism for vasopressin in rats. Acta endocr., Copenh. 53: 644–654 (1966).

Thorn, N. A.; Russell, J. T.; Vilhardt, H.: Hexosamine, calcium, and neurophysin in secretory granules and the rôle of calcium in hormone release. Ann. N.Y. Acad. Sci. 248: 202–217 (1975).

Tweedle, C. D.; Hatton, G. I.: Evidence for dynamic interactions between pituicytes and neurosecretory axons in the rat. Neuroscience 5: 661–667 (1980).

Vanderhaeghen, J. J.; Lotstra, F.; Vandesande, F.; Dierickx, K.: Coexistence of cholecystokinin and oxytocin-neurophysin in some magnocellular hypothalamo-hypophyseal neurons. Cell Tiss. Res. 221: 227–231 (1981).

Vigneaud, V. du: Experiences in the polypeptide field: insulin to oxytocin. Ann. N.Y. Acad. Sci. 88: 537–548 (1960).

Vigneaud, V. du; Ressler, C.; Swan, J. M.; Roberts, C. W.; Katsoyannis, P. G.; Gordon, S.: The synthesis of an octapeptide amide with the hormonal activity of oxytocin. J. Am. chem. Soc. 75: 4879–4880 (1953).

Wakerley, J. B.; Lincoln, D. W.: The milk-ejection reflex of the rat: a 20- to 40-fold acceleration in the firing of paraventricular neurones during oxytocin release. J. Endocr. 57: 477–493 (1973).

Wathes, D. C.; Swann, R. W.: Is oxytocin an ovarian hormone? Nature, Lond. 297: 225–227 (1982).

Wathes, D. C.; Swann, R. W.; Pickering, B. T.; Porter, D. G.; Hull, M. R. G.; Drife, J. O.: Neurohypophysial hormones in the human ovary. Lancet ii: 410–412 (1982).

Wied, D. de; Gispen, W. H.: Behavioral effects of peptides; in Gainer, Peptides in neurobiology (Plenum Press, New York 1977).

Wong, T. M.; Pickering, B. T.: Last in-first out in the neurohypophysis. Gen. compar. Endocr. 29: 242–243 (1976).

Zambrano, D.; De Robertis, E.: The secretory cycle of supraoptic neurons in the rat. Z. Zellforsch. mikrosk. Anat. 73: 414–431 (1966).

Zambrano, D.; De Robertis, E.: Ultrastructure of the hypothalamic neurosecretory system of the dog. Z. zellforsch. mikrosk. Anat. 81: 264–282 (1967).

Zambrano, D.; De Robertis, E.: The effect of castration upon the ultrastructure of the rat hypothalamus. I. Supraoptic and paraventricular nuclei. Z. Zellforsch. mikrosk. Anat. 86: 487–498 (1968).

B. T. Pickering, Department of Anatomy, University of Bristol,
The Medical School, University Walk, Bristol BS8 1TD (England)

Cell Biology of the Secretory Process, pp. 276–308 (Karger, Basel 1984)

Neurophysin: Biology and Chemistry of Its Interactions[1]

Esther Breslow

Department of Biochemistry, Cornell University Medical College, New York, N.Y., USA

Neurophysins comprise the principal proteins of posterior pituitary neurosecretory granules where they occur as noncovalent complexes of oxytocin and vasopressin [2]. The high intragranule concentrations of neurophysin – $10^{-2}M$ to $10^{-1}M$ [43, 52] – and the availability of relatively simple purification methods [111] have encouraged detailed study of neurophysin structure and function. These studies have particularly focused on neurophysin-hormone interaction as a physicochemical model of protein-peptide interaction and on the relationship between neurophysin biosynthesis and biosynthesis of the hormones, but, in addition, have used neurophysins as immunological probes of neuronal pathways involving the hypothalamus and pituitary [67, 117] and as probes of hormone release into the blood [65, 103]. Relatively recent reviews on neurophysin have appeared elsewhere [1, 14, 40, 85, 104]. In this chapter the present status of neurophysin investigations is summarized in the context of the following questions: (1) what are the evolutionary and biosynthetic relationships between individual neurophysins and oxytocin and vasopressin; (2) what is the role of neurophysin in the packaging and transport of hormones within neurosecretory granules, and (3) what is the chemistry of neurophysin-hormone interaction and what does it more generally indicate about the chemistry of protein-peptide interactions?

[1] This work was supported by grant No. GM-17528 from NIH.

Neurophysin Primary Structure in Relation to
Neurophysin Compartmentalization

Neurophysins have been isolated from or detected in species ranging from insects [27] to man [35, 81]. It is likely, but by no means established, that all species containing oxytocin, vasopressin, and related hormones (mesotocin, vasotocin, etc.) contain hormone-associated neurophysins. Each mammalian species that produces both oxytocin and vasopressin also produces at least two neurophysins [14]. Within each species, the two hormones appear to occur in separate neurons [for reviews see 67, 78], compartmentalized within neurosecretory granules with a characteristic neurophysin or group of neurophysins. This latter point was first suggested by *Dean and Hope* and coworkers [42a, b] who showed that the two principal neurophysins of the cow, bovine neurophysins I and II, were found in different granules respectively associated with oxytocin and vasopressin [42a, b], although each of the isolated neurophysins is capable of binding either oxytocin or vasopressin in vitro [16, 20, 26, 53, 59]. The association of the two hormones with different neurophysins in vivo continues to be supported by other studies. Of these, the clearest evidence is found in the rat where Brattleboro rats, which lack vasopressin but which have oxytocin, are missing only one of their characteristic neurophysins [107]. (It is relevant at this stage to point out that the nomenclature of the neurophysins, as I, II, etc., is derived typically from the relative electrophoretic mobilities of the neurophysins within a species. This nomenclature does not describe the structural relationships among neurophysins or their hormonal association (see below).)

Neurophysins from a number of species have been sequenced [34, 36–38, 97, 115]. Representative sequences are shown in figure 1. Several features of the sequences are particularly relevant to the question of neurophysin compartmentalization. First, all neurophysins are remarkably similar, characterized by a central core comprised approximately of residues 10–77 in which very little substitution has occurred with evolutionary time; differences among neurophysins are found principally at the amino end (residues 1–9) and the carboxyl end (residues 78–95). Second, structural differences between the principal neurophysins within a species exceed those between homologous neurophysins from different species. This was first shown by *Capra* et al. [29] who interpreted this as signifying that the gene duplication which gave rise to more than one neurophysin in each species preceded species divergence. The data in figure 1 illustrate the

	1	2	3	4	5	6	7	8	9	10	11	12	13	14	15	16	17	18	19	20	21	22	23	24

Bovine II Ala-Met-Ser-Asp-Leu-Glu-Leu-Arg-Gln-Cys-Leu-PRO-CYS-GLY-PRO-GLY-GLY-LYS-GLY-ARG-CYS-PHE-GLY-PRO

Porcine III _____

Porcine I _____

Bovine I ——Val-Leu———Asp-Val——Thr———————————————————————————————

Porcine II ——Val-Leu———Asp-Val——Lys———————————————————————————————

	25	26	27	28	29	30	31	32	33	34	35	36	37	38	39	40	41	42	43	44	45	46	47	48

Bovine II SER-ILE-CYS-CYS-GLY-ASP-GLU-Leu-Gly-Cys-Phe-Val-Gly-Thr-Ala-Glu-Ala-Leu-Arg-Cys-Gln-Glu-Glu-Asn

Porcine III _____

Porcine I _____

Bovine I _____

Porcine II _____

	49	50	51	52	53	54	55	56	57	58	59	60	61	62	63	64	65	66	67	68	69	70	71	72

Bovine II Tyr-Leu-Pro-Ser-Pro-Cys-Gln-Ser-Gly-Gln-Lys-PRO-CYS-GLY-SER-GLY-GLY-ARG-CYS-ALA-ALA-ALA-GLY-ILE

Porcine III _____

Porcine I _____

Bovine I _____

Porcine II ————————————————————————————Glu————————————————————

	73	74	75	76	77	78	79	80	81	82	83	84	85	86	87	88	89	90	91	92	93	94	95

Bovine II CYS-CYS-ASN-ASP-GLU-Ser-Cys-Val-Thr-Glu-Pro-Glu-Cys-Arg-Glu-Gly-$^{Ile}_{Val}$-Gly-Phe-Pro-Arg-Arg-Val

Porcine III ————————————————————————————————————Ala-Ser———Leu————Ala

Porcine I ————————————————————————————————————Ala-Ser———Leu

Bovine I ————Ser-Pro-Asp-Gly———His-Glu-Asp———Ala———Asp-Pro-Glu-Ala-Ala———Ser-$^{Leu}_{Gln}$

Porcine II ————————Pro-Asp-Gly———Arg-Phe-Asp———Ala———Asp-Pro-Glu-Ala-Thr———Ser-Gln

Fig. 1. Amino acid sequences of representative neurophysins. The complete sequence of only bovine neurophysin II is shown. The sequences of the other neurophysins differ from that of bovine neurophysin II only as indicated. All Cys residues are half-cystines. The duplicated segments are capitalized and should be compared by considering the $Gly_{17}-Lys_{18}$ sequence deleted from the second duplicated segment [29]. Where two residues are identified at the same position, this indicates microheterogeneity (bovine neurophysin II) or ambiguity (bovine neurophysin I). Representative references for the sequences are as follows: bovine II [33]; porcine I, III [115]; bovine I [37; 97]; porcine II [37].

point. Bovine neurophysin II and porcine neurophysin III are almost identical, differing in only 4 of 95 residues (residues 89, 90, 92, and 95). However, bovine neurophysin I differs from bovine neurophysin II by 22 residues (residues 2, 3, 6, 7, 9, and the rest after position 74) and is structurally more similar to porcine neurophysin II from which it differs by only 6 residues [37]. A third conclusion that can be drawn from sequence studies is that, within a species, some of the neurophysins appear to be metabolically derived from the others. This was shown most conclusively by *Wuu and Crumm* [115] in their sequence analysis of porcine neurophysins I and III (fig. 1), the former differing from the latter only in the absence of the last three residues. Other studies suggest a similar relationship between bovine neurophysins C and II [108] and among the different rat neurophysins [78, 80]. Note that the demonstration of these truncated neurophysins occurs under conditions where proteolytic degradation during isolation [15, 42a, 89b] is minimized and appears to be an intrinsic characteristic of posterior pituitary granules. However, additional truncated neurophysin fractions can be obtained if isolation conditions permit proteolytic degradation and these appear to be derived from the 'intrinsic' neurophysins by loss of aminoterminal residues [15] and perhaps also loss of other residues.

Based on early sequence comparisons, *Chauvet* et al. [33] suggested that the different neurophysins can be divided into two principal classes, MSEL and VLDV, identifiable by residues in positions 2, 3, 6, and 7; i.e., Met, Ser, Glu and Leu in the MSEL class (such as bovine neurophysin II) and Val, Leu, Asp, and Val in the VLDV class (see bovine neurophysin I). This view, in its narrowest sense, was challenged by sequence data from other neurophysins in which additional substitutions at these particular positions were found [98]. Nonetheless, as complete sequences of more neurophysins become available, their division into two general classes becomes increasingly apparent. To this reviewer the signal positions which define the class to which an individual neurophysin belongs are positions 6 and 82 (Asp in VLDV and Glu in MSEL), 76–78 (Pro–Asp–Gly in VLDV and Asp–Glu–Ser in MSEL), and 88 (Glu in VLDV and Gly in MSEL). These relationships are documented in table I where the hormonal associations of the different neurophysins, to the extent they have been established, are also shown. The trends in hormonal relationships within the different structural classes suggest the generalization that all VLDV neurophysins are oxytocin associated while all MSEL neurophysins are vasopressin associated. This generalization had been

Table I. Suggested signal positions for establishing relationships between neurophysin sequences, structural classification and hormonal association

Neurophysin	Hormone association	Position					
		6	76	77	78	82	88
Bovine II, (MSEL) [33]	vasopressin [42b]	Glu	Asp	Glu	Ser	Glu	Gly
Porcine I, III (MSEL) [115]	vasopressin [86]	Glu	Asp	Glu	Ser	Glu	Gly
Rat I (MSEL) [79, 98]	vasopressin [79]	Glu	n.d.	n.d.	n.d.	n.d.	n.d.
Horse (MSEL) [36]	n.d.	Glu	Asp	Glu	Ser	Glu	Gly
Whale (MSEL) [38]	n.d.	Glu	Asp	Glu	Ser	Glu	Gly
Ovine (MSEL) [38]	n.d.	Glu	Asp	Glu	Ser	Glu	Gly
Human (MSEL) [38]	vasopressin	Glu	Asp	Glu	Ser	Glu	Gly
Bovine I (VLDV) [37, 97]	oxytocin [42b]	Asp	Pro	Asp	Gly	Asp	Glu
Porcine II (VLDV) [37]	oxytocin [86]	Asp	Pro	Asp	Gly	Asp	Glu
Rat II, III (VLDV) [79, 98]	oxytocin [79]	Asp	n.d.	n.d.	n.d.	n.d.	n.d.
Horse (VDLV) [34]	n.d.	Asp	Pro	Asp	Gly	Asp	Glu
Human I (VLDV) [81, 96b]	n.d.	Asp	Pro	Asp	Gly	Asp	Glu

n.d. = Not determined.

suggested earlier [37, 104], but ran into initial difficulty with rat neuro-physin sequences [98], since the rat neurophysin that fits best into the MSEL class appeared to be oxytocin associated while that which fits into the VLDV class appeared to be vasopressin associated [14]. However, *North and Mitchell* [79] have since demonstrated that the original rat protein sequences [98] were assigned to the wrong neurophysins; their results give renewed support to the generalization that all oxytocin-associated neurophysins have structural features in common that differ from those of vasopressin associated neurophysins. Because the structural relationships among neurophysins from different species are not necessarily paralleled by their relative electrophoretic mobilities and hence by their numerical designation (see above), the designation of neurophysins simply as oxytocin associated (oxytocin NP) or vasopressin associated (vasopressin NP) is generally preferred. This designation also has advantages over the VLDV and MSEL nomenclature since these latter assignments are inexact in selecting the distinguishing features of the two different structural classes.

The evolution of mammalian neurophysins into two classes, one oxytocin associated and the other vasopressin associated, can be explained by the existence of common biosynthetic precursors for each hormone and its associated neurophysin and the separate evolution of the two precursors. Evidence for common biosynthetic precursors is presented below.

Biosynthesis of Neurophysins and the Common Precursor Hypothesis

The cell bodies of the neurons that contain the neurophysins and their associated hormones are located in the supraoptic and paraventricular regions of the hypothalamus, their axons extending from the hypothalamus through the median eminence and terminating in the posterior pituitary. Early studies by *Bargmann and Scharrer* [4] and *Scharrer and Scharrer* [96a] demonstrated that biosynthesis of the hormones and of neurophysins is initiated in the cell bodies of the hypothalamus; the secretory granules in which the neurophysin-hormone complexes are packaged travel down the axons to the posterior pituitary where their contents are released by exocytosis upon stimulation of the nerve [72, 74]. In now classic studies, *Sachs* [94a] and *Sachs* et al. [94b] demonstrated that the nonapeptide hormones were synthesized from higher molecular weight inactive precursors that were processed to the active hormones. Additionally, they suggested that neurophysin biosynthesis, which follows a time course similar to that of the hormones [95], occurs via the same high molecular weight protein that serves as the hormone precursor. The existence of a neurophysin precursor was also argued by the physicochemical properties of neurophysin disulfides which fail to exhibit reversible reduction-reoxidation behavior, suggesting that neurophysins, as isolated, are nonidentical to the species in which the initial disulfide pairing occurred [12, 32, 73]. Conclusive biological evidence has now been obtained, from three different experimental approaches, that neurophysin biosynthesis proceeds via a higher molecular weight precursor. Moreover, the evidence that there is a common precursor for each neurophysin and its associated hormone is all but complete with the very recent sequencing of a common precursor of bovine neurophysin II and arginine vasopressin [63].

Pulse-Chase Studies of Neurophysin Biosynthesis

Possibly the first direct evidence of a neurophysin precursor was obtained by *Walter* et al. [110] using pulse-chase techniques. This was paralleled by a definitive series of studies by *Brownstein and Gainer* [23], *Brownstein* et al. [24, 25], and *Gainer* et al. [49, 50] who pulse-labeled neurophysin and its precursors by direct injection of [^{35} S]-cysteine into the hypothalamus and monitored the molecular weight of labeled proteins that reacted with antineurophysin antibodies as a function of time after the pulse and distance travelled down the axon. Using both electrophoresis and molecular exclusion chromatography, they demonstrated that the first labeled protein to appear in the cell body, that was capable of reacting with antineurophysin antibodies, had a molecular weight of 19,000–20,000 daltons (twice the weight of native neurophysin). As the concentration of label in the 20,000 M_r species diminished with time in the supraoptic nucleus, both this species and a protein of the same size and antibody reactivity as native neurophysin became isolable from the median eminence. 24 h after the pulse, label was found only in protein with the same molecular weight as native neurophysin and was located predominantly in the posterior pituitary. These results were interpreted as indicating biosynthesis of neurophysin via a higher molecular weight precursor, with processing of the precursor during axonal transport. Extensions of these studies indicated that more than one neurophysin high molecular weight precursor was present; these were identified as the precursors for oxytocin NP (pI = 5.4, M_r = 15,000–17,000, nonglycosylated) and vasopressin NP (pI = 6.1, M_r = 20,000–22,000, glycosylated) through comparative studies of normal and Brattleboro rats [23, 25]. Some evidence for a third class of neurophysin precursor was also obtained, possibly associated with vasotocin [25]. Although the intact neurophysin precursors reacted with antibodies against neurophysin, they did not react with antibodies against oxytocin and vasopressin. Limited tryptic digestion of the precursor for vasopressin NP, however, yielded not only a protein of the same size and pI as native NP, but also a heterogeneous set of peptides capable of binding both to vasopressin-directed antibodies and to neurophysin [91], in accord with the common precursor hypothesis. Evidence that the putative oxytocin NP precursor liberated oxytocinlike peptides in the presence of trypsin was also obtained [25, 92].

The tryptic liberation from the precursor of peptides capable both of reacting with antibodies directed against the hormones and of binding to neurophysin is significant. Because peptides that bind to neurophysin must have Tyr (or Phe) in position 2 and a free amino terminus (vide infra), these results can be shown to suggest either that the precursor residue immediately preceding the Cys_1–Tyr_2 sequence of the hormones is a trypsin-sensitive residue (Arg or Lys) or that the Cys_1-Tyr_2 sequence of the hormones represents the amino terminus of the pro-form of the precursor. As discussed below, the latter turns out to be true.

Isolation of Unlabeled High Molecular Weight Neurophysinlike Proteins

Camier et al. [28] and *Nicolas* et al. [76] have isolated putative neurophysin precursors directly from mouse hypothalami and bovine posterior pituitaries using protease inhibitors to prevent further processing or degradation during isolation and molecular exclusion chromatography to select protein of higher molecular weight than neurophysin that reacted with antineurophysin antibodies. Two such classes of bovine protein were found, a minor component with $M_r \cong 140,000$ and a major component with $M_r = 25,000$, and both components were reported to share immunological determinants not only with neurophysin, but also with vasopressin [76]. Removal of protease inhibitors from the partially purified precursors led to their conversion to lower molecular weight species, presumably by serine proteases within the precursor preparations. Of these lower molecular weight species, there was a high proportion of protein with the same molecular weight as neurophysin ($\sim 10,000$) that reacted only with antineurophysin antibodies, in addition to peptides with $M_r \cong 8,000$ that reacted solely with antivasopressin antibodies. Mouse precursor proteins were similarly reactive with antibodies directed against both neurophysin and vasopressin [28].

The higher molecular weight bovine precursor ($M_r \cong 140,000$) found by these investigators has been an enigma. Recent studies [5] indicate a molecular weight under denaturing conditions of 80,000, suggesting that the 140,000 M_r species is a dimer. Two species of $M_r = 80,000$ are actually observed; one appears to be a single polypeptide chain while the other contains two disulfide-linked chains of $M_r \cong 70,000$ and 10,000, respectively. Immunological evidence indicates that only the 80,000 M_r and 10,000 M_r species react with antineurophysin antibodies and accordingly suggests that the two-chain species is derived from the single-chain species by proteolytic cleavage to give a neurophysinlike species of 10,000 and a 70,000 M_r nonneurophysin component. Of particular interest is the very recent demonstration that the 70,000 M_r component contains sequences analogous to those of ACTH and β-endorphin [64]. This has led to the suggestion that the 80,000 M_r component is an exceptionally large 'composite prohormone' for ACTH, β-endorphin, neurophysin, and arginine vasopressin [64]. However, the precursor nature of the 80,000 M_r component remains to be established. Thus, it is generally accepted that cells secreting ACTH and the posterior pituitary hormones are nonidentical [70]. For the 80,000 M_r component to be a true prohormone, processing must destroy the ACTH segment in posterior pituitary neurons and destroy the neurophysin and posterior pituitary hormone segments in ACTH-secreting cells. While differential processing of a single precursor by different tissues is not without precedent, and while the data may signify that the DNA-coding segments for ACTH, β-endorphin, vasopressin, and neurophysin lie adjacent to each other on the same chromosome, it also remains possible that the neurophysinlike immunoreactivity of the 80,000 M_r component does not represent neurophysin. *Lorén* et al. [70] have reported that lipolytic peptide B, an 11,000 M_r protein that is cross-reactive but nonidentical with neurophysin, can be found in cells producing ACTH or α-MSH; and the pars intermedia, which produces α-MSH, is a probable contaminant of posterior pituitary preparations. Additionally, if the disulfide linking the 10,000 M_r neurophysinlike component to the ACTH precursor is supposed to contain a half-Cys from *within* the neurophysin sequence domain, it is difficult to explain how native neurophysin could be derived from such a precursor. In native neurophysin, all half-Cys residues are paired intramolecularly and do not significantly repair to give a functional protein once they are broken [73]; this almost certainly reflects their intraneurophysin pairing in the precursor (vide supra). Finally, we note that neurophysin is able to bind noncovalently to ACTH [41], presumably because it contains the correct aminoterminus (vide infra) for such binding. It is tempting to speculate that ACTH in ACTH-containing cells may bind to a protein (lipolytic peptide B?) structurally similar but not identical to neurophysin.

*Cell-Free Biosynthesis of Neurophysin and Sequencing of a
Neurophysin Precursor*

A third approach to the study of neurophysin biosynthesis has been
the isolation of mRNA that codes for neurophysin and its translation in
cell-free system. *Guidice and Chaiken* [56] isolated mRNA from bovine
hypothalamus, translated it in a wheat germ translation system and isolated proteins that reacted with antibodies directed against bovine neurophysin II. They found a 23,000–25,000 M_r protein that contained tryptic
peptides characteristic of neurophysin II and accordingly identified this as
a neurophysin precursor; it did not react with antibodies directed against
either oxytocin or vasopressin. In subsequent studies [57], this same
approach was used to identify a 17,000–18,000 M_r precursor of bovine
neurophysin I. It is interesting that the difference in molecular weight of
the two precursors follows the same pattern as that observed in the rat, i.e.,
in this case also, the precursor to vasopressin NP (bovine NP II) has a
significantly higher molecular weight than that to oxytocin NP (bovine
NP I).

Schmale et al. [100] and *Schmale and Richter* [101, 102] also utilized
this approach to isolate a precursor to bovine neurophysin I (16,500 M_r)
and a precursor to bovine neurophysin II (21,000 M_r). In these studies,
however, the neurophysin I precursor also reacted specifically with antibodies directed against oxytocin, and the neurophysin II precursor reacted
specifically with antibodies against vasopressin, in accord with a common
precursor. Processing of the 16,500 M_r precursor of oxytocin and NP I by a
microsomal membrane fraction or in Xenopus oocytes converted it to a
15,500 M_r, nonglycosylated species, presumably indicating that the
16,500 M_r species is the 'pre-pro' form, and the 15,500 M_r species is the
'pro' form. Similar processing of the 21,000 M_r precursor of bovine NP II
and vasopressin gave rise to a 'pro' precursor of $M_r = 19,000$ which could be
glycosylated to a species of $M_r = 23,000$. The 'pro' forms of neurophysins I
and II also reacted with antibodies directed against oxytocin and vasopressin, respectively. This approach has also been used with mouse and rat
hypothalami [66]. In both species, the only demonstrable neurophysin
precursor was in the 18,000 M_r class, raising additional questions as to the
significance of the much larger neurophysinlike proteins [5, 64] isolated
from these species by tissue extraction without labeling (vide supra).

The work of *Schmale* et al. [100] and *Schmale and Richter* [101, 102]
has recently culminated in elucidation of the sequence of a precursor to
arginine vasopressin and bovine neurophysin II [63]. This was determined

by cloning of cDNA synthesized using precursor mRNA as the template followed by sequence determination of the cloned cDNA. The amino acid sequence encoded by the nucleotide sequence indicates that the 'pre-pro' form of the precursor contains 166 residues. Of these, the first 19 residues represent the putative signal peptide. The terminal Ala of the signal peptide is immediately followed (in the order given) by: the hormone sequence, the sequence Gly–Lys–Arg, the neurophysin sequence, one Arg residue, and a C terminal peptide of 39 residues that contains the site of precursor glycosylation. The glycosylatable 39-residue C terminal peptide is identical to a peptide isolated earlier from bovine pituitary [113] which, at the time of isolation, was of unknown origin.

Implications of the Precursor Sequence

The structure of the precursor confirms the common precursor hypothesis for vasopressin and its associated neurophysin and indicates that these are synthesized in 1:1 ratio. This latter demonstration is of particular interest in view of the controversial stoichiometry of vasopressin binding to neurophysin (vide infra). The suggested presence of two vasopressin binding sites per neurophysin chain [26, 40] is difficult to explain teleologically if hormone and protein are synthesized and stored in 1:1 ratio. While a second site on neurophysin might represent the vestige of a different hormone-bonding site within the precursor, this now (vide infra) seems unlikely.

A related and key feature of the sequence is that the hormone provides the amino terminal segment of the 'pro' or packaged form of the precursor. The evidence for this rests on the tryptic cleavage pattern of the 'pro' form of the precursor. Treatment of the 'pro' form with trypsin has been shown to liberate a peptide representing vasopressin residues 1–8 [63]. Since the residue immediately preceding the hormone sequence in the 'pro-pro' form is Ala, trypsin would normally be expected to liberate this peptide only if the vasopressin sequence represented the amino terminus of the 'pro' form (note that residue 8 of vasopressin is Arg). The finding by *Russell* et al. [91, 92] that trypsin generates a peptide capable of binding to neurophysin is also clearly in accord with the conclusion that vasopressin is at the amino terminus of the 'pro' form (vide supra). Because a free amino terminus is essential for the noncovalent interaction between hormone and neurophysin in the processed hormone-neurophysin complex (see also below), the demonstration that the hormone represents the amino terminal sequence of the packaged form of the precursor indicates that noncovalent interactions between hormone and neurophysin segments in the precursor may be identical to those in the processed complex.

The structure of the sequenced precursor gives little clue as to the origin of the 80,000 M_r precursor reported by *Béguin* et al. [5]. Interestingly, another report of components with neurophysinlike immunoreactivity and $M_r \geqq 70,000$ has appeared [90], but in this study as well there was no evidence that such components are true precursors.

Another residual question is the identity of the enzymes involved in processing of the precursor to neurophysin and the hormones. All available data support the concept that the precursor is packaged into neurosecretory granules within the cell bodies and that processing occurs within the granules as they are transported down the axon to the posterior pituitary [49]. Thus, the enzymes must be constituents of the granules. However, while intragranule proteolytic enzymes with chymotrypsinlike specificity have been reported [78, 80], information beyond this point is lacking. It is relevant that the sequence which follows the hormone sequence in the precursor is –Gly–Lys–Arg. As discussed by *Land* et al. [63], the occurrence of a Gly immediately after the terminal glycine of the hormone is in accord with the concept that this second Gly may serve, during processing, as the nitrogen donor for the amide of the hormone carboxy-terminal glycineamide.

Function of Neurophysin within Neurosecretory Granules

Processing of the precursor to a neurophysin and its associated hormone is followed by formation of the noncovalent neurophysin-hormone complex. The complex is relatively insoluble at pH 5.5 [15], the intravesicular pH [93]; this allows packaging of the hormones within the granules at very high concentrations without increasing the intragranule osmotic pressure to the point of threatening granule lysis [49]. Additionally, the insolubility of the complex and the fact that the complex is dimeric [77, 109a] reduces leakage of the low molecular weight hormones through the granule membrane. The question arises as to whether neurophysin plays any additional role within the granules. *Russell* et al. [92] have observed that the neurophysin precursor can bind very tightly to neurophysin linked to an agarose column. One potential significance of this is that it might reflect the presence of noncovalent interactions among precursor molecules (or between precursor and newly processed neurophysin if any of the latter is available in the cell body) that would help drive the packaging of the precursor into the granules. Another possible role for neurophysin is to protect the hormones against proteolysis either by insolubilizing them or by directly inhibiting proteolysis. Bovine neurophysin II, for example, can transiently inhibit the activity of pancreatic chymotrypsin [89a], and it is at least remotely possible that this signals its comparable interaction with the chymotrypsin-related enzyme found in posterior pituitary granules (vide supra).

General Features of the Interaction of Neurophysin with Oxytocin and Vasopressin

Neurophysin Domain Structure and Binding Stoichiometry

Neurophysins contain internally duplicated segments [29]. Within each chain, residues 12–31 show approximately 50% homology with residues 60–77 (fig. 1). An interesting question is the relationship of internal duplication to binding and to evolution. Do the different duplicated segments represent autonomous domains, each perhaps capable of providing a separate binding site for hormone or does function depend on interaction between duplicated domains? One answer to this will be found in the pattern of pairing of neurophysin's seven disulfides, but this has not been definitively established [14]. For example, the proposed pairing of Cys_{10} with Cys_{85} [81] would link the amino and carboxyl terminal domains and, taken together with the observation that function is dependent on the integrity of all disulfides [73], would indicate that a functionally viable neurophysin did not arise until evolution of both domains. Alternatively, the suggested pairing of Cys_{10} with Cys_{54} [40, 44] might suggest a greater evolutionary and functional independence of the duplicated segments.

Another clue to the role of internal duplication is the number of hormone-binding sites per chain. This has been studied in several laboratories principally using bovine neurophysins. All studies agree that there are no significant differences between the different isolated bovine proteins in their hormone affinity or specificity [20, 26, 53]. However, there are as yet unresolved differences among laboratories in binding stoichiometry. Equilibrium dialysis studies in our hands consistently indicate that each neurophysin polypeptide chain can bind 1 mol of either oxytocin or vasopressin with equal affinity [10, 20]; we have obtained no evidence for significant secondary sites for either hormone even under conditions (vide infra) where 2:1 stoichiometry has been reported elsewhere. *Hope* et al. [61] and *Wälti* [112b] have also found only a single site for oxytocin [61, 112b] or vasopressin [112b]. The presence of 1:1 hormone:neurophysin stoichiometry for either hormone has been supported by circular dichroism and potentiometric titration studies [22] and, more recently, by NMR studies [10]. Under conditions of standard salt concentrations, there is almost uniform agreement on 1:1 binding stoichiometry for oxytocin [9, 14, 26, 40, 61]. However, studies by *Cohen's* laboratory [26, 40] present compelling evidence for a second binding site for vasopressin, weaker than the first site under most conditions, and this group has also reported the presence of a second oxytocin-binding site in 1.4M LiCl [40, 77]. Another report [53] of a second binding site for vasopressin at standard salt concentration is difficult to evaluate in this context because a second site for oxytocin was also observed.

Although we have found no evidence for significant secondary hormone-binding sites, we have obtained NMR evidence for extremely weak secondary binding sites for small spin-labeled peptides that also (vide infra) bind to the strong site [68, 69]. This evidence is the spin label induced increased relaxation rate of the protons of neurophysin Tyr-49 when the strong site is occupied by nonspin-labeled peptide. It suggests that there is a second site for peptides immediately adjacent to Tyr-49 and coincides with *Cohen's* evidence [40, 114] that the second hormone-binding site (but not the first) is blocked by nitration of Tyr-49. However, the fact that we do not see a second binding site for the hormones leaves open the possibility that the second spin label site may be unrelated to hormone binding. One way to reconcile these diverse findings is to assume a weak peptide-binding site near Tyr-49 to which the intact hormones cannot in fact bind. Since most binding studies in which a second site for hormones was demonstrated utilized highly tritiated hormones [26] that have been found to decompose on storage [88], the apparent second site for hormones might be a site for a tritiated peptide decomposition product of the hormones. Alternatively, the different findings between laboratories might signal the presence of subtle differences between protein preparations or a stoichiometry of three sites per dimer (1.5 per polypeptide chain) as observed in several crystalline neurophysin-hormone complexes [58, 59]. In any event, there is evidence that the strong and putative weak hormone-binding sites are very different with respect to both pH and temperature dependence [53, 75] suggesting that each is unlikely to reside solely within a duplicated segment.

Regardless of stoichiometry ambiguities, most data support the concept that the single oxytocin site is the same as the stronger vasopressin site

[9, 14]. Additionally, small dipeptides and tripeptides resembling the amino terminus of the hormones compete with the hormones for binding to this site and bind with identical qualitative interaction features as do the hormones [10, 11, 15, 22]. These interaction features include the same ligand-induced change in the environment and properties of neurophysin Tyr-49 as measured by absorbance, titration, circular dichroism and NMR [3, 10, 15, 22, 54], the same ligand-induced increase in neurophysin dimerization [22, 40, 77], and the presence of the same bonding interactions between peptide and protein [11, 15, 22, 54]. A more detailed examination of the interaction of small peptides and hormones with the strong site is given below.

Binding Affinity and Function

Studies from different laboratories are also in agreement as to the affinity of the 'strong' or single binding site for the hormones. This affinity is pH dependent, with a maximum near pH 5.5, the same as the intragranular pH [26, 51, 93]. The binding affinity is also dependent on protein concentration because of stronger binding by the protein dimer than by the monomer [40, 77] and diminishes with increasing temperature [53, 75]. Finally, binding to the dimer shows weak positive cooperativity [40, 61, 83]. At pH 5.5 and 25 °C the binding constants of oxytocin or vasopressin to the first subunit of the protein dimer are approximately $1.2 \times 10^5 M^{-1}$ and $2.5 \times 10^5 M^{-1}$ to the second subunit [40, 83]. At 37 °C the binding constants are approximately half of the above values [53, 75], but are clearly sufficient to assure essentially complete binding at the pH and high protein and hormone concentrations found in the granules (vide supra). On the other hand, hormone-neurophysin-binding constants are sufficiently low that complete dissociation should occur when the complexes are released into the blood by exocytosis [20]. Thus, hormone and neurophysin concentrations in the blood are clearly below $10^{-8} M$, and the estimated binding affinity at pH 7.4 and 37 °C is approximately $10^4 M^{-1}$ [20, 53, 75].

Specificity, Thermodynamics, and Kinetics of Hormone Interactions at the Strong Site; General Implications for Protein-Protein Interactions

Nature of the Bonding Interactions

Sites on the hormones that interact with the strong site of neurophysin have been largely, albeit not completely, defined. Figure 2 is a schematic of

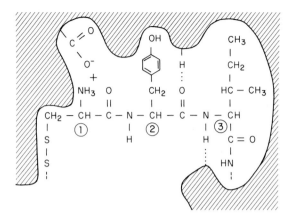

Fig. 2. Interactions of the first three residues of oxytocin with the strong site of neurophysin. The shaded area represents the protein. The first three residues of vasopressin are the same as those of oxytocin with the exception of a phenylalanine in position 3. Residues 4–9 are not shown.

the amino terminus of oxytocin and its putative interactions with neurophysin. Historically, the first proposed feature of the interaction was the salt bridge between the protonated hormone α-amino and a neurophysin carboxyl; the initial evidence for this was the pH dependence of binding and the loss of interaction when the hormone α-amino was substituted by hydrogen [51, 105]. Subsequently, this laboratory demonstrated the critical importance of apolar or π-π interactions involving Tyr-2 of the hormones with the finding that binding was lost if Tyr-2 was substituted by an amino acid other than Phe [16]. This investigation also argued for the presence of apolar interactions involving residue 3; i.e., substitution of Ile-3 of oxytocin by Gly led to a marked binding affinity decrease. Collectively, these results support the concept of an apolar binding site in which a low dielectric constant facilitates salt bridge formation [16]. The importance of the hormone α-amino and residues 2 and 3 to binding led us to look at the binding of dipeptides and tripeptides resembling the amino terminus of the hormones [11, 15, 22] and allowed us to establish that such peptides showed the same binding specificity as the intact hormones and were appropriate models for study of binding of the intact hormones to the strong site (vide supra). With the smaller peptides, two other important features of binding were established. First, it was demonstrated that the side-chain in position 1 contributed to binding, presumably by apolar interactions [15, 22]. Thus, the peptide S-methyl-*L*-cysteinyl-*L*-tyrosyl-*L*-phenylalanine amide was bound with approximately 20-fold greater affinity than glycyl-*L*-tyrosyl-*L*-phenylalanine amide, and the dipeptide *L*-phenylalanyl-*L*-tyrosine amide, which has a particularly hydrophobic side-chain in position 1, is bound very strongly [22]. More recently the importance of the backbone –CONH– between residues 2 and 3 has been shown, and it has been suggested that this region of the hormone backbone, and perhaps others, hydrogen bonds to the protein in the complex [30]; e.g., in dipeptide amides such as *L*-phenylalanyl-*L*-tyrosine amide, substitution of the terminal carboxamide by a hydrogen reduces the binding affinity by a factor of 20.

Almost all of the features of the interaction proposed above have been confirmed or supported by other investigations. The role of Tyr-2 has been confirmed by CD [21], NMR [3, 8, 39], UV absorption [54] and fluorescence [108] spectroscopy; the NMR studies appear to preclude π-π interactions for Tyr-2, suggesting [13] that the apparent requirement for an aromatic side-chain in position 2 signifies apolar bonding of a planar side-chain in a sterically restricted environment. Interestingly, fluorescence studies suggest that the tyrosine –OH might be hydrogen bonded in the complex [108]; however, hydrogen bonding does not contribute to the stability of the complex, since phenylalanine can efficiently substitute for tyrosine [16] and the tyrosine –OH can be alkylated with retention of binding [41]. The importance of the hormone α-amino and its participation in bonding in the protonated form has been confirmed by additional pH [15, 26], chemical modification [19, 60], and NMR studies [7], although thermodynamic dissection of the α-amino interactions (see below) suggest that interactions additional to salt bridge formation are also involved. NMR studies have also provided evidence for participation of the side-chains of residues 1 and 3 [3, 6, 7, 55]. However, this picture of hormone-neurophysin interaction at the strong site may not be complete. For example, the tripeptide S-methyl-*L*-cysteinyl-*L*-tyrosine amide, which participates in all the above interactions, binds with a free energy only 70–80% as negative as that of the hormones [22, 83]. This may signify a role for hormone conformation in binding or the presence of interactions of the protein with regions of the hormone outside of residues 1–3 [13].

Thermodynamics of Bonding and Binding

How important are each of the interactions in figure 2 to binding? This question is relevant to an understanding of factors underlying the specificity of neurophysin-peptide interactions and, as we examine it, reveals some important general principles of protein-ligand interactions.

To formulate a quantitative binding description for the hormones, two different contributions to binding should be recognized [13]. First, the noncovalent bonding of each atom or group of atoms to the protein is characterized by an intrinsic bond strength (ΔG°_{int}). For bonding of any individual residue, this intrinsic bond strength is approximated (such a calculation is only an approximation to the extent, for example, that the protein conformation in the complex is not identical in the presence and absence of the peptide residue in question or that this residue alters the conformation of the peptide in the unbound state) by the difference in stability of the complex when it is formed in the presence of that residue and when it is formed in the absence of that particular residue. Instrinsic bond strengths are characterized by negative free energies. Opposing the negative ΔG°_{int} of bonding are positive free energy contributions associated with energetically uphill conformational changes in the protein and/or peptide and with losses of entropy arising primarily from the decrease in translational and rotational degrees of freedom of the system (variably offset by new vibrational modes) when the complex is formed. We sum all these positive free energy contributions to binding in the term ΔG°_{conf}. *Breslow* [13], and particularly *Jencks* [62a, b], have discussed these terms in greater detail elsewhere. For present purposes it is sufficient to note that the observed free energy of binding of a peptide to neurophysin, ΔG°_{obs}, will typically [62a, b] be less negative than the sum of the intrinsic bonding energies of the individual participant atoms; i.e.,

$$\Delta G^\circ_{obs} = \Delta G^\circ_{conf} + \Sigma \Delta G^\circ_{int}. \tag{1}$$

We have determined the free energy of binding (ΔG°_{obs}) of a series of systematically related peptides to neurophysin and used these values to calculate values of ΔG°_{int} for different segments of these peptides [13]. For example, the difference in binding free energy between the peptide glycyl-*L*-tyrosyl-*L*-phenylalanine amide and glycyl-*L*-tyrosyl amide gives the ΔG°_{int} for the phenylalanine amide residue. The methods used to determine binding constants, the values obtained, and the validity of the assumptions are largely reported elsewhere [13]. For present purposes, the importance of these results is that they help to explain the critical importance to binding of the hormone α-amino and aromatic side-chain in position 2 and indicate very large values of ΔG°_{conf}. Table II shows values of

Table II. Calculation of intrinsic bond strengths (ΔG°_{int}) for the binding of different peptide segments to neurophysin

Peptide	Peptide binding affinity[a] $-\Delta G^\circ_{obs}$ (kcal/mol)	Segment bond contributions[b] $-\Delta G^\circ_{int}$ (kcal/mol)
Gly–*L*–Tyr–*L*–Phe–amide [13]	3.2	–CH(CH$_2$–phenyl) CONH$_2$, in position 3 = 0.9
Gly–*L*–Tyr–amide [13]	2.3	
L–Phe–*L*–Tyr–amide [13, 18]	5.6	–CONH$_2$ in position 2 = 1.9
L–Phe–*L*–tyramine [30]	3.7	
S–CH$_3$–*L*–Cys–*L*–Tyr–*L*–Phe–amide [13]	5.1	–CH$_2$–S–Ch$_3$ in position 1 = 1.9
Gly–*L*–Tyr–*L*–Phe–amide [13]	3.2	
L–Phe–*L*–Tyr–amide [13, 18]	5.6	–CH$_2$–phenyl in position 1 = 3.3
Gly–*L*–Tyr–amide [13]	2.3	
L–Phe–*L*–Tyr–amide [13, 18]	5.6	–hydroxyphenyl in position 2 = >4.4
L–Phe–*L*–Ala–amide[c]	<1.2	
Oxytocin [13]	7.8	α–NH$_3^+$ = >4.6
Deaminooxytocin [13]	<3.2	

[a] Binding affinities, expressed as binding free energies, are calculated from the binding of peptides to $2 \times 10^{-4} M$ mononitrated bovine neurophysin II at 25 °C. Data are expressed in terms of binding protonated peptide to neurophysin in which the nitrotyrosine is protonated and the active site carboxyl is ionized. Additionally, the cooperativity of binding to the protein dimer is treated by expressing binding affinities as the observed affinity at half-saturation [18].
[b] Values of $-\Delta G^\circ_{int}$ for each segment listed are calculated by subtraction of the binding free energy of the peptide that does not contain the segment (second peptide in each group) from that of the analogous peptide (first peptide in each group) in which the segment is present.
[c] Unpublished data [*Virmani-Sardana and Breslow*].

ΔG°_{int} for different peptide segments and the peptide affinities from which these values were calculated. The most negative values of ΔG°_{int} are attributable to interactions of the α-NH_3^+ and the tyrosine (or phenylalanine) side-chain in position 2; peptides lacking either of these bind so weakly that no affinities have been measured and, accordingly, values of ΔG°_{int} reported are only minimum negative values. The results can be used to calculate a minimum value of ΔG°_{conf}. Thus, the peptide glycyl-*L*-tyrosine amide binds with a value of $\Delta G^{\circ}_{obs} = -2.3$ kcal at $25\,^{\circ}C$ [13]. However, the intrinsic bonding contributions of its constituent residues are more negative than -10.9 kcal: $[-(>4.6)$ kcal for the α-NH_3^+, $-(>4.4)$ kcal for the side-chain of Tyr-2, and -1.9 kcal for the carboxamide]. From equation 1, ΔG°_{conf} is, therefore, calculated as $-2.3-[-(>10.9)] => +8.6$ kcal. This value may be the highest value of ΔG°_{conf} calculated for any protein system to date [62a, b], and, as will be seen below, can nonetheless probably be raised upward to >11 kcal/mol. It is likely that large binding-induced conformational changes (vide infra) and a lack of added vibrational modes in the complex [62a] contribute to this high value. The high value of ΔG°_{conf} indicates why both the α-NH_3^+ and the side-chain of Tyr-2 are necessary for binding. These are the only two groups which together have an intrinsic bonding energy sufficiently negative to counterbalance ΔG°_{conf}.

The strongly negative value of ΔG°_{int} for the α-NH_3^+ deserves further comment. This value reflects more than the free energy of charge neutralization associated with α-NH_3^+-carboxylate interactions. For example, there is good evidence that, below pH 2, the neurophysin-active site carboxyl is protonated in the hormone complex [7, 18]. However, the binding free energy in this pH range, where charge neutralization, therefore, does not occur, is only ~ 3 kcal less negative than at the pH optimum, giving an approximate value of ΔG°_{int} for charge neutralization of -2.4 to -3 kcal [18]. This means that more than 1.6 kcal of the total value (table II) of ΔG°_{int} for the α-NH_3^+ is due to interactions other than charge neutralization, a fact also seen by the failure of hormone analogs in which the α-amino is replaced by a proton or hydroxyl group to bind either above or below pH 2 [18, 19]. Moreover, NMR studies [7] have shown that the hormone α-amino (which has a pK_a of 6.3 in the unbound state) is completely protonated in the complex even at pH 10.1, signifying that the pK_a of the hormone α-amino in the complex is shifted upwards from its normal value by at least 5 pH units. This has two implications. First, it raises the minimum total value of ΔG°_{int} for the α-NH_3^+ from -4.6 kcal to above -7 kcal, in turn raising the minimum value of ΔG°_{conf} to 11 kcal. Second, the data indicate that the state of protonation of the hormone α-amino is more important to the binding reaction than the state of protonation of the protein carboxyl with which it forms a salt bridge; i.e., the carboxyl can be protonated in the complex (vide supra), but the α-NH_3^+ cannot be deprotonated in the complex. Although the effect of the protonated α-amino on hormone conformation [19] may contribute somewhat to the necessity for α-NH_2 protonation in the complex, we consider it more likely that this extraordinary dependence reflects the fact that the protonated α-NH_2 participates as a donor in hydrogen-bonding interactions, in addition to its role in ion pair formation. These hydrogen-bonding interactions are likely to involve more than one of the protons of the α-amino [19] and would be unavailable to the deprotonated α-amino which should not be a good hydrogen bond donor.

Thermodynamics, Conformational Change, and Kinetics

Major changes in neurophysin conformation occur when peptide is bound, as evidenced by CD changes [11, 21] and widespread NMR changes involving both Tyr-49 (vide supra) and numerous alkyl protons [109b]. The energy involved in this transition is unknown, but it can be expected to contribute positively to $\Delta G°_{conf}$.

The decreases in rotational and translational freedom that can be expected [62a] to be major contributors to $\Delta G°_{conf}$ are clearly manifest in the overall thermodynamics of neurophysin interaction with peptides. Thus, at 25 °C, this is an enthalpy-driven reaction associated with a large net-negative entropy change [14, 18, 75]. A large negative entropic contribution to $\Delta G°_{conf}$ is also probably one of the factors involved in the kinetics of neurophysin-hormone interaction. At and above neutral pH this interaction has been shown to be slow on the NMR time scale [3, 6, 7, 9, 84]. Although the dissociation rate constant is increased when the protein-active site carboxyl is protonated [7] or when peptides of lower affinity are bound [84], the association rate constant at and below neutral pH is independent of these events [7, 84]. Moreover, the overall exchange rate at and above neutral pH is not markedly temperature dependent [7]. These observations are consistent with the view that the decrease in freedom of the system is a major obstacle to complex formation.

The above studies have implications for protein-peptide interactions in general. In particular, they emphasize that observed binding affinities are typically only the tip of the bonding iceberg; i.e., for substances to bind at all, however weakly, strong intrinsic bond strengths are probably involved but are masked by accompanying energetically uphill events. Additionally, the interactions of a single group may be complex and not necessarily what they appear to be on the surface. In this case, interactions of the hormone α-NH_3^+ with the protein-active site carboxyl involve more than just charge neutralization and these other, relatively unappreciated, interactions may be thermodynamically the more important contribution.

Identification of Neurophysin Residues at the Strong Hormone-Binding Site

The presence of internally homologous segments within each neurophysin chain adds another dimension to the usual question of identifying active site residues in that it raises the question as to the extent to which each of these, as opposed to the nonduplicated segments, participate in bonding. Residues least likely to be involved on evolutionary grounds are those that display evolutionary variability (e.g., 1–7, 77–95, etc.), and limited proteolysis studies have indeed recently shown that residues 1–6 and 91–95 do not directly participate in binding [89a]. Additionally, chemical modification [47], ESR [71], and NMR [39, 69] argue against a role for His-80 of neurophysin I.

On the other hand, the only residues implicated in binding so far have been Glu-31 and more controversially, Tyr-49. Glu-31 has been tentatively identified as the carboxyl involved in salt bridge formation with the hormone α-amino [99, 112a]; i.e., reaction of the hormone-neurophysin complex with water-soluble carbodiimides leads to specific covalent coupling of the hormone α-amino to the carboxyl of Glu-31. While the evidence for Glu-31 involvement is compelling, one caveat is that the ability of the covalently coupled complex to bind hormone was not tested; thus the possibility that Glu-31 is simply a reactive carboxyl not directly involved in binding has not been ruled out.

Glu-31 is in one of the duplicated protein segments. Tyr-49, however, is in the center of the chain in a nonduplicated segment and is particularly easy to study by physicochemical methods because it is the only tyrosine in a protein that is also devoid of tryptophan. As indicated above, Tyr-49 is strongly perturbed by the binding of peptides to the strong site [3, 21, 48, 54, 108]. The change in environment of Tyr-49 associated with binding involves a decrease in its pK_a due to loss of interactions with a neighboring

carboxylate [21, 108], probable loss of stacking interactions with a phenylalanine [3], increased exposure to solvent [54], and possibly greater
motional freedom [3]. In the nitrated protein, binding of peptides is associated with a marked increase in nitrotyrosine optical activity [21]. All
these effects appear to be qualitatively and quantitatively identical irrespective of the identity of the bound peptide, suggesting that they arise
primarily from a conformational change associated with binding. Nonetheless, nuclear Overhauser effect studies suggested a proximity between
the ortho ring protons of Tyr-49 and those of peptide Tyr-2 (or the analogous protons of peptide Phe-2) in the bound state [3], suggesting that
Tyr-49 is at or very close to the binding site.

In order to obtain a better estimate of the distance of Tyr-49 from the
binding site, we initiated a series of chemical and physicochemical studies
which are worthwhile to examine in detail to see what they do or do not tell
us. First, we estimated the efficiency of fluorescence energy transfer
between the ring of Tyr-49 and that of Tyr-2 in the bound state and used
this to calculate probable interring distances [108]. Any such calculation
involves assumptions as to the relative orientations of Tyr-2 and Tyr-49 in
the bound state, but it has recently been argued that distances obtained by
assuming random relative orientations are unlikely to be in error by more
than 20% [106]. Using this assumption therefore, and a more updated
estimate of the critical distance for Tyr-Tyr energy transfer [45], our
original data [108] give a probable distance between the two rings in the
range 7–10.4 Å. We have also used spin label NMR to map the distances
between the nitroxides of peptide spin labels bound to the strong site and
selected neurophysin protons, including those of Tyr-49 [69] as schema-
tized in figure 3. These studies indicate that a nitroxide in a position
similar (but not identical) to that of residue 3 of a bound peptide is >14 Å
from the ortho protons of Tyr-49. If assumptions are made as to the
relative distance between this nitroxide and Tyr-2 of the peptide, the
results argue that the distance between the ortho ring protons of Tyr-49
and Tyr-2 is >5 Å and, therefore, support fluorescence studies in the
conclusion that the two tyrosine rings are not in direct contact in the
complex. By these arguments, the previously observed nuclear Overhauser effect between the protons of the two rings is attributed to spin
diffusion effects [3, 69]. A lack of direct contact between the two tyrosine
rings in the bound state is also in agreement with UV absorbance studies
[54] which place the two rings in different environments. While these
results argue against very close proximity of the Tyr-49 ring to the side-

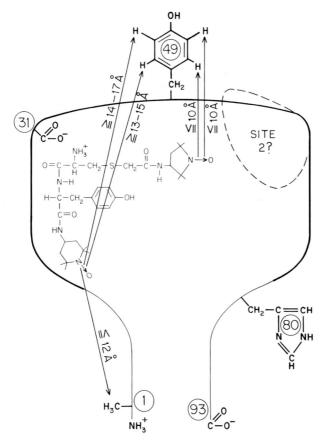

Fig. 3. Hypothetical schematic model of bovine neurophysin I showing distances calculated between nitroxides of spin-labeled peptides bound to the strong site and individual
protein protons. Circled numbers refer to protein residues. For orientation, the peptide
α-NH$_3^+$ and Tyr-2 residues are analogous to the corresponding hormone residues. A possible
position for nitroxides bound to the weak site is also shown [adapted from ref. 69].

chains of either residue 2 or 3 of bound peptides, this may not be true of
Tyr-49 and peptide residue 1. For example, a nitroxide placed on the
extended side-chain of residue 1 gives distances close enough to the ring
protons of Tyr-49 such that some proximity between residue 1 and the ring
of Tyr-49 is allowed, but not demanded, by the data (fig. 3).

Other NMR and chemical modification studies support the conclusion that the ring of Tyr-49 is not a direct binding participant. The
apparently increased motional freedom of Tyr-49 ring protons in the

bound state [3] argues against direct participation. Moreover, Tyr-49 can be nitrated [21, 48, 114], acetylated [46–48], and diiodinated [*Virmani-Sardana and Breslow,* unpublished] without important effects on binding. Dinitration of Tyr-49 [18] leads to a small decrease in binding affinity, as does photooxidation of Tyr-49 [46, 47], but neither blocks binding completely. On the other hand, it should be emphasized that these results do not preclude participation of segments of Tyr-49 *other than the ring* in binding. For example (vide supra) hydrogen-bonding interactions with neurophysin of segments of the hormone appear probable. The carbonyl oxygen and backbone –NH of Tyr-49 are each potential candidates for such hydrogen-bonding interactions, and the carbonyl oxygen can, in principle, be as much as 8 Å from the ortho ring protons. Models can, therefore, be proposed in which the α-NH or carbonyl oxygen of Tyr-49 are hydrogen-bonded to either the peptide backbone or the phenolic –OH of Tyr-2 in the complex. These would not only be completely compatible with *all* measured distances, but would account for the tyrosine ring nuclear Overhauser effect without assuming unreasonable contributions of spin diffusion. Additionally, such models are particularly attractive because they account for the marked perturbation of Tyr-49 by binding.

Until recently, no other residues had been identified as possible binding site participants. However, as shown in figure 3, spin label NMR studies suggest that the amino terminal alanine of neurophysin, although not directly involved in binding (vide supra) may not be distant from the binding site. In this context it is relevant that recent arginine modification studies [17a] argue that one of either Arg_8, Arg_{20}, Arg_{43}, or Arg_{66} (all of which are strictly conserved residues) is close to the binding site, with preliminary data [17b] indicating Arg_8 as the most likely candidate. This is somewhat of a surprise because Arg_8, while itself conserved, is in a region of the protein where oxytocin-associated neurophysins and vasopressin-associated neurophysins show significant differences (fig. 1).

The above results emphasize the importance of neurophysin residues within the sequence 7–49 in binding hormone to the strong site and leave open the possibility that the second duplicated segment participates in weak site interactions. However, in view of the controversial nature of the weak site, it is attractive to speculate that one function of the second duplicated sequence is to participate in supporting interactions with hormone bound to the strong site [14] and/or in interchain contacts in the neurophysin dimer [69]. Interchain contact points in the dimer have not been identified. However, they appear to be largely apolar [75, 82] and not to involve Tyr-49 [40, 82, 108]; a carboxyl group may also participate [31]. These experimental observations and two proposed models for dimerization [69, 89c], all allow participation of either duplicated segment in dimerization in addition to any direct role they might play in hormone binding.

Perspectives

Within the very near future, it is probable that the amino acid sequence of the precursor to oxytocin-associated neurophysin will have been determined, allowing structural comparisons with the precursor to vasopressin-associated neurophysin. Enzymes involved in precursor processing should also have been isolated and characterized, hopefully permitting a complete picture of the biosynthesis of both the hormones and neurophysin. An unambiguous description of the noncovalent interaction between the hormones and neurophysin awaits X-ray crystallographic analyses of neurophysin-peptide complexes that is now in progress [87, 116], but should resolve questions of structure-function relationships in neurophysin and allow for a more precise understanding of the chemical significance of the thermodynamics of individual neurophysin-peptide interactions.

References

1 Acher, R.: Neurophysins: molecular and cellular aspects. Angew. Chem. *18:* 846–860 (1979).

2 Acher, R.; Chauvet, J.; Olivry, G.: Sur l'existence éventuelle d'une hormone unique neurohyphysaire. I. Relations entre l'ocytocine, la vasopressine et la protéine de Van Dyke extraits de la neurohypophyse du bœuf. Biochim. biophys. Acta *16:* 421–427 (1956).

3 Balaram, P.; Bothner-By, A. A.; Breslow, E.: Nuclear magnetic resonance studies of the interaction between peptides and hormones with bovine neurophysin. Biochemistry *12:* 4695–4704 (1973).

4 Bargmann, W.; Scharrer, E.: The site of origin of the hormones of the posterior pituitary. Am. Sci. *39:* 255–259 (1951).

5 Béguin, P.; Nicolas, P.; Boussetta, H.; Fahy, C.; Cohen, P.: Characterization of the 80,000 molecular weight form of neurophysin isolated from bovine neurohypophysis. J. biol. Chem. *256:* 9289–9294 (1981).

6   Blumenstein, M.; Hruby, V.J.: Interaction of oxytocin with bovine neurophysins I and II. Use of ^{13}C nuclear magnetic resonance and hormones specifically enriched with ^{13}C in the glycinamide-9 and half-cystine-1 positions. Biochemistry *16:* 5169–5177 (1977).

7   Blumenstein, M.; Hruby, V.J.; Viswanatha, V.: Investigation of the interactions of oxytocin with neurophysins at low pH using carbon-13 nuclear magnetic resonance and carbon-13 labeled hormones. Biochemistry *18:* 3552–3557 (1979).

8   Blumenstein, M.; Hruby, V.J.; Viswanatha, V.: The tyrosine ring of oxytocin undergoes hindered rotation when the hormone is bound to neurophysin. Biochem. biophys. Res. Commun. *94:* 431–437 (1980).

9   Blumenstein, M.; Hruby, V.J.; Yamamoto, D.M.: Evidence from hydrogen-1 and carbon-13 nuclear magnetic resonance studies that the dissociation rate of oxytocin from bovine neurophysin at neutral pH is slow. Biochemistry *17:* 4971–4977 (1978).

10   Bothner-By, A.A.; LeMarie, B.; Walter, R.; Co, R.T.T.; Rabbani, L.D.; Breslow, E.: NMR and equilibrium dialysis studies of the interaction of bovine neurophysin-I with vasopressin and small peptides. Int. J. Peptide Protein Res. *16:* 450–463 (1980).

11   Breslow, E.: Optical activity of bovine neurophysins and their peptide complexes in the near ultraviolet. Proc. natn. Acad. Sci. USA *67:* 493–500 (1970).

12   Breslow, E.: The neurophysins. Adv. Enzymol. *40:* 271–333 (1974).

13   Breslow, E.: On the mechanism of binding of neurohypophyseal hormones and analogs to neurophysin. Ann. N.Y. Acad. Sci. *248:* 423–441 (1975).

14   Breslow, E.: Chemistry and biology of the neurophysins. A. Rev. Biochem. *48:* 251–274 (1979).

15   Breslow, E.; Aanning, H.L.; Abrash, L.; Schmir, M.: Physical and chemical properties of the bovine neurophysins. J. biol. Chem. *246:* 5179–5188 (1971).

16   Breslow, E.; Abrash, L.: The binding of oxytocin and oxytocin analogues by purified bovine neurophysins. Proc. natn. Acad. Sci. USA *56:* 640–646 (1966).

17a Breslow, E.; Co, T.; Pagnozzi, M.; Rabbani, L.: Effect of arginine modification and proteolytic modification on bovine neurophysins (Abstract No.437). Fed. Proc. *40:* 1615 (1981).

17b Breslow, E.; Pagnozzi, M.; Co, R.T.: Chemical modification or excision of neurophysin arginine-8 is associated with loss of peptide-binding ability. Biochem. biophys. Res. Commun. *106:* 194–201 (1982).

18   Breslow, E.; Gargiulo, P.: Effect of low pH on neurophysin-peptide interactions: implications for the stability of the amino-carboxylate salt bridge. Biochemistry *16:* 3397–3406 (1977).

19   Breslow, E.; Stahl, G.L.; Walter, R.: (*L*-2-Hydroxy-3-mercaptopropionic-acid) oxytocin. Circular dichroism studies of conformation and interaction with neurophysin. Int. J. Peptide Prot. Res. *15:* 314–322 (1980).

20   Breslow, E.; Walter, R.: Binding properties of bovine neurophysins I and II. An equilibrium dialysis study. Molec. Pharmacol. *8:* 75–81 (1972).

21   Breslow, E.; Weis, J.: Contribution of tyrosine to circular dichroism changes accompanying neurophysin-hormone interaction. Biochemistry *11:* 3474–3482 (1972).

22   Breslow, E.; Weis, J.; Menendez-Botet, C.J.: Small peptides as analogs of oxytocin and vasopressin in their interactions with bovine neurophysin-II. Biochemistry *12:* 4644–4653 (1973).

23   Brownstein, M.J.; Gainer, H.: Neurophysin biosynthesis in normal rats and in rats

with hereditary diabetes insipidus. Proc. natn. Acad. Sci. USA *74:* 4046–4049 (1977).

24   Brownstein, M.J.; Robinson, A.G.; Gainer, H.: Immunological identification of rat neurophysin precursors. Nature *269:* 259–261 (1977).

25   Brownstein, M.J.; Russell, J.T.; Gainer, H.: Synthesis, transport and release of posterior pituitary hormones. Science *207:* 373–378 (1980).

26   Camier, M.; Alazard, R.; Cohen, P.; Pradelles, P.; Morgat, J.L.; Fromageot, P.: Hormonal interactions at the molecular level. A study of oxytocin and vasopressin binding to bovine neurophysins. Eur. J. Biochem. *32:* 207–214 (1973).

27   Camier, M.; Girardie, J.; Remy, C.; Girardie, A.; Cohen, P.: Identification of immunoreactive neurophysin-like proteins in the central nervous system of an insect: *Locusta migratoria.* Biochem. biophys. Res. Commun. *93:* 792–796 (1980).

28   Camier, M.; Lauber, M.; Mohring, J.; Cohen, P.: Evidence for higher molecular weight immunoreactive forms of vasopressin in the mouse hypothalamus. FEBS Lett. *108:* 369–373 (1979).

29   Capra, J.D.; Kehoe, J.M.; Kotelchuck, D.; Walter, R.; Breslow, E.: Evolution of neurophysin proteins: the partial sequence of bovine neurophysin-I. Proc. natn. Acad. Sci. USA *69:* 431–434 (1972).

30   Carlson, J.; Breslow, E.: Contribution of the peptide backbone to the binding of peptides and hormones to neurophysin. Biochem. biophys. Res. Commun. *100:* 455–462 (1981).

31   Carlson, J.; Breslow, E.: Interaction of bromophenol blue and related dyes with bovine neurophysin-I: use as a probe of neurophysin chemistry. Biochemistry *20:* 5062–5072 (1981).

32   Chaiken, I.M.; Randolph, R.E.; Taylor, H.C.: Conformational effects associated with the interaction of polypeptide ligands with neurophysins. Ann. N.Y. Acad. Sci. *248:* 442–450 (1975).

33   Chauvet, M.T.; Chauvet, J.; Acher, R.: The neurohypophysial hormone-binding proteins: complete amino acid sequence of ovine and bovine MSEL-neurophysins. Eur. J. Biochem. *69:* 475–485 (1976).

34   Chauvet, M.T.; Chauvet, J.; Acher, R.: Identification of neurophysins: complete amino acid sequence of horse VLDV-neurophysin. Biochem. biophys. Res. Commun. *100:* 600–605 (1981).

35   Chauvet, M.T.; Chauvet, J.; Acher, R.; Robinson, A.G.: Identification of MSEL and VLDV neurophysins in human pituitary gland. FEBS Lett. *101:* 391–394 (1979).

36   Chauvet, M.T.; Codogno, P.; Chauvet, J.; Acher, R.: Phylogeny of the neurophysins: complete amino acid sequence of horse MSEL-neurophysin. FEBS Lett. *80:* 374–376 (1977).

37   Chauvet, M.T.; Codogno, P.; Chauvet, J.; Acher, R.: Comparison between MSEL- and VLDV-neurophysins: complete amino acid sequences of porcine and bovine VLDV neurophysins. FEBS Lett. *98:* 37–40 (1979).

38   Chauvet, M.T.; Hurpet, D.; Chauvet, J.; Acher, R.: The neurophysin domain of human vasopressin precursor. FEBS Lett. *143:* 183–187 (1982).

39   Cohen, P.; Griffin, J.H.; Camier, M.; Caizergues, M.; Fromageot, P.; Cohen, J.S.: Hormonal interactions at the molecular level: a high resolution proton magnetic resonance study of bovine neurophysins and their interactions with oxytocin. FEBS Lett. *25:* 282–285 (1972).

40 Cohen, P.; Nicholas, P.; Camier, M.: Biochemical aspects of neurosecretion: neuro-physin-neurohypophyseal hormone complexes. Current topics in cellular regulation *15:* 263–318 (1979).

41 Cort, J. H.; Sedlakova, E.; Kluh, I.: Neurophysin binding and natriuretic peptides from the posterior pituitary. Ann. N.Y. Acad. Sci. *248:* 336–344 (1975).

42a Dean, C. R.; Hollenberg, M. D.; Hope, D. B.: The relationship between neurophysin and the soluble proteins of pituitary neurosecretory granules. Biochem. J., cell. Aspects *104:* 8–10 (1967).

42b Dean, C. B.; Hope, D. B.; Kazic, T.: Evidence for the storage of oxytocin with neuro-physin-I and of vasopressin with neurophysin-II in separate neurosecretory granules. Br. J. Pharmacol. *34:* 192–193 (1968).

43 Dreifuss, J. J.: A review of neurosecretory granules: their contents and mechanism of release. Ann. N.Y. Acad. Sci. *248:* 184–189 (1975).

44 Drenth, J.: The structure of neurophysin. J. biol. Chem. *256:* 2601–2602 (1981).

45 Eisinger, J.; Feuer, B.; Lamola, A. A.: Intermolecular singlet excitation transfer: application to polypeptides. Biochemistry *8:* 3908–3915 (1969).

46 Fukuda, H.; Hayakawa, T.; Kawamura, J.; Aizawa, Y.: Photooxidation of bovine neurophysin II in the presence of rose bengal. Chem. pharm. Bull. *24:* 36–45 (1976).

47 Fukuda, H.; Hayakawa, T.; Kawamura, J.; Aizawa, Y.: The chemical modifications and their effects on the hormone-binding ability of bovine neurophysin I. Chem. pharm. Bull. *24:* 2043–2051 (1976).

48 Furth, A. J.; Hope, D. B.: Studies on the chemical modification of the tyrosine residue in bovine neurophysin-II. Biochem. J. *116:* 545–553 (1970).

49 Gainer, H.; Peng Loh, Y.; Sarne, Y.: Biosynthesis of neuronal peptides; in Gainer, Peptides in neurobiology, pp. 183–219 (Plenum Press, New York 1977).

50 Gainer, H.; Sarne, Y.; Brownstein, M. J.: Neurophysin biosynthesis: conversion of a putative precursor during axonal transport. Science *195:* 1354–1356 (1977).

51 Ginsburg, M.; Ireland, M.: Binding of vasopressin and oxytocin to protein in extracts of bovine and rabbit neurohypophyses. J. Endocr. *30:* 131–145 (1964).

52 Ginsburg, M.; Ireland, M.: The role of neurophysin in the transport and release of neurohypophyseal hormones. J. Endocr. *35:* 289–298 (1966).

53 Glasel, J. A.; McKelvy, J. F.; Hruby, V.; Spatola, A. F.: Binding studies of polypeptide hormones to bovine neurophysins. J. biol. Chem. *251:* 2929–2937 (1976).

54 Griffin, J. H.; Alazard, R.; Cohen, P.: Complex formation between bovine neurophy-sin-I and oxytocin, vasopressin, and tripeptide analogs of their NH$_2$-terminal region. J. biol. Chem. *250:* 5215–5220 (1973).

55 Griffin, J. H.; DiBello, C.; Alazard, R.; Nicolas, P.; Cohen, P.: Carbon-13 nuclear magnetic resonance studies of the binding of selectively ^{13}C-enriched oxytocins to the neurohypophyseal protein, bovine neurophysin-II. Biochemistry *15:* 4194–4197 (1977).

56 Guidice, L. C.; Chaiken, I. M.: Immunological and chemical identification of a neuro-physin-containing protein coded by messenger RNA from bovine hypothalamus. Proc. natn. Acad. Sci. USA *76:* 3800–3804 (1979).

57 Guidice, L. C.; Chaiken, I. M.: Cell-free biosynthesis of different high molecular weight forms of bovine neurophysins I and II coded by hypothalamic mRNA. J. biol. Chem. *254:* 11767–11770 (1979).

58 Hollenberg, M. D.; Hope, D. B.: The composition of crystalline complexes of neuro-

physin-M with [8-arginine]-vasopressin and oxytocin. Biochem. J. *105:* 921–926 (1967).

59   Hollenberg, M. D.; Hope, D. B.: The isolation of the native hormone-binding proteins from bovine pituitary posterior lobes. Crystallization of neurophysin-I and -II as complexes with [8-arginine]-vasopressin. Biochem. J. *106:* 557–564 (1968).

60   Hope, D. B.; Wälti, M.: [1-(*L*-2-Hydroxy-3-mercaptopropionic acid)]-oxytocin, a highly potent analogue of oxytocin not bound by neurophysin. Biochem. J. *125:* 909–911 (1971).

61   Hope, D. B.; Wälti, M.; Winzor, D. J.: Co-operative binding of oxytocin to bovine neurophysin-II. Biochem. J. *147:* 377–379 (1975).

62a   Jencks, W. P.: Binding energy, specificity and enzymic catalysis – the circe effect. Adv. Enzymol. *43:* 219–410 (1975).

62b   Jencks, W. P.: On the attribution and additivity of binding energies. Proc. natn. Acad. Sci. USA *78:* 4046–4050 (1981).

63   Land, H.; Schütz, G.; Schmale, H.; Richter, D.: Nucleotide sequence of cloned cDNA encoding bovine arginine vasopressin-neurophysin-II precursor. Nature *295:* 299–303 (1982).

64   Lauber, M.; Nicolas, P.; Boussetta, H.; Fahy, C.; Béguin, P.; Camier, M.; Vaudry, H.; Cohen, P.: The M_r 80,000 common forms of neurophysin and vasopressin from bovine neurohypophysis have corticotropin- and β-endorphin-like sequences and liberate by proteolysis biologically active corticotropin. Proc. natn. Acad. Sci. USA *78:* 6086–6090 (1981).

65   Legros, J. J.: Les neurophysines (Masson, Paris 1976).

66   Lin, C.; Joseph-Bravo, P.; Sherman, T.; Chan, L.; McKelvy, J. F.: Cell-free synthesis of putative neurophysin precursors from rat and mouse hypothalamic poly (A)-RNA. Biochem. biophys. Res. Commun. *89:* 943–950 (1979).

67   Livett, B. G.: Immunohistochemical localization of nervous system-specific proteins and peptides. Int. Rev. Cytol., suppl. 7, pp. 53–237 (1978).

68   Lord, S. T.; Breslow, E.: Nuclear magnetic resonance spin label studies of neurophysin: evidence for secondary peptide-binding sites. Biochem. biophys. Res. Commun. *80:* 63–70 (1978).

69   Lord, S. T.; Breslow, E.: Synthesis of peptide spin-labels that bind to neurophysin and their application to distance measurements within neurophysin complexes. Biochemistry *19:* 5593–5602 (1980).

70   Lorén, I.; Schwandt, P.; Alumets, J.; Hokanson, R.; Neureuther, G.; Richter, W.; Sundler, F.: Evidence that lipolytic peptide B occurs in the ACTH/MSH cells of the pituitary and in the brain. Cell Tiss. Res. *205:* 349–359 (1980).

71   Lundt, S. L.; Breslow, E.: Electron spin resonance studies of neurophysin and interaction with spin-labeled peptides. J. phys. Chem. *80:* 1123–1126 (1976).

72   Matthews, E. K.; Legros, J. J.; Grau, J. D.; Nordmann, J. J.; Dreifuss, J. J.: Release of neurohypophyseal hormones by exocytosis. Nature new Biol. *241:* 86–88 (1973).

73   Menendez-Botet, C. J.; Breslow, E.: Chemical and physical properties of the disulfides of bovine neurophysin-II. Biochemistry *14:* 3825–3835 (1975).

74   Nagasawa, J.; Douglas, W. W.; Shulz, R. A.: Ultrastructural evidence of secretion by exocytosis and of 'synaptic vesicle' formation in posterior pituitary glands. Nature *227:* 407–409 (1970).

75   Nicholas, P.; Batelier, G.; Rholam, M.; Cohen, P.: Bovine neurophysin dimeriza-

tion and neurophypophyseal hormone binding. Biochemistry *19:* 3565–3573 (1980).

76   Nicolas, P.; Camier, M.; Lauber, M.; Masse, M.J.O.; Mohring, J.; Cohen, P.: Immunological identification of high molecular weight forms common to bovine neurophysin and vasopressin. Proc. natn. Acad. Sci. USA *77:* 2587–2591 (1980).

77   Nicolas, P.; Wolff, J.; Camier, M.; DiBello, C.; Cohen, P.: Importance of neurophysin dimer and of tyrosine-49 in the binding of neurohypophyseal peptides. J. biol. Chem. *253:* 2633–2639 (1978).

78   North, W.G.; LaRochelle, F.T., Jr.; Morris, J.F.; Sokol, H.W.; Valtin, H.: Biosynthetic specificity of neurons producing neurohypophysial principles; in Lederis, Current studies of hypothalamic function (Karger, Basel 1978).

79   North, W.G.; Mitchell, T.L.: Evolution of neurophysin proteins: partial amino acid sequences of rat neurophysins. FEBS Lett. *126:* 41–44 (1981).

80   North, W.G.; Valtin, H.; Morris, J.F.; LaRochelle, F.T., Jr.: Evidence for metabolic conversions of rat neurophysins within neurosecretory granules of the hypothalamo-neurohyphosial system. Endocrinology *101:* 110–118 (1977).

81   North, W.G.; Walter, R.; Schlesinger, D.H.; Breslow, E.; Capra, J.D.: Structural studies of bovine neurophysin-I. Ann. N.Y. Acad. Sci. *248:* 408–422 (1975).

82   Pearlmutter, A.F.: Bovine neurophysin I dimerization studied by rapid kinetic techniques. Biochemistry *18:* 1672–1676 (1979).

83   Pearlmutter, A.F.; Dalton, E.J.: Thermodynamics and kinetics of bovine neurophysins binding to small peptide analogues of oxytocin and vasopressin. Biochemistry *19:* 3550–3556 (1980).

84   Pearlmutter, A.F.; McMains, C.: Interaction of bovine neurophysin with oxytocin and vasopressin measured by temperature-jump relaxation. Biochemistry *16:* 628–633 (1977).

85   Pickering, B.T.; Jones, C.W.: The neurophysins; in Li, Hormonal proteins and peptides, vol. 5, pp. 103–158 (Academic Press, New York 1978).

86   Pickup, J.C.; Johnston, C.I.; Nakamura, S.; Uttenthal, L.O.; Hope, D.B.: Subcellular organization of neurophysins, [8-lysine]-vasopressin and adenosine triphosphatase in porcine posterior pituitary lobes. Biochem. J. *132:* 361–371 (1973).

87   Pitts, J.E.; Wood, S.P.; Hearn, L.; Tickle, I.J.; Wu, C.W.; Blundell, T.L.; Robinson, I.C.A.F.: Crystallization and preliminary crystallographic data of a porcine neurophysin I-Tyr-Phe-NH$_2$ complex. FEBS Lett. *121:* 41–43 (1980).

88   Pliska, V.; Meyer-Grass, M.; Bersinger, N.; Carlson, L.; Melin, P.; Vilhardt, H.: Long term stability of lysine vasopressin and specifically tritiated lysine vasopressin in weakly acidic aqueous solutions. Experientia *36:* 1145–1146 (1980).

89a Rabbani, L.D.; Pagnozzi, M.; Chang, P.; Breslow, E.: Partial digestion of neurophysins with proteolytic enzymes: unusual interactions between bovine neurophysin-II and chymotrypsin. Biochemistry *21:* 817–826 (1982).

89b Rauch, R.; Hollenberg, M.D.; Hope, D.B.: Isolation of a third bovine neurophysin. Biochem. J. *115:* 473–479 (1969).

89c Rholam, M.; Nicolas, P.: Side-by-side dimerization of neurophysin: sedimentation velocity, viscometry and fluorescence polarization studies. Biochemistry *20:* 5837–5843 (1981).

90   Rosenior, J.C.; North, W.G.; Moore, G.J.: Putative precursors of vasopressin, oxytocin and neurophysins in the rat hypothalamus. Endocrinology *109:* 1067–1072 (1981).

91   Russell, J. T.; Brownstein, M. J.; Gainer, H.: Trypsin liberates an arginine vasopressin-like peptide and neurophysin from a M_r 20,000 putative common precursor. Proc. natn. Acad. Sci. USA 76: 6086–6090 (1979).

92   Russell, J. T.; Brownstein, M. J.; Gainer, H.: Biosynthesis of vasopressin, oxytocin and neurophysins: isolation and characterization of two common precursors (propresso-physin and prooxyphysin). Endocrinology 107: 1880–1891 (1980).

93   Russell, J. T.; Holz, R. W.: Measurement of ΔpH and membrane potential in isolated neurosecretory vesicles from bovine neurohypophyses. J. biol. Chem. 256: 5950–5953 (1981).

94a   Sachs, H.: Neurosecretion. Adv. Enzymol. 32: 327–372 (1969).

94b   Sachs, H.; Fawcett, P.; Takabatake, Y.; Portonova, R.: Biosynthesis and release of vasopressin and neurophysin. Recent Prog. Horm. Res. 25: 447–491 (1969).

95   Sachs, H.; Saito, S.; Sunde, D.: Biochemical studies on the neurosecretory and neuroglial cells of the hypothalamo-neurohypophysial complex; in Heller, Lederis, Sub-cellular organization and function in endocrine tissues, pp. 325–336 (Cambridge Press, New York 1971).

96a   Scharrer, E.; Scharrer, B.: Hormones produced by neurosecretory cells. Recent Prog. Horm. Res. 10: 183–240 (1954).

96b   Schlesinger, D. H.; Audhya, T. K.: A comparative study of mammalian neurophysin protein sequences. FEBS Lett. 128: 325–328 (1981).

97   Schlesinger, D. H.; Audhya, T. K.; Walter, R.: Complete amino acid sequence of bovine neurophysin-I: a major secretory product of the posterior pituitary. J. biol. Chem. 293: 5019–5024 (1978).

98   Schlesinger, D. H.; Pickering, B. T.; Watkins, W. B.; Peek, J. C.; Moore, L. G.; Audhya, T. K.; Walter, R.: A comparative study of partial neurophysin protein sequences of cod, guinea pig, rat and sheep. FEBS Lett. 80: 371–373 (1977).

99   Schlesinger, D. H.; Walter, R.; Audhya, T. K.: A binding site for neurohypophyseal hormones in the neurophysin proteins; in Schlesinger, Proceedings of the symposium on neurohypophyseal hormones and other biologically active peptides (Elsevier, New York 1980).

100   Schmale, H.; Leipold, B.; Richter, D.: Cell-free translation of bovine hypothalamic mRNA. FEBS Lett. 108: 311–316 (1979).

101   Schmale, H.; Richter, D.: In vitro biosynthesis and processing of composite common precursors containing amino acid sequences identified immunologically as neurophysin I/oxytocin and as neurophysin II/arginine vasopressin. FEBS Lett. 121: 358–362 (1980).

102   Schmale, H.; Richter, D.: Immunological identification of a common precursor to arginine vasopressin and neurophysin II synthesized by in vitro translation of bovine hypothalamic mRNA. Proc. natn. Acad. Sci. USA 78: 766–769 (1981).

103   Sinding, C.; Robinson, A. G.: A review of neurophysins. Metabolism 26: 1355–1370 (1977).

104   Soloff, M. S.; Pearlmutter, A. F.: Biochemical actions of neurohypophysial hormones and neurophysin; in Litwak, Biochemical actions of hormones, vol. 6 (Academic Press, New York 1978).

105   Stouffer, J. E.; Hope, D. B.; Vigneaud, V. du: Neurophysin, oxytocin and desamino-oxytocin; in Cori, Foglia, Leloir, Ochoa, Perspectives in biology, pp. 75–80 (Elsevier, Amsterdam 1963).

106 Stryer, L.: Fluorescence energy transfer as a spectroscopic ruler. A. Rev. Biochem. *47:* 819–846 (1978).

107 Sunde, D. A.; Sokol, H. W.: Quantification of rat neurophysins by polyacrylamide gel electrophoresis (PAGE): application to the rat with hereditary hypothalamic diabetes insipidus. Ann. N.Y. Acad. Sci. *248:* 345–364 (1975).

108 Sur, S. S.; Rabbani, L. D.; Libman, L.; Breslow, E.; Fluorescence studies of native and modified neurophysins: effects of peptides and pH. Biochemistry *18:* 1026–1036 (1979).

109a Van Dyke, H. B.; Chow, B. F.; Greep, R. O.; Rothen, A.: The Isolation of a protein from the pars neuralis of the ox pituitary with constant oxytocin, pressor and diuresis-inhibiting activities. J. Pharmac. exp. Ther. *74:* 190–209 (1942).

109b Virmani-Sardana, V.; Breslow, E.: Effects of peptide-binding on the proton NMR spectrum of bovine neurophysin-I. Int. J. Peptide Protein Res. (in press, 1982).

110 Walter, R.; Audhya, T. K.; Schlesinger, D. H.; Shin, S.; Saito, S.; Sachs, H.: Biosynthesis of neurophysin proteins in the dog and their isolation. Endocrinolgy *100:* 162–174 (1977).

111 Walter, R.; Breslow, E.: Methods of isolation and identification of neurophysin proteins; in Marks, Rodnight, Methods in neurochemistry, vol. 2, pp. 247–279 (Plenum Press, New York 1974).

112a Walter, R.; Hoffman, P. L.: Tentative identification of a binding site of arginine vasopressin to neurophysin (Abstract). Fed. Proc. *32:* 567 (1973).

112b Wälti, M.: The binding of polypeptides by neurophysin; thesis, Oxford (1975).

113 Watson, S. J.; Seidah, N. G.; Chrétien, M.: The carboxy terminus of the precursor to vasopressin and neurophysin: immunocytochemistry in rat brain. Science *217:* 853–855 (1982).

114 Wolff, J.; Alazard, R.; Camier, M.; Griffin, J. H.; Cohen, P.: Interactions of bovine neurophysins with neurohypophyseal hormones. On the role of tyrosine-49. J. biol. Chem. *250:* 5215–5220 (1975).

115 Wuu, T. C.; Crumm, S. E.: Characterization of porcine neurophysin-III. Its resemblance and possible relationship to porcine neurophysin-I. J. biol. Chem. *251:* 2735–2739 (1976).

116 Yoo, C. S.; Wang, B. C.; Sax, M.; Breslow, E.: Crystals of a bovine neurophysin-II-dipeptide amine complex. J. molec. Biol. *127:* 241–242 (1979).

117 Zimmerman, E. A.; Defendini, R.; Sokol, H.; Robinson, A. G.: The distribution of neurophysin-secreting pathways in the mammalian brain: light microscopic studies using the immunoperoxidase technique. Ann. N.Y. Acad. Sci. *248:* 92–111 (1975).

Esther Breslow, PhD, Department of Biochemistry, Cornell University Medical College, New York, NY 10021 (USA)

Cell Biology of the Secretory Process, pp. 309–358 (Karger, Basel 1984)

The Secretory Process in
Adrenal Medullary Cells

Bruce G. Livett

Division of Neurology, Montreal General Hospital and McGill University,
Montreal, Quebec, Canada

The mechanisms underlying secretion from the adrenal medulla have
been studied extensively since the initial observations of *Vulpian* [172] that
aqueous extracts of adrenal medulla, and of blood clotted in the adrenal
vein, contain a substance which upon oxidation turned a rose-brown
colour. *Vulpian* [172] proposed that this substance might be released from
the adrenal medulla directly into the blood as its 'produit de sécrétion'.
The first clue that this secretory material might be physiologically active
came 40 years later when *Cybulski and Szyomonowicz* [37], and *Oliver
and Schafer* [126] demonstrated that extracts of the adrenal gland and of
adrenal venous blood caused a rise in blood pressure (pressor activity) and
an increase in heart rate (cardioacceleratory activity). Soon after [48, 73] it
was shown that the secretion of this physiologically active material was
controlled by neural input to the adrenal gland. The active substance was
isolated, identified as a simple organic molecule (a catecholamine), chemically synthesized and termed epinephrine [1] or adrenaline [156].

It was subsequently shown that the immediate physiological stimulus
for the secretion of adrenaline from the adrenal medulla was the release of
acetylcholine (ACH) from the splanchnic nerve terminals that innervate
the catecholamine-containing chromaffin cells [52]. A wide range of physiological stimuli, e.g. emotional and physical stress, asphyxia and anoxia,
cold, heat, pH, hypotension, insulin-induced hypoglycemia, and glucagon, act either indirectly through the splanchnic nerve or directly on the
chromaffin cells themselves to increase the secretion of catecholamines
[for review see 96].

Studies on the cellular mechanisms responsible for the secretion of
catecholamines from the adrenal medulla were greatly facilitated by the
use of the retrogradely perfused adrenal gland maintained in vitro [130].

Extensive biochemical, immunochemical and pharmacological studies with this in vitro preparation by a number of investigators in the early 60s and 70s [for reviews see 44, 168] led to the conclusion that when secretion is evoked by ACH (or related drugs), the release of catecholamines and other chromaffin vesicle constituents occurs quantitatively by exocytosis. Although the ionic mechanisms involved in initiating this process have been reasonably well defined (see below), the molecular mechanisms involved in stimulus-secretion coupling are as yet poorly understood. In addition, the recent discovery that opioid peptides and other neuropeptides, e.g. substance P, somatostatin, vasoactive intestinal polypeptide (VIP), and neurotensin, are present in the adrenal medulla and/or splanchnic nerve terminals, and interact with chromaffin cell secretion in vitro [for review see 103] raises the possibility that these endogenous neuropeptides may have a physiological role in regulating basal and ACH-mediated secretion from the adrenal medulla. In this article, some of these more recent concepts will be discussed in the context of homeostatic mechanisms that may operate at the cellular and molecular levels to regulate secretion from adrenal chromaffin cells.

For further background information on the storage and biosynthesis of catecholamines in the adrenal medulla, the reader is referred to a number of excellent review articles [162, 165, 178, 183–186, 188] and monographs [21, 22, 71, 131, 142a].

Developmental and Physiological Considerations

Developmental Origin and Plasticity of Chromaffin Cells

Adrenal chromaffin cells of the adrenal medulla are derived embryologically from the neutral crest. It is now established that the interaction of the migrating cells with the extracellular matrix in their immediate environment determines both their chemical and morphological phenotype. In addition, a number of derivatives of the neural crest can be interconverted by manipulation of environmental cues [for review see 91]. For example, adrenal chromaffin cells can be induced to express cholinergic properties both in vivo [93] and in vitro [47] (fig. 1). When cultured in the presence of glucocorticoids, chromaffin cells from newborn rats maintain the characteristics of differentiated chromaffin cells. However, in long-term culture in the presence of nerve growth factor and the absence of corticosteroids, many adrenal medullary cells extend neurites [166], lose

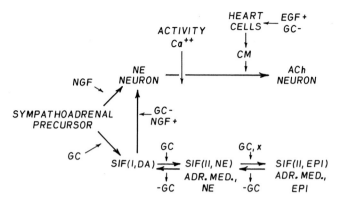

Fig. 1. Possible relationships of the derivatives in the sympathoadrenal lineage. The precursor cells come from the neural crest after migration past the somitic mesenchyme and the neural tube-notochord complex. The neuronal line could arise directly from the sympathoadrenal precursor cells or from a SIF-like intermediate. The experimental evidence supporting many of the steps shown is described in the text. NGF = Nerve growth factor; GC = glucocorticoids; NE = norepinephrine; DA = dopamine; ACH = acetylcholine; EPI = epinephrine; SIF = small intensely fluorescent cell (type I or II); ADR. MED. = adrenal medullary cell; CM = conditioned medium cholinergic factor; EGF = epidermal growth factor; X = unknown factor which may induce the initial appearance of phenylethanolamine N-methyltransferase in the developing adrenal medulla. From [91].

their characteristic large chromaffin granules and acquire small synaptic vesicles in axon varicosities [47] indistinguishable from those seen in sympathetic neurons. In addition, when cultured in medium conditioned by heart cells, the cultures acquire the following cholinergic properties: (1) synthesis and storage of ACH; (2) choline acetyltransferase (CAT), and (3) small agranular vesicles in varicosities [47]. These studies demonstrate that, as previously demonstrated for sympathetic neurons in vivo and in vitro, adrenal chromaffin cells can acquire a cholinergic phenotype.

In addition to epinephrine and norepinephrine cells in the adrenal medulla, another cell type termed 'small granule cells' (SGC) has been described in several species. These cells are thought to contain dopamine. *Coupland* et al. [33] suggested that these cells may form a pool of cells that, under appropriate conditions, differentiate into normal or neoplastic chromaffin cells.

The Concept of Adrenal Paraneurons

Many endocrine cells and secretory neurons develop from common ancestral cells in the neutral crest. Small cell carcinomas, pheochromocytomas (APUDomas) and neuroblastomas often occur concurrently again suggesting a common ancestral origin of these abnormal cell lines [157]. Two characteristics shared by normal and neoplastic cells of the APUD series are (1) the co-storage and co-secretion of various neuropeptides with the classical neurotransmitters, and (2) the extension of cell processes. With the increasing recognition that there exists no sharp boundary between typical 'nerve cells' and some of the peptide- and hormone-secreting endocrine cells, the term 'paraneurons' was proposed by *Fujita* [54a] and *Kobayashi* [86a] to cover both endocrine cells with neuron-like properties and nerve cells with endocrine characteristics. Adrenal chromaffin cells are typical APUD paraneurons: (1) they are of neural crest origin: (2) they secrete catecholamines and opioid peptides by exocytosis; (3) they are electrically excitable and (4) they put out neurite-like processes when maintained in monolayer cultures [for reviews see 103, 167].

The Pheochromocytoma Cell Line (PC12)

The study of the properties of cells derived from the neural crest has been facilitated by the establishment of a clonal cell line (PC12) from a transplantable rat adrenal pheochromocytoma [60, 177]. These cells exhibit differentiated properties of adrenal chromaffin cells such as synthesis [60], storage, uptake and release [57, 58] of catecholamines [for review see 136, 160]. PC12 cells grown in vitro have morphologic and cytochemical features in common with normal chromaffin cells in varying stages of development and with human pheochromocytomas. These features include catecholamine stores demonstrable by formaldehyde-induced fluorescence and argentaffinic secretory granules, measuring 30–350 nm. Dark 'norepinephrine'- and light 'epinephrine'-type granules are both present, despite the absence of epinephrine and its synthesizing enzyme, phenylethanolamine N-methyl transferase.

PC12 cells acquire further neuronal properties after treatment with nerve growth factor (NGF) including a rapid increase in attachment [148], outgrowth of extensive neurites [60], electrical excitability, induction of opiate receptors [70] and high sensitivity to ACH and other nicotinic agonists [42] with depolarization and release of catecholamines [58]. Although the cells bind α-bungarotoxin, as do sympathetic neurons, this binding does not appear to antagonize the nicotinic response of either PC12 cells, adrenal chromaffin cells or sympathetic neurons [127, 135].

As with neural crest derivatives of non-neoplastic origin [e.g. SCG in culture; 55, 128] the PC12 line possesses the enzyme CAT together with the ability to synthesize, store and release ACH [59] and to form cholinergic synapses [147].

Some Differences between Pheochromocytomas and Adrenal Chromaffin Cells

As described well by *Tischler* et al. [158], there are a number of unexplained quantitative and qualitative differences among pheochromocytomas and normal adrenal medullary cells in vivo and in vitro. These differences relate to NGF requirements for outgrowth and maintenance of processes, glucocorticoid modulation of NGF responsiveness, and selective enzyme induction. For example, glucocorticoids do not inhibit the NGF-induced outgrowth of processes from PC12 cells, and NGF-responsive rat pheochromocytoma cells do not synthesize epinephrine but rather contain up to three times as much dopamine as norepinephrine. By contrast, human tumours from which NGF-responsive primary cultures have been derived were originally epinephrine-producing. In contrast to the action of NGF on pheochromocytomas, NGF has no effect on monolayer cultures of adult bovine adrenal medulla [182] or on the morphology of medullary carcinomas of the thyroid or bronchial carcinoid cells maintained as monolayer cultures [159].

Innervation of the Adrenal Medulla

Preganglionic sympathetic neurons in the intermedia lateral cell column (IML) of the thoracic and upper lumbar segments of the spinal cord innervate chromaffin cells in the adrenal medulla via sympathoadrenal preganglionic neurons that comprise the splanchnic nerve in mammals [29–31, 36, 50, 86, 115, 146, 161]. The preganglionic fibres innervating the adrenal medulla are thought to be cholinergic [29, 30, 32, 87, 97]. However, the presence of enkephalin-like immunoreactivity in fibres and terminals in the adrenal medulla of the rat, cat and guinea pig [150], and the presence of substance P-like and VIP-like immunoreactivity in fibres or terminals in the human adrenal medulla [102], suggest that the innervation of the chromaffin cells may not be purely cholinergic.

Neumayr et al. [122] suggested that the control and integration of sympathetic reflexes takes place at the level of preganglionic neurons in the IML and is exerted through a balance of noradrenergic-mediated excitation and serotonergic-mediated inhibition of the IML neurons. Other investigators have demonstrated the importance of descending dopamine, norepinephrine and serotonin pathways in the control of adrenal medullary function and secretion [56, 95, 134]. Descending projections to the thoracic spinal cord from the raphe nuclear group [16], the ventrolateral reticular formation [65], and a group of neurons in the ventrolateral surface of the hindbrain [139] and the nucleus tractus solitarius [109] have been identified. Pathways descending directly to IML have been proposed to originate in several regions of the brainstem and in higher areas of the central nervous system (CNS) [for review see 67], and preganglionic sympathetic neurons to the IML have been studied using both neurophysiological and anatomical methods.

Biochemical and pharmacological evidence which demonstrates the importance of the descending supraspinal input to the sympathoadrenal preganglionic neurons exists [56, 95, 134]. Adrenomedullary ornithine decarboxylase (ODC) activity has been a most useful biochemical marker for mapping the origins of the splanchnic nerve in the rat [3]. It was found that the pathway for regulation of adrenal *medullary* ODC involves nuclei in the diencephalon-telencephalon, and ultimately acts through the sympathetic nervous system. The pathway for regulation of adrenal *cortical* ODC involves the hypothalamus and acts via the anterior pituitary gland. These pathways include serotonergic components which have opposite net effects on the induction of ODC produced by apomorphine; inhibitory for the medulla and facilitatory for the cortex. Other chemically coded afferents to the specific neurons in IML which innervate the adrenal medulla have only recently been identified by making use of specific antibodies to the peptide hormones together with those retrogradely labeled from the adrenal medulla [67].

Holets and Elde [66, 67] utilized the retrograde axonal transport of two fluorescent dyes (Fast Blue and True Blue) to determine the distribution and morphology of the IML neurons which project to the adrenal medulla. They found that the majority (72.3%) of the retrogradely labeled sympathoadrenal preganglionic neurons were located within the T_7 to T_{12} segments of the spinal cord. The T_9 segment contained the largest average number (20.1%) of retrogradely labeled cells in a single segment. Methionine enkephalin, serotonin, and substance-P immunoreactive fibres were prominent in the IML, whereas oxytocin, neurophysin and somatostatin immunoreactive fibres were sparse. The methionine enkephalin, serotonin and substance P fibres were seen to surround both unlabeled and retrogradely labeled neurons; somatostatin fibres appeared to preferentially contact retrogradely labeled neurons whereas the neurophysin and

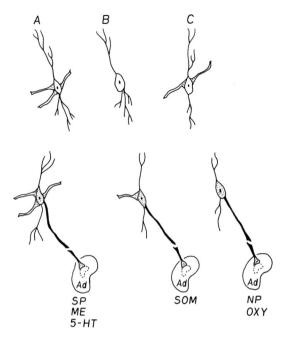

Fig. 2. The distribution and relationship of the chemically coded afferents to sympa-thoadrenal preganglionic neurons (stippled cells) and to unlabeled intermediolateral cell column neurons (unstippled cells) are represented. The Met-enkephalin (ME), substance P (SP) and 5-HT immunoreactive fibers *(A)* appeared to impinge upon both unlabeled and retrogradely labeled cells. Somatostatin (SOM) immunoreactive fibres were seen to surround preferentially retrogradely labeled intermedia lateral cell column (IML) neurons *(B)*. Neu-rophysin (NP) and oxytocin (OXY) fibers, although present in IML, were not found in relationship to retrogradely labeled cells at any level of the thoracic or upper lumbar spinal cord *(C)*. From [67].

oxytocin fibres were not found in proximity to retrogradely labeled neu-rons (fig. 2).

These findings suggest that sympathoadrenal preganglionic neurons, as well as other neurons in the IML, receive methionine enkephalin, serotonin, and substance P afferents; somatostatin afferents preferentially surround sympathoadrenal neurons but neurophysin and oxytocin immu-noreactive fibres do not appear to impinge upon sympathoadrenal pre-ganglionic neurons. The functional significance of these different inputs in chromaffin cell secretion is not known but is a topic of active investigation at the present time.

Ultrastructural changes in the innervation of the rat adrenal medulla at different times of day and after pinealectomy have also been reported. Both small clear synaptic vesicles and large granular synaptic vesicles exhibit two cycles per day, as expressed in numbers per unit area. Sham-operated animals had decreased or insignificant rhythmic changes whereas pinealectomized animals lost the cyclic changes related to the dark cycle. It was also noted that norepinephrine- and epinephrine-containing cells differed consistently and significantly in the quantitative aspects of rhythmic patterns of the two types of synaptic vesicles [75–77].

Two areas of the CNS that have been shown to be involved in the control of secretion from the adrenal medulla are the medulla oblongata, and the hypothalamus. *Matsui* [113] found that stimulation at or near the inferior fovea produced an *increased* secretion of epinephrine, as measured by a fluorimetric assay of adrenal venous blood. Stimulation of other areas in the floor of the 4th ventricle resulted in a decreased secretion of norepinephrine with or without a decrease in epinephrine secretion. In a similar unpublished study [*Robertson, Culverson and Carmichael*, personal commun.], some 344 hypothalamic sites were stimulated in the cat and a preferential release of the two catecholamines was measured.

It is also of interest that lesions in the lateral hypothalamic areas or lateral mamillary nuclei result in a decrease in the mass of the rat adrenal medulla [181] and a reduction in the nuclear volume of the chromaffin cells. In the dog, *Waki* [176] showed that electrical stimulation of the ventral roots caused the release of catecholamines from the adrenal medulla and that this release was mediated by fibres having their origin between T_5 and L_1 for epinephrine and between T_5 and T_{12} for norepinephrine. With both amines the maximal effects were produced by stimulation of the ventral roots around T_{10}. In contrast, stimulation of the dorsal roots was ineffective in releasing the catecholamines. The above studies establish that stimulation of specific sites in the CNS can lead to a preferential secretion of epinephrine and norepinephrine from the adrenal medulla in several species. However, it has recently been shown [*Edwards*, 48a] that changes in the proportions of the two amines that are released from the adrenal gland, in response to direct stimulation of the *whole* of the efferent innervation, can be brought about simply by altering the pattern of stimulation.

Kirby and McCarty [85] have shown that the development of functional sympathetic innervation to the adrenal medulla of rats occurs prior to the development of functional sympathetic innervation to the heart. In another study, *Ross* et al. [138] studied the effect of electrical stimulation of the C1 adrenergic cell group, and of the Kölliker-Fuse nucleus on sympathetic vasomotor activity and adrenal medullary catecholamine secretion in the rat. They concluded: (a) excitation of neurons in the C1 adrenergic group and the Kölliker-Fuse nucleus elicited release of adrenal medullary catecholamines and excited sympathetic vasomotor nerves via direct projections from each nucleus to the intermediolateral cell column, and, (b) within each nucleus, representation of the vasomotor fibres and the adrenal medulla were admixed [138, 144]. In still another study, *Del Bo* et al. [40] showed that the fastigial pressor response in rats with intact sympathetic nerves was mediated by a pathway descending below the midbrain and was associated with an increase in serum norepinephrine and epinephrine; adrenal catecholamines did not contribute to the magnitude of the fastigial pressor response. In the absence of sympathetic nerves, catecholamines of adrenal medullary origin released by fastigial nucleus stimulation contributed only partially to the hypertension. They hypothesized that a circulating pressor factor released by fastigial nucleus stimulation from areas above the midbrain may significantly elevate mean arterial pressure in the absence of sympathetic nerves and adrenal medulla.

An important question concerning innervation of the adrenal medulla cells is whether each chromaffin cell receives splanchnic innervation or whether some cells are activated by electrical coupling via gap junctions. *Grynszpan-Winograd and Nicolas* [61] carried out a freeze fracture study of the intracellular junctions in the adrenal medulla. They found that gap junctions were much less extensive in the medulla than in the cortex and that the junctions were not macular. The number of gap junctions varied with the species; there were more in the guinea pig than in the hamster, and none were seen in the rat. From this study, *Grynszpan-Winograd and Nicolas* [61] suggested that gap junctions were not needed for coupling between chromaffin cells because each cell received direct innervation.

Neurogenic and Non-Neurogenic Control of Secretion from the Adrenal Medulla: Influence of Development

Secretion from the immature adrenal is different from that in the mature animal in that it is independent of splanchnic nerve involvement. Splanchnic denervation [26], nicotinic blocking agents [5] and muscarinic blocking agents [23] are all ineffective in blocking the secretory response in neonatal animals. The adrenal medulla of species such as the rat and calf are not functionally innervated by the splanchnic nerve at birth [152] and non-neurogenic responses have been elicited in response to stresses such as asphyxia [66] or high doses or reserpine [11].

If the acute release of catecholamines in neonates is not mediated by splanchnic nerve stimulation, what other factors might be involved in modulating adrenomedullary secretion in early development? One possibility is the opioid peptides present within the chromaffin cells [104, 149]. Although it is not known if the levels of enkephalins are higher in the neonate than in the adult, it has been shown that upon chronic denervation of mature rat adrenals, a dramatic rise occurs in the level of opioid peptides [98, 150].

Following up this lead, *Chantry* et al. [23] found that naloxone (5 mg/kg s.c.) potentiated the non-neurogenic depletion of catecholamines in the neonate induced by reserpine, while methadone (2.5 mg/kg s.c.) inhibited this non-neurogenic response. By contrast, in adult rats no potentiation of release by naloxone or inhibition by methadone was seen. The onset of neural control of adrenomedullary function in the rat first appears at the end of the first week postnatally and matures fully by approximately 10 days of age [11, 151, 152]. There is thus a narrow window in the rat in which neurogenic non-neurogenic responses co-exist (8–11 days). *Chantry* et al. [23] showed that the non-neurogenic (chlorasondam-ine-resistant) mechanism in the rat disappeared about the same time that splanchnic nerve function matured (8–11 days of age). From 11 days onward, depletion occurred solely through neurogenic mechanisms. By contrast, in the calf, the non-neurogenic mechanism disappears well before splanchnic nerve input becomes capable of eliciting secretion [26].

The finding by *Chantry* et al. [23] suggests that the acute, non-neurogenic effect of reserpine in neonates is not due to uptake inhibition, but rather represents net movement of the amines from the granules to the extracellular compartment (release). Similarly, earlier studies demonstrated that non-neurogenic release from both immature [26, 90] and denervated adult adrenals [187] produced an elevation of circulating catecholamines. A functional role for the endogenous opioids in dampening non-neurogenic release of catecholamines has recently been proposed [23] that may be of survival value to neonates undergoing stress.

Electrophysiological Responses of Adrenal Chromaffin Cells
Although it is now well established (see below) that the release of ACH from the splanchnic nerve terminals causes activation of cholinergic

receptors on the adrenal chromaffin cells followed by entry of Ca^{++} and secretion of the chromaffin vesicle contents by exocytosis, relatively little is known about the electrophysiological changes that accompany this event. Until recently the adrenal medulla has been regarded as electrically inexcitable. However, recent studies in a number of laboratories have shown that the plasma membrane of adrenal chromaffin cells are responsive to both chemical and electrical stimulation.

Several preparations (perfused adrenals, adrenal slices and isolated chromaffin cells) have been used to study the electrical properties of chromaffin cells in vitro. *Wakade* [173, 174] for example showed that transmural stimulation of the isolated adrenal gland with a train of 10-Hz shocks caused excitation of the splanchnic nerve terminals embedded in the adrenal medulla, and that the released ACH activated nicotinic receptors on the chromaffin cells to evoke the secretion of catecholamines. Bovine adrenal medullary slices, a preparation relatively free of splanchnic nerve terminals, also secreted catecholamines following electrical stimulation [18]. Moreover, a single electrical shock to the perfused adrenal gland of the rat resulted in massive secretion of epinephrine and norepinephrine [175]. Unlike the presynaptic cholinergic receptor-activated release, which was saturable and reduced by cholinergic blocking agents and tetrodotoxin, the single shock release was directly related to the strength and duration of the applied stimulus over a wide range, was unaffected by hexamethonium/atropine or tetrodotoxin, but was abolished by removal of Ca^{++} or by addition of 3 mM Mn^{++}. This indicates that adrenal chromaffin cell membranes can undergo nonpropagated electrotonic depolarization upon electrical stimulation resulting in voltage-dependent Ca^{++} channels being opened to initiate secretion. The presence of voltage-dependent Ca^{++} channels have also been demonstrated from work on isolated chromaffin cells (see below).

A limitation of the perfused adrenal gland and of adrenal slices is that these preparations are not really suitable for dose-response studies of agonists and antagonists because the secretory response of the preparation declines steadily. In addition, the complications imposed by the presence of nerve terminals in these preparations encouraged investigators, interested in examining the electrical activity of the chromaffin cells during secretion, to develop techniques for the isolation of single adrenal medullary cells suitable for microelectrode penetration. The first attempts [44–46, 78, 114] were somewhat disappointing because the mixed pancreatic enzymes (Viokase) they used produced cells with low resting mem-

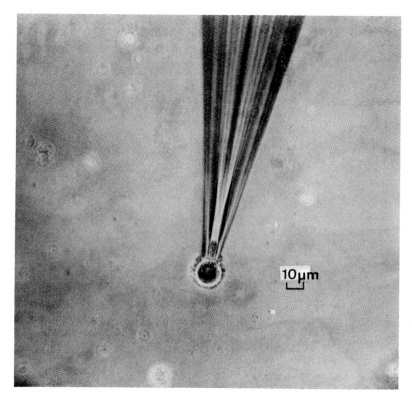

Fig. 3. Photograph showing the microsuction electrode arrangement for extracellular spike recording. A part of a chromaffin cell was sucked into the electrode. A 40 × water-immersion phase-contrast objective (Carl Zeiss, West Germany) was used. The bar indicates 10 µm. From [17].

brane potentials (30 mV). Although depolarization in response to ACH was rapid, they were not able to observe action potentials even when the membrane was depolarized rapidly by passing current across it [78]. *Douglas* [44] was forced to conclude, 'Thus, it appears that spike generation is not involved in stimulus-secretion coupling in the chromaffin cell'. Fortunately, electrophysiologically and functionally viable chromaffin cells can be produced using collagenase, either alone or together with trypsin and DNAase-I [for review see 103].

Using extracellular recording with a suction electrode (fig. 3), *Brandt* et al. [17] demonstrated the presence of spontaneous action potentials in cultured rat adrenal chromaffin cells whose frequency was increased either by electrical depolarization or by elevated concentrations of ACH. In another study [12], action potentials were recorded by intracellular microelectrode penetration of adult human and gerbil adrenal chromaffin cells and in penetrations of fresh slices of gerbil adrenal medullae. In the best recordings, chromaffin cell transmembrane potentials exceeded 50 mV. The cells were capable of generating neuron-like, all-or-none overshooting action potentials in response to depolarizing currents, and in response to applied ACH. These action potentials were blocked by tetrodotoxin but were not blocked by the removal of Ca^{++} or by $CoCl_2$ suggesting that they were probably generated by a Na^+ mechanism with little or no direct Ca^{++} component.

Subsequent studies on chromaffin cell action potentials in perfused rat adrenal medulla [83] have defined three possible modes of regulation of adrenaline secretion by ACH: (1) influx of Ca^{++} through the ACH receptor channel; (2) activation of voltage-dependent Ca^{++} channels by ACH receptor-mediated depolarization, and (3) the voltage-dependent Ca^{++} component of spike activity stimulated by ACH. More recent studies have made use of noise analysis [82] and patch clamp techniques [80, 81, 112] to further define the two components of the action potential (one mediated by Na^+, and the other by Ca^{++} inward currents). One consequence of the fact that these action potentials are mediated by Na^+ and Ca^{++} *inward* currents is that Ca^{++} influx is augmented during each action potential. These studies have revealed marked ACH-induced potential fluctuations (fig. 4) of amplitude 1.6 pA and long duration (34 ms), which may have biological significance since they would favour multiple action potentials (rather than desensitization) during prolonged exposure to ACH [82].

In spite of these elegant studies, the physiological significance of adrenal medullary action potentials in secretion of catecholamines remains elusive. It has been proposed [175] that following stimulation, the adrenal chromaffin cell membrane undergoes non-propagated local depolarization to bring about the voltage-dependent increase in Ca^{++} influx and subsequent secretion of catecholamines. Voltage-dependent inflow of Ca^{++} could serve to activate Ca^+-dependent K^+ channels such as those described in chromaffin cells [112]. These channels display an unusually large unitary conductance, and their properties are affected by the Ca^{++} concentration on the *inner* side of the cell membrane. For low internal Ca^{++}, the Ca^{++}-dependent K^+ channels of chromaffin cells are voltage-dependent. However, their sensitivity to the membrane potential is affected by Ca^{++}, and at $1\,mM$ inner Ca^{++} no voltage-dependence is

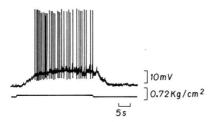

]10mV

]0.72Kg/cm²

5s

Fig. 4. Upon splanchnic stimulation, ACH secretion from splanchnic nerve terminals produces an increase in cytoplasmic Ca⁺⁺ and initiates a sequence of events that ultimately results in catecholamine secretion. Two modes of Ca⁺⁺ entry into the cells have been postulated; one associated with the ACH receptor-ionophore complex, and the other voltage-dependent Ca⁺⁺ channels. The mechanism of action potential formation is thought to be as follows: upon application of ACH the chromaffin cell membrane depolarizes with super-imposing prominent *potential fluctuations* as shown here. Multiple action potentials are generated at the peaks of fluctuation. When noise analysis was applied to determine the underlying mechanism of these potential fluctuations they were estimated to have an amplitude of 0.23 mV and a mean duration of 40 ms. These unitary potential fluctuations could be generated by as little as 0.8 pA of current, which is similar to that reported for the single ACH receptor channel current in skeletal muscle. However, the mean duration was unusually long compared with that in muscle (about 1 ms). Similarly, using single channel recording techniques (patch-clamp) *Kidokoro* [80, 81] estimated the amplitude as 1.6 pA and a duration of 34 ms. From [80].

observed. These results demonstrate that even at very low Ca⁺⁺ concentrations (10^{-8} to 10^{-7} *M*), such as those present intracellularly, unitary conductance of K⁺ channels depends on Ca⁺⁺.

Kirshner [85a] has described tetrodotoxin-sensitive Na⁺ channels in bovine adrenal medullary cells which are functionally linked to Ca⁺⁺-dependent catecholamine secretion. However, the channels are *not* utilized for Na⁺ entry upon activation of nicotinic receptors; rather Na⁺ entry takes place through the nicotinic receptor-ionophore. The role of these TTX-sensitive Na⁺ channels in stimulus-secretion coupling in vivo, if anything, remains to be determined.

While overshooting – short-duration (1–10 ms) – all-or-nothing action potentials are generally considered properties of nerve or muscle, they have been seen in cell lines derived from one mouse neuroblastoma [121] and two primary human neuroblastomas [159]. It has also been reported that rat pheochromocytoma (PC12) cells, and human APUD-derived cell lines (two medullary carcinomas of the thyroid, and two bronchial carcinoids grown in monolayer culture) were electrically excitable and produced action potentials [159]. Whether electrical excitability is involved in hormonal secretion by these neoplastic cells, or by their normal counterparts in the adrenal medulla, thyroid, and bronchii, respectively (as it is for hypothalamic neurosecretory cells), is not yet established.

Storage, Synthesis and Release of Catecholamines

Each of these three topics could easily be a chapter in itself; however, the purpose of this section is to present some of the newer concepts and findings on catecholamine storage, synthesis and release rather than to review each in detail, and to refer the more interested reader to appropiate review articles for details.

Storage
Within the Whole Gland

The two principal catecholamines, adrenaline and noradrenaline, are stored in separate cells, the so-called epinephrine and norepinephrine chromaffin cells, of the adrenal medulla. The two cell types epinephrine and norepinephrine cannot be distinguished when alive (e.g. in culture under the phase microscope). They appear to have approximately the same diameter, shape and surface features. However, they can be distinguished in tissue sections at the light microscope level – formaldehyde reacts selectively with norepinephrine cells to produce fluorescent derivatives [51, 104], antibodies to phenylethanolamine N-methyltransferase (anti-PNMT) react only with epinephrine-synthesizing cells [64, 104] – or at the ultrastructural level [30].

The differential reactivity of epinephrine and norepinephrine cells is seen after staining ultrathin sections of glycol methacrylate-embedded rat and human medullae for carbohydrate (glycoprotein) content [20]. In epinephrine cells, a large part of the Golgi complex stains with phospho-tungstic acid (PTA) and periodic acid-Schiff (PAS), whereas in the norepinephrine cells, the Golgi complex is mostly unstained. The results of these studies indicate that epinephrine vesicles and the Golgi complex of epinephrine cells, but not that of norepinephrine cells, are rich in non-acidic glycoproteins.

The proportion of the two amines (and hence the proportion of the two cell types), and the levels of catecholamines within the adrenal medulla depends on the species and upon the age of the animal [for review see 17a, 153]. In general, within a species, the older the age the higher the proportion of epinephrine. Also, the two adrenals are not identical. In Wistar rats for example, the adrenal on the right hand side has approximately 10% less noradrenaline than that on the left hand side. The gross appearance of the two adrenals and pattern of vasculature is also different in most species. Some species such as the guinea pig and hamster have almost exclusively epinephrine, while others such as the rat, cat, bovine and human have between 60 and 80% epinephrine. The proportion of dopamine is usually less than 2% of the total catecholamines. In man, the tissue concentration of epinephrine in the adrenal medulla is of the order of 0.2–0.8 mg/g and of norepinephrine 0.04–0.16 mg/g.

Within the Chromaffin Granules

Quantitative estimates of the average content of the chromaffin granules that store the catecholamines within the chromaffin granules has come from biochemical analyses [184–186] and more recently from combined catecholamine assay and stereological measurement of the tissue (table II) [124]. The rat adrenal medulla was shown to contain an average 23.1×10^4 epinephrine cells and 5.7×10^4 norepinephrine cells. Each cell contained an average 2.3×10^4 epinephrine-containing granules or 3.3×10^4 norepinephrine-containing granules. In addition, each adrenal gland contained a mean of 44.7 nmol epinephrine and 10.1 nmol of norepinephrine. Thus, it could be calculated that each granule contained 5.07×10^6 molecules ($0.77\,M$) of epinephrine or 3.14×10^6 molecules ($0.62M$) of norepinephrine. These results, which are summarized in table I, make the assumption that all the catecholamines measured are intragranular. However, in other more recent experiments [*Nordmann and Morris,* unpublished] with isolated chromaffin granules, which avoid this assumption, the single chromaffin granule was shown to contain an average 5.33×10^{-12} nmol of catecholamines, which corresponds to 3.2×10^6 molecules – a very similar result. In addition, in rats that were injected with insulin to deplete the catecholamines by reflex sympathetic discharge, the adrenals were shown to have lost 45% of their epinephrine content while the volumetric density of the epinephrine-containing granule decreased by 42%. These results support the idea that most of the catecholamine is stored intragranularly under physiological conditions. Studies on isolated chromaffin cells from various species [for review see 103] have confirmed that the two cell types (fig. 5) synthesize, store and release catecholamines, and maintain high levels of the amines for weeks

Table I. Catecholamine content of the rat adrenal gland

	Epinephrine	Norepinephrine
Number of cells/adrenal medulla	23.1×10^4	5.7×10^4
Number of granules/cell	2.3×10^4	3.3×10^4
Catecholamine content		
nmoles/gland	44.7	10.1
Intragranular concentration		
Molecules/granule	5.07×10^6	3.14
Molar	0.77 *M*	0.62 *M*

Data obtained by a combined stereological and catecholamine assay by *Nordmann* [124].

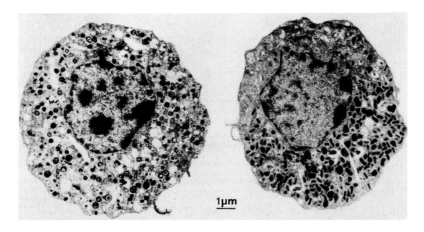

Fig. 5. Ultrastructural features of isolated adult bovine chromaffin cells. Two cell types are present; typical noradrenaline cells (left) containing vesicles with cores contracted from the vesicle membrane, and typical adrenaline cells (right) with evenly filled, pleomorphic vesicles [see 53]. The adrenaline cells predominate in agreement with biochemical analysis of the cell preparation showing an A/NA ratio of approximately 3:1. From [103a].

in culture [84]. Isolated bovine adrenal chromaffin cells such as those in figure 5 contain approximately 80 pg epinephrine cell and 21 pg norepinephrine cell. Similar values can be calculated from the figures given in table I for rat adrenal gland (epinephrine 78 pg/cell; norepinephrine 60 pg/cell). The number of chromaffin granules per cell is similar in both cell types (table I).

Table II. Some quantitative morphological data on the rat adrenal medulla [124]

Adrenal medulla	
Volume of adrenal gland	0.89 ± 0.07 mm³
Chromaffin cells	
Volume of chromaffin cell	2500 µm³
Number of chromaffin cells/adrenal	2.88×10^5
Number of epinephrine cells/adrenal	23.1×10^4
Number of norepinephrine cells/adrenal	5.7×10^4
Calculated $\dfrac{\text{Epinephrine}}{\text{Norepinephrine}}$	4.06
Chromaffin granules	
Volume	1.07×10^7 nm³
Number/gland	7.2×10^9
Number epinephrine granules/gland	5.3×10^9
Number norepinephrine granules/gland	1.9×10^9
Calculated $\dfrac{\text{Epinephrine}}{\text{Norepinephrine}}$	4.11

In addition to the catecholamines, the chromaffin granules contain large amounts of ATP and of soluble acidic proteins, termed chromogranins (table III), as well as a number of peptides, most notably Leu- and Met-enkephalin, enkephalin congeners, and their precursors.

The catecholamines are concentrated within the chromaffin granule by a chemiosmotic mechanism (fig. 6) in which a membrane-bound ATPase drives protons into the chromaffin vesicle thereby creating an electrochemical proton gradient across the chromaffin vesicle membrane. The chemical part (Δ pH) of the gradient supplemented by the electrical potential ($\Delta \psi$) drives the uptake of catecholamines via a specific carrier molecule (the 'catecholamine carrier'). This membrane carrier protein has been solubilized and reconstituted by *Maron* et al. [111], and characterized by a number of groups and shown to be separate from the membrane-bound ATPase [6]. The second component to the electrochemical gradient, the electrical potential, drives the uptake of nucleotides (e.g. ATP) via a separate carrier. The two carriers have been characterized in terms of their differential sensitivity to specific inhibitors [for review see 185].

Table III. Relative number of molecules in the soluble content of chromaffin granules [185]

Constituent	Relative number of molecules
Dopamine β-hydroxylase	1
Chromogranin A	36
Enkephalin-like reactivity	64
(Met- and Leu-enkephalin)	
Catecholamines	22,000
ATP	4,900
ADP	700
AMP	210
GTP	650
UTP	370
Calcium	660
Magnesium	190
Ascorbic acid	880
Sialic acid (glycoproteins)	430
Sulfated hexosamines (glycosaminoglycans)	250
Glucuronic acid (glycosaminoglycans)	250
Acidic groups in protein	19,300
Basic groups in protein	11,400

The results were calculated from published values as described by *Winkler and Westhead* [186].

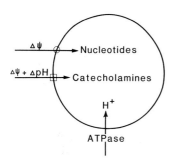

Fig. 6. The membrane-bound ATPase drives protons into the chromaffin vesicle creating an electrochemical proton gradient. The chemical part (ΔpH) of the gradient, supplemented by the electrical potential (Δψ) drives the uptake of catecholamines by a specific carrier (□). Only the electrical potential has been demonstrated to drive the uptake of nucleotides via its carrier (O). The specificity of the two carriers has been established by their sensitivity to different inhibitors.

Synthesis

The well-defined pathway for the synthesis of catecholamines in man from dietary phenylalanine and tyrosine proceeding through dihydroxy-phenylalanine (dopa) and dihydroxyphenylethylamine (dopamine) to norepinephrine and epinephrine, and its alternate (e.g. in the rabbit) via tyramine and octopamine (fig. 7) [for review see 17a] have been studied again recently with a view to defining the mechanism(s) by which the chromaffin cell regulates coordinate control of synthesis of the enzymes responsible for this biosynthetic pathway [74]. The biosynthesis of epinephrine requires four enzymes, tyrosine hydroxylase (TH), aromatic decarboxylase, dopamine-β-hydroxylase (DBH), and phenylethanolamine N-methyltransferase (PNMT). Recent studies have shown that three of these enzymes, TH, DBH and PNMT are regulated together, usually by control of their biosynthesis and that all three have protein domains with similar sequences of amino acids. The finding that all three enzymes share protein domains with a similar primary structure is consistent with the hypothesis that the three enzymes are coded for by a single gene or genes evolved from a common ancestral precursor [74].

The first enzyme in the pathway, TH, is also the rate-limiting enzyme in the biosynthesis of dopamine, norepinephrine and epinephrine and exists in two kinetically distinct forms with differing affinities for pterin cofactor [179]. It is subject to both short-term and long-term regulation by multiple factors. In the rat, a variety of stressful procedures, including electroconvulsive shock, nociceptive stimuli, exposure to cold, and decapitation, are associated with acute activation of adrenal tyrosine hydroxylase. The primary kinetic change appears to involve conversion of a population of less active enzyme molecules to the more active form (higher affinity for the pterin cofactor). Experiments in vitro suggest that the rate of phosphorylation of the soluble enzyme prepared from stressed animals is much slower than that prepared from unstressed animals. However, further kinetic studies suggest that the enzyme is *not* phosphorylated at the cyclic AMP-dependent protein kinase site during stress. Rather, it is thought that a conformational change in the enzyme associated with activation alters the susceptibility of the enzyme to phosphorylation by cyclic AMP-dependent protein kinase. The less active form of TH, having a low affinity for pterin cofactor, is thought to exist 'in a constrained conformation, wherein a cationic group on a regulatory strand of the polypeptide interacts coulombically with an anionic group at the active site of the enzyme at which the pterin cofactor binds. When the cationic group on the

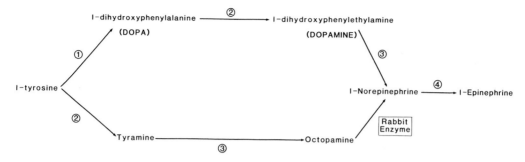

Fig. 7. Biosynthesis of the catecholamines in the adrenal medulla. Enzymes: 1 = tyrosine hydroxylase; 2 = aromatic 1-amino acid decarboxylase; 3 = dopamine-β-hydroxylase; 4 = phenylethanolamine N-methyltransferase. The lower pathway has been identified in the rabbit. The upper is the principle pathway for biosynthesis of norepinephrine and epinephrine in humans [for review see 17a].

regulatory portions of the polypeptide is neutralized by polyanionic inter-actions, interaction with anionic phospholipids or phosphorylation of an adjacent serine hydroxyl (by a cyclic AMP-independent protein kinase), the conformation of the enzyme is altered and the anionic site at which the pterin cofactor binds is rendered more accessible to interaction with the cofactor' [179]. Since the pterin cofactor (tetrahydrobiopterin) appears to be present in concentrations substantially below the K_m for the high affinity form of TH, it has been proposed that the intracellular level of cofactor may regulate TH activity. The levels of pterin cofactor are in turn regulated by the enzyme, guanosine 5'-triphosphate (GTP)-cyclohydrolase. Studies from *Viveros'* laboratory and others have now confirmed that the regulation of availability of the pterin cofactor is one of the key factors that controls TH. These results suggest that TH activity may be coordinately regulated with GTP-cyclohydrolase activity and tetrahydropterin levels both in vivo and in vitro to bring about the most effective increase in the rate of tyrosine hydroxylation in adrenal chromaffin cells [2].

The last two enzymes in the pathway DBH and PNMT are regulated by a combination of neuronal and humoral factors. Hypophysectomy dramatically reduces the levels (and number of immunotitratable molecules) of both enzymes by increasing their rate of proteolysis, and administration of ACTH or dexamethasone restores the situation [24]. The respective cofactors for DBH and PNMT, ascorbate and S-adenosylmethionine, appear to protect the enzymes against proteolysis. On the basis of these findings, *Ciaranello* et al. [24] have proposed that 'intracellular proteolysis is probably the major mode of regulating the steady-state levels of adrenal medullary dopamine-β-hydroxylase and phenylethanolamine N-methyltransferase'. The levels of these two enzymes 'are regulated primarily via adrenal cortical glucocorticoids acting in turn to control the levels of ascorbic acid and S-adenosylmethionine, the cofactors for these enzymes'.

In addition to these biosynthetic considerations, increasing attention has been paid recently to the catabolism of the catecholamines and the possible role(s) of conjugated metabolites (in particular the sulphated derivatives of dopamine, norepinephrine and epinephrine) as biologically active compounds [88].

Release

The concept of stimulus-secretion coupling for Ca^{++}-dependent exocytotic release of catecholamines is now well established and has been more than adequately reviewed by *Douglas* [44], and more recently by *Rubin* [this volume]. In this section three of the newer aspects of the control of secretion will be discussed.

The Role(s) of Nicotinic and Muscarinic Receptors in Control of Secretion

Both nicotinic and muscarinic ACH receptors participate in the secretion of catecholamines from adrenal chromaffin cells. Their relative involvement in the release process depends on the concentration of ACH, and upon the species being studied. The hamster and chick appear to possess only nicotinic or muscarinic receptors, respectively [92, 100], whereas the dog, cat, rat, cow and man possess both nicotinic and muscarinic receptors [154]. Pharmacological studies on isolated bovine adrenal medullary cells have shown, for example, that while both ACH receptor types are present, only the nicotinic receptors (EC_{50} nicotine, $5 \times 10^{-6}\ M$; ACH $5 \times 10^{-5}\ M$) are involved in the active secretion of catecholamines. Muscarinic receptors are present and are fully activated by relatively low concentrations ($10^{-7}\ M$) of ACH resulting in elevated intracellular c-GMP levels that are inhibitory on nicotine-mediated release of catecholamines [41, 145]. At concentrations of ACH greater than $10^{-6}\ M$, this inhibition of release is overcome by activation of the nicotinic receptors. Studies by *Oka* et al. [125] have extended these observations to show that stimulation of the nicotinic ACH receptors causes uptake of Ca^{++}

coupled to catecholamine release and synthesis, while stimulation of the muscarinic ACH receptors causes efflux of Ca^{++} from the chromaffin cells and an increase in c-GMP level and ^{32}P-incorporation into phospholipids that are not directly coupled to catecholamine release and synthesis. At higher concentrations (10^{-5} M ACH) the nicotinic response exhibits desensitization. By contrast with the nicotinic receptor at the neuromuscular junction the nicotinic receptor in chromaffin cells is not inhibited by α-bungarotoxin [135].

A functional role for this nicotinic, muscarinic receptor response in homeostasis of secretion has been proposed [41, 145]. It was suggested that under resting (basal) conditions the low levels of ACH released from the splanchnic nerve terminals may be sufficient to fully activate the muscarinic response, resulting in a net inhibition of catecholamine release. Under stress conditions, when the splanchnic nerve is firing at a higher rate and releasing higher concentrations of ACH, the nicotinic receptors would then be activated and the muscarinic blockade overcome resulting in the exocytotic release of catecholamines. Nicotinic stimulation also releases acetylcholinesterase from the chromaffin cells by a Ca^{++}-dependent mechanism [118, 118a] thereby providing a stimulus-related control to terminate the actions of the released ACH. The efficacy of ACH as a secretory agonist on chromaffin cells may in addition be modulated by endogenous neuropeptides (e.g. substance P and the enkephalins) present in the splanchnic nerve [13, 103].

The Role of Receptor-Mediated vs. Voltage-Dependent
Ion Channels in the Mechanism of Secretion

Studies on isolated bovine chromaffin cells with 'leaky' membranes [8–10] have demonstrated, quite elegantly, that exocytosis of catecholamines can be brought about by micromolar intracellular concentrations of Ca^{++}. The mechanisms by which Ca^{++} enters the cell upon activation by ACH have been the subject of much research. There appear to be three main routes for Ca^{++} entry (fig. 8).

(1) Upon release of ACH from the splanchnic nerve terminals, ACH binds to receptor sites on the chromaffin cell membrane and opens a receptor-linked ion channel that is permeable to both Na^+ and Ca^{++} but is relatively insensitive to tetrodotoxin, Mn^{++} and Co^{++}. Na^+ appears to compete with Ca^{++} for entry via this receptor-linked ion channel. The receptor ionophore complex is inhibited by hexamethonium and d-tubocurarine but not by α-bungarotoxin. However, the amphibian toxin, his-

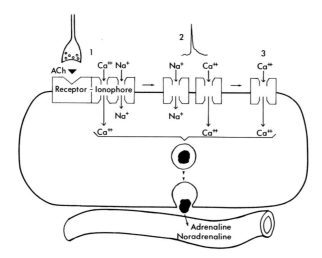

Fig. 8. Normal sequence of events in stimulus-secretion coupling in the bovine chromaffin cell. 1 = Acetylcholine (ACH) released from the splanchnic nerve terminals binds to receptor sites on the chromaffin cell membrane. Binding of ACH to its hexamethonium-sensitive receptor (nicotinic receptor) opens a receptor-linked ion channel that is permeable to both Na^+ and Ca^{++} but is relatively insensitive to TTX, Mn^{++}, and Co^{++}. Na^+ appears to compete with Ca^{++} for entry via the receptor-linked ion channel. ACH also interacts with atropine-sensitive receptors (muscarinic receptors – not shown) to increase the intracellular concentration of c-GMP. 2 = The binding of ACH to the nicotinic receptor-ionophore complex in step 1 stimulates the entry of Na^+ and Ca^{++} (Ca^{++} enters by its own channel inhibited by Co^{++}) and brings about a small depolarization. This depolarization leads in turn to the entry of Na^+ and Ca^{++} via a voltage-sensitive Na^+ channel (that can be inhibited by TTX). The entry of these ions further depolarizes the cell membrane and increases the frequency of action potentials. This depolarization also serves to open D-600-sensitive Ca^{++} channels through which Ca^{++} enters and triggers trifluoperazine-sensitive exocytosis and secretion of the catecholamines and other vesicular components (e.g. ATP and enkephalins) [from 10, 84a, 103, 145a].

trionicotoxin (HTX) (IC_{50} 1.6×10^{-6} *M*), binds to the nicotinic receptor-ionophore complex in chromaffin cells, a site separate from the ACH-binding site, and decreases ion flow through this channel.

(2) The binding of ACH to the nicotinic receptor-ionophore complex in step 1 stimulates the entry of Na^+ and Ca^{++} and brings about a small depolarization. This depolarization leads in turn to the entry of Na^+ and Ca^{++} via another class of channels, the voltage-sensitive Na^+ and Ca^{++} channels. These voltage-sensitive channels are separate for the two ions. The voltage-sensitive Na^+ channel can be inhibited by TTX only to the

extent of 30%, indicating that activation of the voltage-sensitive Na^+ channel is not essential for catecholamine secretion. Ca^{++} enters via its own voltage-sensitive channel inhibited by Co^{++} [4, 68]. The entry of these Na^+ and Ca^{++} ions further depolarizes the cell membrane and increases the frequency of action potentials [80, 81, 82].

(3) The depolarizations caused by Na^+ and Ca^{++} entry through the voltage-sensitive channels also serves to open D-600-sensitive Ca^{++} channels (late Ca^{++} channels) through which Ca^{++} enters to trigger off trifluoperazine/promethazine-sensitive exocytosis of the vesicle contents.

Stimulus-Secretion Coupling and the Intracellular Contractile Apparatus in Chromaffin Cells

Early observations [for review see 44, 142a] that the process of secretion shares a number of features in common with the process of muscle contraction led to the concept of 'stimulus-secretion coupling'. This concept has been reinforced by the finding that chromaffin cells contain contractile proteins (e.g. actin, α-actinin and myosin) [164] and that some of these have been found in association with the secretory vesicles [for reviews see 132, 163]. Further, it is known that the activation of myosin by actin in smooth muscle cells and in non-muscle cells requires the participation of a Ca^{++}-regulatory protein called calmodulin. Calmodulin is also present in chromaffin cells, and may be involved in the process of secretion since trifluoperazine, a drug that inhibits calmodulin-dependent functions [180], also inhibits the secretory response of chromaffin cells [9, 19, 79, 118a]. While the action of this drug must still be interpreted cautiously, e.g. *Izumi* et al. [72] found that trifluoperazine at higher concentrations blocked ACH-induced Ca^{++} influx into chromaffin cells and *Kenigsberg* et al. [79] found that trifluoperazine at a concentration of $2.5 \times 10^{-5}\,M$ produced a 30% inhibition on 56 mM K^+-induced uptake of $^{45}Ca^{++}$, it is of interest that *Kenigsberg* et al. [79] reported that trifluoperazine, unlike the Ca^{++} channel blocker Ni^{++}, in micromolar concentrations that significantly inhibited K^+-induced Ca^{++} release, did not modify K^+-induced $^{45}Ca^{++}$ uptake, $^{45}Ca^{++}$ efflux, or $^{45}Ca^{++}$–Ca^{++} exchange. Taken together with the earlier findings of *Baker and Knight* [9] that trifluoperazine blocks μM Ca^{++}-activated secretion from 'leaky chromaffin cells' (where all ion fluxes associated with membrane depolarization are bypassed), these findings suggest that intracellular Ca^{++} triggers the secretory process of chromaffin cells through a calmodulin-dependent reaction. Which of the many calmodulin-dependent reactions are responsible for

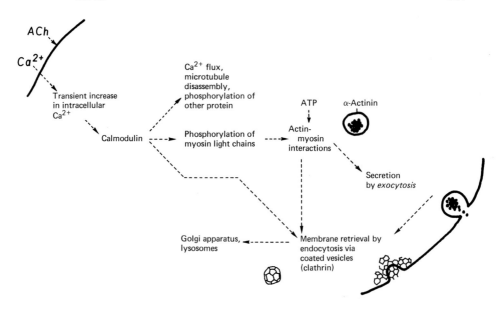

Fig. 9. Proposed role(s) of calmodulin and contractile proteins in the secretory process in chromaffin cells. The molecular events involved in stimulus secretion coupling in chromaffin cells are depicted [from 164].

secretion is not known; however, one possibility is that the transport of chromaffin vesicles to the plasma membrane may involve a sliding filament mechanism similar to that in muscle (fig. 9). In chromaffin cells, K+-induced depolarization not only induces secretion but also increases the phosphorylation of myosin light chains. If, as in other non-muscle tissues, this process is regulated by calmodulin, it might constitute the molecular mechanism of stimulus-secretion coupling.

Two other proteins implicated in the process of secretion and membrane retrieval following exocytosis are synexin [34], and clathrin [164], respectively. Synexin is a Ca++-binding protein molecular weight 47,000 monomer) present in the adrenal medulla and in other tissues. When activated by mM Ca++, synexin polymerizes and can cause isolated chromaffin vesicles in vitro to polymerize into 50 × 150 Å rods that further self-assemble and aggregate irreversibly, forming pentalaminar complexes between adjacent vesicle membranes. This process is reminiscent of compound exocytosis, a common feature of chromaffin cells undergoing exocytosis in vivo [53]. In addition, *Pollard and Scott* [133] have detected Ca^{++}-dependent binding sites for synexin on the plasma membranes of chromaffin cells. In the presence of small amounts (5 μM) of a cis-unsaturated fatty acid such as arachidonic acid, the synexin-aggregated vesicles form large fused membrane sacks, similar to those observed in electron micrographs of chromaffin cells after exocytosis [53]. This amount of arachidonic acid is comparable to the amount of this fatty acid released by phospholipases in stimulated secretory cells. It has therefore been proposed that upon membrane fusion, a membrane-bound Ca^{++} activated phospholipase A_2 could release free arachidonic acid from membrane phospholipids to destabilize the pentalaminar structures and initiate the fusion event. Following this, the chemiosmotic properties of the secretory vesicle would come into play with first the proton-driving ATPase loading protons into the interior of the vesicle, followed closely by extracellular chloride as counter ions, then the osmotic strength of the vesicle would increase resulting in lysis of the vesicle at its point of contact with the plasma membrane and extrusion (exocytosis) of its contents. *Pollard and Scott* [133] have argued that since synexin is very sensitive to certain phenothiazine drugs, e.g. trifluoperazine and promethazine, and that trifluoperazine also blocks calmodulin but promethazine is only poorly active on calmodulin, these findings argue against calmodulin as the target of these drugs in relation to exocytosis. However, the relatively high (mM) concentrations of Ca^{++} required to activate synexin in vitro in contrast with the micromolar concentration of Ca^{++} required intracellularly to evoke secretion [9], and other data that suggests that synexin is non-selective in its ability to increase Ca^{++}-dependent aggregation of biological and artificial membranes in vitro [120], argue against a role for synexin in exocytotic secretion in vivo.

Clathrin, the protein (molecular weight 180,000) that makes up the 'basket' or 'cage'-like structure responsible for vesicle membrane retrieval during endocytosis [35], has been localized by immunofluorescence as a fine granular fluorescence within the cytoplasm of isolated chromaffin cells. This fine granular staining exhibits a strong juxtanuclear localization [164] suggesting an association of clathrin with the Golgi region of the cell. Other morphological observations have shown an association between coated vesicles and actin microfilaments suggesting that this contractile-like protein may be involved in the retrieval mechanism.

To summarize, the finding of contractile proteins in chromaffin cells, their association with chromaffin vesicles, the inhibition of contractile protein phorphorylation and catecholamine release by the calmodulin antagonist trifluoperazine, and the association of α-actinin and clathrin with the vesicles and Golgi region suggest that contractile proteins may be involved in the process of secretion and/or vesicle membrane retrieval. However, the possibility remains that other calmodulin- and/or synexin-mediated processes may facilitate the secretory process unter physiological conditions.

Peptides in the Adrenal Medulla:
How Many, Where Located and What For?

The beginnings of this new field of interest in the adrenal medulla had its roots in some of the initial observations of *Hillarp* [63] on the protein composition of the chromaffin granule. It was soon realized that the chromaffin granules contained, in addition to the catecholamines and nucleotides, large amounts of water-soluble proteins (termed chromogranins) of various molecular sizes. *Helle* [62] pointed out that besides chromogranin A and DBH there were in addition some 22 unidentified proteins and peptides of smaller molecular weight. The first of the peptides to be positively identified as being secreted from the adrenal medulla was, surprisingly, the one that has been the least studied, vasoactive intestinal polypeptide (VIP). *Said* [143] detected increased levels of VIP in the plasma of patients with pheochromocytoma. Soon after, the opioid peptides Leu-enkephalin and Met-enkephalin [69] were demonstrated in the adrenal medulla of various species by immunohistochemical [43, 105, 110, 149], radioimmunoassay [25, 99, 155], and radioreceptor [99, 155, 169, 170] techniques. Within the adrenal medulla enkephalins have been localized within both the splanchnic nerve terminals [150] and the chromaffin granules [99, 155, 169, 170]. The enkephalins are thus stored with the catecholamines in the chromaffin granules [169, 170].

Pelto-Huikko and Hervonen [129] studied the immunohistochemical localization of three neuropeptides (enkephalin, VIP and substance P) in human and rat adrenal medulla. Enkephalin-positive cells were found in both human and rat adrenal medulla; the intensity of immunostaining varied from cell to cell but there was no correlation with histochemically demonstrable catecholamine cells. The relative amount of the stained granules seemed to determine the intensity of the immunostaining as revealed by light microscopy. No positive nerve fibres were found. The VIP-positive fibres formed a tight network around the medullary cells while substance P labelled only a few scattered fibres in the human adrenal medulla. The VIP staining was localized in the large granular vesicles of the medullary axons. The granular localization of enkephalin-like immunoreactivity suggested that enkephalins and related peptides might be released together with the other granular contents by exocytosis, and this has recently been shown to occur (see below). The presence of VIP-containing granular vesicles in the sympathetic pre-ganglionic nerves innervating the medullary cells suggested an active role for VIP in the medullary endocrine function and this is presently under investigation. Two peptides that have been shown to have modulatory roles on catecholamine release from chromaffin cells in vitro are somatostatin and substance P [108, 116, 137].

Preliminary evidence for the somatostatin precursor and two low molecular weight forms in the adrenal medulla was obtained by *Baird* et al. [7]. In this study, bovine adrenal medullae were extracted and the acid defatted material characterized by gel permeation and reverse phase high performance liquid chromatography. Antibodies to somatostatin-14 and somatostatin-28 were used to identify the fractionated material. The ODS-retained material retained greater than 5 ng of somatostatin-like immunoreactivity (SLI) per bovine adrenal medulla which, after G50 chromatography, was resolved into three distinct forms. The greater part of the SLI (75%) eluted in a zone compatible with a peptide of 1.2 kilodalton, while the remaining SLI was divided between a zone compatible with 2.5 kilodalton and another located in the void volume (molecular weight ~ 10 kilodalton). In contrast, no SLI was found in a similar extract of bovine adrenal cortex.

This study, together with an earlier study by *Role* et al. [137] in the guinea pig, and more recent investigations in human adrenal medulla and pheochromocytoma [27a, 28], suggests that there are high molecular weight forms of SLI in the adrenal medulla but none in the cortex. Gel filtration of the bovine extracts indicates that these forms are consistent with the existence of somatostatin-14 and somatostatin-28-like peptides, as well as a putative precursor of higher molecular weight; however, in the human adrenal medulla the somatostatin-14 form predominates. Likewise, in the guinea pig the possibility that the higher molecular weight material recovered by *Role* et al. [137] was a precursor of authentic somatostatin is rendered unlikely by the observation that treatment of these extracts with 8 *M* urea and 5% 2-mecaptomethanol failed to convert this apparently higher molecular weight material to material which co-eluted with synthetic somatostatin. In fact, this treatment had no effect on the mobility of the higher molecular weight adrenal medullary somatostatin-like material. In contrast, in the human pheochromocytoma tissue, significant amounts of a large molecular weight (6 kilodalton) form of somatostatin were present.

Somatostatin-like and substance P-like immunoreactive material in isolated guinea pig chromaffin cells displayed gel filtration profiles identical to those found in adrenal medullary homogenates [137]. In purified chromaffin cells, immunoreactive substance P and somatostatin were present in small amounts. The content of these peptides was 0.8 ± 0.3 (mean$\pm$SD, n = 3) and 0.7 ± 0.1 (mean$\pm$SD, n = 4) pmol/mg protein, respectively. While immunoreactive substance P and somatostatin were present in the purified chromaffin cell extract, more than 90% of which contained catecholamine, it is not clear if these peptides were present inside the chromaffin cells or in a separate population of cells, or if they were contained in nerve terminals which were attached to the chromaffin cells. Immunochistochemical studies in the human suggest that somatostatin may be located in the chromaffin cells themselves [110] whereas substance P may be located in nerve terminals [102].

Substance P has been detected by radioimmunoassay in adrenal medullary tissue and isolated chromaffin cells of several species, and its localization further defined by immunocytochemistry and functional depletion (table IV). Splanchnic nerve terminals containing substance P and enkephalin have been identified making contact with catecholamine- and enkephalin-containing chromaffin cells. It is known that activation of nicotinic cholinergic receptors on the chromaffin cells by ACH released from the splanchnic nerve stimulates release of catecholamines and enkephalin into the circulation. The co-localization of substance P and ACH in splanchnic nerve terminals therefore raises the possibility that substance P might function as a modulator of ACH-induced secretion of catecholamine through an interaction with the nicotinic receptor. To investigate this possibility, *Livett* et al. [103] made use of the model system afforded by primary monolayer cultures of adult dissociated bovine adrenal chromaffin cells (table V). They found that substance P has *two* distinct actions on the nicotinic response: (1) substance P inhibited the ACH-mediated release of catecholamines [108], and (2) substance P protected against ACH-induced desensitization of the nicotinic response [13].

Table IV. Substance P and peptides in the adrenal medulla

Distribution	Species	Content	n	Reference
		pmol/g tissue		
Tissue	man	11.0±2.0	15	*Bucsics* et al. [19a]
		14.8±2.9	15	*Saria* et al. [144a]
	bovine	4.1±0.7	4	*Bucsics* et al. [19a]
	dog	0.37	2	*Nilsson and Brodin* [122a]
	cat	0.36±0.13	5	*Bucsics* et al. [19a]
	rabbit	0.89±0.02	10	*Bucsics* et al. [19b]
	guinea pig	0.71±0.13	11	*Bucsics* et al. [19a]
		1–10	20	*Role* et al. [136a, 137]
	rat	1.19±0.45	5	*Bucsics* et al. [19a]
		0.24±0.06	4–5	*Brodin and Nilsson* [17b]
	mouse	0.44±0.16	4–5	*Brodin and Nilsson* [17b] *Nilsson and Brodin* [122b]
	man (pheochromo- cytoma)	1.5±0.3	3	*Bucsics* et al. [19a]
Isolated cells	guinea pig (chromaffin cells)	pmol/mg protein 0.8±0.3	3	*Role* et al. [136a, 137]
	bovine	fmol/10⁶ cells 38–919	6	*Dean* et al. [38]

With regard to the first mechanism it was shown that substance P: (1) had no direct effect by itself on the secretion of catecholamine but showed contingent action in its inhibition of the secretion of catecholamines evoked by ACH or nicotine (one of the hallmarks of a true neuromodulator); (2) the interaction appeared specific for the nicotinic receptor-ionophore complex since K^+ and veratridine-induced secretion were not inhibited by substance P; (3) a possible interaction of substance P with the Na^+ ionophore of the complex is indicated by the finding that like quina-

Table IV (continued)

Distribution	Species	Content	Reference
Subcellular distribution	bovine human pheo-chromocytoma	subcellular distri-bution of substance P-immunoreactivity pa-ralleled that of catechola-mines; however, not in epinephrine cells	*Bucsics* et al. [19a] and *Floor and Lee-man* [53a]
Histochemical localization	man	restricted to nerve fibres in medulla and cortex	*Linnoila* et al. [102]
Functional depletion	rat	chronic denervation ↓63% substance P-immunoreactivity	*Bucsics* et al. [19a]
	guinea pig	capsaicin pre-treatment ↓80% (adrenal) ↓83% (splanchnic nerve)	*Bucsics* et al. [19a]
		↓96% (splanchnic nerve)	*Gamse* et al. [55a]
	rat	capsaicin nil	*Bucsics* et al. [19a]

crine, the inhibition by substance P was non-competitive with ACH or nicotine, and could not be overcome by raising the Ca^{++} but was less in the presence than in the absence of 125 mM Na^+. (4) In addition, substance P and a series of L-Ala-SP analogs were recognized by a receptor on the chromaffin cells that showed similar structural requirements to sub-stance P receptors in other substance P-responsive tissues [15]. The sub-stance P receptor appeared to be of the low-affinity type similar to that found in vas deferens (receptor sub-type 'E'), and (5) radio-receptor bind-ing studies on isolated chromaffin cell membranes with (4-^{3}H-Phe)-sub-stance P showed specific binding in the nM range displaceable by cold substance P.

In the guinea pig chromaffin cells, *Role* et al. [137] found that at a concentration of 10 μM, somatostatin and substance P partially reversed

Table V. Proposed effects of transmitters released from the splanchnic nerve on catecholamine release from adrenal chromaffin cells [103]

Transmitter released from splanchnic nerve	Mechanism of transmitter action	Effect on catecholamine release from chromaffin cells
Resting conditions		
Enkephalins	specific opioid peptide receptors	stimulates basal release
ACH (acting through muscarinic receptor)	increases intracellular c-GMP	inhibits nicotinic ACH-mediated release
ACH (acting through nicotinic receptor)	increases intracellular Ca^{++}	stimulates nicotinic ACH-mediated release
Substance P	interaction with nicotinic receptor-ionophore complex	inhibits nicotinic ACH-mediated release
Stress conditions (rapid firing of splanchnic nerve)		
ACH (acting through nicotinic receptor)	increases intracellular Ca^{++}	stimulates release (high concentrations of ACH may lead to rapid receptor desensitization)
Substance P	interaction with nicotinic receptor-ionophore complex	inhibits nicotine-mediated release
Substance P	protection against nicotinic desensitization	protects and prolongs nicotinic ACH-mediated release
Enkephalins	'non-specific' interactions with nicotinic receptor-ionophore complex	inhibit nicotine-mediated release

both ACH-induced catecholamine secretion and the inhibition of catecholamine uptake produced by ACH. Replacement of extracellular Ca^{++} by Co^{++} prevented ACH- and veratridine-induced catecholamine secretion but did not block the inhibition of catecholamine uptake caused by these agents. Therefore, the inhibition of catecholamine uptake could not be accounted for by stimulation of catecholamine secretion. Rather, it has been proposed that ACH and veratridine may depolarize the chromaffin cells, thereby decreasing the electrochemical gradient for the entry of Na^+

into the cells; this decrease in the electrochemical gradient for Na^+ may then cause a decrease in catecholamine uptake. The Na^+-dependent catecholamine transport system may be important in maintaining a low basal release of catecholamine from the adrenal medulla in vivo. The inhibition of catecholamine uptake by ACH may prevent the re-uptake of catecholamine during physiological stimulation of catecholamine secretion.

With regard to the second mechanism of action of substance P, *Boksa and Livett* [13] have recently shown that desensitization of nicotine-induced catecholamine release in chromaffin cells develops rapidly (catecholamine release = 80% of control after 15-s desensitization with $10^{-3} M$ nicotine), is not due to depletion of catecholamine stores, and is not Ca^{++}-dependent. Substance P ($10^{-5} M$) completely protected against this ACH- or nicotine-induced desensitization. In addition, the EC_{50} for substance P's protection against desensitization ($10^{-6} M$) and for substance P's inhibition of nicotine-mediated catecholamine release ($3 \times 10^{-6} M$) were similar. Substance P protection against nicotinic desensitization was not Ca^{++}-dependent and, under the conditions of their experiments, substance P had no effect on catecholamine uptake, indicating that substance P's protection against desensitization was a result of facilitation of catecholamine release rather than inhibition of catecholamine re-uptake. In addition, it was shown that K^+ (6–100 mM) could produce desensitization of catecholamine release from chromaffin cells but, in contrast to ACH-induced desensitization, K^+-induced desensitization was completely Ca^{++}-dependent and substance P could not protect against this K^+-induced desensitization.

These novel actions of substance P at the nicotinic receptor may be of physiological importance in maintaining catecholamine output from the adrenal medulla under conditions of acute stress when the splanchnic nerve is releasing large amounts of ACH (fig. 10) [103]. Only experiments in vivo will give answers to the physiological significance of these in vitro observations on cultured chromaffin cells.

The opioid peptides have also been implicated in control of catecholamine secretion from the adrenal medulla [89, 142c]. Immunohistochemical studies [150] have shown that enkephalin-like immunoreactivity is present in the chromaffin cells and also in the splanchnic nerve terminals that innervate the chromaffin cells. In certain species, such as the cow, the two enkephalins appear to be localized exclusively within the epinephrine-synthesizing cells [104]. The significance of this association, however, is not certain since in the same species the opioid peptides are also asso-

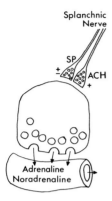

Splanchnic
Nerve

SP

ACH

Adrenaline
Noradrenaline

Fig. 10. Proposed role(s) of substance P (SP) in the control of adrenal medullary cate-cholamine secretion. In vivo and under resting conditions, basal secretion is probably con-trolled by (1) activation of muscarinic receptors and elevation of cyclic GMP (negative control of catecholamine release; see text), and (2) inhibition of ACH-mediated release by substance P released from the splanchnic nerve terminals at the same time as acetylcholine. Upon rapid firing of the splanchnic nerve, e.g. under acute stress situations, release of large amounts of ACH from the splanchnic nerve could rapidly lead to a desensitization of ACH-induced catecholamine release. However, simultaneous release of substance P along with ACH from the nerve might damp down evoked catecholamine release (inhibition) but would also prolong catecholamine release (through protection against nicotinic desensitization). It may be physiologically important that adrenaline and enkephalins continue to be released over a longer period of time following a stressful encounter and this could be achieved by the action of substance P to protect against nicotinic desensitization. Thus, substance P may have a true modulatory function (as a grid in a triode valve) capable of either inhibiting or potentiating the release of catecholamines as required [from 103].

ciated with noradrenergic vesicles in the splenic nerve [182a]. In chromaf-fin cells in monolayer culture the opioid peptides are concentrated within the varicose processes suggestive of transport and processing of the opioids as they move along the axons [105]. It is not known whether these enke-phalin-containing cells are exclusively adrenaline-synthesizing as seen in tissue sections, but release studies [106] indicate a co-release of enkephalin with adrenaline. The enkephalins can be shown to inhibit nicotine-evoked release of ^{3}H-norepinephrine but this occurs only at high concentrations of opioid (greater than $10^{-5} M$). This inhibition is probably not via the stereospecific high affinity opiate receptors present on these cells since it lacks stereospecificity and is not reversible by naloxone or by naltrexone [38, 94]. The controversy concerning this issue has been presented by *Costa* et al. [28a] and *Lemaire* et al. [94]. If the opioids are involved in modulating catecholamine release from chromaffin cells, it is likely that they act through a specific receptor, one with a relatively low affinity for morphine and a high affinity for β-endorphin [28a]. Such a unique class of opiate receptor may exist since *Saiani and Guidotti* [142b, 142c] have shown that etorphine binding by chromaffin cells is displaceable by low concentrations of diprenorphine that also reversed the etorphine inhibition of nicotine-evoked release of catecholamines. Nevertheless, it is disturbing that a number of pro-opiomelancortin fragments which are not endogenous components of the adrenal medulla showed a similar inhibition of nicotine-mediated release that had no effect on basal release [14]. In contrast, morphine and the opioid peptides at concentrations (10^{-6} to $10^{-3} M$) enhanced the *basal* release of endogenous catecholamines. The enhancement by morphine ($5 \times 10^{-4} M$) was Ca^{++}-dependent and stereospecific and reversed by naloxone and naltrexone [39]. The physiological significance of the enhancement of basal release by enkephalins and opioids is not fully understood. However, it is interesting that as early as 1912, *Elliott* [49] noted that morphine depleted the epinephrine content of the adrenal medulla. Since then several studies have demonstrated the direct and neurally mediated effects of both acute and chronic morphine administration on adrenal catecholamine synthesis, storage and release. *Slotkin* et al. [152] have shown that in the mature rat, morphine and methadone evoke a reflex sympatho-adrenal discharge which results in an initial depletion of adrenal catecholamines. Following chronic treatment, TH and DBH activities increase resulting in an elevation of catecholamine levels. The presence of high levels of enkephalins within the adrenal medulla and within the splanchnic nerve suggests that these opioid peptides may have a role in controlling basal secretion. This stimulatory role would be opposite to the inhibitory role proposed above for substance P. The maintenance of homeostasis in the adrenal medulla might therefore be controlled by opposing actions of enkephalin and substance P. Again, further studies are required in vivo to evaluate the physiological significance of these proposals. It is of interest that the opioid peptide most potent in these actions [103b] is the peptide BAM-22P [119]. This is one of a number of higher molecular weight opioid fragments that have been isolated from the adrenal medulla as endogenous components. In addition to Leu- and Met-enkephalin, the adrenal medulla of several species contains large amounts of pro-enkephalin and smaller amounts of processed derivatives (Met-enkephalin Arg^6-Phe^7 and Met-enkephalin Arg^6-Gly^7-Leu^8) [123]. For a review of enkephalin synthesis in the adrenal medulla see

Rossier [140] and *Viveros* et al. [171]. The higher molecular weight precursors and congeners are also released from the bovine adrenal medulla upon depolarization [141]. It is thought that these higher molecular weight precursors may be more resistant than Leu- and Met-enkephalin to proteolytic degradation in the blood stream, and physiological and pharmacological effects of the released opioids may be mediated to some extent by the enkephalin congeners and by the higher molecular weight precursor forms [142c]. A recent demonstration of the simultaneous release of a number of neuropeptides (neurotensin, somatostatin and the enkephalins) from perfused cat adrenal glands together with the catecholamines [27b] raises the question of the co-interaction of neuropeptides and their possible potentiation by each other following release.

With the demonstration of high molecular weight precursors and of biologically active enkephalin congeners in the adrenal medulla of several species [98], increasing attention has recently been paid to the processing of enkephalins within the chromaffin cells. Work from several laboratories has now shown, for example, that the chromaffin granules contain enzymes capable of hydrolyzing the precursor molecule into smaller active fragments and much work is currently being devoted to purifying and characterizing these enzymes [54, 101]. Another area of research yielding interesting results is the study of the regional distribution of enkephalin congeners. For example, *Yang and Costa* [186a] have shown that Met^5-enkephalin Arg^6-Phe^7-like material in rat brain is unevenly distributed throughout the brain with the highest contents in striatum and hypothalamus and the lowest in the cerebellum and cortex. Various brain structures have different amounts of high molecular weight Met-enkephalin Arg^6-Phe^7-like activity and it is possible through differential processing that some neuronal pathways may express different amounts of the congeners relative to the enkephalins. For example, there is some suggestion from the work of *Panula* et al. [126a] of a differential localization of Met-enkephalin Arg^6-Phe^7-like immunoreactivity and of Met-enkephalin-like immunoreactivity in nerve terminals of the spinal cord. Their results suggest that at least in this area, the heptapeptide may function as a neuromodulator in addition to, or instead of, being a precursor for Met-enkephalin.

What then are the functional roles proposed for enkephalins released from the adrenal medulla? From the work of *Lewis* et al. [99a] adrenal medullary enkephalin-like peptides have a role in mediating opioid but not non-opioid stress-induced analgesia. Stress-induced analgesia was reduced by adrenal demedullation and denervation and was potentiated by reserpine, a drug known to increase concentrations of adrenal medullary enkephalin-like peptide. The results of these studies also present an interesting paradox: adrenalectomy blocks an opioid form of stress analgesia but potentiates the analgesic response to morphine. To account for these apparently discordant results, *Lewis* et al. [99a] proposed that 'adrenalectomy potentiates opiate analgesia by increasing the affinity of receptors for opiate drugs in response to the loss of adrenocortical hormones, but decreases opioid stress analgesia by eliminating an important source of opioid peptides, the adrenal medulla'. The apocryphal tale that 'a soldier going into battle feels no pain' may therefore have some basis in this new understanding of the role of adrenal medullary secretion in activating long-term stress-induced analgesia.

Acknowledgement

I thank Dr. *Patricia Boksa,* Mr. *Larry Whelan* and Mrs. *Gwen Peard* for their assistance in preparing this chapter which was written during support by a Canadian MRC Program Grant (PG 20) to B.G.L.

References

1   Abel, J. J.: On the blood-pressure-raising constituent of the suprarenal capsule. Johns Hopkins Hosp. Bull. *8:* 151–157 (1897).

2   Abou-Donia, M. M.; Wilson, S. P.; Nichol, C. A.; Viveros, O. H.: Regulation of tyrosine hydroxylation by co-ordinate control of tyrosine hydroxylase, tetrahydrobiopterin, and GTP-Cyclohydrolase. Int. Conf. Molecular Neurobiology of Peripheral Catecholaminergic Systems, Ibiza 1982, pp. 64–65.

3   Almazan, G.; Pacheco, P.; Vassilieff, V. S.; Sourkes, T. L.: Adrenomedullary ornithine decarboxylase activity; its use in biochemical mapping of the origins of the splanchnic nerve in the rat. Brain Res. *237:* 397–404 (1982).

4   Amy, C.; Kirshner, N.: Na^+ uptake and catecholamine secretion by primary cultures of adrenal medulla cells. J. Neurochem. *39:* 132–142 (1982).

5   Anderson, T. R.; Slotkin, T. A.: The role of neural input in the effects of morphine on the rat adrenal medulla. Biochem. Pharmac. *25:* 1071–1074 (1976).

6   Apps, D. K.; Pryde, J. G.; Sutton, R.; Phillips, J. H.: Inhibition of adenosine triphosphatase, 5-hydroxytryptamine transport and proton translocation activities of released chromaffin-granule 'ghosts'. Biochem. J. *190:* 273–282 (1980).

7   Baird, A.; Benoit, R.; Ling, R.; Bohlen, P.; Guillemin, R.: Partial characterization of somatostatin-like material in the bovine adrenal medulla. Soc. Neurosci. Abstr. *7:* 606 (1981).

8   Baker, P. F.; Knight, D. E.: Calcium-dependent exocytosis in bovine adrenal medullary cells with leaky plasma membranes. Nature, Lond. *276:* 620–622 (1978).

9   Baker, P. F.; Knight, D. E.: Calcium-dependent exocytosis in 'leaky' bovine adrenal medullary cells has a specific requirement for magnesium adenosine triphosphate. J. Physiol., Lond. *295:* 89 (1979).

10   Baker, P. F.; Knight, D. E.: Gaining access to the site of exocytosis in bovine adrenal medullary cells. J. Physiol., Paris *76:* 497–504 (1980).

10a   Baker, P. F.; Knight, D. E.: Calcium control of exocytosis and endocytosis in bovine adrenal medullary cells. Phil. Trans. R. Soc. B*96:* 83–103 (1981).

11   Bartolome, J.; Slotkin, T. A.: Effects of postnatal reserpine administration on sympatho-adrenal development in the rat. Biochem. Pharmac. *25:* 1513–1519 (1976).

12   Biales, B.; Dichter, M.; Tischler, A.: Electrical excitability of cultured adrenal chromaffin cells. J. Physiol., Lond. *262:* 743–753 (1976).

13   Boksa, P.; Livett, B. G.: Substance P protects against nicotinic desensitization of cultured adrenal chromaffin cells. Soc. Neurosci. Abstr.*8:* 986 (1982).

14   Boksa, P.; Seidah, N. G.; Chrétien, M.; Livett, B. G.: Effects of pro-opiomelanocortin fragments on release of catecholamines from adrenal chromaffin cells. Neurosci. Lett. *28:* 199–204 (1982).

15   Boksa, P.; St. Pierre, S.; Livett, B. G.: Characterization of substance P and somatostatin receptors on adrenal chromaffin cells using structural analogues. Brain Res. *245:* 275–283 (1982).

16   Bowker, R. M.; Steinbusch, H. W. M.; Coulter, J. D.: Serotonergic and peptidergic projections to the spinal cord demonstrated by a combined retrograde HRP histochemical and immunocytochemical staining method. Brain Res. *211:* 412–417 (1981).

17   Brandt, B. L.; Hagiwara, S.; Kodokoro, Y.; Miyazaki, S.: Action potentials in the rat chromaffin cell and effects of acetylcholine. J. Physiol., Lond. *263:* 417–439 (1976).

17a   Bray, G. A.; DeQuattro, V.; Fisher, D. A.: Catecholamines: a symposium teaching conference. Calif. Med. *117:* 32–62 (1972).

17b   Brodin, F.; Nilsson, G.: Concentration of substance P-like immunoreactivity (SPLI) in tissues of dog, rat and mouse. Acta physiol. scand. *112:* 305–312 (1981).

18   Brookes, J. C.; Burke, D. H.; Treml, S.: Effect of various stimulus parameters on electrically induced catecholamine secretion by thin slices of bovine adrenal medulla. J. neural Transm. *50:* 201–208 (1981).

19   Brookes, J. C.; Treml, S.: Trifluoperazine-induced inhibition of catecholamine secretion by isolated adrenal chromaffin cells. Soc. Neurosci. Abstr. *7:* 211 (1981).

19a   Bucsics, A.; Saria, A.; Lembeck, F.: Substance P in the adrenal gland: origin and species distribution. Neuropeptides *1:* 329–341 (1981).

20   Cantin, M.; Benchimol, S.: Localization, synthesis and transport of proteins, glycoproteins and catecholamines in the adrenal medulla: ultrastructural, cytochemical and radioautographic studies. Adv. Biosci. *36:* 119–126 (1982).

21   Carmichael, S. W.: The adrenal medulla, vol.1. Annu. Res. Revs (Eden Press, Montreal 1979).

22   Carmichael, S. W.: The adrenal medulla, vol.2. Annu. Res. Revs (Eden Press, Montreal 1981).

23   Chantry, C. J.; Seidler, F. J.; Slotkin, T. A.: Non-neurogenic mechanism for reserpine-

induced release of catecholamines from the adrenal medulla of neonatal rats: Possible modulation by opiate receptors. Neuroscience 7: 673–678 (1982).

24 Ciaranello, R. D.; Berenbeim, D. M.; Masover, S.; Wong, D. L.: Synthesis and degradation of adrenal medullary catecholamine biosynthetic enzymes. Int. Conf. Molecular Neurobiology of Peripheral Catecholaminergic Systems, Ibiza 1982, pp. 66–67.

25 Clement-Jones, V.; Corder, R.; Lowry, P. J.: Isolation of human met-enkephalin and two groups of putative precursors (2K-pro-met-enkephalin) from an adrenal medullary tumour. Biochem. biophys. Res. Commun. 95: 665–673 (1980).

26 Comline, R. S.; Silver, M.: The development of the adrenal medulla of the foetal and new-born calf. J. Physiol., Lond. 183: 305–340 (1966).

27a Corder, R.; Lowry, P. J.: Large-molecular-weight somatostatin in human adrenal medullary tissue. Bioscience Reports 2: 397–403 (1982).

27b Corder, R.; Mason, D. F. J.; Perrett, D.; Clement-Jones, V.; Rees, L. H.; Besser, G. M.; Lowry, P. J.: Neurotensin, somatostatin enkephalins and catecholamines are co-secreted from perfused cat adrenal glands. Int. Conf. Molecular Neurobiology of Peripheral Catecholaminergic Systems, Ibiza 1982, p. 125.

28 Corder, R.; Sykes, J. E.; Lowry, P. J.: Characterization of the somatostatin-like immunoreactivity extracted from an adrenal medullary tumour. Biosci. Rep. 2: 147–154 (1982).

28a Costa, E.; Guidotti, A.; Saiani, L.: Opiate receptors and adrenal medullary junction. Reply to Lemaire, S., Lemaire, I., Dean, D. M. and Livett, B. G. Nature, Lond. 288: 304 (1980).

29 Coupland, R. E.: The natural history of the chromaffin cell, pp. 1–279 (Longmans, London 1965).

30 Coupland, R. E.: Electron microscopic observations on the structure of the rat adrenal medulla. I. The ultrastructure and organization of chromaffin cells in the normal adrenal medulla. J. Anat. 99: 231–254 (1965).

31 Coupland, R. E.: The chromaffin system; in Blaschko, Muscholl, Catecholamines, pp. 16–45 (Springer, Berlin 1972).

32 Coupland, R. E.; Holmes, R. L.: The distribution of cholinesterase in the adrenal glands of the rat, cat and rabbit. J. Physiol., Lond. 141: 97–106 (1958).

33 Coupland, R. E.; Kent, C.; Kobayashi, S.: Amine turnover and the effects of insulin hypoglycemia on small-granule chromaffin (SGC) cells of the mouse adrenal medulla; in Coupland, Forssmann, Peripheral neuroendocrine interaction, pp. 86–96 (Springer, Berlin 1978).

34 Creutz, C. E.; Pazoles, C. J.; Pollard, H. B.: Identification and purification of an adrenal medullary protein (synexin) that causes calcium-dependent aggregation of isolated chromaffin granules. J. biol. Chem. 253: 2858–2866 (1978).

35 Crowther, R.; Pearse, B.: Assembly and packing of clathrin into coats. J. Cell Biol. 91: 790–797 (1981).

36 Cummings, J. F.: Thoracolumbar preganglionic neurons and adrenal innervation in the dog. Acta anat. 73: 27–37 (1969).

37 Cybulski; Szyomonowicz: Gazeta Lekarska, p. 299 (1895); quoted by Dreyer [48].

38 Dean, D. M.; Lemaire, S.; Livett, B. G.: Evidence that inhibition of nicotine-mediated catecholamine secretion from adrenal chromaffin cells by enkephalin, β-endorphin, dynorphin (1–13), and opiates is not mediated via specific opiate receptors. J. Neurochem. 38: 606–614 (1982).

39 Dean, D. M.; Livett, B. G.: Opioid peptides and narcotic analgesics modulate the basal release of catecholamines from adrenal chromaffin cells in culture. Soc. Neurosci. Abstr. *7:* 212 (1981).

40 Del Bo, A.; Ross, C.; Pardal, J.; Saavedra, J.; Reis, D. J.: Neural and humoral components of the pressor response elicited by electrical stimulation of fastigial nucleus (FN) in rats before and after sympathectomy. Soc. Neurosci. Abstr. *7:* 632 (1981).

41 Derome, G.; Tseng, R.; Mercier, P.; Lemaire, I.; Lemaire, S.: Possible muscarinic regulation of catecholamine secretion mediated by cyclic GMP in isolated bovine adrenal chromaffin cells. Biochem. Pharmac. *30:* 855–860 (1981).

42 Dichter, M. A.; Tischler, A. S.; Greene, L. A.: Nerve growth factor-induced increase in electrical excitability and acetylcholine sensitivity of a rat pheochromocytoma cell line. Nature, Lond. *268:* 501–504 (1977).

43 DiGiulio, A. M.; Yang, H.-Y. T.; Fratta, W.; Costa, E.: Decreased content of immunoreactive enkephalin-like peptide in peripheral tissues of spontaneously hypertensive rats. Nature, Lond. *278:* 646–647 (1979).

44 Douglas, W. W.: Stimulus-secretion coupling: the concept and clues from chromaffin and other cells. Br. J. Pharmacol. *34:* 451–474 (1968).

45 Douglas, W. W.; Kanno, T.; Sampson, S. R.: Effects of acetylcholine and other medullary secretagogues and antagonists on the membrane potential of adrenal chromaffin cells: an analysis employing techniques of tissue culture. J. Physiol., Lond. *188:* 107–120 (1967).

46 Douglas, W. W.; Kanno, T.; Sampson, S. R.: Influence of the ionic environment on the membrane potential of adrenal chromaffin cells and on the depolarizing effect of acetylcholine. J. Physiol., Lond. *191:* 107–121 (1967).

47 Doupe, A. J.; Patterson, P. H.; Landis, S. C.: Induction of cholinergic neuronal properties in cultured rat adrenal chromaffin cells. Soc. Neurosci. Abstr. *8:* 257 (1982).

48 Dreyer, G. P.: On secretory nerves to the suprarenal capsules. Am. J. Physiol. *2:* 203–219 (1895).

48a Edwards, A. V.: Adrenal catecholamine output in response to stimulation of the splanchnic nerve in bursts in the conscious calf. J. Physiol., Lond. *327:* 409–419 (1982).

49 Elliott, T. R.: The control of the suprarenal glands by the splanchnic nerves. J. Physiol., Lond. *44:* 374–409 (1912).

50 Ellison, J. P.; Clarke, G. M.: Retrograde axonal transport of horseradish peroxidase in peripheral autonomic nerves. J. comp. Neurol. *161:* 103–114 (1975).

51 Falck, N. A.; Hillarp, G.; Thieme, G.; Torp, A.: Fluorescence of catecholamines and related compounds with formaldehyde. J. Histochem. Cytochem. *10:* 348–354 (1962).

52 Feldberg, W.; Minz, B.; Tsudzimura, H.: The mechanism of the nervous discharge of adrenaline. J. Physiol., Lond. *81:* 286–304 (1934).

53 Fenwick, E. M.; Fajdiga, P. B.; Howe, N. B. S.; Livett, B. G.: Functional and morphological characterization of isolated bovine adrenal medullary cells. J. Cell Biol. *76:* 12–30 (1978).

53a Floor, E.; Leeman, S. E.: Depletion of substance P in synaptic vesicle fraction derived from veratridine-depolarized synaptasomes. Soc. Neurosci. Abstr. *6.:* 573 (1980).

54 Fricker, L. D.; Snyder, S. H.: Enkephalin convertase: purification and characterization

of a specific enkephalin-synthesizing carboxypeptidase localized to adrenal chromaffin granules. Proc. natn. Acad. Sci. USA *79:* 3886–3890 (1982).

54a Fujita, T.: The gastro-enteric endocrine cell and its paraneuronic nature; in Coupland, Fujita (eds), Chromaffin, enterochromaffin and related cells, pp. 191–208 (Elsevier, Amsterdam 1976).

55   Furshpan, E. J.; MacLeish, P. R.; O'Lague, P. H.; Potter, D. D.: Chemical transmission between rat sympathetic neurons and cardiac myocytes developing in microcultures: evidence for cholinergic, adrenergic and dual-function neurons. Proc. natn. Acad. Sci. USA *73:* 4225–4229 (1976).

55a Gamse, R.; Wax, A.; Zigmond, R. E.; Leeman, S. E.: Immunoreactive substance P in sympathetic ganglia: distribution and sensitivity towards capsaicin. Neuroscience *6:* 437–441 (1981).

56   Gauthier, S.; Gagner, J.-R.; Sourkes, T. L.: Role of descending spinal pathways in the regulation of adrenal tyrosine hydroxylase. Expl Neurol. *66:* 42–54 (1979).

57   Greene, L. A.; Rein, G.: Release, storage and uptake of catecholamines by a clonal cell line of nerve growth factor (NGF) responsive pheochromocytoma cells. Brain Res. *129:* 247–263 (1977).

58   Greene, L. A.; Rein, G.: Release of (^{3}H) norepinephrine from a clonal line of pheochromocytoma cells (PC12) by nicotinic cholinergic stimulation. Brain Res. *138:* 521–528 (1977).

59   Greene, L. A.; Rein, G.: Synthesis, storage and release of acetylcholine by a noradrenergic pheochromocytoma cell line. Nature, Lond. *268:* 349–351 (1977).

60   Greene, L. A.; Tischler, A. S.: Establishment of a noradrenergic clonal line of rat adrenal pheochromocytoma cells which respond to nerve growth factor. Proc. natn. Acad. Sci. USA *73:* 2424–2428 (1976).

61   Grynszpan-Winograd, O.; Nicolas, G.: Intercellular junctions in the adrenal medulla: a comparative freeze-fracture study. Tissue Cell *12:* 661–672 (1980).

62   Helle, K.: Some chemical and physical properties of the soluble protein fraction of bovine adrenal chromaffin granules. Molec. Pharmacol. *2:* 298–310 (1966).

63   Hillarp, N. A.: Isolation and some biochemical properties of the catecholamine granules in the cow adrenal medulla. Acta physiol. scand. *43:* 82–96 (1958).

64   Hökfelt, T.; Fuxe, K.; Goldstein, M.; Joh, T. H.: Immunohistochemical studies of three catecholamine synthesizing enzymes: aspects on methodology. Histochemie *33:* 231–254 (1973).

65   Hökfelt, T.; Terenius, L.; Kuypers, H. G. J. M.; Dann, O.: Evidence for enkephalin immunoreactive neurons in medulla oblongata projecting to the spinal cord. Neuroscience *14:* 55–60 (1979).

66   Holets, V.; Elde, R.: Ultrastructural localization of met-enkephalin, serotonin and substance P in the cat intermediolateral cell column. Soc. Neurosci. Abstr. *7:* 117 (1981).

67   Holets, V.; Elde, R.: The differential distribution and relationship of serotonergic and peptidergic fibers to sympathoadrenal neurons in the intermediolateral cell column of the rat: A combined retrograde axonal transport and immunofluorescence study. Neuroscience *7:* 1155–1174 (1982).

68   Holz, R. W.; Senter, R. A.; Frye, R. A.: Relationship between Ca^{++} uptake and catecholamine secretion in primary dissociated cultures of adrenal medulla. J. Neurochem. *39:* 635–646.

69   Hughes, J.; Smith, T. W.; Kosterlitz, H. W.; Fothergill, L. A.; Morgan, B. A.; Morris,
     H. R.: Identification of two related pentapeptides from the brain with potent opiate
     agonist activity. Nature, Lond. *258:* 577–579 (1975).
70   Inoue, N.; Hatanaka, H.: Nerve growth factor induces specific enkephalin binding sites
     in a nerve cell line. J. biol. Chem. *257:* 9238–9241 (1982).
71   Izumi, F.; Oka, M.; Kumakura, K.: Synthesis, storage and secretion of adrenal cate-
     cholamines. Adv. Biosci. *36:* 1–301 (1982).
72   Izumi, F.; Wada, W.; Yanagihara, N.; Toyohira, Y.; Okuda, K.; Kashimoto, T.:
     Intracellular mechanism for the release of adrenal catecholamine. Protein nucleic Acid
     Enzyme *26:* 1651–1656 (1981).
73   Jacobi, C.: Beiträge zur physiologischen und pharmakologischen Kenntnis der Darm-
     bewegungen mit besonderer Berücksichtigung der Beziehung der Nebenniere zu den-
     selben. Arch. exp. Path. Phamak. *29:* 171–211 (1892).
74   Joh, T. H.; Baetge, E. E.; Kaplan, B. B.; Ross, M. E.; Brodsky, M. J.; Albert, V. P.; Park,
     D. H.; Reis, D. J.: Do catecholamine synthesizing enzymes share common gene coding
     squences? Soc. Neurosci. Abstr. *7:* 206 (1981).
75   Kachi, T.; Benerji, T. K.; Quay, W. B.: Daily rhythmic changes in synaptic vesicle
     contents of nerve endings on adrenomedullary adrenaline cells and their modification
     by pinealectomy and sham operations. Neuroendocrinology *28:* 201–211 (1979).
76   Kachi, T. K.; Banerji, T. K.; Quay, W. B.: Circadian and ultradian changes in synaptic
     vesicle numbers in nerve endings on adrenomedullary noradrenaline cells and their
     modifications by pinealectomy and sham operations. Neuroendocrinology *30:*
     291–299 (1980).
77   Kachi, T.; Banerji, T. K.; Quay, W. B.: The invagination complex in nerve endings on
     adrenomedullary adrenaline cells: quantitative ultrastructural description and analysis
     of changes with time-of-day and their modification by sham surgery and pinealectomy.
     J. autonom. Nerv. System *2:* 241–258 (1980).
78   Kanno, T.; Douglas, W. W.: Effect of rapid application of acetylcholine or depolarizing
     current on transmembrane potentials of adrenal chromaffin cells. Proc. Can. Fed. biol.
     Soc. *10:* 39 (1967).
79   Kenigsberg, R. L.; Cote, A.; Trifaro, J. M.: Trifluoperazine, a calmodulin inhibitor,
     blocks secretion in cultured chromaffin cells at a step distal from calcium entry. Neu-
     roscience *7:* 2277–2286 (1982).
80   Kidokoro, Y.: The effect of acetylcholine on the adrenal chromaffin cell: two pathways
     for Ca entry. Adv. Biosci. *36:* 11–20 (1982).
81   Kidokoro, Y.: Properties of acetylcholine receptor channels in the rat adrenal chro-
     maffin cell. Int. Conf. Molecular Neurobiology of Peripheral Catecholaminergic Sys-
     tems, 1982, p.135.
82   Kidokoro, Y.; Miyazaki, S.; Ozawa, S.: Acetylcholine-induced membrane depolari-
     zation and potential fluctuations in the rat adrenal chromaffin cell. J. Physiol., Lond.
     324: 203–220 (1982).
83   Kidokoro, Y.; Ritchie, A. K.: Chromaffin cell action potentials and their possible role
     in adrenaline secretion from rat adrenal medulla. J. Physiol., Lond. *307:* 199–216
     (1980).
84   Kilpatrick, D. L.; Ledbetter, F. H.; Carson, K. A.; Krishner, A. G.; Slepetis, R.; Kirsh-
     ner, N.: Stability of bovine adrenal medulla cells in culture. J. Neurochem. *35:*
     679–692 (1980).

84a Kilpatrick, D.L.; Slepetis, R.J.; Corcoran, J.J.; Kirshner, N.: Calcium uptake and catecholamine secretion by cultured bovine adrenal medulla cells. J. Neurochem. *38:* 427–435 (1982).

85 Kirby, R.F.; McCarty, R.: Development of adrenal medullary and cardiac responses to sympathetic stimulation in rats. Soc. Neurosci. Abstr. *7:* 401 (1981).

85a Kirshner, N.; Kilpatrick, D.L.; Amy, C.M.; Slepetis, R.J.: Ion channels and stimulus secretion coupling in cultured bovine adrenal medulla cells; in Molecular biology of peripheral catecholamine systems, Ibiza 1982, p.133 (1982).

86 Kiss, T.: Experimental-morphologische Analyse der Nebenniereninnervation. Acta anat. *13:* 81–89 (1951).

86a Kobayashi, S.: An autoradiographic study of the mouse carotid body using tritiated leucine, dopa, dopamine and ATP with special reference to the chief cell as a paraneuron. Arch. Histol. Jap. *39:* 295–317 (1976).

87 Koelle, G.: The histochemical differentiation of types of cholinesterase and their localization in tissue of the cat. J. Pharmac. exp. Ther. *100:* 158–179 (1950).

88 Kuchel, O.; Buu, N.T.: Conjugation of norepinephrine and other catecholamines; in Ziegler, Lake, Frontiers of clinical neuroscience Vol.4: Norepinephrine: clinical aspects (Williams & Wilkins, Baltimore 1983).

89 Kumakura, K.; Karoum, F.: Guidotti, A.; Costa, E.: Modulation of nicotinic receptors by opiate receptor agonists in cultured adrenal chromaffin cells. Nature, Lond. *283:* 489–492 (1980).

90 Lagercrantz, H.; Pistoletti, P.; Catecholamine release in the newborn infant at birth. Pediat. Res. *11:* 889–893 (1973).

91 Landis, S.C.; Patterson, P.H.: Neural crest cell lineages. Trends Neurosci. *4:* 172–175 (1981).

92 Ledbetter, F.H.; Kirshner, N.: Studies of chick adrenal medulla in organ culture. Biochem. Pharmac. *24:* 967–974 (1975).

93 Le Douarin, N.M.: The ontogeny of the neural crest in avian embryo chimaeras. Nature, Lond. *286:* 663–669 (1980).

94 Lemaire, S.; Livett, B.; Tseng, R.; Mercier, P.; Lemaire, I.: Studies on the inhibitory action of opiate compounds on isolated bovine adrenal chromaffin cells: noninvolvement of stereospecific opiate binding sites. J. Neurochem. *36:* 886–892 (1981).

95 Lewander, T.; Joh, T.H.; Reis, D.: Tyrosine hydroxylase: delayed activation in central noradrenergic neurons and induction in adrenal medulla elicited by stimulation of central cholinergic receptors. J. Pharmac. exp. Ther. *200:* 523–534 (1977).

96 Lewis, G.P.: Physiological mechanism controlling secretory activity of adrenal medulla; in Handbook of physiology, vol.VI, Sect. 7, pp.309–319 (1975).

97 Lewis, R.P.; Shute, C.C.D.: An electronmicroscopic study of cholinesterase distribution in the rat adrenal medulla. J. Microscopy *89:* 181–194 (1969).

98 Lewis, R.V.; Stern, A.S.; Kilpatrick, D.L.; Gerber, L.D.; Rossier, J.; Stein S.; Udenfriend, S.: Marked increases in large enkephalin-containing polypeptides in the rat adrenal gland following denervation. J. Neurosci. *1:* 80–82 (1981).

99 Lewis, R.V.; Stern, A.S.; Rossier, J.; Stein, S.; Udenfriend, S.: Putative enkephalin precursors in bovine adrenal medulla. Biochem. biophys. Res. Commun. *89:* 822–829 (1979).

99a Lewis, J.W.; Tordoff, M.G.; Sherman, J.E.; Liebeskind, J.C.: Adrenal medullary enkephalin-like peptides may mediate opioid stress analgesia. Science *217:* 557–559 (1982).

100 Liang, B.T.; Perlman, R.L.: Catecholamine secretion by hamster adrenal cells. J. Neurochem. *32:* 927–933 (1979).

101 Lindberg, I.; Yang, H.-Y.T.; Costa, E.: An enkephalin-generating enzyme in bovine adrenal medulla. Biochem. biophys. Res. Commun. *106:* 186–193 (1982).

102 Linnoila, R.I.; Diaugustine, R.P.; Hervonen, A.; Miller, R.J.: Distribution of (Met[5])- and (Leu[5])-enkephalin, vasoactive-intestinal polypeptide, and substance P-like immunoreactivities in human adrenal glands. Neuroscience *5:* 2247–2259 (1980).

103 Livett, B.G.; Boksa, P.; Dean, D.M.; Mizobe, F.; Lindenbaum, M.H.: Use of isolated chromaffin cells to study basic release mechanisms. J. autonom. Nerv. System *7:* 59–86 (1983).

103a Livett, B.G.; Fenwick, E.M.; Fajdiga, P.G.; Have, N.B.S.: A retrograde perfusion technique for high-yield production of single chromaffin cells from the bovine adrenal gland. Proc. Aust. physiol. pharmacol. Soc. *7:* 108P (1976).

103b Livett, B.G.; Boksa, P.; White, T.D.: On-line detection of exocytosis from adrenal chromaffin cells. Soc. Neurosci. Abstr. *8:* 120 (1982).

104 Livett, B.G.; Day, R.: Elde, R.P.; Howe, P.R.C.: Co-storage of enkephalins and adrenaline in the bovine adrenal medulla. Neuroscience *7:* 1323–1332 (1982).

105 Livett, B.G.; Dean, D.M.: Distribution of immunoreactive enkephalins in adrenal paraneurons: preferential localization in varicose processes and terminals. Neuropeptides *1:* 3–13 (1980).

106 Livett, B.G.; Dean, D.M.; Whelan, L.G.; Udenfriend, S.; Rossier, J.: Co-release of Leu-enkephalin and catecholamines from adrenal chromaffin cells in culture. Nature, Lond. *289:* 317–319 (1981).

107 Livett, B.G.; Kozousek, V.; Dean, D.M.: Substance P inhibits the nicotinic activation of adrenal medullary cells in culture. Proc. Int. IBRO Workshop Neuronal Secretion in Physiological and Pathological States. Antwerp, p.15 (1978).

108 Livett, B.G.; Kozousek, V.; Mizobe, F.; Dean, D.M.: Substance P inhibits nicotinic activation of chromaffin cells. Nature, Lond. *278:* 256–257 (1979).

109 Loewy, A.D.; Burton, H.: Nuclei of the solitary tract: efferent projection to the lower brain stem and spinal cord of the cat. J. comp. Neurol. *181:* 421–450 (1978).

110 Lundberg, J.M.; Hamberger, B.; Schultzberg, M.; Hokfelt, T.; Granberg, P.-O.; Efendic, S.; Terenius, L.; Goldstein, M.; Luft, R.: Enkephalin- and somatostatin-like immunoreactivities in human adrenal medulla and pheochromocytoma. Proc. natn. Acad. Sci. USA *76:* 4079–4083 (1979).

111 Maron, R.H.; Fishkes, H.; Kanner, B.I.; Schuldiner, S.: Solubilization and reconstitution of the catecholamine transporter from bovine chromaffin granules. Biochemistry *18:* 4781–4785 (1979).

112 Marty, A.: Ca-dependent K channels with large unitary conductance in chromaffin cell membranes. Nature, Lond. *219:* 497–500 (1981).

113 Matsui, H.: Adrenal medullary secretory response to stimulation of the medulla oblongata in the rat. Neuroendocrinology *29:* 385–390 (1979).

114 Matthews, E.K.: Membrane potential measurement in cells of the adrenal gland. J. Physiol., Lond. *189:* 139–148 (1967).

115 Maycock, W.A.; Heslop, T.S.: An experimental investigation of the nerve supply of the adrenal medulla of the cat. J. Anat. *73:* 551–558 (1939).

116 Mizobe, F.; Kozousek, F.; Dean, D. M.; Livett, B. G.: Pharmcological characterization of adrenal paraneurons: substance P and somatostatin as inhibitory modulators of the nicotine response. Brain Res. *178:* 555–566 (1979).

117 Mizobe, F.; Livett, B. G.: Production and release of acetylcholinesterase by a primary cell culture of bovine adrenal medullary chromaffin cells. J. Neurochem. *35:* 1469–1472 (1981).

118 Mizobe, F.; Livett, B. G.: Nicotine stimulates secretion of both catecholamines and acetylcholinesterase from cultured adrenal chromaffin cells. J. Neurosci. *3:* 871–876 (1983).

118a Mizobe, F.; Iwamoto, M.; Livett, B. G.: Parallel but separate release of catecholamines and acetylcholinesterase from stimulated adrenal chromaffin cells in culture. J. Neurochem. (submitted; 1983).

119 Mizuno, K.; Minamino, N.; Kangawa, K.; Matsuo, H.: A new family of endogenous 'big' Met-enkephalins from bovine adrenal medulla: purification and structure of docosa-(BAM-22P) and eicosapeptide (BAM-20P) with very potent opiate activity. Biochem. biophys. Res. Commun. *97:* 1283–1290 (1980).

120 Morris, S. J.; Hughes, J. M. X.: Synexin protein is nonselective in its ability to increase Ca^{++}-dependent aggregation of biological and artificial membranes. Biochem. biophys. Res. Commun. *91:* 345–350 (1979).

121 Nelson, P. G.: Neuroblastoma: Electrophysiological properties. Response of clonal cell lines. Neurosci. Res. Prog. Bull. *11:* 422–427 (1973).

122 Neumayr, R. J.; Hare, B. D.; Franz, D. N.: Evidence for bulbospinal control of sympathetic preganglionic neurons by monoaminergic pathways. Life Sci. *14:* 793–806 (1974).

122a Nilsson, G.; Brodin, E.: Tissue distribution of substance P-like immunoreactivity in dog, cat, rat and mouse: in Von Euler, Pernone, Substance P, pp. 49–54 (Raven Press, New York 1977).

123 Noda, M.; Furutani, Y.; Takahashi, H.; Toyosato, M.; Hirose, T.; Inayama, S.; Nakanishi, S.; Numa, S.: Cloning and sequence analysis of cDNA for bovine adrenal preproenkephalin. Nature, Lond. *295:* 202–206 (1982).

124 Nordmann, J. J.: Content and release of chromaffin granules: a combined stereological and assay analysis. Int. Conf. Molecular Neurobiology of Peripheral Catecholaminergic Systems, Ibiza 1982, p. 91.

125 Oka, M.; Isosaki, J.; Watanabe, H.; Houchi, H.; Tsunematsu, T.; Minaguchi, K.: Nicotinic and msucarinic receptors in chromaffin cells; their role in the regulation of Ca^{++} flux and catecholamine release and synthesis. Int. Conf. Molecular Neurobiology of Peripheral Catecholaminergic Systems, Ibiza 1982, p. 70.

126 Oliver, G.; Schafer, E. A.: On the physiological action of extract of the suprarenal capsules. J. Physiol., Lond. *16:* 1–4 (1895).

126a Panula, P.; Yang, H. Y. T.; Cheney, D.; Costa, E.: Localization of Met-enkephalin Arg^6-Phe^7-like immunoreactivity in the rat brain and spinal cord. Fed. Proc. *41*(5): 1469, abstr. 7010 (1982).

127 Patrick, J.; Stallcup, W. B.: α-Bungarotoxin binding and cholinergic receptor function on a rat sympathetic nerve line. J. biol. Chem. *252:* 8629–8633 (1977).

128 Patterson, P. H.: Environmental determination of autonomic neurotransmitter functions. Annu. Rev. Neurosci. *1:* 1–17 (1978).

129 Pelto-Huikko, M.; Hervonen, A.: Immunohistochemical demonstration of enkephal-

ins, VIP and SP in Adrenal Medulla (Abstract). 8th Meet. Int. Soc. Neurochemistry, Nottingham, p.442 (1981).

130 Philippu, A.; Schumann, H.J.: Der Einfluss von Calcium auf die Brenzcatechin-aminfreisetzung. Experientia 18: 138–140 (1962).

131 Poisner, A.M.; Trifaro, J.M.: The secretory granule (Elsevier Biomedical Press, Amsterdam 1982).

132 Pollard, H.; Pazoles, C.J.; Creutz, C.E.; Zinder, O.: The chromaffin granule and possible mechanisms of exocytosis. Int. Rev. Cytol. 58: 160–198 (1979).

133 Pollard, H.B.; Scott, J.H.: Control of Ca^{++}-dependent membrane contact and fusion in chromaffin cells by synexin. Int. Conf. Molecular Neurobiology of Peripheral Cate-cholaminergic Systems, Ibiza 1982, p.175.

134 Quik, M.; Sourkes, T.L.: Central dopaminergic and serotonergic systems in the reg-ulation of adrenal tyrosine hydroxylase. J. Neurochem. 28: 137–147 (1977).

135 Quik, M.; Trifaró, J.M.: The α-bungarotoxin site and its relation to the cholinergic and nerve growth factor mediated increases in tyrosine hydroxylase activity in cultures of sympathetic ganglia and chromaffin cells. Brain Res. 244: 332–336 (1982).

136 Richelson, E.: The use of cultured cells in neurobiological studies. Int. Rev. Biochem. 26: 81–120 (1979).

136a Role, L.W.; Perlman, R.L.; Leeman, S.E.: Somatostatin and substance P inhibit cat-echolamine secretion from guinea pig chromaffin cells. Soc. Neurosci. Abstr. 5: 597 (1979).

137 Role, L.W.; Leeman, S.E.; Perlman, R.L.: Somatostatin and substance P inhibit cat-echolamine secretion from isolated cells of guinea-pig adrenal medulla. Neuroscience 6: 1813–1821 (1981).

138 Ross, C.A.; Del Bo, A.; Reis, D.J.: Effect of electrical stimulation of the C$_1$ adrenergic cell group and the Kolliker-Fuse nucleus on sympathetic vasomotor activity and adrenal medullary catecholamine secretion in the rat. Soc. Neurosci. Abstr. 7: 210 (1981).

139 Ross, C.A.; Ruggiero, D.A.; Reis, D.J.: Projections from neurons close to the ventral surface of the hindbrain to the spinal cord in the rat. Neurosci. Lett. 21: 143–148 (1981).

140 Rossier, J.: Enkephalin biosynthesis. Trends Neurosci. 4: 94–96 (1981).

141 Rossier, J.; Dean, D.M.; Livett, B.G.; Udenfriend, S.: Enkephalin congeners and precursors are synthesized and released by primary cultures of adrenal chromaffin cells. Life Sci. 28: 781–789 (1981).

142a Rubin, R.P.: Calcium and cellular secretion (Plenum Press, New York 1982).

142b Saiani, L.; Guidotti, A.: Suggestions for a new class of opiate receptors in bovine chromaffin cells. Soc. Neurosci. Abstr. 6: 35 (1980).

142c Saiani, L. and Guidotti, A.: Opiate receptor-mediated inhibition of catecholamine release in primary cultures of bovine adrenal chromaffin cells. J. Neurochem. 39: 1669–1670 (1982).

143 Said, S.I.: Evidence for secretion of vasoactive intestinal peptide by tumours of pan-creas, adrenal medulla, thyroid and lung: Support for the unifying apud concept. Clin. Endocr. 5: 201–204 (1976).

144 Saper, C.B.; Loewy, A.D.: Efferent connections of the parabrachial nucleus in the rat. Brain Res. 197: 291–317 (1980).

144a Saria, A.; Wilson, S.P.; Molnar, A.; Viveros, O.H.; Lembeck, F.: Substance P and opiate-like peptides in human adrenal medulla. Neurosci. Lett. 20: 195–200 (1980).

145 Schneider, A. S.; Cline, H. T.; Lemaire, S.: Rapid rise in cyclic GMP accompanies catecholamine secretion in suspensions of adrenal chromaffin cells. Life Sci. *24:* 1389–1394 (1979).

145a Schneider, A. S.; Cline, H. T.; Rosenbeck, K.; Sonenberg, M.: Stimulus-secretion coupling in isolated adrenal chromaffin cells: calcium channel activation and possible role of cytoskeletal elements. J. Neurochem. *37:* 567–575 (1981).

146 Schramm, L. P.; Adair, J. R.; Stribling, J. M.; Gray, L. P.: Preganglionic innervation of the adrenal gland of the rat: a study using horseradish peroxidase. Expl. Neurol. *49:* 540–553 (1975).

147 Schubert, D.; Heinemann, S.; Kidokoro, Y.: Cholinergic metabolism and synapse formation by a rat nerve cell line. Proc. natn. Acad. Sci. USA *74:* 2579–2583 (1977).

148 Schubert, D.; Whitlock, C.: Alteration of cellular adhesion by nerve growth factor. Proc. natn. Acad. Sci. USA *74:* 4055–4058.

149 Schultzberg, M.; Hökfelt, T.; Lundberg, J. M.; Terenius, L.; Elfvin, L. G.; Elde, R.: Enkephalin-like immunoreactivity in nerve terminals in sympathetic ganglia and adrenal medulla and in adrenal medullary gland cells. Acta physiol. scand. *103:* 475–477 (1978).

150 Schultzberg, M.; Lundberg, J. M.; Hökfelt, T.; Terenius, L.; Brandt, J.; Elde, R. P.; Goldstein, M.: Enkephalin-like immunoreactivity in gland cells and nerve terminals of the adrenal medulla. Neuroscience *3:* 1169–1186 (1978).

151 Slotkin, T. A.: Maturation of the adrenal medulla. II. Content and properties of catecholamine storage vesicles of the rat. Biochem. Pharmac. *22:* 2033–2044 (1973).

152 Slotkin, T. A.; Smith, P. G.; Lau, C.; Bareis, D. L.: Functional aspects of development of catecholamine biosynthesis and release in the sympathetic nervous system; in Parvez, Parvez, Biogenic amines in development, pp. 29–48 (Elsevier/North Holland, Amsterdam 1980).

153 Smith, A. D.; Winkler, H.: Fundamental mechanisms in the release of catecholamines; in Blaschko, Muscholl, Handbook of experimental pharmacology. Catecholamines, vol. 33, pp. 538–617 (Springer, New York 1972).

154 Sorimachi, M.; Yoshida, K.: Exocytotic release of catecholamines and dopamine-β-hydroxylase from the perfused adrenal gland of the rabbit and cat. Br. J. Pharmacol. *65:* 117–125 (1979).

155 Stern, A. S.; Lewis, R. V.; Kimura, S.; Rossier, J.; Gerber, L. D.; Brink, L.; Stein, S.; Udenfriend, S.: Isolation of the opioid heptapeptide (Arg6-Phe7-) met-enkephalin from bovine adrenal medullary granules and striatum. Proc. natn. Acad. Sci. USA *76:* 6680–6683 (1979).

156 Takamine, J.: The isolation of the active principle of the suprarenal gland. J. Physiol. Lond. *27:* 24–30 (1901).

157 Tischler, A. S.: Small cell carcinoma of the lung: cellular origin and relationship to other neoplasms. Semin. Oncol. *5:* 244–252 (1978).

158 Tischler, A. S.; De Lellis, R. A.; Biales, B.; Nunnemacher, G.; Carabba, V.; Wolfe, H. J.: Nerve growth factor-induced neurite outgrowth from normal human chromaffin cells. Lab. Invest. *43:* 399–409 (1980).

159 Tischler, A. S.; Greene, L. A.: Nerve growth factor-induced process formation by cultured rat pheochromocytoma cells. Nature, Lond. *258:* 341–342 (1975).

160 Tischler, A. S.; Greene, L. A.: Morphologic and cytochemical properties of a clonal line of rat adrenal pheochromocytoma cells which respond to nerve growth factor. Lab. Invest. *39:* 77–89 (1978).

161 Tournade, A.; Chabrol, M.; Wagner, P. E.: Le système nerveux adrenaline-sécréteur. C. r. Séanc. Soc. Biol. *93:* 93 (1925).

162 Trifaró, J. M.: The secretory process of the adrenal medulla (review). Endocrinol. exp. *4:* 225–251 (1970).

163 Trifaró, J. M.: Contractile proteins in tissues originating in the neural crest. Neuroscience *3:* 1–24 (1978).

164 Trifaró, J. M.; Lee, R. W. H.; Kenigsberg, R. L.; Cote, A.: Contractile proteins and chromaffin cell function; in Izumi et al., Synthesis, storage and secretion of adrenal catecholamines. Dynamic integration of functions. Adv. Biosci., vol. 36, pp. 151–158 (Pergamon Press, Oxford 1982).

165 Trifaró, J. M.; Poisner, A. M.: Common properties in the mechanisms of synthesis, processing and storage of secretory products; in Poisner, Trifaro, The secretory granule, pp. 387–407 (Elsevier, Amsterdam 1982).

166 Unsicker, K.; Krisch, B.; Otten, U.; Thoenen, H.: Nerve growth factor induced fiber outgrowth from isolated rat adrenal chromaffin cells: impairment by glucocorticoids. Proc. natn. Acad. Sci. USA *75:* 3498–3502 (1978).

167 Unsicker, K.; Rieffert, B.; Ziegler, W.: Effects of cell culture conditions, nerve growth factor, dexamethazone, and cyclic AMP on adrenal chromaffin cells in vitro. Adv. biochem. Psychopharmacol. *25:* 51–59 (1980).

168 Viveros, O. H.: Mechanism of secretion of catecholamines from adrenal medulla. Handbook of Physiology, vol. VI, sect. 7, pp. 389–426 (1975).

169 Viveros, O. H.; Diliberto, E. J., Jr.; Hazum, E.; Chang, K. J.: Opiate-like materials in the adrenal medulla: evidence for storage and secretion with catecholamines. Molec. Pharmacol. *16:* 310–317 (1979).

170 Viveros, O. H.; Diliberto, E. J.; Hazum, E.; Chang, K. J.: Enkephalins as possible adrenomedullary hormones: storage, secretion and regulation of synthesis. Adv. biochem. Psychopharmacol. *22:* 191–204 (1980).

171 Viveros, O. H.; Wilson, S. P.; Chang, K.-J.: Regulation of synthesis and secretion of enkephalins and related peptides in adrenomedullary chromaffin cells and human pheochromocytoma; in Costa, Trabucci, Regulatory peptides: from molecular biology to function, pp. 217–224 (Raven Press, New York 1982).

172 Vulpian, M.: Note sur quelques réactions propres à la substance des capsules surrénales. C. r. hebd. Séanc. Acad. Sci., Paris *43:* 663–665 (1856).

173 Wakade, A. R.: Studies on secretion of catecholamines evoked by acetylcholine or transmural stimulation of the rat adrenal gland. J. Physiol., Lond. *313:* 463–480 (1981).

174 Wakade, A. R.: Facilitation of secretion of catecholamines from rat and guinea-pig adrenal glands in potassium-free medium or after ouabain. J. physiol., Lond. *313:* 481–498 (1981).

175 Wakade, A. R.; Wakade, T. D.: Secretion of catecholamines from adrenal gland by a single electrical shock: electronic depolarization of medullary cell membrane. Proc. natn. Acad. Sci. USA *79:* 3071–3074 (1982).

176 Waki, S.: Spinal root pathway of the adrenal medullary secretory nerve fibres in the dog. Tohoku J. exp. Med. *104:* 341–348 (1971).

177 Warren, S.; Chute, R. N.: Pheochromocytoma. Cancer *29:* 327–331 (1972).

178 Weiner, N.; Masserano, J. M.; Meligeni, J.; Tank, A. W.: Activation of adrenal tyrosine hydroxylase following acute and chronic stress. Adv. Biosci. *36:* 37–46 (1982).

179 Weiner, N.; Masserano, J. M.; Tank, A. W.: Studies on the mechanism of activation of tyrosine hydroxylase during acute nerve stimulation or stress. Int. Conf. Molecular Neurobiology of Peripheral Catecholaminergic Systems, Ibiza 1982, p.63.

180 Weiss, B.; Levin, R. M.: Mechanism for selectively inhibiting the activation of cyclic nucleotide phosphodiesterase and adenylate cyclase by antipsychotic agents. Adv. cyclic Nucleotide Res. *9:* 285–303 (1978).

181 Weiss, I.; Kulla, B.; Muller, B.: Morphometrical studies of the adrenals in rats after hypothalamic lesions. Folia morph. *28:* 137–142 (1980).

182 Whelan, L. G.; Livett, B. G.: Time-lapse video observations on chromaffin cells in culture. Int. Conf. Molecular Neurobiology of Peripheral Catecholaminergic Systems. Ibiza 1982, p.183.

182a Wilson, S. P.; Klein, R. L.; Chang, K.-J.; Gasparis, M. S.; Viveros, O. H.; Yang, W. H.: Are opioid peptides co-transmitters in noradrenergic vesicles of sympathetic nerves? Nature, Lond. *288:* 707–709 (1980).

183 Winkler, H.: The biogenesis of adrenal chromaffin granules. Neuroscience *2:* 657–683 (1977).

184 Winkler, H.: The proteins of catecholamine-storing organelles. Scand. J. Immunol. *15:* suppl.9, pp.75–96 (1982).

185 Winkler, H.; Carmichael, S. W.: The chromaffin granule; in Poisner, Trifaró, The secretory granule, pp.3–79 (Elsevier Biomedical Press, Amsterdam 1982).

186 Winkler, H.; Westhead, E. W.; The molecular organization of adrenal chromaffin granules. Neuroscience *5:* 1803–1823 (1980).

186a Yang, H. Y. T.; Costa, E.: Regional distribution of Met[5]-enkephalin-Arg[6]-Phe' immunoreactivity in rat brain. Fed. Proc. *41*(5): 1469, Abstr.7009 (1982).

187 Yoshizaki, T.: Participation of muscarinic receptors on splanchnic adrenal transmission in the rat. Jap. J. Pharmac. *23:* 813–816 (1973).

188 Zinder, O.; Pollard, H. B.: The chromaffin granule – recent studies leading to a functional model for exocytosis. Essays Neurochem. Neuropharmacol. *4:* 125–162 (1980).

B. G. Livett, PhD, Department of Biochemistry, University of Melbourne, Parkville, Vic. 3052 (Australia)

Cell Biology of the Secretory Process, pp. 359–388 (Karger, Basel 1984)

The Secretory Process in B Cells of the Pancreas

Gerald Gold, Gerold M. Grodsky

Metabolic Research Unit and Department of Biochemistry and Biophysics, University of California, San Francisco, Calif., USA

Introduction

Insulin promotes glucose uptake by tissues such as fat and muscle at the same time that it inhibits glucose output by the liver. As a result, rising insulin levels depress the concentration of blood glucose, and insulin levels are essentially determined by secretory rates because this hormone is cleared from the circulation with a $t_{1/2}$ of less than 10 min. Most studies of glucose monitoring and associated metabolic and ionic signals that regulate insulin synthesis, storage and secretion have used isolated islets. Islets represent only 1–2% of total pancreatic tissue, and in most species only 70–80% of the approximately 3,000 cells per islet are B cells. Thus, large quantities of homogeneous B cells are generally unavailable for experimentation. Isolation of islets without contamination by acinar tissue entails tedious microdissection or collagenase digestion plus manual selection. At best only a few milligrams of islet tissue are obtained each day. Insulinomas, which contain more easily isolated insulin-secreting cells, may have both erratic storage characteristics and responses to regulators.

Regulation of Insulin Biosynthesis

Glucose is the most important regulator of both insulin biosynthesis and secretion. Physiologic concentrations stimulate biosynthesis of insulin in preference to other islet proteins, and more than 50% of the increase in total protein synthesis is represented by insulin [120]. Compounds chemically related to glucose or its triose metabolites stimulate either insulin biosynthesis or secretion in proportion to their rate of metabolism in islets [6]. Therefore, an event during the metabolism of glucose has been postulated to be a common point of origin for secondary signals that probably accelerate independently both insulin biosynthesis and secretion [6, 44, 71, 103].

Biosynthesis and secretion of insulin, though coordinated and coregulated by many compounds, are subject to independent regulation. Some agents that stimulate secretion inhibit synthesis (e.g. sulfonylureas) [89]. Cyclic AMP, which potentiates secretion, has been found to stimulate, inhibit, or have no effect on insulin biosynthesis [133]. In addition, inhibition of insulin synthesis by puromycin [72] or cyclohexamide [89] does not directly impair short-term release of stored insulin. Secretion during intense glucose or sulfonylurea stimulation also exceeds insulin replacement. Thus, the normal pancreas can accommodate periods of temporary deficit, and it contains enough preformed insulin to provide for more than 10 h of stimulated secretion. Insulin biosynthesis has a threshold of $3 \times 10^{-3} M$ glucose and a K_m of approximately $6 \times 10^{-3} M$, whereas insulin secretion has a threshold of $5 \times 10^{-3} M$ and a K_m of approximately $8 \times 10^{-3} M$ glucose [25, 80]. Thus, stored insulin increases when glucose concentrations are below or only slightly above the secretory threshold for insulin. At concentrations also below the secretory threshold, glucose increases islet ATP [75] and causes initial depolarization of the plasma membrane (to be discussed later). In addition, ionic requirements for insulin biosynthesis vs. secretion are different. Glucose-stimulated biosynthesis requires extracellular magnesium [70], but not calcium or electrical depolarizing events at the plasma membrane [103]. Insulin secretion, on the other hand, is inhibited by magnesium, but requires extracellular calcium [45] and depolarization of the plasma membrane.

Within minutes, glucose produces a 5- to 10-fold increase in the rates of both insulin biosynthesis and secretion [58]. Upon cessation of a strong glucose stimulus, insulin secretion rapidly returns to basal. Insulin biosynthesis, however, only rapidly returns to basal levels if messenger RNA synthesis has been previously blocked with actinomycin D; otherwise it continues at elevated rates [62]. Thus, glucose has a dual effect on insulin biosynthesis: first, translational efficiency increases [100] with a fixed amount of insulin messenger RNA [59], and enhanced biosynthesis is initially insensitive to actinomycin D [102]; secondly, the specific insulin messenger RNA increases by 40–60 min after the stimulus [34, 35] and produces an additional and relatively sustained enhancement of insulin biosynthesis that is sensitive to actinomycin D [101].

Insulin Genetics

The gene for human insulin is on the short arm of chromosome 11 [95] and is approximately 500 base pairs downstream from a polymorphic insertion region [9]. After digestion of human DNA with the endonuclease Bgl I, the insulin gene is found on large (4.6, 4.8 kb) and/or small (2.8, 2.9 kb) nucleotide fragments of DNA that segregate as alleles. Two studies reported that a noninsulin-dependent diabetic population has a higher incidence of people homozygous for the large Bgl I fragment than a control population [96, 108], but no difference was found in a third [9]. Divergent conclusions may reflect the racial mixtures in the control populations studied; the negative report had a higher percentage of Blacks in the control group and, consequently, a higher incidence of nondiabetic people homozygous for the large Bgl I genotype than the other reports. With a study using several generations of a single large Caucasian family, both homozygous and heterozygous carriers of the large Bgl I fragment had a higher incidence of elevated glycosylated hemoglobin (HbA$_{1c}$), which suggests a modest but chronic elevation of blood glucose concentrations in the nondiabetic carriers of these genotypes [97]. How or if this insertion sequence affects transcription of insulin messenger RNA or otherwise regulates insulin biosynthesis or secretion is unknown at present.

Sequestration and Processing of Insulin Precursors

Insulin messenger RNA codes in vitro for preproinsulin [73], which also has been identified in islets with rapid labeling techniques [2, 99]. In rat there are two different preproinsulins, and glucose preferentially stimulates production of preproinsulin I vs. II [63]. Rapid loss of the approximately 23-amino acid, hydrophobic NH$_2$-terminal region as preproinsulin is discharged into the lumen of the rough endoplasmic reticulum is consistent with the initiating signal hypothesis [11]. The resulting product, proinsulin [119] is continuously sequestered within cellular membranes as it transits the B cell. It is a peptide of approximately 9,000 daltons and can be thought of as a composite of three smaller peptides of nearly equal size

Table I. Distribution of radioautographic grains over cell components

	% of radioautographic grains			
	pulse (5 min)	chase incubation		
		+15 min	+30 min	+60 min
Rough endoplasmic reticulum and cytoplasm	78.7	36.6	33.3	25.8
Golgi complex	4.5	41.7	37.4	12.8
Secretory granules	5.2	5.1	12.8	44.3
Mitochondria	3.1	3.4	5.4	5.5
Nuclei	8.5	13.2	11.1	11.6
Number of grains counted	224	295	433	342

Taken from reference 92 with permission from the publisher.

[121]. The middle connecting peptide (C-peptide) is flanked by a pair of basic amino acids and the A-chain of insulin on the NH_2-terminus; another pair of basic amino acids and the B-chain of insulin flanks the COOH-terminus [43]. The C-peptide, which is clinically useful for measuring endogenous pancreatic function during insulin therapy [69], may correctly fold or align proinsulin during biosynthesis so that, after the pairs of basic amino acids are hydrolyzed and C-peptide falls away, the A- and B-chains of insulin have the correct disulfide bonding. Although the amino acid composition of C-peptide varies considerably from species to species, both the A- and B-chains of insulin have amino acid sequences that have been highly conserved in most mammals.

Proteolytic removal of the C-peptide is a relatively slow process and occurs concomitantly with transit of proinsulin through the B cell via the same route that was originally mapped for secretory proteins in pancreatic acini [98]. Table I summarizes pulse-labeling experiments with ^{3}H-leucine in fetal rat pancreas [92]. Radioactive traces from labeled secretory proteins first appear over membranes of the rough endoplasmic reticulum and, via transitional vesicles, move to the Golgi apparatus by 15–30 min, then concentrate over secretory vesicles 60 min after the pulse. Concomitant with transport, conversion of proinsulin to insulin occurs with little or no loss, as shown in figure 1 [37]. Conversion is unaffected by the glucose concentration [36, 120], and occurs, after a lag time of approximately 30 min (corresponding to the transit time to the secretory vesicles and

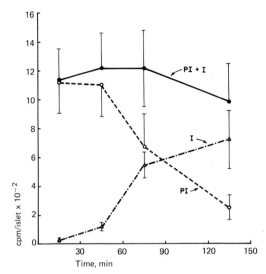

Fig. 1. Effect of time on the recovery of labeled proinsulin plus insulin in the islets plus incubation buffer [taken from ref. 37 with permission from the publisher].

Golgi), as a pseudo-first-order process with a $t_{1/2}$ of 1 h. Thus, kinetic experiments indicate that the enzymatic conversion of proinsulin to insulin occurs in secretory vesicles and perhaps the Golgi apparatus [55].

Inhibitors of oxidative phosphorylation, such as antimycin A [118], or inhibitors of microtubules, such as colchicine [79], block both transport of radioactive proteins to the Golgi apparatus and conversion of proinsulin to insulin. Converting enzymes coisolate with a crude secretory vesicle fraction prepared from islets, and are localized within vesicles [65]. Thus, inhibitor and cell fractionation studies also indicate that secretory vesicles are the site of proinsulin-to-insulin conversion, and suggest that conversion is inhibited by preventing the distribution of proinsulin into the secretory vesicles. The normal distribution of total proinsulin in glucose-stimulated islets has been estimated by computer analysis of a pulse of labeled proinsulin passing through the various cellular compartments. The rough endoplasmic reticulum and Golgi apparatus may contain approximately 25% of the total proinsulin in nearly equal amounts, and the secretory vesicles may contain the remaining 75% [*Landahl, Grodsky, and Gold,* unreported observation].

Electron microscopy indicates that, at the same time that chemical modification of proinsulin to insulin occurs in the cell, membranes sequestering this process also undergo chemical and structural modification.

With the freeze-fracture technique, a progressive loss of integral membrane particles is visible, starting with membranes of the Golgi apparatus nearest the transitional elements of the endoplasmic reticulum, through the membrane stacks to the vesicle-forming face. Loss continues during formation and maturation of secretory vesicles, and some of the remaining proteins may be destined for insertion into the plasma membrane [91]. Cholesterol, visualized as aggregates with philipin, is richer in membranes of secretory vesicles than in membranes of the Golgi apparatus and also is richer than in segments of the plasma membrane that are specifically reutilized for pinocytosis [91]. Thus, membranes of secretory vesicles interacting with the plasma membrane are more protein-poor and cholesterol-rich than other intracellular membranes. In addition, a clathrin-like coating is frequently present on the forming face of the Golgi apparatus and is included on adjacent secretory vesicles but is lost from the surface of secretory vesicles with time [91].

Total maturation of a secretory vesicle or completion of conversion of proinsulin to insulin is not a prerequisite for secretion. Approximately 30 min after the pulse, both secretion of labeled hormone and onset of conversion of labeled proinsulin to insulin begin. Thus, labeled proteins enter into secretory vesicles by 30 min. With time, labeled proinsulin is secreted at a fractional rate comparable with that of labeled insulin; however, the proportion of labeled proinsulin is initially somewhat higher in islets than in the secreted fraction because cells contain nonsecreted, prevesicle fractions (endoplasmic reticulum and Golgi) that are initially rich in labeled proinsulin [37]. Thus, despite the occurrence of considerable chemical modification of the membranes and contents, secretion is relatively independent of the completeness of conversion of labeled proinsulin to insulin and does not require a lengthy maturation process of vesicles.

Characteristics of Stored Insulin

Assuming that secretory vesicles make up 10–12% of B-cell volume [21, 105] and that B-cell insulin content is approximately $1 \times 10^{-3} M$ [29, 49], the insulin within secretory vesicles is about $10^{-2} M$ (or 60 mg/ml). Dye fluorescence studies indicate pH within secretory vesicles is approximately 6.0 [1] – close to the isoelectric point of insulin. Because insulin is extremely insoluble under these conditions, the hormone is found in the

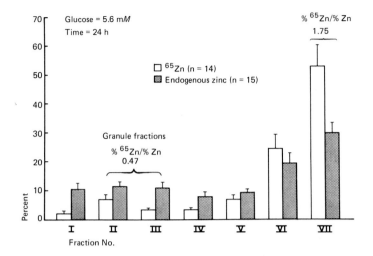

Fig. 2. ^{65}Zn and endogenous zinc distribution in sucrose gradient fractions of homogenized islets [taken from ref. 29 with permission from the publisher].

secretory vesicles of most species as a central insoluble aggregate, usually with crystalline structure.

In contrast to the lower pH of the secretory vesicles, cytosolic pH is approximately neutral [52]. Thus, the vesicles probably contain a proton pump system coupled with low proton permeability. In addition to insulin, ions (zinc, calcium, potassium, phosphate), nucleotides (AMP, cyclic AMP), histamines, and the enzymes involved in their binding or metabolism are also probably components of secretory vesicles of the B cell [14, 54, 124]. Of these components, zinc is found in most, but not all, species and has been most extensively studied. Islets concentrate zinc to more than 50 times the level found in blood [29], and uptake is carrier-mediated or occurs by diffusion, depending on extracellular concentrations [74].

Histologic data have consistently shown B-cell zinc to be exclusively concentrated in the protein core of the secretory vesicles [13, 27, 82, 134], indicating that it is associated with the insulin molecule. This zinc content decreases with increased secretion and degranulation induced by glucose or sulfonylureas. However, standard fixation and staining methods do not detect soluble or loosely bound zinc. Recent studies measuring the distribution of endogenous ^{65}Zn in islets cultured with this label have shown that the majority of islet zinc is extragranular and is, therefore, not stored or secreted with insulin [28, 29] (fig. 2). Most of the endogenous, nongran-

Table II. Evidence for islet zinc in excess of that required for two-zinc-insulin hexamer

Insulin content/islet	13.1 pmol
Zinc content/islet	11.1 pmol
Zinc required for two-zinc hexamer	4.4 pmol

Taken from reference 28 with permission from the publisher.

ular zinc distributes with the upper fraction of sucrose gradients so that this zinc is either free, bound to soluble proteins or associated with low-density particles. The function of this large amount of extragranular zinc requires clarification. Zinc does not affect the rate of proinsulin synthesis or conversion to insulin, although it has been reported to stabilize isolated granules [57, 81]. However, stability of granules in intact islets is dependent on calcium more than zinc [54].

By X-ray crystallographic analysis, the classic unit crystal consists of two zinc molecules in association with three insulin dimers (two-zinc hexamer), with the zinc binding at the b-10 histadines (23, 112). However, crystals containing four zinc per hexamer or in fact no zinc (24, 41, 60) can be produced. Crystalline structures for insulin in granules from rats occur with a 50-A periodicity and polyhedral structure [41], but this histological observation alone cannot distinguish the subtle differences between two-zinc, four-zinc, or zinc-free insulin crystals, and does not establish the nature of the endogenous insulin complex. As shown in table II [28], zinc and insulin are present in almost equal concentrations in islets, so three times more total zinc is present than is required for the two-zinc hexamer [113]. However, because secretory vesicles contain approximately one-fourth of the B-cell zinc, granular zinc is, at best, barely adequate to complex stored insulin as a two-zinc hexamer. Therefore, although large amounts of zinc can be bound to insulin under specific conditions in vitro [26], it is unlikely that zinc-insulin complexes exceeding the two-zinc hexamer exist in the normal B cell. Two observations suggest, although they do not prove, that zinc is incorporated into the secretory vesicle during its initial or early formation. Zinc forms two-zinc proinsulin hexamers with high affinity and with a structure similar to that of zinc-insulin [40]. Zinc-proinsulin, in contrast to zinc-insulin, is highly soluble and does not form crystals. In addition, kinetic studies with ^{65}Zn indicate that cytosolic zinc only slowly equilibrates with granular zinc, suggesting that this ion does not rapidly exchange across formed granular membranes but enters the secretion vesicles during their initial formation [29]. Thus, current evidence indicates zinc initially binds to proinsulin (possibly as the two-zinc hexamer); as proinsulin is slowly cleaved to insulin in the maturing granule, the resulting zinc-insulin complex rapidly aggregates or crystalizes out of solution. Soluble insulin may exist only transiently in maturing vesicles. Crystallization allows storage of large amounts of insulin and may reduce its sensitivity to further digestion by enzymes within the secretory vesicle. Therefore, with time, and roughly paralleling the conversion of proinsulin to insulin, the pale immature granule containing its diffused contents may develop into the characteristic vesicle containing crystalline insulin.

Intact crystalline insulin is secreted by exocytosis [68, 93], but it is unlikely that soluble zinc-insulin complexes are stable for long periods of time in the portal blood. Insulin complexes are diluted approximately 10^7-fold from vesicles to portal vein blood. At such dilutions, zinc-insulin-soluble complexes are almost completely dissociated within seconds [29, 39, 40]. Factors in the blood, including phosphate and other zinc-complexing ions or proteins, could facilitate dissociation even further. Nevertheless, if zinc-insulin complexes released from the pancreas can retain some stability in the portal blood for the few seconds required to reach the liver, our concept of the hepatic effects of endogenous insulin based on experiments using exogenous insulin will require revision.

Movement of Secretory Vesicles and Role of Microtubules

As is typical of secretory proteins, insulin is released from the B cell by emiocytosis. Secretory vesicles move to and fuse with the plasma membrane. The membranous junction rapidly dissolves and the contents of a vesicle are externalized. Movements of secretory vesicles in the B cell occur by saltatory rather than random motion. Motions start and stop abruptly; vesicles move faster than is possible with forces associated with random motion; and subsequent movements of a single vesicle occur principally in a straight line or at a 180° angle [76]. Only 2% per min of the vesicles in cultured B cells demonstrate this motion, which indicates that special activation and compartmentalization occur [64].

Internal cellular structures, such as microtubules, could direct saltatory motion and could also be required for transport of stored insulin and plasma membrane proteins, such as the somatostatin receptor [125], to the plasma membrane. Glucose stimulates polymerization of the microtubule subunit protein, tubulin, into microtubules [88], and secretory vesicles appear in association with microtubules in the B cell by electron microscopy [77]. Secretory vesicles from islets bind specifically in vitro to microtubules formed with isolated brain tubulin and microtubule-associated proteins in the presence of calcium [123]. The microtubule inhibitor, colchicine, also blocks saltatory motion, insulin secretion and glucose-stimulated externalization of the somatostatin receptor in a dose-dependent manner [64, 66, 125]. However, with cultured B cells [15] or intact rats [114], microtubule inhibitors block insulin release at concentrations below those that produce demonstrable changes in microtubular structures. In addition, a 2-hour incubation of cultured B cells with $10^{-6} M$ colchicine totally disassembles cytoplasmic microtubules; however, glucose-stimulated insulin secretion is initially normal but diminishes with time [15]. Therefore, intact microtubules may not be required for short-term insulin secretion, but may assist the migration of secretory vesicles toward the plasma membrane, which is necessary for sustained secretion.

Kinetics of Insulin Secretion

Beta-cells, and probably many other secretory cells [reviewed in ref. 90], are sensitive to both the static concentration of a secretagogue and its rate of change in concentration. Figure 3 shows some characteristics of insulin secretion from the in vitro perfused rat pancreas in response to changing glucose levels [90]. When the pancreas is suddenly exposed to stimulating concentrations of glucose, there is an immediate first phase of insulin secretion of 2–5 min (No. 1) followed by a second ascending secretion to equilibrium with time (No. 2) [4] (with comparable 'optional' stimulation of secretion, glucagon and somatostatin also are released multiphasically). When the glucose concentration is suddenly reduced to a lower stimulating concentration, a transient negative spike (No. 3) is observed [42]. Also, when restimulated, a potentiating action of glucose can be duplicated with similarly metabolized carbohydrates (such as glyceraldehyde) but not by secretagogues (such as cyclic AMP) [4, 42]. When glucose concentrations are increased in a slowly rising ramp, the transient negative and positive spikes of insulin release are not as prominent, but may still exist as underlying phenomena. Slow oscillations of basal insulin secretion are observed both in vitro and in vivo, indicating that such

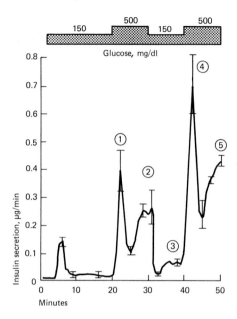

Fig. 3. Characteristic patterns of insulin secretion during step stimulations of the per-fused rat pancreas with glucose [taken from ref. 90 with permission from the publisher].

oscillations are initiated from the B cell [116]. Other secretagogues that cause sustained insulin release also produce similar multiphasic patterns of insulin secretion with some quantitative and qualitative differences [90].

Several mathematical models have been devised to account for mul-tiphasic insulin secretion [reviewed in ref. 67, 90]. In one form or another, they fall within two categories. The first is a 'storage limited model', in which hormone is stored heterogeneously in compartments of differing lability to secretagogues. This causes one compartment to be more rapidly released (and depleted) than the other by a step-stimulation of rapid onset, thereby producing rate sensitivity and first-phase secretion. In the second category, the 'signal limited model', a negative feedback or an exciter-inhibitor (metabolic or ionic) relationship is invoked to permit a transient first-phase release. Second-phase release (and the related potentiation) is a separate phenomenon in most models. The capabilities and limitations of these models have been reviewed [90] and a combination of the two [67] has been shown recently to duplicate most closely a variety of stimulation-secretion patterns for insulin secretion.

The closed-loop artificial pancreas permits studies whereby blood glucose is continuously monitored and regulated by computer-directed insulin infusion. These studies show that the ability of normal B cells to respond transiently and promptly to changes in glucose levels results in a reduction in the total insulin required to provide proper regulation and minimizes to occurrence of any subsequent hypoglycemia [3, 104]. Thus, in a rapid, closed-loop endocrine system, positive rate sensitivity may provide a priming dose to the peripheral active site of the hormone, which initiates rapid regulation before the secretagogue reaches a high level and reduces the total hormone requirement. On the other hand, a negative rate sensitivity prevents excessive hormone secretion, even though secretagogue levels, if considered only as static regulators, may still be in the stimulatory range.

Heterogeneous Storage and Secretion of Insulin

Several laboratories have concluded that insulin storage may be heterogeneous and that newly synthesized insulin is secreted preferentially from glucose-stimulated pancreatic slices or islets [20, 36–38, 46, 47, 56, 109, 110, 128]. Typical evidence is shown in figure 4b. Rat islets were incubated for 45 min in high glucose, and pulse-labeled for 15 min with ^{3}H-leucine, also in high glucose. Secreted insulin was then collected in noncumulative, 20-min observation periods (windows) at various times after the pulse. Preferential release of newly synthesized vs. preexisting insulin is indicated at several time points by secreted insulin's having a higher specific activity than that of the average cellular insulin. It is also indicated by a higher fractional secretory rate of labeled vs. immunoreactive insulin [37]. With these experimental conditions, the specific activity of the average cellular insulin is relatively stable between 1 and 2.5 h after the pulse. If there is a homogeneous storage compartment where both newly formed and preexisting secretory vesicles equilibrate, the specific activity of secreted and cellular insulin would always be identical. Thus, preferential secretion indicates that newly formed secretory vesicles equilibrate with only a percentage of the preexisting vesicles, which represent a readily secreted insulin compartment. Hypothetical curves predicting specific activity relationships for different sizes of readily secreted compartments are shown in figure 4a [37] and were drawn by computer analysis of steady-state equations describing rates of insulin synthesis, storage and

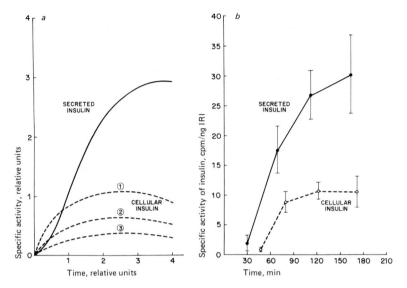

Fig. 4. Effect of time on the predicted *(a)* and experimentally determined *(b)* specific activity of secreted and cellular insulin. In figure 4a the glucose-labile compartment is 33% in No. 1, 20% in No. 2, 11% in No. 3. [Taken from ref. 37 with permission from the publisher.]

secretion. Results indicate that as much as 33% of the total islet insulin may be in a storage compartment readily secreted in response to glucose.

Regional localization of islets within either the dorsal or ventral pancreas has been reported to affect hormone storage and secretory characteristics [129]. Pulse-labeling experiments with islets specifically isolated from these regions of the pancreas show identical preferential secretion of newly synthesized insulin [37]. Furthermore, there is no correlation between the degree of preferential secretion and the amount of insulin recovered in the islets (a measure of islet size). Therefore, in islets continuously stimulated by glucose, preferential secretion of newly labeled insulin does not reflect regional or size differences.

Previous investigations indicated that newly synthesized insulin is secreted with a greater degree of preference when stimulatory concentrations of arginine or sulfonylureas are present during labeling [111]. With pulse-labeled islets in culture, the specific activity of insulin secreted in

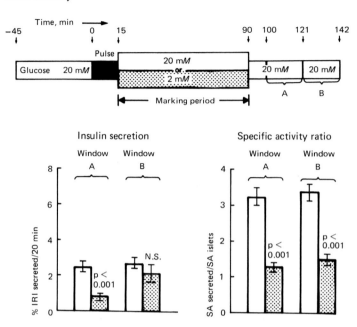

Fig. 5. The effect of glucose concentration during the marking period on the secretion rate of immunoreactive insulin (IRI) and the specific activity (SA) ratio between secreted and cellular insulin [taken from ref. 36 with permission from the publisher].

response to $8.3 \times 10^{-3} M$ glucose was greater than if a high concentration of glucose or 3-isobutyl-1-methylxanthine (IBMX) plus a nonstimulatory concentration of glucose was the secretagogue [47]. Thus, preferential secretion may be regulated. We recently accumulated evidence that preferential secretion of newly synthesized insulin may be regulated by glucose, but only during a specific period of time ('marking' period) directly after labeling. In figure 5 [36], two groups of islets were identically incubated and labeled with ^{3}H-leucine. After the pulse, one group of islets was maintained in high glucose and the other was switched to a nonstimulatory concentration. On subsequent stimulation of secretion, islets 'marked' with high glucose preferentially released the newer insulin, as indicated by a specific activity ratio (secreted vs. average cellular insulin) of more than 3. However, islets exposed to low glucose during the critical period after the pulse, but before the secretory event, released labeled and unlabeled insulin near-randomly, as indicated by a specific activity ratio near unity.

Table III. Effect of glucose concentration during the marking period on the rate of insulin secretion and the specific activity ratio between secreted and cellular insulin (taken from ref. 36 with permission from the publisher)

	Number of experiments	Unmarked in 2 mM glucose		Marked in 20 mM glucose	
		percentage secreted per 20 min	SA secreted ⎯⎯⎯⎯ SA in islets	percentage secreted per 20 min	SA secreted ⎯⎯⎯⎯ SA in islets
Glucose (20 mM) plus IBMX (1 mM)[1]	11	7.72±2.2	1.54±0.18	5.81±1.6	3.69±0.32*
Glucose (2 mM) plus potassium (50 mM)[1]	8	2.54±0.40	1.29±0.12	2.76±0.35	3.47±0.44*
Glucose (5 mM) plus tolbutamide (100 μg/ml)[1]	8	0.50±0.11	1.56±0.12	1.60±0.26**	3.25±0.42**
Glucose (5 mM)[2]	8	0.13±0.01	0.79±0.29	1.40±0.25*	2.60±0.33*

* $p<0.001$; ** $p<0.005$, compared to unmarked islets.
[1] Samples taken during the second observation period (window B). Because of the increased secretion rate, 20 mM glucose plus 1 mM IBMX was used in a 7-min rather than a 20-min window.
[2] Sample taken during the first observation period (window A).

Total proinsulin synthesis and conversion of labeled proinsulin to insulin progressed identically in both groups [36].

Once marked, the glucose concentration, secretagogue or secretion rate achieved during the subsequent short stimulatory period had little effect (in our hands) on the specific activity ratio, as shown in table III. Thus, preferential secretion of newly labeled insulin seems to be regulated by glucose at a critical time only. This time corresponds to the period that labeled material is transiting the Golgi apparatus, entering forming secretory vesicles and starting to be converted to insulin. Glucose during the marking period may alter vesicles forming in either the rough endoplasmic reticulum or the Golgi apparatus. Perhaps glucose regulates a sorting function of the Golgi apparatus or opens a novel secretory route as it increases the rate of secretory vesicle formation [22]. How the cell marks and identifies marked vesicles, and if marking plays a role in time-dependent modulation of insulin secretion, remains to be determined. Marking also may affect a variety of cell types that secrete nonrandomly in response to their physiologic regulators.

Internal Degradation of Insulin

The quantity of insulin stored in the B cell is affected by intracellular degradation in addition to biosynthesis and secretion. Explants of fetal rat pancreas, when transferred from an elevated to a nonstimulatory concentration of glucose, contained less insulin than secretion alone could account for; crinophagy was indicated after granulolytic lysosomes were observed to contain secretory vesicles and their contents [106]. Increased numbers of granulolytic lysosomes were also observed when the inhibitor of insulin secretion, diazoxide, was present [19]. Similarly, spiny mice, which store adequate amounts of insulin but have a defect in insulin secretion, have prevalent granulolytic lysosomes [117]. In experiments following the fate of a ^{3}H-leucine pulse, intracellular loss of labeled insulin by degradation is inversely correlated with glucose concentration. Islets cultured with high concentrations of glucose and actively secreting insulin degrade negligible amounts of labeled insulin [48] or proinsulin [37], whereas, at low concentrations of glucose, as much as 37% of the labeled insulin per day is degraded within the cell [48]. Thus, with nonstimulated or blocked B cells, degradation may cull an overproduction of insulin and secretory vesicles from the storage compartment, perhaps by crinophagy.

Effects of Islet Morphology on Insulin Secretion

Within the islet, cells containing insulin, glucagon and somatostatin are in close proximity to one another. Therefore, B cells may be regulated by high local concentrations of glucagon or somatostatin in the interstitial space (paracrine effect) or in the proximal islet vascular circulation (short-loop endocrine effect) [31, 91]. Glucagon stimulates insulin release by increasing the concentration of cyclic AMP in the B cell and, therefore, like cyclic AMP, works only in the presence of glucose. On the other hand, low concentrations of somatostatin inhibit insulin secretion, either by blocking the accumulation of cyclic AMP or interfering with normal cellular handling of potassium or calcium [133]. Evidence that local interactions by other islet hormones affect insulin secretion was obtained by: (1) correlation of insulin secretion to non-B-cell content of dorsal and ventral islets [129]; (2) effects of perfusion or perifusion of non-B-cell islet hormones on insulin secretion [30]; and (3) effects of antibodies against these islet hormones on insulin secretion [50]. Microscopic reconstruction of the vasculature of many rat islets indicates that glucagon and somatostatin are frequently secreted into blood, leaving (rather than entering) regions rich in B cells, thereby minimizing or eliminating short-loop endocrine effects [12].

Mechanism of Action of Glucose on Insulin Secretion

Beta-cells are regulated by numerous compounds; however, glucose is probably the most important because it is a major primary stimulant and also is required at low concentrations for many of the secondary regulators to be effective. Glucose regulation of insulin secretion will be discussed only briefly here because excellent reviews of the extensive literature are available [see ref. 5, 50]. Beta-cells have a uniquely high capacity for facilitated diffusion of glucose; thus, circulating and cytoplasmic glucose are essentially in equilibrium. A conventional receptor-ligand interaction has been postulated between a putative protein receptor and intact glucose, but most evidence indicates that secretion is caused by signals elucidated as a result of glucose metabolism. As the glucose concentration is raised, the rate of glucose metabolism increases in parallel with increases in the rate of insulin secretion [136]. Alpha-D-glucose is a more potent secretagogue than beta-D-glucose, and also is more readily metabolized by the B cell [78]. Mannose, N-acetyl-glucosamine, inosine, dihydroxyacetone, glyceraldehyde, fructose and galactose are metabolized at different rates by the B cell [51, 61], and the rate of metabolism of these compounds is proportional to their potency as insulin secretagogues [5, 44]. Nonmetabolized carbohydrates do not stimulate insulin secretion. Finally, inhibition of glucose or mannose phosphorylation with mannoheptulose renders these sugars ineffective as stimulators, but substrates, such as glyceraldehyde, dihydroxyacetone or inosine, which bypass the site of mannoheptulose inhibition, retain their ability to stimulate insulin secretion [5, 135]. No consensus of opinion exists about which step or steps of sugar metabolism are critical, but changes in cellular pH or the ratio of reduced to oxidized pyridine nucleotides may trigger secondary signals that regulate both insulin secretion and biosynthesis [50].

Role of Cytosolic Calcium as a Signal for Insulin Secretion

The requirement of calcium for regulated insulin secretion also has been studied extensively [for a recent comprehensive review see ref. 133]. Regardless of the type of secretory stimulant, sustained insulin release requires a minimum of $10^{-4} M$ extracellular calcium, and maximum insulin secretory rates require between 1 and $5 \times 10^{-3} M$ [45, 133]. Within minutes after stimulation, total calcium in the islet increases by approximately 50% from 4.7 to 6.9×10^{-12} mol [107]. Glucose increases calcium influx and inhibits calcium efflux; however, most of the increase of cellular calcium is due to a tripling of the influx rate [32, 131]. Influx, for the most part, probably occurs by diffusion through a voltage-sensitive calcium channel from the millimolar concentrations outside to the submicromolar concentrations of calcium in the cytosol. Rapid calcium influx by itself causes insulin secretion, and calcium influx and the resulting secretion of insulin can be induced by partially depolarizing the plasma membrane of the B cell with high concentrations of extracellular potassium. The important oral hypoglycemic agents, such as the sulfonylureas, elevate the insulin secretory response to glucose, at least in part, by facilitating depolarization and, thereby, increasing calcium influx. Calcium influx via the voltage-sensitive channel (and most of insulin secretion) can be blocked with drugs such as verapamil or elevated concentrations of extracellular magnesium [10, 132]. A small increase in cellular calcium also occurs by partial inhibition of external sodium/internal calcium exchange, which decreases the overall rate of calcium efflux [53].

Glucose also is thought to affect the relationship between ionized and sequestered calcium in the B cell. Almost all cellular calcium is sequestered. Ionized calcium has been estimated at approximately $10^{-7} M$, which represents only 0.03% of the total islet calcium [133]. This small pool of ionized calcium also has been estimated to turn over more than 20 times a second; thus, any agent that affects calcium handling can rapidly change the concentration of ionized calcium in the cytosol and signal for insulin secretion [133].

*Role of Electric and Morphologic Events at the
Plasma Membrane as Signals for Insulin Secretion*

Electrical events at the plasma membrane are part of the signal process, and correlations between electrical depolarizations and insulin secre-

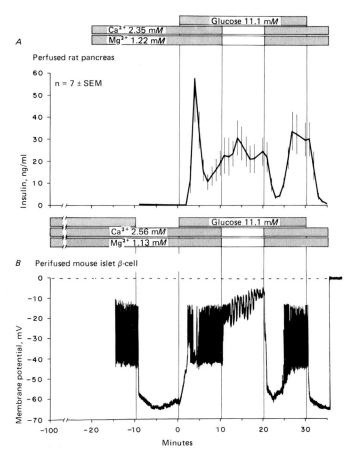

Fig. 6. Effects of rapid changes in the concentration of glucose, calcium, and magnesium on insulin secretion *(A)* and membrane potentials *(B)* [taken from ref. 30 with permission from the publisher].

tion have been reviewed [7, 86]. In resting B cells incubated with nonstimulatory concentrations of glucose, the plasma membrane maintains a -65×10^{-3} V potential (see min −10 to 0 in figure 6b [30]. This potential is maintained by a continuous efflux of potassium [85]. There are two stages in depolarization: at concentrations below the secretory threshold, a slow depolarization of approximately 15×10^{-3} V occurs, which is independent of calcium influx and is associated with inhibition of potassium efflux by glucose [133]; a further increase in the glucose or calcium concentration results in calcium influx through the voltage-sensitive calcium channels and further rapid depolarization, as seen in figure 6b, and insulin secretion, as seen in figure 6a. Concentrations of glucose that are insufficient to stimulate secretion, but cause partial depolarization, also stimulate islet ATP production and insulin synthesis and potentiate the glucose-dependent secretagogues, such as cyclic AMP, gut hormones or amino acids. Therefore, all secondary secretagogues may require low concentrations of glucose to inhibit potassium efflux and initiate depolarization of the plasma membrane.

As a result of the depolarized state, a series of action potentials or spikes is produced. Both the frequency of this spiking and the duration of the period of depolarization increase with increasing concentrations of glucose [8, 86]. Previously, it was uncertain if spikes were signals for insulin secretion or the result of emiocytotic events themselves. With studies at 25 °C, however, action potentials were still induced, but glucose-stimulated insulin secretion was blocked [*Atwater,* personal commun.] Thus, action potentials are early signals or regulators of insulin secretion in the B cell.

Insulin secretion also may be regulated by structural alterations of the plasma membrane. Gap junctions have been identified between B cells and between B and A cells [94]. Injection of fluorescent dyes indicated that dye passes from one cell to another over several cellular diameters without entering the extracellular space [87]. Thus, small molecules, such as those that signal for insulin secretion, may pass from the cytoplasm of one cell to the cytoplasm of the next. Glucose stimulation of insulin secretion is accompanied by an increase in the number of gap junctions between B cells [84]. However, elevated insulin secretion is not required to increase the number of gap junctions because the number increases after exposure to tetraethylammonium plus a low concentration of glucose, which together do not elevate insulin secretory rates [115]. Thus, the number or location of gap junctions may affect insulin secretion, but gap junctions do not increase in parallel with insulin secretory rates.

Role of Cyclic AMP in Insulin Secretion

Elevated concentrations of islet cyclic AMP are also signals for insulin secretion. Unlike calcium, cyclic AMP cannot by itself cause insulin release; a stimulatory concentration of glucose also is required. Cyclic AMP greatly enhances the magnitude of the secretory response to glucose alone. Enhancement is particularly effective at maximal glucose concentrations, suggesting that cyclic AMP affects the V_{max} of the secretory process rather than only the K_m for glucose [16]. Glucose raises the cellular content of cyclic AMP, but extracellular calcium is also required [18]. Thus, calcium influx, rather than glucose per se, probably causes the rise in cyclic AMP. Several dependent insulin secretagogues that require only a threshold concentration of glucose are thought to act by increasing cellular cyclic AMP. These include exogenous theophylline, IBMX, beta-adrenergic agonists, glucagon, secretin, gut hormones, cholera toxin and cyclic AMP and its derivatives. Mechanistically, the actions of cyclic AMP and calcium are thought to be distinct, but both signals are interrelated. Cyclic AMP increases ionized calcium in the cytosol probably by mobilizing internal stores rather than by affecting calcium influx and efflux, and elevated cellular calcium increases the rate of cyclic AMP formation.

Role of Calmodulin in Insulin Secretion

A calmodulin-like activity has been identified at concentrations between 36 and $50 \times 10^{-3} M$ in normal islets, and calmodulin has been isolated from abundant insulinoma tissue [133]. This ubiquitous calcium-binding protein affects many cellular processes that can be inhibited by antipsychotic drugs, such as trifluoperazine and its analogs, which bind to the calmodulin-calcium complex [130]. Insulin secretion is inhibited at the same concentrations of trifluoperazine that inhibit a calmodulin-dependent phosphodiesterase in vitro; thus, a calmodulin-dependent process probably is one of the mechanisms whereby calcium signals for insulin release [127]. Elevated glucose increases the amount of calcium-activated calmodulin without increasing the actual amount of calmodulin itself [17].

Both ionized cytosolic calcium via calmodulin and cyclic AMP independently may signal for insulin release by protein phosphorylation, as shown diagrammatically in figure 7. The universal receptor for cyclic AMP is cyclic AMP-dependent protein kinase(s), and the calcium-calmodulin complex activates similar, but probably different, protein kinases. Several endogenous islet proteins are phosphorylated with endogenous kinases and, thus, are labeled when islet homogenates are incubated with gamma-^{32}P-ATP [122]; the addition of exogenous calmodulin increases the rate of phosphorylation [33]. With intact islets incubated with ^{32}P, glucose-stimulated insulin secretion is accompanied by enhanced phos-

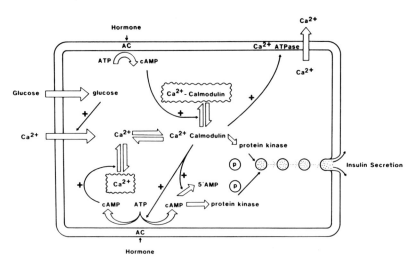

Fig. 7. The possible biochemical events linking secretion of insulin to regulation by calcium, cyclic AMP, glucose, and other secretagogues [taken from ref. 127 with permission from the publisher].

phorylation of six proteins ranging in molecular weight from 15,000 to 138,000 daltons [126]. With intact hamster insulinoma cells, phosphorylations of 16,000 and 28,000 dalton proteins are specifically stimulated by increasing cellular cyclic AMP; a 60,000-dalton protein is specifically phosphorylated after induction of calcium influx [113]. This latter phosphorylation is blocked by trifluoperazine, and is probably mediated by calmodulin. Some of the phosphorylated proteins may be structural or contractile proteins, and recently a calmodulin-dependent myosin light chain kinase activity has been demonstrated in islets [83]. The scheme in figure 7 assumes that the signalling actions of glucose, calcium and cyclic AMP are ultimately mediated by kinases, and, although not proven, provides an attractive and plausible mechanism for regulation of the secretory process in the B cell.

References

1   Abrahamsson, H.; Gylfe, E.: Demonstration of a proton gradient across the insulin granule membrane. Acta physiol. scand. *109:* 113–114 (1980).
2   Albert, S.; Chyn, R.; Goldford, M.; Permutt, A.: Insulin biosynthesis. Evidence for the existence of a precursor to proinsulin in cells. Diabetes *26:* suppl. 1, p. 378 (1977).

3   Albisser, A. M.; Leibel, B. S.; Ewart, T. G.; Davidovac, Z.; Botz, C. K.; Zingg, W.: An artificial endocrine pancreas. Diabetes 23: 389–396 (1974).

4   Ashby, J. P.; Shirling, D.: The priming effect of glucose on insulin secretion from isolated islets of Langerhans. Diabetologia 21: 230–234 (1981).

5   Ashcroft, S. J.: Glucoreceptor mechanisms and the control of insulin release and biosynthesis. Diabetologia 18: 5–15 (1980).

6   Ashcroft, S. J. H.; Bunce, J.; Lawry, M.; Hansen, S. E.; Hederkov, C. J.: The effect of sugars on (pro)insulin biosynthesis. Biochem. J. 174: 517–526 (1978).

7   Atwater, I.; Beigelman, P. M.: Dynamic characteristics of electrical activity in pancreatic B-cells. I. Effects of calcium and magnesium removal. J. Physiol., Paris 72: 769–786 (1976).

8   Beigelman, P. M.; Ribalet, B.; Atwater, I.: Electric activity of mouse pancreatic beta-cells. II. Effects of glucose and arginine. J. Physiol., Paris 73: 201–217 (1977).

9   Bell, G. I.; Karam, J. H.; Rutter, W. J.: Polymorphic DNA region adjacent to the 5' end of the human insulin gene. Proc. natn. Acad. Sci. USA 78: 5759–5763 (1981).

10   Bennett, L. L.; Curry, D. L.; Grodsky, G. M.: Calcium-magnesium antagonism in insulin secretion by the perfused rat pancreas. Endocrinology 85: 594–596 (1969).

11   Blobel, G.; Dobberstein, B.: Transfer of protein across membranes. I. Presence of proteolytically processed and unprocessed nascent immunoglobulin light chains on membrane-bound ribosomes of murine myeloma. J. Cell Biol. 67: 835–851 (1975).

12   Bonner-Weir, S.; Orci, L.: New perspectives on the microvasculature of the islets of Langerhans in the rat. Diabetes 31: 883–889 (1982).

13   Boquist, L.; Lernmark, A.: Effects on the endocrine pancreas in Chinese hamsters fed zinc deficient diets. Acta path. microbiol. scand. 76: 215–228 (1969).

14   Borowitz, J. L.; Mathews, E. K.: Calcium exchangeability in subcellular fractions of pancreatic islet cells. J. Cell Sci. 41: 233–243 (1980).

15   Boyd, A. E., III; Bolton, W. E.; Brinkley, B. R.: Microtubules and beta cell function. Effect of colchicine on microtubules and insulin secretion in vitro by mouse beta cells. J. Cell Biol. 92: 425–434 (1982).

16   Brisson, G. R.; Malaisse-Lagae, F.; Malaisse, W. J.: The stimulus-secretion coupling of glucose-induced insulin release. VII. A proposed site of action for adenosine-3',5'-cyclic monophosphate. J. clin. Invest. 51: 232–241 (1972).

17   Chafouleas, J. G.; Dedman, J. R.; Munjaal, R. P.; Means, A. R.: Calmodulin. Development and application of a sensitive radioimmunoassay. J. biol. Chem. 254: 10262–10267 (1979).

18   Charles, M. A.; Lawecki, I.; Pictet, R.; Grodsky, G. M.: Insulin secretion. Interrelationships of glucose, cyclic adenosine 3:5-monophosphate, and calcium. J. biol. Chem. 250: 6134–6140 (1975).

19   Creutzfeldt, W.; Creutzfeldt, C.; Frericks, H.; Perings, E.; Sickinger, K.: The morphological substrate of the inhibition of insulin secretion by diazoxide. Hormone metabol. Res. 1: 53–64 (1969).

20   Creutzfeldt, C.; Track, N. S.; Creutzfeldt, W.: In vitro studies of the rate of proinsulin and insulin turnover in seven human insulinomas. Eur. J. clin. Invest. 3: 371–384 (1973).

21   Dean, P. M.: Ultrastructural morphometry of the pancreatic B-cell. Diabetologia 9: 115–119 (1973).

22 Dean, P.M.: The kinetics of beta-granule formation. A morphometric study. Diabetologia *12:* 111–114 (1976).

23 Dodson, E.J.; Dodson, G.G.; Hodgkin, D.C.; Reynolds, C.D.: Structural relationships in the two-zinc-insulin hesamer. Can. J. Biochem. *57:* 469–479 (1979).

24 Dodson, E.J.; Dodson, G.G.; Lewitova, A.; Sabesan, M.: Zinc-free cubic pig insulin. Crystallization and structure determination. J. molec. Biol. *125:* 387–396 (1978).

25 Duran Garcia, S.; Jarrousse, C.; Rosselin, G.: Biosynthesis of proinsulin and insulin in newborn rat pancreas. Interaction of glucose, cyclic AMP, somatostatin, and sulfonylureas on the [^{3}H] leucine incorporation into immunoreactive insulin. J. clin. Invest. *57:* 230–243 (1976).

26 Emdin, S.O.; Dodson, G.G.; Cutfield, J.M.; Cutfield, S.M.: Role of zinc in insulin biosynthesis. Some possible zinc-insulin interactions in the pancreatic B-cell. Diabetologia *19:* 174–182 (1980).

27 Falkmer, S.; Pihl, E.: Structural lability of zinc-containing secretion granules of pancreatic B-cells after exposure to hydrogen sulphide. Diabetologia *4:* 239–242 (1968).

28 Figlewicz, D.P.; Formby, B.; Hodgson, A.T.; Schmid, F.G.; Grodsky, G.M.: Kinetics of ^{65}zinc uptake and distribution in fractions from cultured rat islets of Langerhans. Diabetes *19:* 767–773 (1980).

29 Figlewicz, D.P.; Formby, B.; Hodgson, A.T.; Schmid, F.G.; Grodsky, G.M.: Uptake and distribution of ^{65}zinc in cultured rat islets; in Proc. 10th Congr. Int. Diabetes Federation, Vienna 1979, pp.146–153. Excerpta medica International Congress Series No. 500 (Excerpta medica, Amsterdam 1980).

30 Frankel, B.J.; Atwater, I.; Grodsky, G.M.: Calcium affects insulin release and membrane potential in islet B-cells. Am. J. Physiol. *240:* C64–C72 (1981).

31 Frankel, B.J.; Heldt, A.M.; Grodsky, G.M.: Effects of K$^+$ and arginine on insulin, glucagon, and somatostatin release from the in vitro perfused rat pancreas. Endocrinology *110:* 428–431 (1982).

32 Frankel, B.J.; Kromhout, J.A.; Imagawa, W.; Landahl, H.D.; Grodsky, G.M.: Glucose-stimulated ^{45}Ca uptake in isolated rat islets. Diabetes *27:* 365–369 (1978).

33 Gagliardino, J.J.; Harrison, D.E.; Christie, M.R.; Gagliardino, E.E.; Ashcroft, S.J.H.: Evidence for participation of calmodulin in stimulus-secretion coupling in the pancreatic B-cell. Biochem. J. *192:* 919–927 (1980).

34 Giddings, S.J.; Chirgwin, J.; Permutt, M.A.: Effects of glucose on proinsulin messenger RNA in rats in vivo. Diabetes *31:* 624–629 (1982).

35 Giddings, S.J.; Chirgwin, J.; Permutt, M.A.: The effect of fasting and feeding on preproinsulin messenger RNA in rats. J. clin. Invest. *67:* 952–960 (1981).

36 Gold, G.; Gishizky, M.L.; Grodsky, G.M.: Evidence that glucose 'marks' B cells resulting in preferential release of newly synthesized insulin. Science *218:* 56–58 (1982).

37 Gold, G.; Landahl, H.D.; Gishizky, M.L.; Grodsky, G.M.: Heterogeneity and compartmental properties of insulin storage and secretion in rat islets. J. clin. Invest. *69:* 554–563 (1982).

38 Gold, G.; Reaven, G.M.; Reaven, E.P.: Effect of age on proinsulin and insulin secretory patterns in isolated rat islets. Diabetes *30:* 77–82 (1981).

39   Goldman, I.; Carpenter, F. H.: Zinc binding, circular dichroism, and equilibrium sedimentation studies on insulin (bovine) and several of its derivatives. Biochemistry *13:* 4566–4574 (1974).

40   Grant, P. T.; Coombs, T. L.; Frank, B. H.: Differences in the nature of the interaction of insulin and proinsulin with zinc. Biochem. J. *126:* 433–440 (1972).

41   Greider, M. H.; Howell, S. L.; Lacy, P. E.: Isolation and properties of secretory granules from rat islets of Langerhans. II. Ultrastructure of the beta cell granule. J. Cell Biol. *41:* 162–166 (1969).

42   Grill, V.; Adamson, U.; Cerasi, E.: Immediate and time-dependent effects of glucose on insulin release from rat pancreatic tissue. Evidence for different mechanisms of action. J. clin. Invest. *61:* 1034–1043 (1978).

43   Grodsky, G. M.: Chemistry and function of the hormones. Thyroid, pancreas, adrenal and gastrointestinal tract; in Harper, Rudwell, Mayes, Review of physiological chemistry, pp. 511–555 (Lange, Los Altos 1979).

44   Grodsky, G. M.; Batts, A. A.; Bennett, L. L.; Vcella, C.; McWilliams, N. B.; Smith, D. F.: Effects of carbohydrates on secretion of insulin from isolated rat pancreas. Am. J. Physiol. *205:* 638–644 (1963).

45   Grodsky, G. M.; Bennett, L. L.: Cation requirements for insulin secretion in the isolated perfused pancreas. Diabetes *15:* 910–913 (1966).

46   Gutman, R. A.; Fink, G.; Shapiro, J. R.; Selawry, H.; Recant, L.: Proinsulin and insulin release with a human insulinoma and adjacent nonadenomatous pancreas. J. clin. Endocr. Metab. *36:* 978–987 (1973).

47   Halban, P.: Differential rates of release of newly synthesized and of stored insulin from pancreatic islets. Endocrinology *110:* 1183–1188 (1982).

48   Halban, P. A.; Wollheim, C. B.: Intracellular degradation of insulin stored by rat pancreatic islets in vitro. An alternative pathway for homeostasis of pancreatic insulin content. J. biol. Chem. *255:* 6003–6006 (1980).

49   Havu, N.; Lundgren, G.; Falkmer, S.: Zinc and manganese contents for micro-dissected pancreatic islets of some rodents. A microchemical study in adult and newborn guinea pigs, rats, Chinese hamsters and spiny mice. Acta endocr., Copenh. *86:* 570–577 (1977).

50   Hedeskov, C. J.: Mechanism of glucose-induced insulin secretion. Physiol. Rev. *60:* 442–509 (1980).

51   Hellman, B.; Idahl, L. A.; Lernmark, A.; Sehlin, J.; Taljedal, I. B.: The pancreatic B-cell recognition of insulin secretagogues. Comparisons of glucose with glyceraldehyde isomers and dihydroxyacetone. Archs biochem. Biophys. *162:* 448–457 (1974).

52   Hellman, B.; Sehlin, J.; Taljedal, I. B.: The intracellular pH of mammalian pancreatic B-cells. Endocrinology *90:* 335–337 (1972).

53   Herchuelz, A.; Sener, A.; Malaisse, W. J.: Regulation of calcium fluxes in rat pancreatic islets. Calcium extrusion by sodium-calcium countertransport. J. membr. Biol. *15/57:* 1–12 (1980).

54   Howell, S. L.: In Ceccarelli, Clementi, Meldolesi, Advances in cytopharmacology, vol. 2, pp. 319–327 (Raven Press, New York 1974).

55   Howell, S. L.; Kostianovsky, M.; Lacy, P. E.: Beta granule formation in isolated islets of Langerhans. A study by electron microscopic radioautography. J. Cell Biol. *42:* 695–705 (1969).

56 Howell, S. L.; Parry, D. G.; Taylor, K. W.: Secretion of newly synthesized insulin in vitro. Nature, Lond. *208:* 487 (1965).

57 Howell, S. L.; Tyhurst, M.; Duvefeldt, H.; Andersson, A.; Hellerstrom, C.: Role of zinc and calcium in the formation and storage of insulin in the pancreatic beta-cell. Cell Tiss. Res. *188:* 107–118 (1978).

58 Itoh, N.; Okamoto, H.: Translational control of proinsulin synthesis by glucose. Nature, Lond. *283:* 100–102 (1980).

59 Itoh, N.; Sei, T.; Nose, K.; Okamoto, H.: Glucose stimulation of the proinsulin synthesis in isolated pancreatic islets without increasing amounts of proinsulin in RNA. FEBS Lett. *93:* 343–347 (1978).

60 Jackson, R. L.: In Holt, Lefebvre, Butterfield, Vallance-Owen, International Congress Series No.280, pp.225–226 (Excerpta medica, Amsterdam 1973).

61 Jarrett, R. J.; Keen, H.: Oxidation of sugars, other than glucose, by isolated mammalian islets of Langerhans. Metabolism *17:* 155–157 (1968).

62 Kaelin, D.; Renold, A. E.; Sharp, G. W. G.: Glucose stimulated proinsulin biosynthesis. Rates of turn off after cessation of stimulus. Diabetologia *14:* 329–335 (1978).

63 Kakita, K.; O'Connel, K.; Giddings, S.; Permutt, M. A.: Biosynthesis of insulins I and II in rodents. Evidence for differential expression of the two genes. Diabetes *31:* suppl. 2, p. 7a (1982).

64 Kanazawa, Y.; Kawaza, S.; Ikeuchi, M.; Kosaka, K.: The relationship of intracytoplasmic movement of beta granules to insulin release in monolayer-cultured pancreatic beta-cells. Diabetes *29:* 953–959 (1980).

65 Kemmler, W.; Steiner, D. F.; Borg, J.: Studies on the conversion of proinsulin to insulin. III. Studies in vitro with a crude secretion granule fraction isolated from rat islets of Langerhans. J. biol. Chem. *248:* 4544–4551 (1973).

66 Lacy, P. E.; Howell, S. L.; Young, D. A.; Fink, C. J.: New hypothesis of insulin secretion. Nature, Lond. *219:* 1177–1179 (1968).

67 Landahl, H. D.; Grodsky, G. M.: Comparison of models of insulin release. Bull. Math. Biol. *44:* 399–409 (1982).

68 Lee, J. C.; Grodsky, G. M.; Bennett, L. L.; Smith-Kyle, D. F.; Craw, L.: Ultrastructure of B-cells during the dynamic response to glucose and tolbutamide in vitro. Diabetologia *6:* 542–549 (1980).

69 Liljenquist, J. E.; Horwitz, D. L.; Jennings, A. S.; Chaisson, J. L.; Keller, U.; Rubenstein, A. H.: Inhibition of insulin secretion by exogenous insulin in normal man as demonstrated by C-peptide assay. Diabetes *27:* 563–570 (1978).

70 Lin, B. J.; Haist, R. E.: Effect of some modifiers of insulin secretion on insulin biosynthesis. Endocrinology *92:* 735–742 (1973).

71 Lin, B. J.; Haist, R. E.: Insulin biosynthesis. The monoaminergic mechanisms and specificity of 'glucoreceptor'. Endocrinology *96:* 1247–1253 (1975).

72 Lin, B. J.; Nagy, B. R.; Haist, R. E.: Effect of various concentrations of glucose on insulin biosynthesis. Endocrinology *91:* 309–311 (1972).

73 Lomedico, P. T.; Chan, S. J.; Steiner, D. F.; Saunders, G. F.: Immunological and chemical characterization of bovine preproinsulin. J. biol. Chem. *252:* 7971–7978 (1977).

74 Ludvigsen, C.; McDaniel, M.; Lacy, P. E.: The mechanism of zinc uptake in isolated islets of Langerhans. Diabetes *28:* 570–576 (1979).

75   Malaisse, W. J.; Hutton, J. C.; Kawazu, L.; Herchuelz, A.; Valverde, I.; Sener, A.: The stimulus-secretion coupling of glucose-induced insulin release. XXXV. The links between metabolic and cationic events. Diabetologia *16:* 331–341 (1979).

76   Malaisse, W. J.; Leclercq-Mayer, V.; Van Obberghan, E.; Somers, G.; Devis, G.; Ravazzola, M.; Malaisse-Lagae, F.; Orci, L.: The role of the microtubular-microfilamentous system in insulin and glucagon release by the endocrine pancreas; in Borgers, de Brabander, Microtubules and microtubule inhibitors, pp. 143–152 (North-Holland, Amsterdam 1975).

77   Malaisse, W. J.; Malaisse-Lagae, F.; Van Obberghen, E.; Somers, G.; Devis, G.; Ravazzola, M.; Orci, L.: Role of microtubules in the phasic pattern of insulin release. Ann. N.Y. Acad. Sci. *253:* 630–652 (1975).

78   Malaisse, W. J.; Sener, A.; Koser, M.; Herchuelz, A.: Stimulus secretion coupling of glucose-induced insulin release. Metabolism of α- and β-D-glucose in isolated islets. J. biol. Chem. *251:* 5936–5943 (1976).

79   Malaisse-Lagae, F.; Amherdt, M.; Ravazzola, M.; Sener, A.; Hutton, J. C.; Orci, L.; Malaisse, W. J.: Role of microtubules in the synthesis, conversion and release of (pro)insulin. A biochemical and radioautographic study in rat islets. J. clin. Invest. *63:* 1284–1296 (1979).

80   Maldonato, A.; Renold, A. E.; Sharp, G. W. G.; Cerasi, E.: Glucose-induced proinsulin biosynthesis. Role of islet cyclic AMP. Diabetes *26:* 538–545 (1977).

81   Maldonato, A.; Truehart, P. A.; Renold, A. E.; Sharp, G. W. G.: Effect of streptozotocin in vitro on proinsulin biosynthesis, insulin release and ATP content of isolated rat islets of Langerhans. Diabetologia *12:* 471–481 (1976).

82   Maske, H.: Interaction between insulin and zinc in the islets of Langerhans. Diabetes *6:* 335–341 (1957).

83   MacDonald, M. J.; Kowluru, A.: Calcium-calmodulin-dependent myosin phosphorylation by pancreatic islets. Diabetes *31:* 566–570 (1982).

84   Meda, P.; Perrelet, A.; Orci, L.: Increase of gap junctions between pancreatic B-cells during stimulation of insulin secretion. J. Cell Biol. *82:* 441–448 (1979).

85   Meissner, H. P.; Henquin, J. C.; Preissler, M.: Potassium dependence of the membrane potential of pancreatic B-cells. FEBS Lett. *94:* 87–89 (1978).

86   Meissner, H. P.; Preissler, M.: Treatment of early diabetes; in Camerini-Davalos, Hanover, Glucose-induced changes of the membrane potential of pancreatic B cells. Their significance for the regulation of insulin release, p. 97107 (Plenum Press, New York 1979).

87   Michaels, R. L.; Sheridan, J. D.: Islets of Langerhans dye coupling among immunocytochemically distinct cell types. Science *214:* 801–803 (1981).

88   Monotague, W.; Howell, S. L.; Green, I. C.: Insulin release and the microtubular system of the islets of Langerhans. Effect of insulin secretagogues on microtubule subunit pool size. Hormone metabol. Res. *8:* 166–169 (1976).

89   Morris, G. E.; Korner, A.: The effect of glucose on insulin biosynthesis by isolated islets of Langerhans of the rat. Biochim. biophys. Acta *208:* 404–413 (1970).

90   O'Connor, M. D. L.; Landahl, H.; Grodsky, G. M.: Comparison of storage- and signal-limited models of pancreatic insulin secretion. Am. J. Physiol. *238:* R378–R389 (1980).

91   Orci, L.: Macro- and micro-domains in the endocrine pancreas. Diabetes *31:* 538–565 (1980).

92 Orci, L.; Lambert, A.E.; Kanazawa, T.; Amherdt, M.; Rouiller, C.; Renold, A.E.: Morphological and biochemical studies of B cells of fetal rat endocrine pancreas in organ culture. Evidence for (pro)insulin biosynthesis. J. Cell Biol. *50:* 565–582 (1971).

93 Orci, L.; Perrelet, A.; Friend, D.S.: Freeze-fracture of membrane fusions during exocytosis in pancreatic B-cells. J. Cell Biol. *75:* 23–30 (1977).

94 Orci, L.; Unger, R.H.; Renold, A.E.: Structural coupling between pancreatic islet cells. Experientia *29:* 1015–1018 (1973).

95 Owerbach, D.; Bell, G.I.; Rutter, W.J.; Brown, J.A.; Shows, T.B.: The insulin gene is located on the short arm of chromosome 11 in humans. Diabetes *30:* 267–270 (1981).

96 Owerbach, D.; Nerup, J.: Restriction fragment length polymorphism of the insulin gene in diabetes mellitus. Diabetes *31:* 275–277 (1982).

97 Owerbach, D.; Poulson, S.; Billesbolle, P.; Nerup, J.: DNA insertion sequences near the insulin gene affect glucose regulation. Lancet *i:* 880–882 (1982).

98 Palade, G.: Intracellular aspects of the process of protein synthesis. Science *189:* 347–358 (1975).

99 Patzelt, C.; Chan, S.J.; Duguid, J.; Hortin, G.; Keim, P.; Heinsikson, R.L.; Steiner, D.F.: In Magnusson, Ottesen, Foltmann, Dano, Neurath, Regulating proteolytic enzymes and their inhibitors, vol. 47, Symp. A6, pp. 69–78 (Pergamon Press, Oxford 1978).

100 Permutt, M.A.: Insulin biosynthesis. IV. Effect of glucose on initiation and elongation rates in isolated rat pancreatic islets. J. biol. Chem. *249:* 2738–2742 (1974).

101 Permutt, M.A.; Kipnis, D.M.: Insulin biosynthesis. I. On the mechanism of glucose stimulation. J. biol. Chem. *247:* 1194–1199 (1972).

102 Permutt, M.A.; Kipnis, D.M.: Insulin biosynthesis. Studies of islet polyribosomes (nascent peptides-sucrose gradient analysis-gel filtration). Proc. natn. Acad. Sci. USA *69:* 505–509 (1972).

103 Pipeleers, D.G.; Mariehal, M.; Malaisse, W.J.: The stimulus-secretion coupling of glucose-induced insulin release. Glucose regulation of insulin biosynthesis activity. Endocrinology *93:* 1001–1011 (1973).

104 Pfeiffer, E.F.; Thum, C.H.; Clemens, A.H.: The artificial beta cell. A continuous control of blood sugar by external regulation of insulin infusion (glucose controlled insulin infusion system). Hormone metabol. Res. *487:* 339–342 (1974).

105 Reaven, E.P.; Gold, G.; Reaven, G.M.: Effect of age on glucose-stimulated insulin release by the B cell. J. clin. Invest. *64:* 591–599 (1979).

106 Renold, A.E.: The beta cell and its responses: summarizing remarks and some contributions from Geneva. Diabetes *21:* suppl. 2, pp. 619–631 (1972).

107 Ribes, G.; Siegel, E.G.; Wollheim, C.B.; Renold, A.E.; Sharp, G.W.: Rapid changes in calcium content of rat pancreatic islets in response to glucose. Diabetes *30:* 52–55 (1981).

108 Rotwein, P.; Chyn, R.; Chirgwin, J.; Cordell, B.; Goodman, H.M.; Permutt, M.A.: Polymorphism in the 5′ flanking region of the human insulin gene and its possible relation to type 2 diabetes. Science *213:* 1117–1120 (1981).

109 Sando, H.; Borg, H.; Steiner, D.F.: Studies on the secretion of newly synthesized proinsulin and insulin from isolated rat islets of Langerhans. J. clin. Invest. *51:* 1476–1485 (1972).

110 Sando, H.; Grodsky, G.M.: Dynamic synthesis and release of insulin and proinsulin from perifused islets. Diabetes *22:* 354–360 (1973).

111 Schatz, H.; Nierle, C.; Pfeiffer, E. F.: (Pro)insulin biosynthesis and release of newly synthesized (pro)insulin from isolated islets of rat pancreas in the presence of amino acids and sulphonylurea. Eur. J. clin. Invest. 51: 477–485 (1975).

112 Schlichtkrull, K.: Insulin crystals. I. The minimum mole-fraction of metal in insulin crystals prepared with Zn^{++}, Ca^{++}, Co^{++}, Ni^{++}, Mn^{++}, or Fe^{++}. Acta chem. scand. 10: 1455–1458 (1956).

113 Schubart, U. R.; Erlichman, J.; Fleischer, N.: Insulin release and protein phosphorylation. Possible role of calmodulin. Fed. Proc. 41: 2278–2282 (1982).

114 Shah, J. H.; Stevens, B.; Sorensen, B. J.: Dissociation of the effects of vincristine on stimulated insulin release and the pancreatic B-cell microtubular structures in the intact rat. Diabetes 30: 539–544 (1981).

115 Sheppard, M. S.; Meda, P.: Tetraethylammonium modifies gap junctions between pancreatic B-cell. Am. J. Physiol. 240: C116–120 (1981).

116 Stagner, J. I.; Samols, E.; Weir, G. C.: Sustained oscillations of insulin, glucagon, and somatostatin from the isolated canine pancreas during exposure to a constant glucose concentration. J. clin. Invest. 65: 939–942 (1980).

117 Stauffacher, W.; Orci, L.; Cameron, D. P.; Burr, J. M.; Renold, A. E.: Spontaneous hyperglycemia and/or obesity in laboratory rodents: an example of the possible usefulness of animal disease models with both genetic and environmental components. Recent Prog. Horm. Res. 27: 41–95 (1971).

118 Steiner, D. F.; Clark, J. L.; Nolan, C.; et al.: The pathogenesis of diabetes mellitus; in Cerasi, Luft, Proc 13th Nobel Symposium, p.123 (Almquist & Wiksell, Stockholm 1970).

119 Steiner, D. F.; Cunningham, D.; Spigelman, L.; Aten, B.: Insulin biosynthesis. Evidence of a precursor. Science 157: 697–700 (1967).

120 Steiner, D. F.; Kemmler, W.; Clark, J. L.; Oyer, P. E.; Rubenstein, A. H.: The biosynthesis of insulin; in Steiner, Freinkel, Handbook of physiology, section 7: Endocrinology, vol. I, pp.175–198 (Waverly, Baltimore 1972).

121 Steiner, D. F.; Kemmler, W.; Tager, H. S.; Peterson, J. D.: Proteolytic processing in the biosynthesis of insulin and other proteins. Fed. Proc. 33: 2105–2115 (1974).

122 Sugden, M. C.; Christie, M. R.; Ashcroft, S. J. H.: Presence and possible role of calcium-dependent regulator (calmodulin) in rat islets of Langerhans. FEBS Lett. 105: 95–100 (1979).

123 Suprenant, K. S.; Dentler, W. L.: Association between endocrine pancreatic secretory granules and in-vitro-assembled microtubules is dependent upon microtubule-associated proteins. J. Cell Biol. 93: 164–174 (1982).

124 Sussman, K. E.; Leitner, J. W.: In Bajaj, Insulin and metabolism, pp. 39–56 (Elsevier, Amsterdam 1977).

125 Sussman, K. E.; Mehler, P. S.; Leitner, J. W.; Draznin, B.: Role of the secretion vesicle in the transport of receptors. Endocrinology 111: 316–323 (1982).

126 Suzuki, S.; Hiroshi, O.; Yasuda, H.; Ikeda, M.; Cheng, P. Y.; Oda, T.: Effect of glucose on protein phosphorylation in rat pancreatic islets. Biochem. biophys. Res. Commun. 99: 987–993 (1981).

127 Tomlinson, S.; Walker, S. W.; Brown, B. L.: Calmodulin and insulin secretion. Diabetologia 22: 1–5 (1982).

128 Track, N. S.: Insulin biosynthesis; in Bajaj, Insulin and metabolism, pp. 13–38 (Excerpta Medica, Amsterdam 1977).

129 Trimble, E.R.; Renold, A.E.: Ventral and dorsal areas of rat pancreas. Islet hormone content and secretion. Am. J. Physiol. *240:* E422–427 (1981).

130 Wolff, D.J.; Brostrom, C.O.: Properties and functions of the calcium dependent regulator protein. Adv. cyclic Nucl. Res. *11:* 27–88 (1979).

131 Wollheim, C.B.; Kikuchi, M.; Renold, A.E.; Sharp, G.W.G.: Somatostatin and epinephrine-induced modifications of $^{45}Ca^{++}$ fluxes in insulin release in rat pancreatic islets maintained in tissue culture. J. clin. Invest. *60:* 1165–1173 (1977).

132 Wollheim, C.B.; Kikuchi, B.M.; Renold, A.E.; Sharp, G.W.G.: The role of intracellular and extracellular Ca^{++} in glucose-stimulated biphasic insulin release by rat islets. J. clin. Invest. *62:* 451–458 (1978).

133 Wollheim, C.B.; Sharp, G.W.G.: Regulation of insulin release by calcium. Physiol. Res. *61:* 914–973 (1981).

134 Wolters, G.H.J.; Pasma, A.; Konijnendijk, W.; Brom, G.: Calcium, zinc and other elements in islet and exocrine tissue of the rat pancreas as measured by histochemical methods and electron-probe micro-analysis. Effects of fasting and tolbutamide. Histochemistry *62:* 1–17 (1979).

135 Zawalich, W.S.; Dye, E.S.; Rognstad, R.; Matschinsky, F.M.: On the biochemical nature of triose- and hexose-stimulated insulin secretion. Endocrinology *103:* 2027–2034 (1978).

136 Zawalich, W.S.; Matschinsky, F.M.: Sequential analysis of the releasing and fuel function of glucose in isolated perifused pancreatic islets. Endocrinology *100:* 1–8 (1977).

G. Gold, PhD, Research Biochemist, Metabolic Research Unit, University of California, San Francisco, CA 94143 (USA)

Cell Biology of the Secretory Process, pp. 389–422 (Karger, Basel 1984)

Actions of Cholecystokinin and Insulin on the Acinar Pancreas

Functional and Morphological Studies of Hormone Receptors Regulating Secretion and Metabolism[1]

Ira D. Goldfine, John A. Williams[2]

Cell Biology Laboratory, Harold Brunn Institute, and Department of Medicine, Mount Zion Hospital and Medical Center, and Departments of Medicine and Physiology, University of California, San Francisco, Calif., USA

Introduction

The major function of the exocrine pancreas is to secrete zymogen and bicarbonate into the intestine in response to the ingestion of foodstuffs. The regulation of the exocrine pancreas is largely via muscarinic cholinergic neurons and polypeptide hormones. While the biochemical and ultrastructural events involved in acinar cell secretion have been studied for several decades [60], the hormonal regulation of secretion has only been studied in detail for a few years. Two developments have made these hormonal studies possible. First, hormone-sensitive preparations of pancreatic acini have been developed. Second, biologically active radiolabeled hormones have been prepared. This review will survey the recent studies carried out in our laboratory to probe the receptors and mechanism of action of two polypeptide hormones, cholecystokinin (CCK) and insulin, in the exocrine pancreas.

[1] This research was supported by NIH grants AM 21089, AM26422, AM26667 and AM 29971, and the Elise Stern Haas Research Fund, Harold Brunn Institute, Mount Zion Hospital and Medical Center.
[2] With the editorial assistance of *J. Kalbach.*

Isolated Pancreatic Acini in the Study of Exocrine Function

Amsterdam and Jamieson [1] were the first to devise a procedure for preparing isolated pancreatic acinar cells that employed digestion of the pancreas with collagenase and chymotrypsin, chelation of divalent cations with EDTA, and mechanical shearing. Similar preparations have now been used for studies of acinar cell hormone receptors, ion fluxes, cyclic nucleotide levels, and amylase release [16, 17, 24, 44, 89, 92]. The ability of secretagogues to induce $^{45}Ca^{2+}$ efflux in isolated cells suggests that both hormone receptors and the initial steps in stimulus-secretion coupling are intact. In most investigations, however, the measurement of enzyme secretion by isolated acinar cells has been difficult. For example, amylase release from the perfused rat pancreas is increased 8- to 20-fold by both acetylcholine and CCK [39], whereas it is increased only 2-fold or less from isolated rat acinar cells [44].

The poor secretory response of isolated cells does not appear to reflect overt cellular damage, as isolated cells exclude trypan blue, contain normal concentrations of Na^+ and K^+ [92], and appear ultrastructurally intact under the electron microscope (fig. 1, upper). The rise in intracellular second messengers after hormonal stimulation also suggests that receptors are not defective in isolated cells. Furthermore, the Ca^{2+} ionophore A23187, an agent that bypasses the receptor and increases cytoplasmic Ca^{2+}, also fails to stimulate secretion in cells responding poorly to either acetylcholine or CCK [*J. A. Williams,* unpubl. data]. Thus, the cellular defect in isolated cells appears to be in the process of secretion itself. Secretion consists of the movement of zymogen granules to the membrane, their fusion with it, and the subsequent release of zymogen granule contents [60]. The hypothesis that secretion is defective in isolated cells is supported by ultrastructural analyses. In the intact pancreas, the luminal membrane contains microvilli along with a network of actin-like microfilaments underlying the membrane and projecting into the microvilli (fig. 2). In isolated cells, however, these specializations are lost from the secretory pole (fig. 1, lower). Furthermore, due to the ability of membrane proteins to diffuse laterally in the plane of the membrane, any structural or protein specialization of the luminal membrane [51] would be expected to be lost following destruction of junctional complexes during the isolation procedure.

Supporting the importance of the acinar configuration in secretion is the finding that when the dissociation procedure is modified to produce

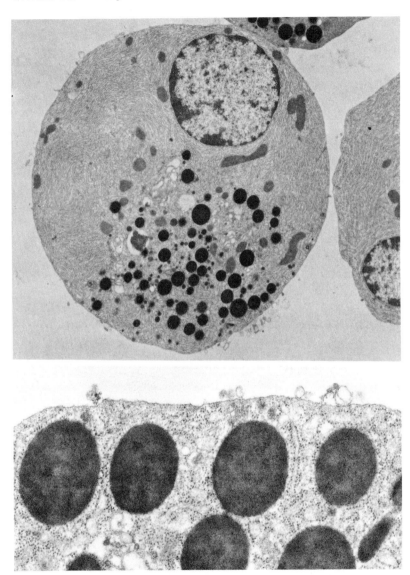

Fig.1. Upper Electron micrograph of isolated mouse pancreatic acinar cells. ×8,000.
Lower Electron micrograph of the secretory pole of isolated mouse pancreatic acinar cell [89].
×25,600.

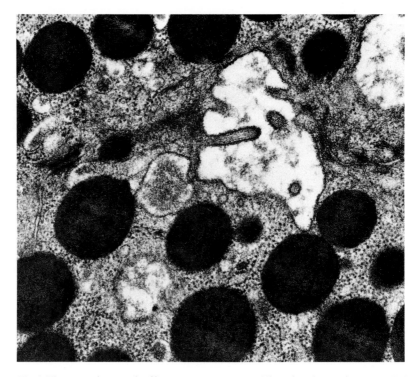

Fig. 2. Electron micrograph of intact mouse pancreas. The acinar lumen is surrounded by a network of microfilaments [89]. ×29,000.

isolated acini, a considerably improved secretory response is consistently observed [62, 93]. Isolated acini are prepared in a manner similar to isolated cells but without the calcium chelation step necessary to break junctional complexes [2]. Isolated acini consist of groups of acinar (and on occasion centroacinar) cells arranged around an intact lumen (fig. 3). Ultrastructural evaluation reveals that the tight junctions connecting adjacent acinar cells are maintained along with the specialized microvilli and their underlying microfilament network located at the apical border of the acinar cell (fig. 4). Since isolated acini can be studied as a homogeneous

Fig. 3. Light micrograph of isolated mouse pancreatic acini [93]. ×855.

Fig. 4. Electron micrograph of the luminal area of an isolated mouse pancreatic acinus [93]. ×24,500.

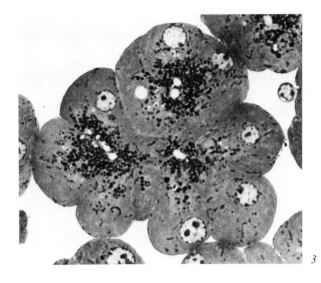

3

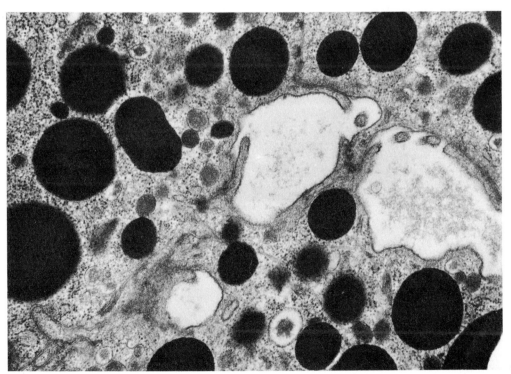

4

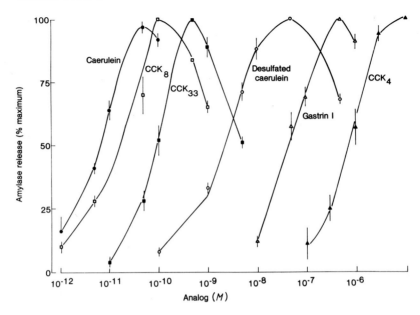

Fig. 5. Dose-response relationship for amylase release from isolated rat pancreatic acini induced by CCK and its analogs. In each case basal release was subtracted and secretagogue-induced release calculated as percent of maximal release. All analogs induced a similar maximal release. ● =Caerulein; □ =CCK$_8$; ■ =CCK$_{33}$; ○ =desulfated caerulein; △ =gastrin; ▲ =CCK$_4$ [94].

suspension and have their basolateral plasma membrane exposed to the incubation medium, they possess all the advantages of the isolated cells, and it is also possible to measure and correlate enzyme release with other cellular functions [90]. Isolated pancreatic acini have proved especially useful in evaluating the complex dose-response curves for secretagogues such as the neurotransmitter acetylcholine and the hormone CCK (fig. 5). Isolated acini have also been utilized to study secretagogue inhibitors, including atropine for the cholinergic agents and dibutyryl cyclic GMP for CCK [63, 93].

Cholecystokinin

Background

CCK is generally considered one of the three classical gut hormones along with gastrin and secretin [38]. Understanding the physiology and

Table I. Amino acid sequence of CCK, gastrin and caerulein

CCK$_{33}$	Lys–Ala–Pro–Ser–Gly–Arg–Val– SO$_3$H Ser–Met–Ile–Lys–Asn–Leu–Glu–Ser– | Leu–Asp–Pro–Ser–His–Arg–Ile–Ser–Asp–Arg–Asp–Tyr–Met–Gly–Trp–Met– Asp–Phe–NH$_2$
Gastrin$_{17}$	Glp–Gly–Pro–Trp–Met–Glu–Glu–Glu–Glu–Glu–Ala–Tyr*–Gly–Trp–Met– Asp–Phe–NH$_2$

* Gastrin exists in two forms, I (non-sulfated tyrosine) and II (sulfated tyrosine). Glp = Pyroglutamic acid.

pathology of CCK, however, has lagged behind that of these other gut hormones due to the difficulty in developing reliable and sensitive assays and, until recently, the relative unavailability of the purified hormone. Over the last few years, however, there has been an explosion of interest in CCK from the disciplines of chemistry, physiology, neurobiology, and psychology.

CCK, which was originally isolated from the porcine intestine based on its ability to stimulate gallbladder contraction and pancreatic secretion, is a straight chain 33 amino acid peptide (CCK$_{33}$) with an amidated C terminus [45]. A prominent feature is the presence of a sulfated tyrosine at position 27 (7 from C terminus) (table I, fig. 5). The C-terminal octapeptide (CCK$_8$) possesses a high degree of biological activity and is in fact more potent than CCK$_{33}$ (fig. 5) [53, 57, 58]. The unsulfated octapeptide, however, has only 1/150 of the activity of the sulfated form [24, 57, 71, 94]. The C-terminal tetrapeptide amide appears to contain all the biological activity of CCK although it is 1- to 30-thousand-fold weaker than the octapeptide, whereas both the C-terminal tripeptide amide and the deamidated tetrapeptide have no activity [52, 67, 78].

Another notable feature of the C-terminal portion of CCK is its homology to the similar portion of the gastrin molecule. The C-terminal pentapeptides of both molecules are identical and both CCK and some forms of gastrin contain a sulfated tyrosine, although this tyrosine in gastrin is not in an identical position to the one in CCK (table I). Separate CCK and gastrin-like molecules are only found in reptiles, birds, and mammals [50], and not lower animals. Thus, both functional and structural evidence indicates that CCK and gastrin have developed from a single caerulein-like molecule [19, 69]. The basic biological activity of

CCK is contained in the C-terminal tetrapeptide amide whereas the additional amino acids are essential to increase specificity for action on pancreas and gallbladder.

CCK was originally isolated from porcine intestine and subsequent radioimmunoassay investigation has shown it to be most concentrated in the duodenum and proximal jejunum [7, 18, 68]. A specific class of gut endocrine cells containing CCK has been identified by immunohistochemistry [9, 65]. These cells are dispersed throughout the duodenal and jejunal mucosa and are elongated or flask-shaped with an apical process in contact with the intestinal lumen. The cells are not argyrophilic and can also be distinguished from enterochromaffin cells by their staining characteristics [9]. By comparison of antibody-stained sections with adjacent thin section examined by electron microscopy, these cells were found to be identical with the intestinal I cell, previously identified in ultrastructural studies [9, 65]. I cells contain a number of dense secretory granules with an average diameter of 250 nm and were originally classified as such because of the intermediate size of their granules which are between those of S and L cells, believed to produce secretin and gut glucagon, respectively [8, 65]. Both immunoreactive CCK and I cells are relatively abundant in portions of the intestine which physiological studies have shown to be involved in nutrient-induced pancreatic secretion and gallbladder contraction. The geometry of the CCK cell has led to the generally accepted schema that this cell senses nutrients at its luminal process followed by exocytotic release of granules on the opposite or capillary facing side of the cell [40].

Since CCK binds initially to receptors localized on the basolateral plasma membrane and rapidly initiates zymogen release at the luminal membrane, it has long been apparent that that action of CCK must be mediated by a second messenger. Although early work focused on a possible role for cyclic AMP, it is now clear that this is not the case. Under physiological conditions CCK does not increase pancreatic cyclic AMP content and exogenous derivatives of cyclic AMP or phosphodiesterase inhibitors do not mimic the action of CCK [13, 42]. CCK does bring about an increase in cyclic GMP, but that rise appears secondary to the rise in Ca^{2+} discussed below [17]. In contrast to the results with cyclic AMP, considerable evidence exists for the role of cytoplasmic Ca^{2+} as the intracellular mediator of CCK [13, 82, 90]. The major points are that (1) CCK increases the movement of Ca^{2+} into and out of acinar cells, (2) removal of Ca^{2+} from the medium reduces or abolishes the action of CCK, and (3) the action of CCK can be mimicked by artificial introduction of Ca^{2+} into acinar cells by means of calcium ionophores, particularly A23187. A recent report using Ca^{2+}-sensitive microelectrodes has determined that the concentration of ionized Ca^{2+} in unstimulated pancreas is $3 \times 10^{-7} M$ and that it increases upon stimulation with acetylcholine [56]. Since CCK is known to act similarly to acetylcholine, although via a distinct receptor, it seems likely that it will have a similar effect. Some controversy exists over the source of Ca^{2+} as both entry from the extracellular fluid and release from intracellular stores have been proposed [82, 90]. It seems clear, however, that release from intracellular stores is the predominant event since initial enzyme secretion can take place in the complete absence of extracellular Ca^{2+}, even in the presence of chelators such as EGTA. It is not completely clear which organelle(s) releases Ca^{2+} upon stimulation with CCK; mitochondria, plasma membrane, and endoplasmic reticulum have been proposed [14, 21, 82]. The nature of the signal from CCK receptors located on the plasma membrane to the intracellular Ca^{2+} stores is also a matter for further investigation.

Little is presently known about the mechanism by which the rise in cytoplasmic Ca^{2+} brings about the increase in amylase release [72, 73]. This process is energy-dependent as inhibitors which lower cellular ATP levels block the action of Ca^{2+} on secretion [36, 91]. It seems likely that the calcium receptor calmodulin may be involved. Calmodulin is present in pancreas [86]. In other cell types after binding Ca^{2+}, calmodulin is known to activate a number of Ca^{2+}-activated protein kinases. Ca^{2+}-activated kinases are present in pancreatic cytosol and CCK stimulation leads to alteration in phosphorylation of a number of acinar proteins [10]. Hypothetically, the CCK-induced rise in cytoplasmic Ca^{2+} might lead to phosphorylation of zymogen granules or plasma membrane proteins leading to exocytotic release of zymogen.

CCK Receptors

In order to directly study the initial steps in pancreatic stimulus-secretion coupling that are initiated by CCK, we have employed isolated pancreatic acini to study CCK receptors. For these studies, we prepared radioiodinated CCK by its conjugation to ^{125}I-labeled Bolton-Hunter (BH) reagent [76]. This procedure avoids the use of oxidizing conditions that destroy the biological activity of CCK, presumably by oxidation of essential methionine residues. The ^{125}I-BH-CCK conjugate is of high specific activity (215–260 µCi/µg) and retains full biological and immunological activity (fig. 6) [76]. Using this ligand, we were able to study hormone

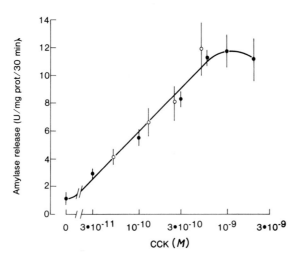

Fig. 6. Comparative biological activities (±1 SD) of CCK (●) and ^{125}I-BH-CCK (○) in the mouse amylase assay [76].

binding to acini at 37 °C under the same conditions employed to study amylase release [77]. At a ligand concentration of 2.5 pM, total binding was rapid, being one-half maximal after 3 min and maximal after 30 min (fig. 7a). Nonspecific binding (determined with ligand in the presence of 125 nM unlabeled CCK) was unchanged throughout the incubation period. Degradation of tracer in the extracellular fluid was monitored by measuring the trichloroacetic acid (TCA) precipitability of the ligand in supernatants of the acinar suspensions. Degradation (appearance of TCA-soluble radioactivity) was minimal for up to 30 min of incubation and then increased linearly. The fall in binding observed after 60 min of incubation presumably was due to hormone degradation.

Binding of ^{125}I-BH-CCK to acini was competitively inhibited by increasing concentrations of CCK but not by unrelated hormones. The affinity and capacity of binding were determined by a computer fitting of bound versus free hormone. The best fit was obtained in terms of two orders of saturable binding sites. These saturable components, replotted by the method of Scatchard, had affinities (Kds) of approximately 70 pM and 30 nM (fig. 7b).

The relationship between receptor binding and amylase release was studied with CCK and its analogs. For both parameters, the potency ratios

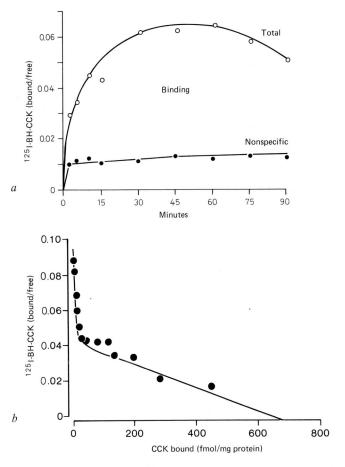

Fig. 7. a Time course of ^{125}I-BH-CCK binding in isolated rat pancreatic acini [77].
b Scatchard plot of the specific binding of CCK to rat pancreatic acini. These data can be
interpreted as indicating either the presence of two orders of binding sites or negative cooperativity [77].

were caerulein[3] > CCK$_8$ > CCK$_{33}$ > desulfated caerulein > gastrin I
> CCK$_4$. There was a high correlation between receptor binding and bio-
logical activity (r = 0.99).

The relationship of CCK receptor occupancy to CCK regulation of
amylase release was then studied. The progressive stimulation of amylase

---

[3] Caerulein is a frog skin decapeptide that is identical in this c terminus with
CCK$_8$.

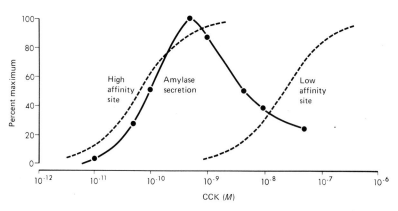

Fig. 8. Concentration dependence of CCK-stimulated amylase secretion and the occupation of high and low affinity CCK receptor sites in rat pancreatic acini [modified from ref. 77].

release by CCK correlated well with the occupancy of the high affinity CCK binding sites, suggesting a close coupling of these two processes. The subsequent fall in amylase release, however, occurred when the lower affinity binding sites began to be occupied (fig. 8). It is possible, therefore, that the two binding sites for CCK regulate different biological responses (i.e., stimulation and then inhibition of amylase). In concert with this hypothesis, we find that other actions of CCK on isolated pancreatic acini, such as stimulation of sugar transport [46] or inhibition of α-aminoisobutyric acid uptake [35], correlate with occupancy of the lower affinity binding site [80].

CCK Autoradiographs

Recently it has been possible to study the location of hormones in target cells with quantitative electron microscope autoradiographs. Such studies have been carried out for the polypeptide hormones insulin, growth hormone, glucagon and parathyroid hormone [4–6, 28, 29, 55]. In general, these hormones initially bind to the plasma membrane and then become internalized and associated with specific organelles. While the location over the plasma membrane suggests that this organelle is a site of hormone action, localization over intracellular organelles has also raised the possibility of intracellular loci of hormone action and/or degradation [29, 32, 61]. However, no autoradiographic studies of CCK had previously been carried out.

Table II. Observed and theoretical grain distribution in acini incubated with ^{125}I–BH–CCK

Organelle	Grain distribution, % total		
	observed		theoretical
	2 min	30 min	
Plasma membrane	64.4[a]	31.1[a]	9.1
Mitochondria	2.4	2.9	5.9
Zymogen granules	2.4	4.1	4.2
Golgi	0.5	4.2	2.1
Multivesicular bodies	0	2.6[a]	0
Endoplasmic reticulum	29.1	49.9	73.6
Nucleus	1.1	3.8	4.8
Smooth vesicles	0	1.3	0.2

Preparation of Sections for Autoradiography. Acini were intubated in Hepes-Ringer buffer with 0.5 nM[125]I–BH–CCK. Acini were collected, washed, and fixed in glutaraldehyde and paraformaldehyde. The fixed acini were postfixed in unbuffered 1% osmium tetroxide containing potassium ferrocyanide [43] and embedded in Epon 812 epoxy resin. Gold sections were coated with a monolayer of Ilford L-4 emulsion [75]. Grids were stained with lead citrate.

Morphological Analysis. Electron micrographs of all the autoradiographic grains present over intact pancreatic acini were photographed. The organelle distribution of the grains was determined using a circle analysis [30, 97]. A transparent overlay having a circle with a radius equal to the resolution half -distance (0.085 µm) was placed over the center of each grain and the underlying organelles recorded. When grains were localized over two or more organelles, credit was equally divided. Any circle touching the plasma membrane, however, was scored solely as plasma membrane. The theoretical random distribution of grains over the organelles in the same micrographs was then determined using a plastic overlay with randomly placed circles of the same radius [97]. 370 observed grains at 2 min, 425 observed grains at 30 min and 1,557 theoretical grains were analyzed.

[a] Hormone concentration (observed/theoretical grains > 1.5).

Accordingly, we used quantitative electron microscopic autoradiography of ^{125}I-BH-CCK in isolated mouse pancreatic acini to localize the initial site of interaction and the subsequent fate of this ligand [95]. Acini were incubated with 0.5 nM ^{125}I-BH-CCK for 2 and 30 min and the distribution of ^{125}I grains over the various cellular organelles determined. In acini incubated 2 min, 64% of the observed grains were localized over the plasma membrane (table II, fig. 9); all grains were over the basolateral

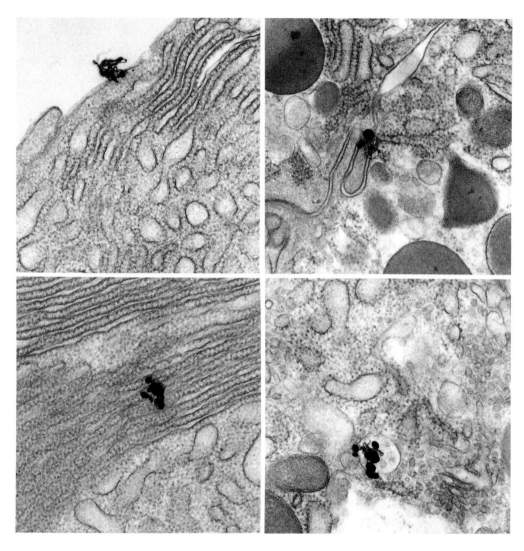

Fig. 9. Electron microscope autoradiographs of pancreatic acini incubated with ^{125}I-BH-CCK for 2 or 30 min showing localization of silver grains over typical organelles [95]. Final magnification ×36,000.

Table III. Radioactivity remaining as intact CCK versus percent remaining with tissue after processing for electron microscopy

Incubation, min	^{125}I–BH–CCK radioactivity, % total	
	remaining intact	remaining with processed tissue
2	90 ± 2 (3)	84 ± 2 (3)
30	65 ± 1 (5)	78 ± 3 (6)

^{125}I–BH–CCK, 0.5 nM, was incubated with pancreatic acini for 2 and 30 min, and the amount of radioactivity bound to acini was measured. For each experiment, one portion of the preparation was processed for electron microscopy and the amount of radioactivity remaining with the tissue calculated. Another portion was extracted and the ^{125}I radioactivity analyzed by gel filtration to determine the percent of radioactivity remaining as intact ^{125}I–BH–CCK. All values are the mean ± SE for the number of experiments shown in parenthesis. From [95].

plasma membrane and no grains were observed over the luminal plasma membrane. The plasma membrane over which grains were localized did not have areas of specialization such as coated pits. The remainder of the grains were located over the endoplasmic reticulum, particularly near the basolateral plasma membrane.

The micrographs were then analyzed by the random circle technique [97] to obtain a theoretical grain distribution and this theoretical distribution was then compared to the observed distribution. With this analysis at 2 min, seven times as many grains were found over the plasma membrane than would be expected on a random basis; at this time, no other cellular organelle showed a similar concentration of hormone. After a 30-min incubation of acini with ^{125}I-BH-CCK, the percent of grains over the plasma membrane (fig. 9) had fallen to 31% of total, but this value was four times that expected on a random basis (table II). At 30 min, the percent of grains over other cellular organelles had increased, but concentration of hormone was observed only over multivesicular bodies.

To study the amount of CCK that had fixed to acinar cells, the acini were incubated for either 2 or 30 min with 0.5 nM ^{125}I-BH-CCK and then washed, fixed, and processed for electron microscopy. After fixation, postfixation and dehydration in ethanol, 84% of the radioactivity bound during a 2-min incubation remained with the acini. After a 30-min incubation, the percentage of bound radioactivity fixing to acini had decreased to 78% (table III). After a 2-min incubation, the fraction of radioactivity fixing to acini correlated well with the fraction remaining as intact CCK (table III). After 30 min, slightly more radioactivity fixed to the acini than remained intact, as determined by gel filtration (table III).

This autoradiograph study, demonstrating the concentration of CCK grains over the plasma membrane, is in concert with the concept that this organelle is the major site of CCK action. In particular, the effects of CCK to both increase Ca^{2+} uptake into acini and to stimulate the release of Ca^{2+} from the plasma membrane most likely are a direct result of CCK binding to this organelle and thus reflect the very rapid effects of the hormone. However, we and others have shown that CCK also mobilizes Ca^{2+} from the endoplasmic reticulum and other intracellular organelles [21]. Thus, it is possible that the presence of CCK in the cell interior could reflect the regulation of the endoplasmic reticulum and other organelles by CCK. This observation could explain how CCK exerts long-term effects on pancreatic acini such as those on protein synthesis [48].

Insulin

Background

The mammalian pancreas contains clusters of heterogenous endocrine cells, the islets of Langerhans, whose function is to secrete insulin, glucagon, and other peptides into the blood. These islets are distributed throughout the pancreatic exocrine tissue which secretes both digestive enzymes and bicarbonate-rich fluid into the intestine via the pancreatic duct. This unique anatomical arrangement is also found in birds, but is either less extensively developed or is missing altogether in lower vertebrates where endocrine cells are clustered as a separate gland. In the most primitive vertebrates, these particular endocrine cells are found only in the mucosa of the gastrointestinal tract [85].

The reasons for the evolutionary migration of endocrine cells from the intestine into numerous islets are unclear. It has been suggested, however, that the formation of islets may allow for hormonal interaction between the endocrine cells, while the dispersal of these islets within the pancreas allows for an interaction between islet and acinar cells. The possibility that hormone-secreting islet cells may interact with their neighboring exocrine cells and form an islet-acinar axis has thus given rise to the hypothesis that islet cell hormones may regulate the exocrine function of the pancreas [34].

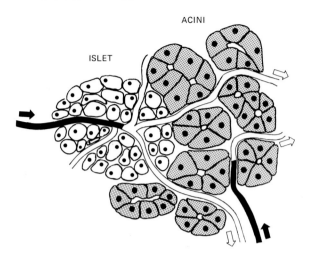

Fig. 10. Schematic representation of intrapancreatic blood flow. Intralobular arterial blood supply is depicted by solid arrows. The arterial vessel entering directly into an islet drains via multiple branches into capillaries surrounding the neighboring acini. Venous drainage is depicted by open arrows.

There is histological evidence for the existence of a capillary portal circulation within the exocrine pancreas [23]. About 75% of the pancreatic arterial blood flows directly to the exocrine pancreas while one quarter flows to the islets. Venous blood from the islets then enters capillaries surrounding the acinar cells prior to returning to the heart via the systemic arterial blood supply. This portal system thus carries hormone-rich blood from the islets directly to the acinar cells (fig. 10), creating a milieu in which the levels of islet hormones are considerably higher than levels found in the peripheral circulation.

These high hormone levels within the pancreas could exert trophic effects on the acinar cells, and thus explain the long-standing observation that acinar cells, adjacent to islets are both larger and richer in zymogen granules than other acini. Histochemical staining for digestive enzymes reveals these larger acini as dark rings or 'haloes' surrounding the paler islets.

The experimental basis for the existence of an islet-acinar axis is strongest for insulin, and was initially deduced from dietary studies. *Grossman* et al. [33], along with other investigators, showed that a glucose-rich diet was effective in inducing an increase in pancreatic amylase con-

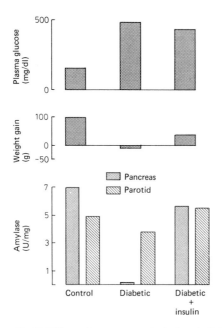

Fig. 11. Effects of streptozotocin-induced diabetes and insulin administration on rat pancreas and parotid amylase levels. Streptozotocin (75 µg/g) was injected into the tail vein. Insulin was administered by regular twice-daily intraperitoneal injection (6 U/day) for 8 days, beginning 2 days after the injection of streptozotocin.

tent. Similar results for both enhanced protein synthesis and raised amylase content are observed following the refeeding of starved animals.

Diabetes mellitus in man and animals may be associated with disturbances in pancreatic exocrine function, including decreased secretion of bicarbonate-rich fluid and decreased enzyme output. Although clinically overt pancreatic insufficiency is not a common clinical manifestation of diabetes, several studies have demonstrated that many insulin-dependent diabetic patients have abnormal pancreatic responses to secretin and CCK, the hormones that regulate the secretion of pancreatic fluid and enzymes [15, 20]. Pathologically, the pancreas of diabetics is typically small and the exocrine tissue may show fatty degeneration and fibrosis. Animals that develop spontaneous diabetes have similar abnormalities [3].

Three observations support the hypothesis that deficiency of insulin is responsible for these abnormalities. First, pancreatic exocrine dysfunction is only found in insulinopenic diabetes [64]. Second, the extent to which pancreatic exocrine function is preserved in insulin-dependent diabetic patients correlates directly with the persistence of beta-cell secretory activity, measured by C-peptide levels [22]. This peptide is secreted with insulin and its plasma level can be used, therefore, as an index of endogenous insulin secretion. Third, in experimental animals the induction of diabetes by beta-cell toxins, such as alloxan or streptozotocin, results in a time-dependent fall in pancreatic amylase levels that can be reversed by the in vivo administration of insulin [83].

We have also confirmed that streptozotocin-induced diabetes is associated with a fall in pancreatic amylase levels, and that this decrease can be prevented by the in vivo administration of insulin. Moreover, we have shown that the marked fall in pancreatic amylase levels was associated with only a mild decrease in parotid gland amylase levels (fig. 11), indicating that the synthesis of pancreatic amylase is considerably more sensitive to insulin than parotid gland amylase. Although parotid gland amylase levels did respond to the in vivo administration of insulin, the response was not of the same magnitude as that observed in the pancreas.

Insulin Receptors

Earlier studies have reported effects of insulin on the exocrine pancreas in vivo; however, it has been difficult to study effects of insulin on the exocrine pancreas in vitro. Accordingly, we investigated both insulin receptors and insulin action on isolated pancreatic acini.

Insulin receptors in pancreatic acini were studied using biologically active porcine ^{125}I-labeled insulin prepared by the stoichiometric chloramine-T method [26]. When mouse acini were incubated with 167 pM ^{125}I-labeled insulin at 37 °C (fig. 12a), binding was one-half maximal after 10 min of incubation and maximal after 30 min of incubation. Scatchard plots of insulin binding, like CCK binding, were not linear and could be resolved into two orders of binding sites, a high affinity site with a Kd of 1.6 nM and a low affinity site with a Kd of 84 nM (fig. 12b).

Subsequently we carried out experiments with pancreatic acini obtained from rats. ^{125}I-labeled insulin binding to receptors in acini from normal rats was similar to that seen with acini from normal mice. We then investigated the effects of insulin on [3H]-leucine incorporation into protein, a well-known effect of insulin in other tissues [37]. Insulin had only a small effect on this function in acini from normal rats but had a much larger effect on acini from rats rendered diabetic with streptozotocin [49].

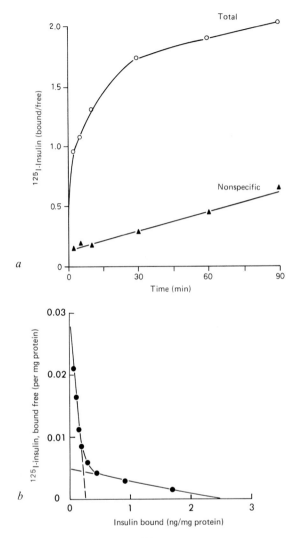

Fig. 12. a Time course of ^{125}I-labeled insulin binding to receptors on isolated pancreatic acini. *b* Scatchard plot of specific binding of insulin to acini [modified from ref. 45].

In acini from diabetic rats, the effect of insulin was one-half maximal at 0.6 nM and maximal at 5 nM (fig. 13a). In these acini from diabetic rats, two orders of insulin binding sites were also detected and stimulation of [3H]-leucine incorporation by insulin correlated closely with occupancy of the high affinity site [79].

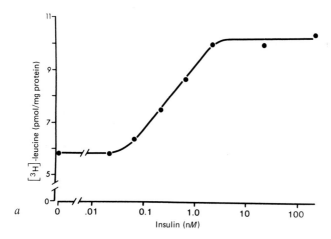

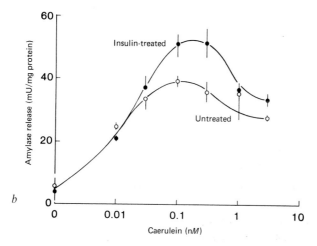

Fig. 13. a Stimulation of [³H]-leucine incorporation into protein by insulin in pancreatic acini prepared from diabetic rats. b Effect of a 2-hour treatment with 167 nM insulin on the subsequent stimulation of amylase release by the CCK analog caerulein in pancreatic acini prepared from diabetic rats [94].

Insulin Autoradiographs

Insulin regulates cellular functions in the exocrine pancreas [41, 45, 49, 74]. In turn, the pancreas degrades insulin [79]. However, the cellular sites for these processes are unknown. Recently, several electron microscope autoradiographic studies of liver and lymphocytes [6, 12, 28, 31, 70]

Table IV. Observed versus theoretical grain distribution in pancreatic acini incubated with ^{125}I-insulin

Organelle	Grain distribution, % total		
	observed		theoretical
	3 min	30 min	
Plasma membrane	45.0[a]	19.7	8.5
Rough endoplasmic reticulum	29.7	37.5	50.4
100-nm vesicles	11.9[a]	15.6[a]	5.8
Zymogen granules	2.4	10.0	7.7
Golgi	2.4[a]	7.7[a]	1.1
Mitochondria	5.8	3.8	12.5
Nuclei	0.8	2.2	7.0
Multivesicular bodies and autophagic vacules	1.4	2.3	1.0
Other	0.6	1.2	5.9

In this study, 316 observed grains and 915 theoretical grains at 3 min and 921 grains at 30 min were analyzed; $\chi^2 = 190.85$; $\chi = 8$; $p < 0.001$.
[a]   Hormone concentration (observed/theoretical grain >1.5).

have indicated that insulin initially binds to receptors on the cell surface, enters the interior of cells, and then interacts with several intracellular structures [11, 26, 27, 66, 87]. These findings have suggested the possibility, therefore, that insulin may either act on or be degraded in the cell interior [25, 84].

In order to better understand the relationship between the intracellular distribution of insulin and its action and/or degradation, we have quantitatively examined electron microscope autoradiographs of isolated pancreatic acini that have been incubated for 3 and 30 min with ^{125}I-insulin in order to determine the initial site of insulin-acinar cell interaction and whether the hormone was translocated into the cell interior.

Studies were undertaken to determine whether the distribution of ^{125}I-insulin grains among the various cellular organelles was a random process. First, using a circle with a radius of one-half distance, the organelles underlying the grains were noted (table IV). Second, by use of the same micrographs and circles of the same radius, a theoretical random grain distribution was determined [97]. Chi-square analyses of the ob-

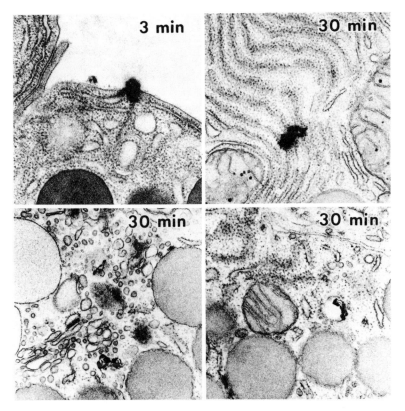

Fig. 14. Electron microscope autoradiographs of pancreatic acini incubated with ^{125}I-insulin for 2 or 30 min showing localization of silver grains over typical organelles. Final magnification ×36,000. From [30].

served versus the theoretical grain distribution patterns indicated that at 3 and 30 min the observed grain distributions were nonrandom (p < 0.001).

After 3 min of incubation, the percent of observed grains over both the plasma membrane and vesicles with an average diameter of 100 nm was higher than the percent of grains over these two structures in a hypothetical random distribution (i.e., the grains were concentrated) (table IV, fig. 14). At 30 min of incubation, grains were concentrated over the plasma membrane, 100 nm vesicles, and Golgi (table IV, fig. 14). Although at both

3 and 30 min a large fraction of the grains was observed over the rough endoplasmic reticulum, there was no relative concentration of grains over this structure.

At neither 3 nor 30 min were grains observed over the luminal acinar membrane. Particularly at 3 min, grains were observed over both basal and lateral cell membranes. Pancreatic acinar cells are connected by tight junctions at their luminal margins, and these junctions presumably restricted the diffusion of either insulin, its receptors, or both.

The distribution of ^{125}I-insulin among intracellular structures was not a random process. Compared with a theoretical random grain distribution, there was a relative concentration of ^{125}I-insulin grains in vesicles with an average diameter of 100 nm at both 3 and 30 min of incubation. Presumably these vesicles originated from the cell surface. Prior studies in hepatocytes and fibroblasts suggest that insulin can be internalized from the cell surface via endocytotic vesicles [6, 11, 12, 31, 70, 81]. In an in vivo study with rat liver, we injected both ^{125}I-insulin and horseradish peroxidase (a marker of endocytosis) into the portal vein and concomitantly localized the peroxidase by histochemical staining and the ^{125}I-insulin by autoradiography [70]. We were able to calculate that approximately 30% of the cellular ^{125}I-insulin grains were present in 100-nm endocytotic vesicles. These vesicles were then observed either in the Golgi area or fused with the bile canalicular membrane [70]. Two other electron microscope autoradiographic studies of ^{125}I-insulin in liver have also suggested that a fraction of the internalized ^{125}I-insulin is present in small vesicles [6, 12]. In contrast to the studies in liver and pancreas, an in vitro electron microscopic study of ^{125}I-insulin entry into human cultured lymphocytes demonstrated internalization but did not detect ^{125}I-insulin grains in vesicular structures [27]. These findings suggest, therefore, that vesicular transport is one mechanism whereby insulin is internalized into certain cell types, but in addition there may be other mechanisms through which the hormone is internalized.

Grains representing ^{125}I-insulin were also concentrated in the Golgi of pancreatic acini. We and others have observed that ^{125}I-insulin grains accumulate in the Golgi of hepatocytes in vivo [6, 12, 70]. In pancreatic acini, one function of the Golgi is to concentrate and package newly synthesized proteins [60]. In pancreatic acini isolated from diabetic rats we find that although incubation with insulin increases the synthesis of amylase, it does not directly stimulate amylase release from the cell. Preliminary studies indicate, however, that incubation of these acini with insulin does potentiate amylase release that is evoked by CCK [47]. These data, therefore, raise the possibility that the Golgi could be an intracellular regulatory site of insulin action. Alternatively, the interaction of insulin with the Golgi may represent either a step in a pathway of insulin degradation [70] or the recycling of insulin and its receptor to the cell surface.

Interactions between CCK and Insulin

To directly evaluate the nature and mechanism of the pancreatic exocrine dysfunction that occurs in diabetes mellitus, we have studied in vitro isolated pancreatic acini prepared from streptozotocin-induced diabetic rats [59]. In these studies, the content of the digestive enzymes amylase and ribonuclease, the effect of secretagogues on their release, and the intracellular transport of newly synthesized proteins were examined. Two defects were observed [59]. First, acini from diabetic rats contained less than 1% of the normal content of amylase, and approximately 50% of the normal content of ribonuclease (fig. 15, 16). Further, reduced amounts of both enzymes were secreted by diabetic acini in response to both CCK and the cholinergic analog carbamylcholine. Second, the sensitivity to CCK was altered in acini from diabetic rats (fig. 15, 16). In acini from normal rats, the effect of CCK was maximal at a concentration of 100 pM; higher concentrations led to submaximal enzyme release. In acini from diabetic rats, the dose-response curve was similarly shaped, but shifted 3-fold towards higher concentrations of CCK. When another effect of secretagogue action, the mobilization of cellular Ca^{2+}, was evaluated the dose response curve for CCK was also selectively shifted towards higher concentrations. In contrast to these results with CCK, the dose-response curve to carbamylcholine was unaltered by diabetes [59].

Control experiments were carried out to confirm that the effects observed were due to insulin deficiency and not due to a direct toxic effect of streptozotocin on acinar cells. First, streptozotocin treatment had no acute effect on acini when measured 24 h after administration. Second, alloxan, another beta-cell toxin, induced similar changes in acinar enzyme content and the secretory response. Third, the administration of exogenous insulin to diabetic rats returned the content of pancreatic amylase and the secretory response to CCK towards normal (fig. 17). The altered pancreatic function in diabetes did not appear due to malnutrition since

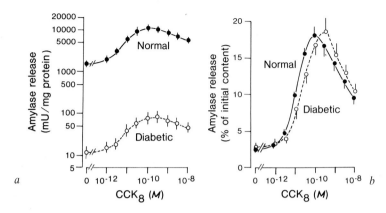

Fig. 15. Concentration dependence of amylase release stimulated by CCK$_8$ from normal (●) and streptozotocin-induced diabetic (○) rat acini. *a* Amylase activity released into the medium expressed relative to the acinar protein concentration. *b* Amylase release expressed as the percentage of the total amylase activity initially present in the acini [59].

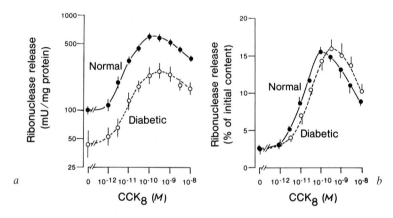

Fig. 16. Concentration dependence of ribonuclease release stimulated by CCK$_8$ from normal (●) and streptozotocin-induced diabetic (○) rat acini. *a* Ribonuclease release is expressed relative to the acinar protein concentration. *b* Ribonuclease release expressed as percent of the total ribonuclease initially present in the acini [59].

starvation for 48 h, although inducing a significant weight loss, did not mimic the effects of diabetes.

Since pancreatic amylase content in diabetic rats is very low and levels can be restored to normal by replacement doses of insulin, we studied the effect of insulin in vitro. We also found that treatment of these acini with

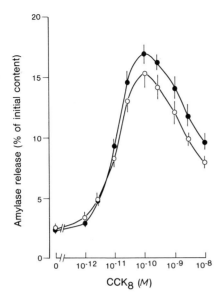

Fig. 17. Concentration dependence of amylase release stimulated by CCK_8 from normal (●) and insulin-treated diabetic (○) rat acini. Amylase release over 30 min expressed as percent of the total amylase activity initially present in the acini is plotted as a function of the added concentration of CCK_8 [59].

insulin in vitro did not directly increase amylase secretion, but after a 2-hour preincubation period insulin potentiated the secretory effects of CCK and its analog caerulein (fig. 13b).

These studies demonstrate, therefore, that two major abnormalities in pancreatic exocrine secretion exist in the diabetic rat due to insulin lack. First, the content of certain digestive enzymes is markedly altered, leading to an altered amount of zymogen secretion. Second, the sensitivity to the secretagogue CCK is selectively reduced, most likely related to a defect in receptor-activated transmembrane signaling. These findings also raise the possibility that the gastrointestinal abnormalities occurring in diabetes mellitus may be a direct result of insulinopenia.

In addition, since both CCK and insulin stimulate protein synthesis in isolated acini, the effects of both were tested in acini from diabetic rats. The major effects of insulin were to enhance the action of CCK at all hormone concentrations [48]. In addition to synergistic effects on protein synthesis, both CCK and insulin stimulate glucose uptake into acinar cells. Each hormone, however, uses a different intracellular mechanism [96].

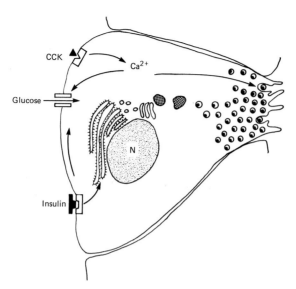

Fig. 18. Schematic diagram of how insulin and CCK influence the exocrine pancreas. CCK via Ca^{2+} acts to increase zymogen secretion directly. Insulin acting via an unknown mechanism increases protein synthesis to both change the acinar zymogen content and potentiate the actions of CCK. In addition, both hormones increase glucose transport.

Conclusions

These studies, therefore, allow a model of how CCK and insulin regulate the acinar pancreas in a coordinated manner (fig. 18). CCK, after its secretion by gut cells, interacts with a specific receptor on the cell surface and then increases intracellular free Ca^{2+}. Ca^{2+} in turn interacts with the secretory granules leading to zymogen release. The model is supported by the finding of specific high affinity CCK receptors on acini and by the localization of CCK to the plasma membrane in electron microscope autoradiographs.

Insulin, secreted from the pancreatic islets, also interacts with a specific receptor on the cell surface. Either via a messenger generated by this reaction or via its subsequent direct interaction with intracellular organelles, such as the Golgi endoplasmic reticulum, protein synthesis is initiated. Then a series of events is initiated to increase cell growth, amylase content, and sensitivity to CCK. These studies, therefore, indicate that the control of acinar cell function is a product of intrahormonal interaction.

References

1 Amsterdam, A.; Jamieson, J. D.: Structural and functional characterization of isolated pancreatic exocrine cells. Proc. natn. Acad. Sci. USA *69:* 3028–3032 (1972).

2 Amsterdam, A.; Jamieson, J. D.: Studies on dispersed pancreatic exocrine cells. I. Dissociation technique and morphological characteristics of separated cells. J. Cell Biol. *63:* 1037–1056 (1974).

3 Balk, M. W.; Lang, C. M.; White, W. J.; Munger, B. L.: Exocrine pancreatic dysfunction in guinea pigs with diabetes mellitus. Lab. Invest. *32:* 28–32 (1975).

4 Barazzone, P.; Gorden, P.; Carpentier, J.-L.; Orci, L.; Freychet, P.; Canivet, B.: Binding, internalization and lysosomal association of ^{125}I-glucagon in isolated rat hepatocytes. J. clin. Invest. *66:* 1081–1093 (1980).

5 Barazzone, P.; Lesniak, M. A.; Gorden, P.; Van Obberghen, E.; Carpentier, J.-L.; Orci, L.: Binding, internalization, and lysosomal association of ^{125}I-human growth hormone in cultured human lymphocytes: a quantitative morphological and biochemical study. J. Cell Biol. *87:* 360–369 (1980).

6 Bergeron, J.; Sikstrom, R.; Hand, A.; Posner, B. I.: Binding and uptake of ^{125}I-insulin into rat liver hepatocytes and endothelium. An in vivo radioautographic study. J. Cell Biol. *80:* 427–443 (1979).

7 Bloom, S. R.: Hormones of the gastrointestinal tract. Br. med. Bull. *30:* 62–67 (1974).

8 Buchan, A. M. J.; Polak, J. M.; Solcia, E.; Capella, C.; Hudson, D.; Pearse, A. G. E.: Electron immunohistochemical evidence for the human intestinal I cell as the source of CCK. Gut *19:* 403–407 (1978).

9 Buffa, R.; Solcia, E.; Go, V. L. W.: Immunohistochemical identification of the cholecystokinin cell in the intestinal mucosa. Gastroenterology *70:* 528–532 (1976).

10 Burnham, D. B.; Williams, J. A.: Effects of carbachol, cholecystokinin, and insulin on protein phosphorylation in isolated pancreatic acini. J. biol. Chem. *257:* 10523–10528 (1982).

11 Carpentier, J.-L.; Gorden, P.; Amherdt, M.; Van Obberghen, E.; Kahn, C. R.; Orci, L.: ^{125}I-insulin binding to cultured human lymphocytes. J. clin. Invest. *61:* 1057–1070 (1978).

12 Carpentier, J.-L.; Gorden, P.; Barazzone, P.; Freychet, P.; Le Cam, A.; Orci, L.: Intracellular localization of ^{125}I-labeled insulin in hepatocytes from intact rat liver. Proc. natn. Acad. Sci. USA *76:* 2803–2807 (1979).

13 Case, R. M.: Synthesis, intracellular transport and discharge of exportable proteins in the pancreatic acinar cell and other cells. Biol. Rev. *53:* 211–354 (1978).

14 Chandler, D. E.; Williams, J. A.: Intracellular divalent cation release in pancreatic acinar cells during stimulus-secretion coupling. II. Subcellular localization of the fluorescent probe chlorotetracycline. J. Cell Biol. *76:* 386–399 (1978).

15 Chey, W. Y.; Shay, H.; Shuman, C. R.: External pancreatic secretion in diabetes mellitus. Ann. intern. Med. *59:* 812–815 (1964).

16 Christophe, J. P.; Conlon, T. P.; Gardner, J. D.: Interaction of porcine vasoactive intestinal peptide with dispersed acinar cells from the guinea pig. J. biol. Chem. *251:* 4629–4634 (1976).

17 Christophe, J. P.; Frandsen, E. K.; Conlon, T. P.; Krishna, G.; Gardner, J. D.: Action of

cholecystokinin, cholinergic agents, and A23187 on accumulation of guanosine 3′:5′-monophosphate in dispersed guinea pig pancreatic acinar cells. J. biol. Chem. *251:* 4640–4645 (1976).

18 Dockray, G.J.: Immunoreactive components resembling cholecystokinin octapeptide in intestine. Nature, Lond. *270:* 359–361 (1977).

19 Dockray, G.J.: Evolutionary relationships of the gut hormones. Fed. Proc. *38:* 2295–2301 (1979).

20 Domschke, W.; Tympner, F.; Domschke, S.; Demling, L.: Exocrine pancreatic function in juvenile diabetics. Dig. Dis. *20:* 309–315 (1975).

21 Dormer, R.L.; Williams, J.A.: Secretagogue-induced changes in subcellular Ca^{2+} distribution in isolated pancreatic acini. Am. J. Physiol. G *240:* 130–140 (1981).

22 Frier, B.M.; Faber, O.K.; Binder, C.; Elliott, H.L.: The effect of residual insulin secretion on exocrine pancreatic function in juvenile-onset diabetes mellitus. Diabetologia *14:* 301–306 (1978).

23 Fujita, T.; Murakami, T.: Microcirculation of monkey pancreas with special reference to the insulo-acinar portal system. A scanning electron microscope study of vascular casts. Archvm histol. jap. *35:* 255–267 (1973).

24 Gardner, J.D.; Conlon, T.P.; Klaeveman, H.L.; Adams, T.D.; Ondetti, M.A.: Action of cholecystokinin and cholinergic agents on calcium transport in isolated pancreatic acinar cells. J. clin. Invest. *56:* 366–375 (1975).

25 Goldfine, I.D.: Does insulin need a second messenger? Diabetes *26:* 148–155 (1977).

26 Goldfine, I.D.; Smith, G.J.: Binding of insulin to isolated nuclei. Proc. natn. Acad. Sci. USA *73:* 1427–1431 (1976).

27 Goldfine, I.D.; Smith, G.; Wong, K.Y.; Jones, A.L.: Cellular uptake and nuclear binding of insulin in human cultured lymphocytes: evidence for potential intracellular sites of insulin action. Proc. natn. Acad. Sci. USA *74:* 1368–1372 (1977).

28 Goldfine, I.D.; Jones, A.L.; Hradek, G.T.; Wong, K.Y.; Mooney, J.S.: Entry of insulin into human cultured lymphocytes: electron microscope autoradiographic analysis. Science *202:* 760–763 (1978).

29 Goldfine, I.D.; Kriz, B.M.; Wong, K.Y.; Jones, A.L.; Renston, R.; Hradek, G.T.: Entry of insulin into target cells in vitro and in vivo; in Middlebrook, Receptor-mediated binding and internalization of toxins and hormones, pp.233–247 (Academic Press, New York 1981).

30 Goldfine, I.D.; Kriz, B.M.; Wong, K.Y.; Hradek, G.; Jones, A.L.; Williams, J.A.: Insulin action in pancreatic acini from streptozotocin-treated rats. III. Electron microscope autoradiography of ^{125}I-insulin. Am. J. Physiol. G *240:* 69–75 (1981).

31 Gorden, P.; Carpentier, J.-L.; Freychet, P.; Le Cam, A.; Orci, L.: Intracellular translocation of iodine-125-labeled insulin: direct demonstration in isolated hepatocytes. Science *200:* 782–785 (1978).

32 Gorden, P.; Carpentier, J.-L.; Freychet, P.; Orci, L.: Morphological probes of polypeptide hormone receptor interactions. J. Histochem. Cytochem. *28:* 811–817 (1980).

33 Grossman, M.I.; Greengard, H.; Ivy, A.C.: The effect of dietary composition on pancreatic enzymes. Am. J. Physiol. *138:* 676–682 (1942).

34 Henderson, J.R.: Why are the islets of Langerhans? Lancet *ii:* 469 (1969).

35 Iwamoto, Y.; Williams, J.A.: Inhibition of pancreatic α-aminoisobutyric acid uptake

by cholecystokinin and other secretagogues. Am. J. Physiol. G *238:* 440–444 (1980).

36 Jamieson, J.D.; Palade, G.E.: Condensing vacuole conversion and zymogen granule discharge in pancreatic exocrine cells: metabolic studies. J. Cell Biol. *48:* 503–522 (1971).

37 Jefferson, L.S.: Role of insulin in the regulation of protein synthesis. Diabetes *29:* 487–496 (1980).

38 Jorpes, J.E.; Mutt, V.: Secretin and cholecystokinin (CCK); in Jorpes, Mutt, Secretin, cholecystokinin, pancreozymin and gastrin, pp.1–179 (Springer, Berlin 1973).

39 Kanno, T.: Calcium-dependent amylase release and electrophysiological measurements in cells of the pancreas. J. Physiol., Lond. *226:* 353–371 (1972).

40 Kanno, T.: Process of reception, conduction and extrusion in paraneurons. Biomed. Res. *1:* suppl., pp.10–16 (1980).

41 Kanno, T.; Saito, A.: The potentiating influence of insulin on pancreozymin-induced hyperpolarization and amylase release in the pancreatic acinar cell. J. Physiol., Lond. *261:* 505–521 (1976).

42 Kanno, T.; Yamamoto, M.: Differentiation between the calcium-dependent effects of cholecystokinin-pancreozymin and the bicarbonate-dependent effects of secretin in exocrine secretion of the rat pancreas. J. Physiol. *264:* 787–799 (1977).

43 Karnovsky, M.: Use of ferrocyanide-reduced osmium tetroxide in electron microscopy (Abstract). 11th Annu. Meet. Am. Soc. Cell Biol., New Orleans 1971, p.412.

44 Kondo, S.; Schulz, I.: Calcium ion uptake in isolated pancreas cells induced by secretagogues. Biochim. biophys. Acta *419:* 76–92 (1976).

45 Korc, M.; Sankaran, H.; Wong, K.Y.; Williams, J.A.; Goldfine, I.D.: Insulin receptors in isolated mouse pancreatic acini. Biochem. biophys. Res. Commun. *84:* 293–299 (1978).

46 Korc, M.; Williams, J.A.; Goldfine, I.D.: Stimulation of the glucose transport system in isolated mouse pancreatic acini by cholecystokinin and analogues. J. biol. Chem. *254:* 7624–7629 (1979).

47 Korc, M.; Williams, J.A.; Goldfine, I.D.: Why the islets of Langerhans? Clin. Res. *27:* 86 (1979).

48 Korc, M.; Bailey, A.; Williams, J.A.: Regulation of protein synthesis in normal and diabetic rat pancreas by cholecystokinin. Am. J. Physiol. G *241:* 116–121 (1981).

49 Korc, M.; Iwamoto, Y.; Sankaran, H.; Williams, J.A.; Goldfine, I.D.: Insulin action in pancreatic acini from streptozotocin-treated rats. I. Stimulation of protein synthesis. Am. J. Physiol. G *240:* 56–62 (1981).

50 Larsson, L.-I.; Rehfeld, J.F.: Evidence of community evolutionary origin of gastrin and cholecystokinin. Nature, Lond. *269:* 335–338 (1977).

51 Meldolesi, J.; Borgese, N.; De Camilli, P.; Ceccarelli, B.: Cytoplasmic membranes and the secretory process. in Poste, Nicholson, vol. 5, pp. 510–598 (North Holland, Amsterdam 1978).

52 Morley, J.S.; Tracy, J.H.; Gregory, R.A.: Structure-function relationships in the active C-terminal tetrapeptide sequence of gastrin. Nature, Lond. *207:* 1356–1359 (1965).

53 Mutt, V.; Jorpes, J.E.: Structure of porcine cholecystokinin-pancreozymin. I. Cleavage with thrombin and with trypsin. Eur. J. Biochem. *6:* 156–162 (1968).

54 Mutt, V.; Jorpes, J.E.: Hormonal polypeptides of the upper intestine. Biochem. J. *125:* 57–58 (1971).

55 Nordquist, R. E.; Palmieri, G. M. A.: Intracellular localization of parathyroid hormone in the kidney. Endocrinology 95: 229–237 (1974).

56 O'Doherty, J.; Stark, R. J.: Ca^{2+} mediation of acetylcholine-induced pancreatic acinar secretion: direct measurements of cytosolic Ca^{2+}. Physiologist 24: 56 (1981).

57 Ondetti, M. A.; Pluscec, J.; Sabo, E. F.; Sheehan, J. T.; Williams, N.: Synthesis of cholecystokinin-pancreozymin. I. The C-terminal dodecapeptide. J. Am. chem. Soc. 92: 195–199 (1970).

58 Ondetti, M. A.; Rubin, B.; Engel, S. L.; Pluscec, J.; Sheehan, J. T.: Cholecystokinin-pancreozymin. Recent developments. Digest. Dis. Sci. 15: 149–156 (1970).

59 Otsuki, M.; Williams, J. A.: Effect of diabetes mellitus on the regulation of enzyme secretion by isolated rat pancreatic acini. J. clin. Invest. 70: 148–156 (1982).

60 Palade, G.: Intracellular aspects of the process of protein synthesis. Science 189: 347–358 (1975).

61 Pastan, I. H.; Willingham, M. C.: Receptor-mediated endocytosis of hormones in cultured cells. A. Rev. Physiol. 43: 239–250 (1981).

62 Peikin, S. R.; Rottman, A. J.; Batzri, S.; Gardner, J. D.: Kinetics of amylase release by dispersed acini prepared from guinea pig pancreas. Am. J. Physiol. E 235: 743–749 (1978).

63 Peikin, S. R.; Costenbader, C. L.; Gardner, J. D.: Actions of derivations of cyclic nucleotides on dispersed acini from guinea pig pancreas: discovery of a competitive antagonist of the action of cholecystokinin. J. biol. Chem. 254: 5321–5327 (1979).

64 Peters, N.; Dick, A. P.; Hales, C. N.; Orrell, D. H.; Sarner, M.: Exocrine and endocrine pancreatic function in diabetes mellitus and chronic pancreatitis. Gut 7: 277–281 (1966).

65 Polak, J. M.; Bloom, S. R.; Rayford, P. L.; Pearse, A. G. E.; Buchan, A. M. J.; Thompson, J. C.: Identification of cholecystokinin-secreting cells. Lancet ii: 1016–1018 (1975).

66 Posner, B. I.; Josefsberg, Z.; Bergeron, J. J. M.: Intracellular polypeptide hormone receptors. J. biol. Chem. 253: 4067–4073 (1978).

67 Rajh, H. M.; Smyth, M. J.; Renckens, B. A. M.; Jansen, J. W. C. M.; DePont, J. J. H. H. M.; Bonting, S. L.; Tesser, G. I.; Nivard, R. J. F.: Role of the tryptophan residue in the interaction of pancreozymin with its receptor. Biochim. biophys. Acta 632: 386–398 (1980).

68 Rehfeld, J. F.: Immunochemical studies on cholecystokinin. II. Distribution and molecular heterogeneity in the central nervous system and small intestine of man and hog. J. biol. Chem. 253: 4022–4030 (1978).

69 Rehfeld, J. F.: Four basic characteristics of the gastrin-cholecystokinin system. Am. J. Physiol. G 240: 255–266 (1981).

70 Renston, R. H.; Maloney, D. G.; Jones, A. L.; Hradek, G. T.; Wong, K. Y.; Goldfine, I. D.: The bile secretory apparatus: evidence for a vesicular transport mechanism for proteins in the rat, using horseradish peroxidase and ^{125}I-insulin. Gastroenterology 78: 1373–1388 (1980).

71 Robberecht, P.; Deschodt-Lanckman, M.; Morgat, J.-L.; Christophe, J.: The interaction of caerulein with the rat pancreas. 3. Structural requirements for in vitro binding of caerulein-like peptides and its relationship to increased calcium outflux, adenylate cyclase activation and secretion. Eur. J. Biochem. 91: 39–48 (1978).

72 Rothman, S. S.: Protein transport by the pancreas. Science *190:* 747–753 (1975).

73 Rothman, S. S.: Passage of proteins through membranes – old assumptions and new perspectives. Am. J. Physiol. G *238:* 391–402 (1980).

74 Saito, A.; Williams, J. A.; Kanno, T.: Potentiation of cholecystokinin-induced exocrine secretion by both exogenous and endogenous insulin in isolated and perfused rat pancreata. J. clin. Invest. *65:* 777–782 (1980).

75 Salpeter, M.; Fertuck, H.; Salpeter, E.: Resolution in electron microscope autoradiography. III. Iodine-125, the effect of heavy metal staining and reassessment of critical parameters. J. Cell Biol. *72:* 161–173 (1977).

76 Sankaran, H.; Deveney, C. W.; Goldfine, I. D.; Williams, J. A.: Preparation of biologically active radioiodinated cholecystokinin for radioreceptor assay and radioimmunoassay. J. biol. Chem. *254:* 9349–9351 (1979).

77 Sankaran, H.; Goldfine, I. D.; Deveney, C. W.; Wong, K. Y.; Williams, J. A.: Binding of cholecystokinin to high affinity receptors on isolated rat pancreatic acini. J. biol. Chem. *255:* 1849–1853 (1980).

78 Sankaran, H.; Bailey, A.; Williams, J. A.: CCK_4 contains the full hormonal information of cholecystokinin in isolated pancreatic acini. Biochem. biophys. Res. Commun. *103:* 1356–1362 (1981).

79 Sankaran, H.; Iwamoto, Y.; Korc, M.; Williams, J. A.; Goldfine, I. D.: Insulin action in pancreatic acini from streptozotocin-treated rats. II. Binding of ^{125}I-insulin to receptors. Am. J. Physiol. G *240:* 63–68 (1981).

80 Sankaran, H.; Goldfine, I. D.; Bailey, A.; Licko, V.; Williams, J. A.: Relationship of CCK binding to regulation of biological functions in pancreatic acini. Am. J. Physiol. G *242:* 250–257 (1982).

81 Schlessinger, J.; Shechter, Y.; Willingham, M.; Pastan, I.: Direct visualization of binding, aggregation, and internalization of insulin and epidermal growth factor on living fibroblastic cells. Proc. natn. Acad. Sci. USA *75:* 2659–2663 (1978).

82 Schulz, I.: Messenger role of calcium in function of pancreatic acinar cells. Am. J. Physiol. G *239:* 335–347 (1980).

83 Söling, H. D.; Unger, K. O.: The role of insulin in the regulation of amylase synthesis in the rat pancreas. Eur. J. clin. Invest. *2:* 199 (1972).

84 Steiner, D. F.: Insulin today. Diabetes *26:* 322–340 (1976).

85 Van Noorden, S.; Falkmer, S.: Gut-islet endocrinology. Some evolutionary aspects. Invest. Cell Pathol. *3:* 21–28 (1980).

86 Vandermeers, A.; Vandermeers-Piret, M. C.; Rathe, J.; Kutzner, R.; Delforge, A.; Christophe, J.: A calcium-dependent protein activator of guanosine 3′:5′-monophosphate phosphodiesterase in bovine and rat pancreas. Eur. J. Biochem. *81:* 377–386 (1977).

87 Vigneri, R.; Goldfine, I. D.; Wong, K. Y.; Smith, G. J.; Pezzino, V.: The nuclear envelope. J. biol. Chem. *253:* 2098–2103 (1978).

88 Webster, P. D.; Black, O.; Mainz, D.; Singh, M.: Pancreatic acinar cell metabolism and function. Gastroenterology *73:* 1434–1449 (1977).

89 Williams, J. A.: Effects of cytochalasin B on pancreatic acinar cell structure and secretion. Cell Tiss. Res. *179:* 453–466 (1977).

90 Williams, J. A.: Regulation of pancreatic acinar cell function by intracellular calcium. Am. J. Physiol. G *238:* 269–279 (1980).

91 Williams, J. A.; Lee, M.: Pancreatic acinar cells: use of a calcium ionophore to separate

enzyme release from the earlier steps in stimulus-secretion coupling. Biochem. bio-phys. Res. Commun. *60:* 542–548 (1974).

92 Williams, J. A.; Cary, P.; Moffat, B.: Effects of ions on amylase release by dissociated pancreatic acinar cells. Am. J. Physiol. *231:* 1562–1567 (1976).

93 Williams, J. A.; Jorc, M.; Dormer, R. L.: Action of secretagogues on a new preparation of functionally intact, isolated pancreatic acini. Am. J. Physiol. E *235:* 517–524 (1978).

94 Williams, J. A.; Sankaran, H.; Korc, M.; Goldfine, I. D.: Receptors for cholecystokinin and insulin in isolated pancreatic acini: hormonal control of secretion and metabolism. Fed. Proc. *40:* 2497–2502 (1981).

95 Williams, J. A.; Sankaran, H.; Roach, E.; Goldfine, I. D.: Quantitative EM autoradio-graphs of radioiodinated CCK in isolated mouse pancreatic acini. Am. J. Physiol. *243:* G291–G296 (1982).

96 Williams, J. A.; Bailey, A.; Preissler, M.; Goldfine, I. D.: Insulin regulation of sugar transport in isolated pancreatic acini from diabetic rats. Diabetes *31:* 674–682 (1982).

97 Williams, M. A.: Advances in optical and electron microscopy, pp. 219–272 (Academic Press, New York 1969).

I. D. Goldfine, MD, Cell Biology Laboratory, Harold Brunn Institute, Department of Medicine, Mount Zion Hospital and Medical Center, PO Box 7921, San Francisco, CA 94120 (USA)

Cell Biology of the Secretory Process, pp. 423–442 (Karger, Basel 1984)

Secretory Process in Thyroid Cells

N. J. Nadler[1]

Department of Anatomy, McGill University, Montreal, Que., Canada

The thyroid gland produces two principal classes of hormones, the iodinated thyroid hormones and calcitonin. This article deals only with the secretory process involving the iodinated hormones. While other iodinated substances may be released by the thyroid gland, l-thyroxine (3,5,3′,5′-tetraiodo-l-thyronine) and 3,5,3′-triiodo-l-thyronine constitute the two main functional iodinated hormones. There is actually no precise information as to how these iodothyronines pass out of the thyroid cells; it is generally assumed that after their production the amino acids simply diffuse from the base of the cells into the extracellular space. Ignoring, then, this final step in the elaboration of the iodinated hormones, it is not easy to mark precisely where synthesis stops and secretion begins. Moreover, during intermediate stages of synthesis, thyroid cells actively secrete products from the apex of the cells, so that to discuss properly the secretory process demands to include the synthesis process as well.

In the thyroid gland, synthesis of the iodinated hormones takes place in the whole of follicles. These follicles are composed of epithelial cells that surround a lumen containing a viscous fluid termed colloid [26]. The epithelial cells are of two types: predominant are the follicular cells (involved in the formation of the iodinated hormones), while less numerous are the C cells (involved in the formation of calcitonin). The main steps in the elaboration of the iodinated hormones are: (1) synthesis of the glyco-

[1] The author wishes to acknowledge the assistance of Dr. *Beatrix Kopriwa* for the preparation of the radioautographs. The work was supported by a grant from the Medical Research Council of Canada.

protein moiety of thyroglobulin; (2) addition of iodine to the tyrosyl residues of thyroglobulin to yield iodotyrosyl residues, and, by combination, iodothyronyl residues, and (3) proteolysis of thyroglobulin to liberate free iodothyronines which are normally deiodinated except for the hormones which pass out of the follicles.

Synthesis of the Glycoprotein Moiety of Thyroglobulin

Thyroglobulin is a large molecule with a normal molecular weight of 660,000 daltons and a sedimentation coefficient of 19; but, under certain conditions, 12S half-molecules and 27S dimers are also found [8–10, 34, 36]. The long-held quaternary structure of thyroglobulin has been challenged and several alternatives proposed [9, 13, 16, 31, 37, 42, 46]. Recent studies suggest that thyroglobulin isolated from various species is composed of two polypeptide subunits, each with a molecular weight of about 300,000 [35]. Tyrosine makes up about 2% of the molecule [38]. Three types of carbohydrate units, which make up about 10% of the bulk of the thyroglobulin molecule, are covalently linked to asparagine, serine and threonine residues [1, 40].

The synthesis of thyroglobulin accounts for more than half the total protein production by the follicular cell [41]. Little is known about the nature of the thyroglobulin gene or genes nor of the transcription. The half-life of thyroglobulin messenger RNA has been estimated to be in excess of 20 h, so that it must be quite stable; accordingly, despite the rapid rate of synthesis of thyroglobulin, the rate of transcription of the thyroglobulin gene is not required to be too active under steady state conditions [39, 45]. Messenger RNA containing a polyadenylated sequence has been purified from thyroglobulin-synthesizing ribosomes isolated from cow thyroid gland and found to have a sedimentation coefficient of 33S and a molecular weight of 330,000 daltons [48]. This is about ten times larger than the average eukaryotic messenger RNA and allows it to code for a polypeptide with a maximal molecular weight of 330,000 daltons. After injection into oocytes of *Xenopus laevis,* it was found that the batrachian oocyte translated the bovine 33S messenger RNA into protein corresponding to 12S half-thyroglobulin molecules. These spontaneously formed 19S dimers with a molecular weight of 300,000 daltons, related chemically and immunologically with cow thyroglobulin [4, 47]. Notwithstanding, there is still controversy as to whether smaller putative thyroglobulin messenger

RNA and smaller thyroglobulin subunits that have been found are cleavage products or valid entities [6].

The structures in the thyroid follicular cells which correlate with the synthesis of thyroglobulin have been elucidated by a large number of investigations using the technique of radioautography. After introduction of appropriate radiolabeled precursors, the location of the developed silver grains in the photographic emulsion showed, at early times, the sites in the thyroid follicles of incorporation of the various labels into newly synthesized products, and at sequential times later, the migration of these products through the follicles. Thus, since leucine is an abundant amino acid of thyroglobulin, ^{3}H-leucine was used to label the polypeptide moiety, while ^{3}H-mannose, -galactose, -fucose, -N-acetylmannosamine were used for the carbohydrate moieties and ^{125}I for the iodine component. The overall activity of the entire gland is the aggregate contribution by all the follicular cells of all the follicles. By these means, it was discovered that a product is synthesized which is exported by the follicular cell into the lumen of the thyroid follicle and which is consistent chemically and immunologically with thyroglobulin [2, 12, 14, 15, 20, 28, 30]. The synthesis of the polypeptide, presumably as thyroglobulin subunits, starts at the ribosomes which stud the membranes of the rough endoplasmic reticulum in the cytoplasm of these cells. After this initial event, thyroglobulin molecules are discharged into the normally dilated cisternae of the endoplasmic reticulum of the follicular cells. The probable mechanism involves a signal peptide translated from the 5′ end to the messenger RNA that codes for hydrophobic amino acids [3]. Glycosylation of thyroglobulin occurs during its transfer through the cisternae into the Golgi apparatus. This involves first the synthesis, independent of polypeptide, of a core carbohydrate at the membranes of the rough endoplasmic reticulum as a dolichol derivative [49]. The core carbohydrate is then somehow transferred to the thyroglobulin chain in an unexplained manner, and it then undergoes a multistage process involving also the addition of final residues, such as sialic acid. At the Golgi apparatus, in the normal rat thyroid gland about 0.5–1 h after the initiation of the molecule, the final glycoprotein moiety is secreted into the apical zone of the follicular cells in membrane-enclosed vesicles. These vesicles have a mean diameter of about 0.17 μm, possess a regular, round outline, and display an inner structure of uniform gray threads alternating with light areas after glutaraldehyde fixation. The density of the content is similar to that of the condensing vacuoles which appear to be yielded from the mature (trans) surface of the Golgi apparatus. These vesicles migrate towards the apical membrane reaching the apical surface about 0.5 h later. There, the membranes presumably fuse, and the contents are discharged into the lumen of the thyroid follicles contributing to the colloid contained therein [7, 11].

Addition of Iodine

It has long been demonstrated that inorganic iodide is trapped by thyroid follicles, achieving a gradient of more than 100:1. It is expected that the passage of iodide from the extracellular space into the cytoplasm of the thyroid follicular cells occurs against an electrical gradient. Newly trapped iodide ions can also flow out of the thyroid follicular cells, and it is probable that the ions can just as easily flow apically into the lumen of the follicles as basally back into the extracellular space [17, 18, 21, 22].

In the thyroid follicle, iodide reacts with a peroxidase forming an oxidized species of iodine which is added to the tyrosyl residues of thyroglobulin. This results in monoiodotyrosine and diiodotyrosine residues in the glycoprotein moiety. The mechanism is still unclear by which hydrogen peroxide required by the peroxidase is supplied. Within the thyroglobulin molecule, iodotyrosines undergo oxidative coupling to form the iodothyronines, the process seemingly catalyzed by the same peroxidase [5, 33, 43].

The elaboration of this thyroperoxidase, a hematoprotein, has been investigated by means of radioautography by *Nadler and Cassol* [25, 27, 28]. Inasmuch, as these results have been published only in abstract form, the text will be expanded to include results obtained from these experiments. The description will also serve to depict the technical mode by which histological analyses have served to elucidate the secretory process in general. Labeled iron (Fe) and labeled delta-aminolevulinic acid (ALA) were determined to be suitable precursors to label the heme moiety of thyroperoxidase. ^{55}Fe-transferrin was prepared by slowly mixing 50 μ-Ci ^{55}FeCl$_3$ in 10 ml 0.001 N HCl with 20 ml of rat serum and incubating at room temerature for 30 min. After incubating rat thyroid lobes with either this ^{55}Fe-transferrin of with ^{14}C-ALA, thyroperoxidase was then solubilized from membranes of the particulate fraction with cholate and trypsin. Following removal of major contaminants, the preparations were analyzed by gel electrophoresis. Two corresponding bands were detected, one labeled with either ^{55}Fe or ^{14}C and the other stainable with diaminobenzidine at pH 7.6 and low H_2O_2 concentration. This would indicate that both labels were incorporated in the thyroid gland into an enzyme with the characteristics of thyroperoxidase. The ability of these preparations to catalyze the oxidation of iodide and the coupling of guaiacol provided additional evidence that the enzyme was thyroperoxidase. Moreover, it was demonstrated that the incorporation of labeled ALA was stimulated by erythropoietin which has been known to stimulate also heme synthesis. The incorporation of labeled ALA was also stimulated by thyroid-stimulating hormone (TSH), suggesting that this hormone may regulate the rate of elaboration of thyroperoxidase. Puromycin had no effect on the incorporation of labeled ALA, suggesting that heme synthesis is independent of a simultaneous synthesis of acceptor apoprotein. Perhaps the thyroid follicle contains a preformed pool of apothyroperoxidase, in which case the inhibitory effects of puromycin on heme synthesis might not be expressed until the apothyroperoxidase pool were depleted. Iodoacetic acid ($6.7 \times 10^{-4} M$) produced a marked inhibition in labeled ALA incorporation. The reason for this response is not clear, but, since this chemical is an inhibitor of sulfhydryl enzymes, it is reasonable to expect that the inhibition reflects inactivation of sulfhydryl groups. Perhaps, the incorporation of ALA into thyroperoxidase involves ALA synthetase, as sulfhydryl enzyme which catalyzes an essential step in the biosynthesis of heme.

After 15 min of incubation of rat thyroid lobes with ^{3}H-ALA, light microscope radioautographs demonstrated that the label was primarily in the follicular cells. However, after 3 h of continuous incubation, label was found in the lumen of the thyroid follicles. These results indicate that thyroperoxidase is synthesized in the cytoplasm of the thyroid follicular cells and that the newly synthesized product is exported into the lumens. Investigations as to the ultrastructural site of synthesis and the pathway of secretion was explored further. 0.9 mCi ^{55}FeCl$_3$ in 0.18 ml 0.001 N HCl was injected slowly through the jugular vein of rats, and pairs were killed at sequential times. Thyroid tissue was examined by electron microscope radioautography, employing quantitative grain-counting techniques [12, 29]. At 15 min and 1 h, most label was associated with extrafollicular connective tissue or with the basal plasma membranes of the follicular cells (fig. 1). Radiolabel in the cells was attributed mainly to mitochondria and rough endoplasmic reticulum (fig. 2). Between 1 and 4 h, label associated with extracellular components declined, while more accumulated in mitochondria and rough endoplasmic reticulum. Later, between 4 h and 4 days, label accumulated in membrane-bound vesicles near the Golgi

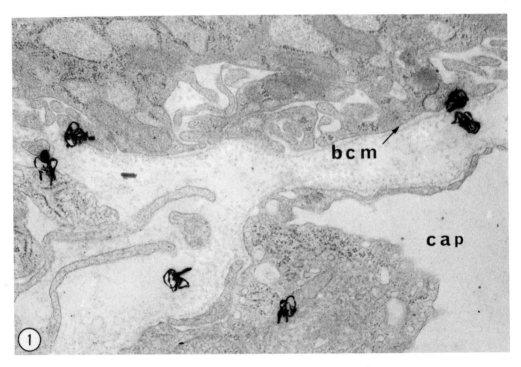

Fig. 1. Electron microscope radioautograph of the basal zone of a rat thyroid follicular cell 15 min after injection of $^{55}FeCl_3$. This and subsequent radioautographs were stained with uranyl acetate and lead citrate. In the lower right corner a capillary (cap) is evident and in the lower left corner a portion of a fibrocyte is visible. At least two of the silver grains present in the picture are centered over the basal plasma membrane (bcm) of the follicular cell. Another two grains lie over the interfollicular connective tissue space, while a fifth grain lies over the endothelial cell of the capillary. The sixth grain is difficult to localize to any one structure. Grain counts showed 40% of the total to lie directly over the basal plasma membrane, interpreted to represent the binding of ^{55}Fe-transferrin to receptor sites in this location prior to incorporation by follicular cells. ×28,500.

apparatus and in the apical zone of the follicular cells, in the follicular lumens, in intracellular colloid droplets, and in lysosomes (fig. 3–8).

Figures 4–6 show electron microscope radioautographs of the apical region of thyroid follicular cells 4 days after intravenous injection of $^{55}FeCl_3$. These results are interpreted to represent the secretion of thyroperoxidase from the follicular cells into the colloid of the lumens.

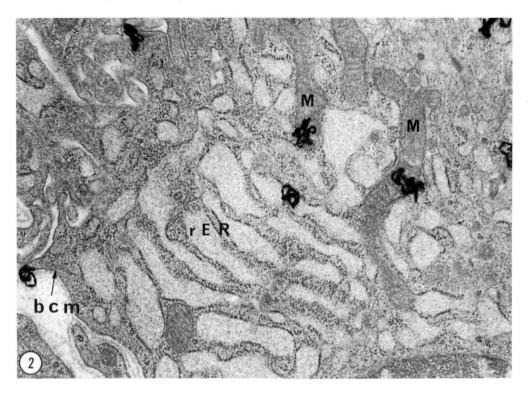

Fig. 2. Electron microscope radioautograph of a thyroid follicular cell 1h after intra-venous injection of $^{55}FeCl_3$. Silver grains are present over mitochondria (M) and over the rough endoplasmic reticulum (r ER) as well as over the basal plasma membrane (bcm). Early labeling of mitochondria and of rough endoplasmic reticulum suggests that these organelles are the intracellular sites of synthesis of iron-containing molecules. Mitochondria are inter-preted to be the site of iron incorporation into heme; while the rough endoplasmic reticulum is interpreted to represent the site of assembly of heme with apoproteins. ×28,500.

The quantititative results are expressed in figure 9. It is expected that the early labeling of extracellular connective tissue is due to circulating ^{55}Fe-transferrin. Initially, labeling of the basal plasma membrane probably represents the binding of labeled transferrin to membrane receptor sites. The appearance of label in mitochondria and rough endoplasmic reticu-lum suggest these organelles as the intracellular sites of synthesis of ^{55}Fe-

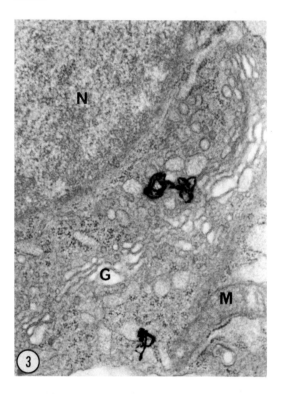

Fig. 3. Electron microscope radioautograph of a thyroid follicular cell 4 h after intravenous injection of ^{55}FeCl$_3$ showing two grains over condensing vacuoles near the mature (trans) surface of the Golgi apparatus (G). This suggests that the Golgi apparatus is involved in the elaboration of thyroperoxidase designated for secretion. M=Mitochondria; N=Nucleus. ×28,500.

containing molecules which include thyroperoxidase. The subsequent appearance of label in the cytoplasmic vesicles in the apical zone and in follicular lumens are interpreted to represent the events in the secretion of thyroperoxidase. It is concluded from these studies that thyroperoxidase, synthesized in mitochondria and in rough endoplasmic reticulum of the thyroid follicular cells, is exported by means of apical vesicles into the lumen of thyroid follicles. It is not known whether these vesicles are the same or are different from the vesicles which contain the glycoprotein moiety of thyroglobulin.

The final step in the synthesis of thyroglobulin involves the iodination of the glycoprotein moiety. To demonstrate the site of iodination in the thyroid follicle, the typical experimental approach is to inject a tracer dose

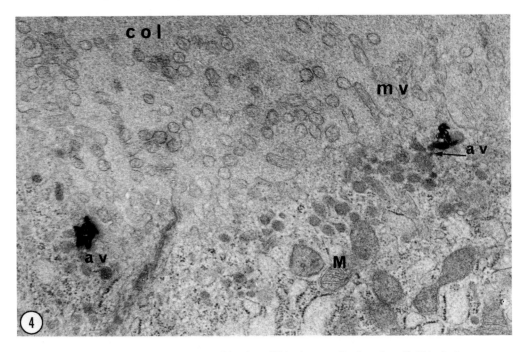

Fig. 4. Part of two follicular cells with microvilli (mv) protruding into the colloid (col) of the follicular lumen. Two silver grains are present over apical vasicles (av) located adjacent to the apical border. M=Mitochondria. ×28,500.

of ^{125}I by intravenous route and follow as soon as possible (30 s) with intravenous perfusion of fixative so as to arrest almost immediately the migration of the newly labeled molecules. Exercising every effort to obtain the best possible resolution that can be afforded by electron microscope radioautography, experimental results obtained in the rat reveal that the peak photographic reaction is located over the colloid in the follicular lumens in a compartment within 1 μm from the apical border of the follicular cells (fig. 10). When the photographic exposure is prolonged in order to attempt detection of relatively weaker sources of radioactive label in the tissue, radioautographs generally show that over 90% of silver grains are located over the lumenal colloid of follicles (fig. 11). The remaining grains are over the follicular cells, and the significance of this relatively

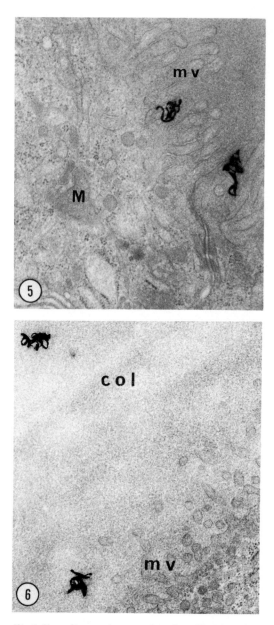

Fig. 5. Two silver grains over the microvillus (mv)-lumenal colloid interface. M=Mi-
tochondria. ×28,500.

    Fig. 6. Two silver grains over the colloid (col) in the follicular lumen. mv=Microvilli.
×28,500.

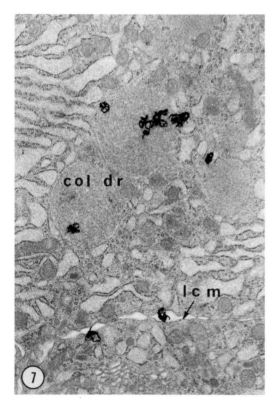

Fig. 7. Electron microscope radioautograph 4 days after intravenous injection of $^{55}FeCl_3$. At least three silver grains are located over the internal matrix of intracellular colloid droplets (col dr), two other grains straddle the membranes of a colloid droplet, and two other grains are over the lateral plasma membrane (l cm). Label in intracellular colloid droplets implies that iodination conceivably could continue intracellulary within colloid droplets. ×14,000.

small but definite cellular reaction was the subject of controversy for many years. Now, it is almost universally agreed that silver grains over the cells can be attributed almost wholly to scatter of radiation emanating from radioactive label in the lumenal colloids, and not from the presence of newly iodinated thyroglobulin in the cells. Analysis of radioautographic grain counts demonstrated that the distribution of silver grains over the various structures in the follicular cells was proportional with the cross-sectional area occupied by these structures in tissue sections, so that grains over cells were located at random and not concentrated over any organelle [24, 26]. Moreover, it was observed after ^{3}H-fucose labeling in the rat that it

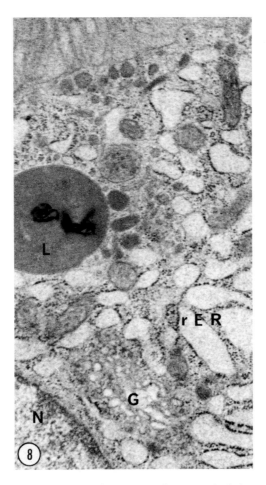

Fig. 8. Electron microscope radioautograph 4 days after intravenous injection of [55]FeCl$_3$. Two silver grains are present over the internal matrix of a large dense body in the thyroid follicular cell, consistent with a lysosome (L). This observation implicates lysosomes, perhaps to recycle iron in its metabolism. r ER = Rough endoplasmic reticulum; G = surface of the Golgi apparatus; N = Nucleus. ×28,500.

normally takes at least 0.5 h for vesicles containing the glycoprotein moiety of thyroglobulin to migrate from the Golgi apparatus into the follicular lumen; implying that newly iodinated thyroglobulin present in the colloid of the follicular lumens could not have originated in the cells and transported to the lumens in 30 s (the time between ^{125}I injection and fixation). Thus, it is concluded that the site of binding of iodine to the

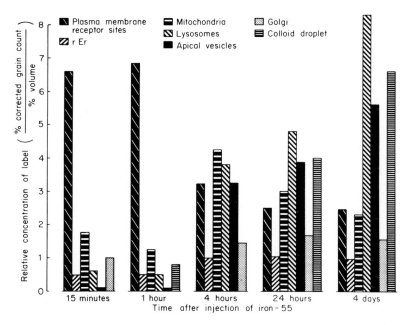

Fig. 9. Results of analysis of grain counts over thyroid follicular cells after intravenous injection of $^{55}FeCl_3$, expressing the relative concentration of label in various cellular structures. The data emphasize the importance of a role for mitochondria, Golgi apparatus, apical vesicles, lysosomes and intracellular colloid droplets. r Er=Rough endoplasmic reticulum.

glycoprotein moiety of thyroglobulin in rat thyroid follicles takes place in the follicular lumens just at the cell apical border. It is realized of course that the ingredients for iodination are present in the follicular cells: the glycoprotein molecule, peroxidase, and iodide. Why the reaction does not take place in the cytoplasm is not known. It can be speculated that perhaps the vesicles which contain the glycoprotein moiety differ from those which contain peroxidase; or perhaps the chemical environment in the cells is inhibitive.

When thyroid tissues are fixed at longer intervals after the injection of ^{125}I, the silver grains in radioautographs distribute over the whole of the colloid in the follicular lumens [17, 21, 23, 24]. This implies that thyroglobulin molecules, whether iodinated or not, mix by diffusion in the follicular lumens. Iodination of tyrosine residues and oxidative coupling would appear to occur when a molecule approaches the region adjacent to the apical membrane of the follicular cell.

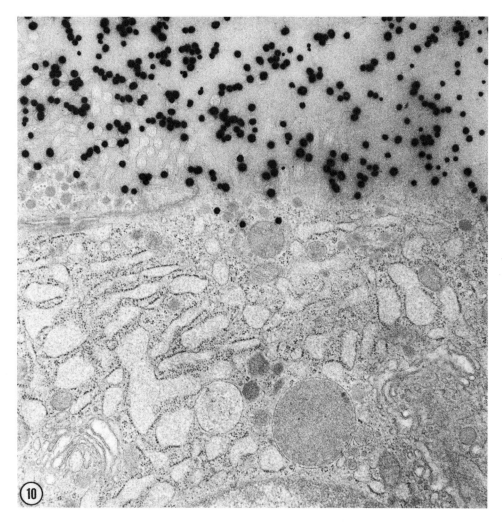

Fig. 10. Electron microscope radioautograph 30 s after intravenous injection of [125]I. The so-called solution physical development technique was used to produce small compact silver grains which afford better resolution than the filamentous type obtained after conventional chemical development. The technique demonstrates the site of iodination to protein.

Analysis of radidoautographs depicting the various stages in the synthesis and secretion of thyroglobulin in the normal rat thyroid gland indicated that nearly all follicular cells of nearly all follicles operate at about the same rate, performing each of the separate functions required for synthesis and secretion simultaneously at all times. As a result, there

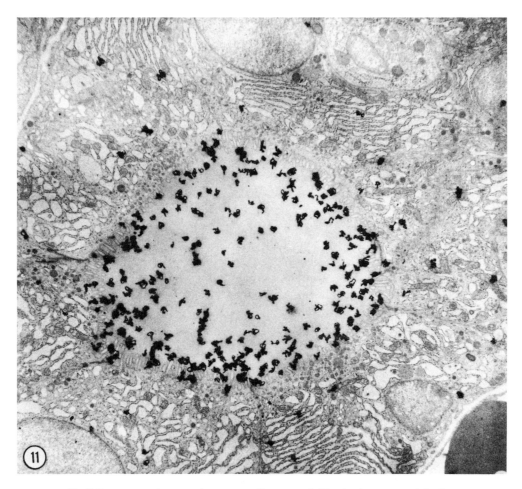

Fig. 11. Low-power electron microscope radioautograph 30 s after intravenous injection of [125]I. Of the 310 silver grains present over the thyroid follicle, 91% are located over the lumen including microvilli projections, and 9% are located over the cytoplasm of follicular cells.

appears to be a correlation between the rates at which the individual functions proceed in any follicle and the number of cells in that follicle [17]. This means that although the phenomenon of iodination of thyroglobulin takes place outside of the follicular cells, the rate at which it occurs in any follicle depends on the number of follicular cells in the epithelium. Since, by simple mathematical derivation, smaller follicles have more cells per unit volume of follicle than do larger follicles, it follows that smaller follicles are more efficient, that is, produce more thyroglobulin per unit volume of follicle.

Proteolysis of Thyroglobulin

Radioautographs at successive time intervals after labeling by any means of thyroglobulin in the colloid of the follicular lumens show a gradual disappearance of the label. This depicts the continuous turnover of thyroglobulin in this compartment, accomplished by the movement of colloid from the lumens into the follicular cells by a pinocytotic process which occurs at the apical membrane of the cells. This process is accelerated after a single injection of TSH [19]. Starting about 10 min, follicular cells in the rat thyroid gland send out streamers of cytoplasm from their apical surfaces into the lumens of the follicles and transfer droplets of colloid into the body of the cells. When thyroglobulin in the colloid of the lumens had been prelabeled with ^{125}I, radioautographs revealed silver grains over the newly formed colloid droplets, proving that the intracellular droplets contained iodinated thyroglobulin. It is of interest to recall that in the experiments to label thyroperoxidase with ^{55}Fe (see above), it was noted that much later after label had appeared in the follicular lumens, label was also detected in the intracellular colloid droplets. This suggests that since intracellular colloid droplets contain all the ingredients of follicular lumen colloid surrounded by a membrane which must be identical with the apical membrane of the follicular cells, it is conceivable that the process of iodination of glycoprotein moieties could continue, and, in this sense, the activity would be intracellular. Intracellular colloid droplets are still observed even in the absence of exogenous TSH stimulation, so that presumably the pinocytotic activity proceeds normally. However, without excessive stimulation by TSH, the intracellular colloid droplets are smaller and less numerous.

It is general belief that digestion of thyroglobulin is achieved by lysosomal action on colloid droplets in the cytoplasm of the follicular cells [50]. Evidence is that reduction of disulfide bonds precedes proteolysis [32].

Intracellular colloid droplets normally vary in size from 0.2 to 2 µm in diameter, and their content displays a spectrum of electron densities between gray and white. When the density content is almost white, it suggests that the original store of thyroglobulin has been depleted. The observation of intracellular colloid droplets fusing with and/or being invaginated by dense bodies or lysosomes is compatible with the concept that the breakdown of colloid droplets might occur by the interaction in the cytoplasm between these droplets and the enzyme content of lysosomes. However, although appearances of an actual union between colloid droplets and lysosomes do occur, they are observed only rarely [26]. Indeed, the relative infrequency and nonspecificity of this association, and the absence of consistent electron density changes in the colloid droplet content even when fused with a dense body, suggest that the lysosome may not necessarily represent the exclusive source of proteolytic enzyme for colloid droplet dissolution. Nevertheless, it is true that colloid droplets which contain the mature molecules of thyroglobulin do eventually disintegrate in the follicular cells. Therefore, it is reasonable to assume that this action must be facilitated by intracellular enzymatic activity of some kind, but the precise morphodynamics remain unclear.

After proteolysis of thyroglobulin molecules, iodotyrosines which are released appear to be deiodinated in the follicular cells by another enzyme, deiodinase. There is evidence that some thyroxine may be deiodinated to triiodothyronine as well. Otherwise, as described in the 'Introduction', the thyroid hormones, thyroxine and triiodothyronine, may simply diffuse from the base of the follicular cells to be secreted into the extracellular space. The iodide which is released mixes with newly trapped iodide and is consequently recycled, as presumably do the released amino acids and carbohydrates.

Conclusions

The secretory process in thyroid follicular cells is unlike other cells in that it is not a single operation. TSH regulates the cellular structures responsible for each step and controls almost every aspect of cell function through modulation of the intracellular concentrations of cyclic adenosine monophosphate, although other physiologic and pathologic extracellular signals may have a positive or negative effect on the same cyclic adenosine monophosphate system [44]. The ultimate secretion of the iodinated thyroid hormones is in one direction, basally; the intermediate secretion of the glycoprotein moiety of thyroglobulin from which these hormones are derived is in the other direction, apically. Complicating, too, is the fact that the ingredients required for iodination of the glycoprotein molecule, thyroperoxidase and iodide, are also exported apically into the follicular lumens, so that the maturation of thyroglobulin occurs external to the cells. As often pointed out, the storage capacity of the follicular lumens for thyroglobulin permits the thyroid gland to sustain an uninterrupted secretion of the hormones even after synthesis of thyroglobulin is blocked for whatever reasons. Pinocytotic activity at the apical surface of the cells allows for the turnover of the contents of the follicular lumens. Despite the complexity, and in some ways because of it, it has been possible to describe for many of the steps the correlation of function with structure.

References

1 Arima, T.; Spiro, M.J.; Spiro, G.G.: Studies on the carbohydrate units of thyroglobulin: evaluation of their microheterogeneity in the human and calf proteins. J. biol. Chem. *247:* 1825–1835 (1972).

2 Bennett, G.; Kan, F.W.K.; O'Shaughnessy, D.: The site of incorporation of sialic acid residues into glycoproteins and the subsequent fates of these molecules in various rat and mouse cell types as shown by radioautography after injection of ^{3}H N acetylmannosamine II. J. Cell Biol. *88:* 16–28 (1981).

3 Blobel, G.; Dobberstein, B.: Transfer of proteins across membranes. I. Presence of proteolytically processed and unprocessed nascent immunoglobulin light chains on membrane-bound ribosomes of murine myeloma. J. Cell Biol. *67:* 835–851 (1975).

4 Chebath, J.; Chabaud, O.; Becarivic, A., et al.: Thyroglobulin messenger ribonucleic acid translation in vitro. Eur. J. Biochem. *77:* 243–252 (1977).

5 DeGroot, L.J.; Niepomniszcze, H.: Biosynthesis of thyroid dhormone: basic and clinical aspects. Metabolism *26:* 665–718 (1977).

6 DeNayer, P.; Cucheteux, L.B.: Identification of RNA species with messenger activity in the thyroid gland., FEBS Lett. *76:* 316–319 (1977).

7 Dinsart, C.; Lecocq, R.; Dumont, J.E., et al.: Control by TSH of protein turnover in thyroid subcellular fractions. Hormone metabol. Res. *8:* 140–145 (1976).

8 Edelhoch, H.; Lippoldt, R.F.: The properties of thyroglobulin II. The effects of dodecyl sulfate. J. biol. Chem. *235:* 1335–1440 (1960).

9 Edelhoch, H.; Lippoldt, R.F.: The properties of thyroglobulin X. The effects of urea. Biochim. biophys. Acta *79:* 64–79 (1964).

10 Edelhoch, H.: The structure of thyroglobulin and its role in iodination. Recent Prog. Horm. Res. *21:* 1–31 (1965).

11 Feeney, L.; Wissig, S.L.: A biochemical and radioautographic analysis of protein secretion by thyroid lobes incubated in vitro. J. Cell Biol. *53:* 510–522 (1972).

12 Haddad, A.; Smith, M.; Herscovics, A., et al.: Radioautographic study of in vivo and in vitro incorporation of fucose-^{3}H into thyroglobulin by rat thyroid follicular cells. J. Cell Biol. *49:* 856–882 (1971).

13 Haeberli, A.; Bilstad, J.; Edelhoch, H., et al.: Elementary chain composition of guinea pig thyroglobulin. J. biol. Chem. *250:* 7294–7299 (1975).

14   Herscovics, A.: Biosynthesis of thyroglobulin. Incorporation of 1-^{14}C-galactose, 1-^{14}C-mannose and 4,5-^{3}H-leucine into soluble proteins by rat thyroids in vitro. Biochem. J. *112:* 709 (1969).

15   Herscovics, A.: Biosynthesis of thyroglobulin. Incorporation of ^{3}H fucose into proteins by rat thyroids in vitro. Biochem. J. *117:* 411 (1970).

16   Lissitzky, S.; Rolland, M.; Reynaud, J., et al.: Structure sous-unitaire de la thyroglobuine de mouton. Eur. J. Biochem. *4:* 464–471 (1968).

17   Nadler, N.J.; Leblond, C.P.: The site and rate of formation of thyroid hormones. Brookhaven Symp. Biol. *7:* 40–60 (1955).

18   Nadler, N.J.; Leblond, C.P.: Rates of passage of iodine into and out of the thyroid gland of the rat under various conditions of dietary iodine intake and body weight. Endocrinology *62:* 768–782 (1958).

19   Nadler, N.J.; Sarkar, S.K.; Leblond, C.P.: Origin of intracellular colloid droplets in the rat thyroid. Endocrinology *71:* 120–129 (1962).

20   Nadler, N.J.; Young, B.A.; Leblond, C.P., et al.: Elaboration of thyroglobulin in the thyroid follicle. Endocrinology *74:* 333–354 (1964).

21   Nadler, N.J.: Iodination of thyroglobulin in the thyroid follicle; in Cassano, Andreoli, Current topics in thyroid research, pp. 73–76 (Academic Press, New York 1965).

22   Nadler, N.J.; Benard, B.; Fitzsimons, G., et al.: An autoradiographic technique to demonstrate inorganic radioiodide in the thyroid gland; in Roth, Stumpf, Autoradiography of diffusible substances, pp. 121–130 (Academic Press, New York 1969).

23   Nadler, N.J.: The application of radioautography to the localization of the binding of iodine to thyroglobulin in rat thyroid follicles. Anat. Rec. *169:* 384–385 (1971).

24   Nadler, N.J.; Chajut, N.R.: Site of bindig of iodine to thyroglobulin in rat thyroid follicles using em radioautography. IVth Int. Congr. of Endocrinology, Washington 1972.

25   Nadler, N.J.; Cassol, S.A.: The radioautographic localization of peroxidase in the rat thyroid follicle after labeling with ^{3}H-delta-aminolevulinic acid and [55]Fe. Anat. Rec. *178:* 324 (1974).

26   Nadler, N.J.: Anatomical features; in Greep, Astwood, Greer, Solomon, Geiger, Handbook of physiology – endocrinology III, Section 7, chapter 4, pp. 39–54 (Waverly Press, Baltimore 1974).

27   Nadler, N.J.; Cassol, S.A.: Radioautographic studies on thyroperoxidase. Anat Rec. *187:* 662–663 (1977).

28   Nadler, N.J.; Cassol, S.A.: Utilization of labeled heme precursors ^{14}C-delta-aminolevulinic acid and [55]Fe to investigate thyroperoxidase, a hematoprotein (Abstract). Endocrinology (1977).

29   Nadler, N.J.: Quantitation and resolution in electron microscope radioautography. J. Histochem. Cytochem. *27:* 1531–1533 (1979).

30   Paiement, J.M.; Leblond, C.P.: Localization of thyroglobulin antigenicity in rat thyroid follicles using antibodies labeled with peroxidase or radioiodine. J. Cell Biol. *74:* 992–1013 (1977).

31   Pierce, J.G.; Rawitch, A.B.; Brown, D.M., et al.: Human thyrogobulin studies on the native and S-carboxymethylated protein. Biochim. biophgys. Acta *3:* 247–256 (1965).

32   Pisarev, M.A.; Dumont, J.E.: The role of reduced glutathione in thyroglobulin proteolysis in vitro. Acta endocr. *79:* 76–85 (1975).

33   Pommier, J.: Structure-function relationship thyroglobulin; in Dumont, Nunez, Hor-
      mones and cell regulation, vol. 2, pp. 180–190 (North Holland Publishing, Amsterdam
      1978).

34   Robbins, J.; Rall, J. E.: Proteins associated with the thyroid hormones. Physiol. Rev.
      40: 415–489 (1960).

35   Rolland, M.; Lissitzky, S.: Endogenous proteolytic activity and constituent polypep-
      tide chains of sheep and pig 19S thyroglobulin. Biochim. biophys. Acta *427:* 696–707
      (1976).

36   Salvatore, G.; Vecchio, G.; Salvatore, M., et al.: 27S thyroid iodoprotein: isolation and
      properties. J. biol. Chem. *240:* 2935–2943 (1965).

37   Schneider, A. B.; Bornet, H.; Edelhoch, H.: The effects of low temperature on the
      conformation of thyroglobulin. J. biol. Chem. *246:* 2835–2841 (1971).

38   Seed, R. W.; Goldberg, I. H.: Biosynthesis of thyroglobulin: relationship to RNA-
      template and precursor protein. Proc. natn. Acad. Sci. USA *50:* 275–282 (1963).

39   Seed, R. W.; Goldberg, I. H.: Biosynthesis of thyroglobulin. II. Role of sub-units, iod-
      ination, and ribonucleic acid synthesis. J. biol. Chem. *240:* 764–773 (1965).

40   Spiro, R. G.; Spiro, M. J.: The carbohydrate composition of the thyroglobulins from
      several species. J. biol. Chem. *240:* 997–1001 (1965).

41   Spiro, M. J.: Studies on the protein portion of thyroglobulin: amino acid compositions
      and terminal amino acids of several thyroglobulins. J. biol. Chem. *245:* 5820–5826
      (1970).

42   Spiro, M. J.: Sub-unit heterogeneity of thyroglobulin. J. biol. Chem. *248:* 4446–4460
      (1973).

43   Taurog, A.: Thyroid peroxidase and thyroxine biosynthesis. Recent Prog. Horm. Res.
      26: 189–247 (1970).

44   Van Herle, A. J.; Vassart, G.; Dumont, J.: Control of thyroglobulin synthesis and
      secretion. New Engl. J. med. *301:* 239–249 (1979).

45   Vassart, G.; Dumont, J. E.; Identification of polysomes synthesizing thyroglobulin.
      Eur. J. Biochem. *32:* 322–330 (1973).

46   Vassart, G.; Refetoff, S.; Brocas, H., et al.: Translation of thyroglobulin 33S messenger
      RNA as a means of determining thyroglobulin quaternary structure. Proc. natn. Acad.
      Sci. USA *72:* 3839–3843 (1975).

47   Vassart, G.; Brocas, H.; Lecocq, R., et al.: Thyroglobulin messenger RNA: translation
      of a 33S mRNA into a peptide immunologically related to thyroglobulin. Eur. J.
      Biochem. *55:* 15–22 (1975).

48   Vassart, G.; Verstreken, L.; Dinsart, C.: Molecular weight of thyroglobulin 33S mes-
      senger RNA as determined by polyacrylamide gel electrophoresis in the presence of
      formamide. FEBS Lett. *79:* 15–18 (1977).

49   Waechter, C. J.; Lennarz, W. J.: The role of polyprenol-linked sugars in glycoprotein
      synthesis. A. Rev. Biochem. *45:* 95–112 (1976).

50   Wollman, S. H.: Secretion of thyroid hormones; in Dingle, Fell, Lysosomes in biology
      and pathology, vol. 2, pp. 483–512 (North-Holland Publishing, Amsterdam 1969).

Dr. N. J. Nadler, McGill University, Department of Anatomy, Strathcona Anatomy
and Dentistry Building, 3640 University Street, Montreal, Que. H3A 2B2 (Canada)

Cell Biology of the Secretory Process, pp. 443–480 (Karger, Basel 1984)

Biosynthesis and Secretion of Parathyroid Hormone

Byron Kemper

Department of Physiology and School of Basic Medical Sciences, University of Illinois, Urbana, Ill., USA

Introduction

The parathyroid glands were first described by *Sandstrom* in 1880 but their significance was not appreciated until 1891 when *Gley* [57] observed that removal of the parathyroids resulted in tetany. The role of calcium in the action of PTH was established in 1909 by *MacCallum and Voegtlin* [113] who showed that parathyroidectomy resulted in low concentrations of calcium in the plasma. Extracts of parathyroid tissue that influenced calcium metabolism were reported in 1924 and 1925 by several laboratories [11, 45, 86]. In these studies to prevent proteolysis during isolation, the tissue was treated with hot acid which cleaved intact PTH to a biologically active fragment [103]. The intact hormone was isolated only after less drastic methods were developed using phenol [7] or urea acid [153, 154] to initially extract polypeptides from the tissue. PTH plays an important role in maintaining the extracellular concentration of calcium within a narrow range and in particular to prevent hypocalcemia. In turn decreaing concentrations of extracellular calcium stimulate the secretion of PTH [145]. PTH, in concert with vitamin D, causes increased absorption of calcium from the gastrointestinal tract [48], reabsorption of bone and retention of calcium by the kidney [8], all of which tend to increase plasma calcium concentrations.

The purpose of this review is to summarize the intracellular mechanisms involved in the production and secretion of appropriate amounts of PTH to maintain calcium homeostasis. Other recent reviews of this subject include an excellent comprehensive review [38] as well as reviews that deal more selectively with the biosynthesis of PTH [68, 72, 73] or the factors regulating secretion of PTH [23, 60].

Primary Structure of PTH and Its Precursors

The complete amino acid sequences for bovine [19, 140], porcine [156] and human PTH [18, 105] have been determined and are shown in figure 1. In each case PTH is a linear basic polypeptide containing 84 amino acids with no cysteine residues. Partial structures of PTH from chicken [129] and dog [38] are known and the proteins of these two species and rat are similar in size and change to the other PTH molecules [35, 38, 129]. Two additional forms or isohormones of bovine PTH have been isolated which differ slightly in amino acid composition, each containing a threonine which is not present in the major species [102].

Differences from the amino acid sequences shown in figure 1 have been reported and include glutamine instead of glutamic acid$_{22}$ in bovine, pig and human PTH, lysine instead of leucine$_{28}$ and leucine instead of aspartic acid$_{30}$ in human PTH [17]. Determination of the sequence by biosynthetic incorporation of radioactive amino acids into PTH in vitro [104] and the predicted amino acid sequences based on the nucleic acid sequences of cloned cDNAs of human and bovine PTH mRNA establish the correct structure as shown in figure 1 [89, 100, 107, 169]. In addition Asn$_{76}$ in the human sequence was reported as aspartic acid in the protein-sequencing studies but sequences of three separate cloned cDNAs indicate that asparagine is correct [89].

The activity of PTH resides in the amino terminal third of the molecule [101, 127, 167]. The activity of fragments of PTH measured by stimulation of adenylate cyclase in rat renal membranes in vitro indicate that the molecule PTH(1–34) retains essentially 100% activity and that residues 2–27 are minimal requirements for activity [101, 167]. Qualitatively similar results are obtained by measurement of hypercalcemia induced in chickens in vivo although the potencies of some of the fragments differ from the adenylate cyclase assay and PTH(1–28) is not detectably active [101, 167].

PTH is synthesized initially as a larger polypeptide precursor, pre-ProPTH, which is converted by two sequential proteolytic cleavages to ProPTH and then to PTH. ProPTH was the first of the precursors detected in studies of protein synthesis in slices of parathyroid tissue incubated in vitro [39, 41, 82, 97]. Initially two major species of radioactive small basic polypeptides were observed, one of which was identified as PTH. The other species had a higher specific radioactivity, consistent with a role as a precursor [82], and later pulse chase studies conclusively demonstrated the

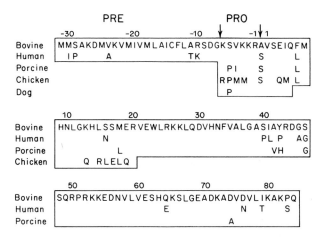

Fig. 1. Comparison of the amino acid sequences of preProPTH from bovine [19, 140], human [18, 105], porcine [156], chicken [129] and dog [38]. Regions of the molecules for which the sequence is known are enclosed within the lines. Only amino acids that differ from the bovine are shown for the other species. The two arrows indicate the proteolytic cleavage sites for the conversion of pre-ProPTH to ProPTH and ProPTH to PTH. The codes for the amino acids are: A=alanine; C=cysteine; D=aspartic acid; E=glutamic acid; F=phenylalanine; G=glycine; H=histidine; I=isoleucine; K=lysine; L=leucine; M=methionine; N=asparagine; P=proline; Q=glutamine; R=arginine; S=serine; V=valine; W=tryptophan; Y=tyrosine.

precursor nature of this peptide [41, 97]. The sequence of the pro-sequence of bovine ProPTH was determined by conventional automated Edman degradation [83] and the sequences of the bovine [92], human [40, 90, 92], pig [34], dog [38] and chicken [129] pro-sequences were determined by biosynthetic incorporation of selected radioactive amino acids into ProPTH followed by Edman degradation of the radioactive proteins. A polypeptide with properties similar to ProPTH has been detected in the rat [35]. As shown in figure 1 all the pro-sequences contain an additional hexapeptide at the amino terminus. All the mammalian sequences have an amino terminal lysine and three basic amino acids at the carboxyl terminus. The chicken sequence, while similar, contains an amino terminal arginine and only two basic amino acids at the carboxyl end. The sequence of the cDNA for bovine [100, 107] and human [89] PTH mRNA have confirmed the amino acid sequence and have established definitively that no additional amino acids are present in ProPTH at the carboxyl terminal which had been suggested earlier [81] but was considered unlikely on the basis of peptides of ProPTH produced by tryptic and chemical cleavage

[64, 126]. The two basic amino acids at the site of cleavage are similar to the sequences at the cleavage site for a number of different pro-proteins [162] suggesting that a general trypsin-like enzyme, which catalyzes the conversion of pro-forms to mature proteins, is present in secretory cells. The isolation of the PTH gene and the present technology which allows the introduction and expression of PTH genes in nonparathyroid cells should permit this hypothesis to be tested.

PreProPTH was identified initially by the translation of bovine parathyroid mRNA in wheat germ extracts [96]. A major translational product was observed that reacted with antisera to PTH but migrated more slowly than PTH or ProPTH upon electrophoresis in either acidic urea acrylamide gels or acrylamide gels containing sodium dodecyl sulfate. Analysis of tryptic peptides and cyanogen bromide fragments demonstrated that the carboxyl terminus of preProPTH was the same as ProPTH and that an initiator methionine was present at the amino terminus of the cell-free product. The complete sequences of the bovine [77, 95] and a partial sequence for the human [67] pre-sequences were determined by microsequencing techniques on radioactive preProPTH. The sequence was confirmed by the determination of the sequence of cloned bovine PTH cDNA [100, 107, 169] and the complete sequence of the human pre-sequence was derived from cloned human PTH cDNA [89] as shown in figure 1. The pre-sequences of both human and bovine preProPTH each contain 25 amino acids and exhibit considerable homology with 5 of 25 differences which is similar to the 12 differences out of 84 in PTH itself. The pre-sequence, like pre-sequences of most secretory and membrane proteins [9], is characterized by a hydrophobic string of amino acids in the center of the sequence and several charged amino acids at either end. In contrast to many pre-sequences large numbers of leucine residues are not present in the hydrophobic region. The secondary structure of the pre-sequence, predicted as described by *Chou and Fasman* [33], contains a β turn at the cleavage site and about 57% β sheet and 20% α helix [9, 44, 150] in one high probability structure and 0% β sheet and 83% α helix in a second high probability structure [150]. The results of an analysis of the structure of a synthetic prepro-sequence by circular dichroism spectra were reasonably consistent with these predictions with a high β sheet structure present in an aqueous environment and a high α helical structure present in a nonpolar solvent [150]. This observation suggests that the structure of the pre-sequence may be different in the aqueous environment of the cytoplasm from the structure in the membrane which may prove to be functionally significant. The hydrophobic nature of the pre-sequence is consistent with its proposed signal function [12] in binding of ribosomes to the membranes of the endoplasmic reticulum.

Structure of PTH mRNA and PTH Genes

The mRNA for bovine PTH has been partially purified to about 50% purity by centrifugation on sucrose gradients and affinity chromatography on oligo(dT) cellulose [163]. The size of the mRNA was estimated to be about 258,000 daltons or 750 nucleotides of which about 60 were present in the poly(A) sequence at the 3' terminus [163]. The mRNA is about twice as large as necessary to code for preProPTH. Little is known about the role of the noncoding sequences in the mRNA. Like most other eukaryotic mRNAs, bovine PTH mRNA probably contains a 7-methylguanosine residue in 5'-5' linkage through a triphosphate which is required for maximal translational activity in vitro [94, 163]. PTH mRNA is therefore a typical mammalian mRNA containing a poly(A) sequence at the 3' terminus and a 7-methylguanosine at the 5' terminus and is larger than required to code for its protein products.

The entire sequence of the bovine PTH mRNA (fig. 2) and nearly the entire sequence of the human PTH mRNA have been determined, primarily, by sequencing cloned PTH cDNA. *Kronenberg* et al. [107] cloned cDNA corresponding to about 60% of the bovine PTH mRNA into the Pst I site of pBR322 DNA (pPTHml) and determined the sequence of the cDNA. Analysis of restriction sites present in bovine PTH cDNA indicated that approximately 200 nucleotides at the 3' terminus were missing in pPTHml [59]. An essentially full-length bovine PTH cDNA was cloned (pPTHi4) by techniques similar to those used for pPTHml and the DNA sequence was determined [100, 169]. The sequence for the first 50 nucleotides of pPTHml were different from those of pPTHi4. Analysis of several other cloned cDNAs in this region and determination of the sequence of restriction fragments hybridized to bovine PTH mRNA and extended toward the 5' terminus of PTH mRNA with reverse transcriptase showed that the sequence of the cDNA in pPTHi4 was correct [168]. The first 50 nucleotides of pPTHml were an exact inverted complement of the first 50 nucleotides of pPTHi4 and apparently were the result of errors introduced during the in vitro cloning procedures. The remaining sequence of bovine PTH mRNA was determined by sequencing a restriction fragment that had been hybridized to PTH mRNA and extended toward the 5' region

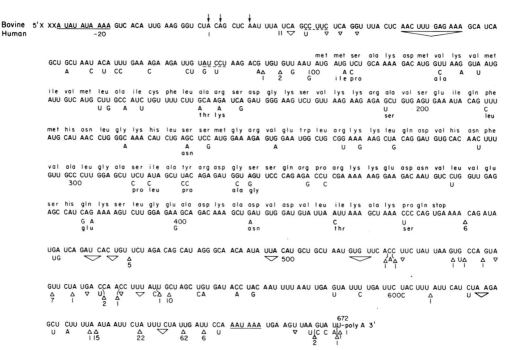

Fig. 2. Nucleotide sequence of bovine PTH mRNA [107, 169] and comparison with the nucleotide sequence of human PTH mRNA [89]. The transcription of PTH mRNA probably begins at the nucleotide numbered 1 but transcription starts at the nucleotides indicated by the arrows cannot be excluded at present. The sequence from −1 to −27 was determined by reverse transcription of parathyroid RNA and is probably present in a minor species of PTH mRNA for which transcription initiates about 30 nucleotides upstream of the normal initiation site. The possible TATA region (putative eukaryotic promotor) is underlined and regions in the 5′ noncoding region which could potentially hybridize to the 3′ region of the 18S ribosomal RNA are underlined with broken lines. In the 3′ noncoding region the sequence, AAUAAA, common to most other animal mRNAs and possibly involved in polyadenylation is underlined. The amino acids for bovine preProPTH are indicated above their appropriate codons. A comparison of the nucleotide sequence of human PTH mRNA with the bovine sequence is also shown. The human sequence, determined from cloned cDNA, begins at the nucleotide corresponding to nucleotide 14 of the bovine sequence and probably is missing some sequence from the 5′ end. Only differences between the sequences are shown. To maximize homology between the two sequences, deletions (indicated by triangles pointing down) or gaps (indicated by triangles pointing up with the number of nucleotides in the gap indicated below) had to be introduced in the bovine mRNA [89].

with reverse transcriptase. A single fragment was not observed in these experiments as would have been the case if all the mRNA molecules had the same 5′ terminus [169]. Instead three major fragments were observed with a maximum difference in size of about 8 nucleotides and a minor fragment about 30 bases longer than the major fragments. Sequencing of the largest fragments revealed that at the 5′ terminus (relative to mRNA) a perfect 'TATA' box was present. This sequence is thought to be a eukaryotic promoter [16] and is normally 25–30 nucleotides upstream from the initiator site for mRNA transcription. These data are consistent with the hypothesis that transcription of PTH is initiated at more than one site. A minor site is near the 'TATA' box and the major site(s) are about 25 nucleotides downstream from this site. Further studies will be required to support this hypothesis.

Bovine PTH mRNA contains 664–672 nucleotides, depending on which major initiation site is chosen, of which about 100 are in the 5′ noncoding region, 348 in the coding region and 224 in the 3′ noncoding region (fig. 2). The 3′ coding region is particularly rich in A and U. The dinucleotide sequence CG occurs only 3 times, consistent with the under representation of this sequence in other mRNAs [141]. This sequence is a site of methylation in eukaryotic DNA and a potential mutational hot spot and thus may be selected against in genes [155]. The methionine codon, AUG, is not present before the initiator AUG codon for preProPTH [168], in contrast to the earlier report which was based on the erroneous sequence in pPTHm1 [107]. This finding is consistent with the scanning hypothesis of protein synthesis initiation by *Kozak* [106]. Two regions of sequence, CCUUC and AUCCU, about 80 and 15 nucleotides from the initiator codon, respectively, are present which are complementary to the 3′ end of 18S rRNA. As in other mammalian mRNAs, the usage of codons is not always distributed evenly among codons coding for the same amino acids. In contrast to other mRNAs, no preference is found for G or C in the third position. The 3′ noncoding region contains the sequence AAUAAA 19 bases from the 3′ terminus. This sequence is present in most eukaryotic mRNAs [146] and may be involved in polyadenylation of the mRNA at the 3′ terminus.

The sequence of human PTH mRNA has also been determined by sequence analysis of several cloned human PTH cDNAs except for probably less than 50 nucleotides at the 5′ terminus [89]. As noted above, this sequence predicted the amino acid sequence of human PTH and established several controversial assignments of amino acids. In the coding region of PTH the two mRNAs exhibit about 90% homology similar to the strength of homology of the amino acids (fig. 2). The number of nucleotide differences in codons which do not change the amino acid (silent mutations) are less than expected on the basis of the evolutionary distance between human and bovine and suggests changes in mRNA sequence in this region may be conserved for reasons other than amino acid coding, for example secondary structure of the mRNA. The sequence of the 5' noncoding regions are conserved also, but not as strongly as the coding region. In the 3' noncoding region only 48% of the nucleotides are the same and extensive gaps must be introduced in the bovine sequence, which is 124 nucleotides shorter than the human sequence, to maximize homology (fig. 2). Nevertheless several extensive regions of homology are seen including 16 of 18 nucleotides around the sequence AAUAAA near the 3' terminus. Two interesting aspects of the human PTH mRNA structure are an AUG codon before the true initator codon which rarely occurs in eukaryotic mRNAs and a second AAUAAA sequence 87 nucleotides before the AAUAAA sequence near the 3' terminus.

The structures of the human and bovine PTH genes have been partially determined and it is likely that a detailed sequence analysis of these two genes will be available in the near future. A 14,000 base pair fragment of human DNA which contains some or all of the gene for PTH has been isolated from a human DNA library in lambda phage by hybridization with radioactive bovine PTH cDNA probes [108]. The human gene was initially shown to have no intervening sequences in the region coding for amino acid 28 of PTH to the first nucleotides of the 3' noncoding region. Additional sequencing studies have shown that an intervening sequence of 109 base pairs in the gene interrupts the mRNA between amino acid(−2) and (−3) which lie within the pro region of preProPTH. A second intervening sequence occurs 100 base pairs upstream from the first one in the 5' noncoding region and contains about 3,700 base pairs [Kronenberg, personal commun]. The organization of the bovine PTH gene has been studied by analysis of restriction fragments of bovine DNA by the Southern blot technique using restriction fragments of cloned bovine PTH cDNA as probes [Gordon and Kemper, manuscript submitted]. This analysis suggested that no intervening sequences are present from amino acid 15 of PTH through nearly all the 3' noncoding sequence and that at least 1,500 base pairs are present in one or more intervening sequences in the 5' region which is consistent with the positions of the intervening sequences in the human gene.

Cohn et al. [44] have suggested that the PTH gene evolved by gene duplication on the basis of internal homologies detected within the amino acid sequence of preProPTH. Significant homology was detected between the first half of the molecule, including the pre and pro regions and PTH(1–29) and the rest of PTH. The sequences of the mRNAs did not add additional support to this hypothesis but were consistent with it. The positions of the intervening sequences in the gene also do not add support to this hypothesis since no intervening sequences are observed in the portion of the gene coding for the second half of the preProPTH but two or more are probably present in the first half. The possibility that intervening sequences were deleted or inserted after gene duplication of course cannot be excluded.

The isolation of PTH genes and PTH cDNAs will permit an understanding of the transcription of the PTH gene and potentially the regulation of the gene. Within a few years both cell-free transcription studies and the introduction of PTH genes in yeast and other mammalian cells combined with in vitro mutagenesis should define the regions of the gene important in regulated transcription.

Transcription and Processing of PTH mRNA

There is as yet no direct experimental evidence on the transcription of the PTH gene or on the potential processing and transport of the initial nuclear transcript to produce active mRNA in the cytoplasm. The evidence demonstrating that intervening sequences are present certainly suggests that processing occurs and potentially could be a site for regulation of the biosynthesis of PTH.

Conversion of preProPTH to ProPTH

Cell-free translation of PTH mRNA in either wheat germ extracts or reticulocyte lysates which contain little or no microsomal membranes results in the synthesis of preProPTH [51, 96]. In ascites cell extracts which contain some membranes both preProPTH and ProPTH are produced [66]. Addition of dog pancreatic microsomal membranes to either the reticulocyte or wheat germ systems results in the formation of ProPTH indicating that the enzyme(s) required for conversion of preProPTH to ProPTH are present in the microsomal membranes [51]. As noted for other pre-proteins the cleavage activity is neither tissue nor species specific [12].

The ProPTH that is formed is resistant to cleavage by proteolytic enzymes indicating that the hormone has been transported across the membrane into the lumen of the membrane vesicle [51]. Both the processing and transport of preProPTH are cotranslational and do not occur if microsomal membranes are added to the reactions after preProPTH has been formed. It has also been shown in this cell-free system that ribosomes associated with radioactive PTH mRNA become bound with the membrane [52]. Active protein synthesis is required for this association to occur and the ribosome-membrane complex is resistant to dissociation by high ionic strength in a manner similar to membrane-bound polysomes formed in vivo.

The pre-piece, thus, appears to function as a classical signal sequence in the model originally proposed by *Blobel and Sabatini* [13]. Direct evidence to support this hypothesis has been obtained in studies with a synthetic precursor-specific sequence of preProPTH which contained the hydrophobic pre-sequence as well as the hydrophilic pro-sequence [117]. In the reticulocyte lysate cell-free system containing dog pancreatic microsomal membranes the synthetic prePro-sequence inhibited the processing and membrane transport of preProPTH and three other pre-proteins, bovine pre-growth hormone, bovine pre-prolactin and human pre-placental lactagen. The data suggest that a pre-specific saturable site is present and required in microsomal membranes for the transport and processing of pre-proteins. It will be of interest to determine whether the signal activity resides solely in the pre-region or if the pro-region contributes to the activity.

For several proteins the pre-sequence appears to be removed before synthesis of the protein is complete [12, 15]. *Habener* et al. [75], however, detected preProPTH in parathyroid slices incubated for 1 min or more with ^{35}S-methionine and the radioactive preProPTH disappeared after a chase incubation suggesting a precursor-product relationship with ProPTH. In more recent studies the amounts of preProPTH and ProPTH were determined by SDS slab gel electrophoresis and autoradiography [69] which is a more sensitive method than the cylindrical gel systems used earlier. In these studies radioactive preProPTH was again observed after a 1-min incubation of parathyroid slices with ^{35}S-methionine. In contrast to the first study after a 0.5-min incubation ProPTH but no preProPTH was detected. The preProPTH formed in the cells is susceptible to proteolytic digestion indicating that it is either adsorbed to the outside of the microsomal membrane or only partially embedded in the membrane and thus represents aberrant synthesis of preProPTH. This strongly suggests that the preProPTH observed in the cell is not a direct precursor and that the ProPTH is rapidly formed by cotranslational processing of nascent chains.

The pre-sequence that is removed appears to be rapidly degraded. In in vitro studies in spite of efficient conversion of preProPTH, to ProPTH, no radioactivity corresponding to an intact pre-sequence could be detected on the electrophoretic gels [51]. Since the pre-sequence contains 5 of the 7 methionines in the preProPTH ^{35}S-methonine-labelled pre-sequence should have been easily detected. Likewise in slices of parathyroid tissue no radioactive pre-sequence could be detected while a synthetic pre-sequence added to the tissue extracts was recovered and detected with good efficiency [76].

Conversion of ProPTH to PTH

The presence of lys-lys-arg at the site of proteolytic cleavage to convert proPTH to PTH indicates that an enzyme with a specificity like trypsin is involved. In fact, limited digestion of proPTH with trypsin selectively cleaves this bond producing intact PTH [41, 58]. This occurs in spite of the fact that there are numerous potential sites for tryptic cleavage at arginines and lysines in PTH itself including two clusters of three basic amino acids which suggests that the secondary or tertiary structure of the protein at the site of cleavage is important.

The search for the enzyme in parathyroid tissue that is responsible for the conversion of ProPTH to PTH has been hindered by the fact that tryptic-like enzymes are rather ubiquitous in tissues, including serum [79], and can efficiently catalyze the conversion. Nevertheless, activity which converts ProPTH to PTH has been detected in all subcellular fractions of parathyroid tissue [62, 123] with the highest specific activity in particulate fractions. While the activity had some properties similar to trypsin, for example, pH optimum and inhibition by benzamidine, it was clearly different from trypsin. Pancreatic and soybean trypsin inhibitors and tosyl-L-lysine chloromethylketone did not inhibit the activity while EDTA and chloroquine, a cathepsin B_1 inhibitor, did inhibit the activity. The particulate activity appears to be tightly bound to membranes since relatively high concentrations of Triton X100 were required to solubilize the activity [123]. Additional work is required to establish that this activity is responsible for the in vivo conversion of ProPTH to PTH.

The hexapeptide that is produced during the cleavage process is rapidly degraded by exopeptidases [62, 126]. Potential aminopeptidase activity was suggested by the detection of ProPTH missing the amino terminal lysine and the probable identification of tetra- and pentapeptides missing lysine after conversion of ProPTH to PTH in vitro [126]. Carboxypeptidase activity was initially suggested by the observation that amino acids from the carboxyl terminus of a synthetic hexapeptide Pro-sequence were removed by particulate fractions containing converting activity [62]. Tetrapeptides missing the carboxyl terminal arginine have also been identified in in vitro conversion systems. The combination of trypsin and carboxypeptidase has been shown to be involved in the conversion of proinsulin to insulin and may be a general mechanism of conversion for many proproteins [162]. Proteins like insulin require the removal of the basic amino acids at the carboxyl terminus to generate the active hormone but the function of carboxypeptidase digestion of the hexapeptide Pro-sequence of ProPTH is obscure. This may simply be the reflection of the general existence of both trypsin and carboxypeptidase-like enzymes in the Golgi apparatus and secretory granules that function in processing of precursors.

Conversion of ProPTH to PTH in slices of bovine parathyroid tissue occurs about 15–20 min after the synthesis of the hormone which is about the time required for the proteins to reach the Golgi apparatus [41, 97]. The hypothesis that the ProPTH must reach the Golgi apparatus before conversion to PTH occurs is supported by studies with Tris buffer, other amines [37, 121, 122] and the ionophore X537A [78, 122, 148] which inhibit conversion of ProPTH to PTH and at the same time cause alterations in the structure of the Golgi apparatus. In contrast, A23187, another ionophore which is selective for divalent cations, and lower concentrations of X537A also affected the ratio of ProPTH to PTH, but by increasing degradation of PTH instead of through effects on conversion. No significant changes in the structure of the Golgi apparatus were observed in these cases [122]. These experiments clearly correlate the pathology of the Golgi apparatus with the conversion of ProPTH to PTH and thus strongly suggest that the enzymes involved in the conversion are localized there.

Storage and Secretion of PTH

PTH is the predominant form of hormone that is stored in the gland and secreted. There is evidence that fragments of PTH are generated in the gland and also secreted. Fragments of PTH are clearly produced in the circulation after secretion [157] but the half-life of fragments in the circulation is also consistent with the possibility that some of the fragments of PTH are secreted from the gland [161]. Subsequently, several studies of the secretion of PTH in situ or in vitro have shown that fragments of PTH are secreted from the glands and that the ratio of fragments to intact hormone increased under hypercalcemic conditions [55, 85, 120]. This is consistent with the proposal that degradation of PTH is increased in glands suppressed by high extracellular calcium. The mechanism of PTH degradation is not understood but two carboxyl-terminal fragments, PTH(37–84) and PTH(34–84), have been detected in dispersed porcine parathyroid cells in amounts that were equal to about half of the PTH that had been degraded in the gland [137]. Cathepsin B from parathyroid tissue selectively cleaves PTH to produce PTH(37–84) and may be involved in this degradation [128]. The physiological importance of these fragments both from a quantitative and qualitative standpoint is not yet established.

The discharge of PTH from the cell probably occurs by exocytosis although evidence supporting two pools of stored hormone suggests the possibility of two mechanisms of secretion. The number of secretory granules present in parathyroid tissue is generally less than that observed in many other secretory tissues (fig. 3) and varies greatly between species [151]. Bovine tissue contains more secretory granules than human or rat tissue. In human parathyroid adenomas, secretion granules are generally observed near the margins of the cell [32, 166]. Secretory granules in these hyperactive tissues are often arranged tandem-like and examples of close attachment to the plasma membrane are observed [166]. Freeze fracture studies reveal that fusion between the granule membrane and the plasma membrane is marked by a clearing of particles from the membrane [166]. The observation of large invaginations in the plasma membrane and endocytic vesicles with bristle coats suggests that membrane recycling is associated with exocytosis [166]. Biochemical evidence to support this hypothesis is obtained from the secretion of parathyroid secretory protein (PSP) and plasminogen activator both of which appear to be associated with PTH in the secretory granules. The secretion of both proteins as measured by radioactivity [98, 135, 136] or radioimmunoactivity [43] is regulated in parallel with that of PTH [43, 98, 135, 136], although in one study the ratio of radioactive PSP to PTH has been reported to increase as secretion is suppressed [132]. PSP has also been shown to be present in the same cells as PTH and presumably the same secretory granules by immu-

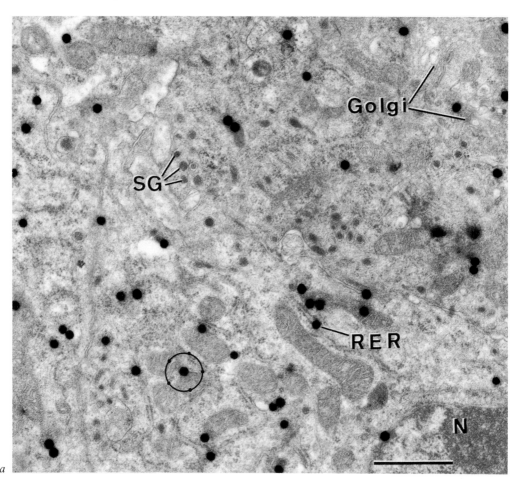

Fig. 3. Autoradiographs of bovine parathyroid cells after incubation with ^{3}H-leucine. Cellular organelles labelled are: rough endoplasmic reticulum (RER); Golgi apparatus (Golgi); secretory granules (SG); nucleus (N); mitochondria (M), and lysosomes (Ly). *a* Cells after a 5-min pulse with ^{3}H-leucine. At this time many autoradiographic grains (black dots) relate to the RER cisternae. Around one developed grain, the probability circle, used for quantitative evaluation of the labeling has been drawn. On this circle, the five equidistant points are represented; three of five points fall on RER profiles whereas two of five are over mitochondrial profiles. Bar=1µm. ×21,700. *b* Cells after a 5-min pulse with ^{3}H-leucine and a 25-min chase. At this time period, Golgi cisternae still contain substantial labeling; in addition, several labeled secretory granules appear outside Golgi areas. Bar=1µm. ×23,000. Reproduced from *Habener* et al. [61].

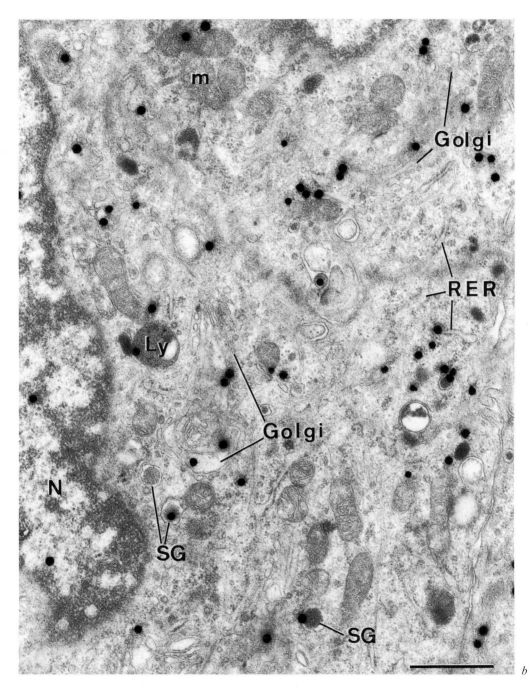

b

nofluorescent localization [149]. Thus the morphological studies and the parallel secretion of multiple proteins support the hypothesis that secretion of PTH occurs by exocytosis.

The relationship of the secretory granules to secretory activity has been studied in parathyroid tissue in vitro and in vivo. In general the number of secretory granules did not correlate well with the stimulation or suppression of the tissue although the amount of other organelles involved in biosynthesis such as the rough endoplasmic reticulum and Golgi apparatus did correlate well [2, 3]. In human adenomas a lack of correlation of the amount of Golgi apparatus and secretory granules with the rate of secretion of PTH has led to a proposal of a 'bypass' secretion pathway [49]. Studies in porcine glands and human adenomas on the recovery of activity after long-term suppression indicate that PTH secretion increases before the Golgi apparatus and secretory granules reappear, further supporting the 'bypass' mechanism [50]. These results are similar to those seen in hyperstimulated pancreatic cells in which secretion continues at high rates after zymogen granules have been depleted [93] and suggest that the storage vesicles for PTH may be heterogeneous ranging from mature secretory granules to a form not easily recognized morphologically. Biochemical evidence does indicate, however, that even newly synthesized PTH is in a membrane-limited compartment [74, 112, 124, 138].

MacGregor et al. [125] initially proposed two pathways of secretion on the basis of the observation that the specific activity of secreted radioactive PTH was much greater than that of cellular PTH in bovine parathyroid slices incubated in vitro. Furthermore a 'pool' of PTH that could be extracted from the tissue with detergent had a specific activity similar to the secreted PTH. Similar results were obtained in dispersed porcine cells [133]. These results could have been obtained if only a fraction of the cells was metabolically active under the in vitro condition used, so that the inactive cells would contribute to the low specific activity of the cellular PTH. More convincing evidence for two pools was the observation in dispersed porcine cells that stimulation of PTH secretion by low extracellular calcium stimulated both newly synthesized as well as old PTH but that secretagogues that stimulate PTH secretion by increasing cAMP stimulate only the secretion of old PTH [133, 134]. Newly synthesized PTH becomes responsive to cAMP stimulation within about 1 h and responsiveness increases for several hours with 50% responsiveness occurring after 3 h [133]. Similar data for parathyroid cells from other species have not been reported.

The basis for the two pools of PTH is not clear. *Cohn and MacGregor* [38] have proposed three basic models. In the sequential model a single pathway exists and heterogeneity results from different stages of maturation of the storage vesicles. In a divergent pathway, the pathway has a branch point presumably at the level of the Golgi apparatus and then two separate pathways of maturation. In the third model the heterogeneity results from two cells either fundamentally different or in different states of activity. The studies by *Dietel and Dorn-Quint* [49] and *Dietel* et al. [50], suggesting that the Golgi apparatus does not correlate with secretion under certain conditions, would require that heterogeneity in the pool could develop before the PTH reaches the Golgi. Further work will clearly be necessary to establish the basis for the two pools of PTH, in particular a structural correlate for the newly synthesized pool of PTH which is not secreted in response to cAMP would be very useful.

Kinetics of Intracellular Transport of PTH

The movement of PTH and its precursor through the cells is consistent with the model proposed by *Palade* [143]. Support for *Palade's* [143] model was obtained from biochemical studies in which the appearance of a pulse of radioactively labelled protein into various subcellular fractions was observed and in morphological studies in which the association of a similar pulse of radioactive protein with cellular organelles was determined by autoradiography. In acinar pancreatic tissue greater than 90% of protein synthesized was secreted so total radioactive protein could essentially be equated with proteins destined for secretion. In parathyroid tissue PTH accounts for only about 20–25% of radioactively labelled protein but the other major secreted protein, PSP, contains about $2^{1}/_{2}$ times as much radioactivity as PTH and the two together represent more than 75% of total radioactive protein [98]. Studies detecting total protein in parathyroid tissues as in autoradiographic approaches, thus, are valid for the secreted proteins but are not specific for PTH.

Biochemical studies of the subcellular distribution of PTH have been hindered by the inability to obtain reasonably distinct subcellular fractions and by nonspecific adsorption of ProPTH and PTH to cellular membranes. Nevertheless, it has been shown that both newly synthesized radioactive and immunoreactive ProPTH and PTH are associated with particulate fractions obtained from bovine glands [74, 112, 138]. About

50% of the particulate ProPTH and PTH is resistant to proteolytic digestion indicating they are within the membrane vesicles [74, 138]. The percentage of ProPTH was greater in the 105,000-g pellet than in the 10,000-g
pellet and the distribution of PTH was the opposite [74]. This is consistent
with the presence of ProPTH in the rough endoplasmic reticulum and
Golgi apparatus and the PTH present in the Golgi apparatus and secretory
granules as would be expected on the basis of *Palade's* [143] model.

Autoradiographic studies of the movement of radioactive proteins
have been done in vivo and in vitro. *Nakagami* et al. [139] injected [3]H-
tyrosine into rats and determined the distribution of radioactive grains
over parathyroid cells. Initially the highest concentration of grains were
over the rough endoplasmic reticulum and maximum relative amounts of
grains were over the Golgi apparatus and secretory granules at about and
20–30 min, respectively. *Habener* et al. [61] studied the distribution of
grains over bovine parathyroid cells after incubation with radioactive
amino acids in vitro and correlated these results with the relative amounts
of radioactive ProPTH and PTH in the cells and medium (fig. 3, 4). Initial
conversion of ProPTH was detected at about 15 min and the maximal rate
of conversion of ProPTH occurred between 15 and 30 min which was the
period of maximum labelling of the Golgi apparatus. It was concluded that
conversion of ProPTH to PTH occurred in the Golgi apparatus although
significant labelling of secretory granules (50% of maximum) was also
observed at 15 min so that conversion occurring in these organelles is
certainly possible. Secretion into the medium was observed between 20
and 30 min and coincided with the appearance of labelled secretory granules in the periphery of the cell near the plasma membranes. Both the
biochemical studies and the autoradiographic studies, thus, indicate that
the major intracellular pathway in the secretion of PTH is the same as that
proposed by Palade for secretory proteins in the exocrine pancreas.

The observation that ProPTH probably has to reach the Golgi apparatus before conversion to PTH provides a relatively simple assay for
transit time to the Golgi by measuring the time at which PTH appears in
the cell. In the presence of vinblastine, colchicine and D_2O the ratio of
ProPTH to PTH was increased and the appearance of PTH was delayed
[99]. Lumicolchicine, an analog of colchicine without effects on microtubules, did not alter the ratio. These results suggested that microtubules
were probably involved in the intracellular transport of ProPTH. More
detailed kinetic studies with colchicine indicated that time required for the
first appearance of PTH increased from 15 to 25 min indicating clearly that

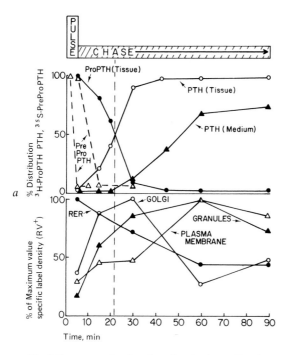

Fig. 4. Summary of pulse-chase incubations of bovine parathyroid slices. Parathyroid gland slices were pulse-labeled with [³H]leucine for 5 min, after which incubations were continued in the presence of unlabeled leucine (chase incubations) for the times indicated. *a* Distributions of [³H]leucine-labeled ProPTH and PTH in parathyroid tissues and in incubation medium as determined by polyacrylamide-gel electrophoresis. Open and closed circles indicate percent distributions of PTH and ProPTH within tissues, where the sum of PTH and ProPTH at each time is taken as 100%. Closed triangles indicate percent of PTH in the incubation medium, where PTH in medium plus PTH in tissue is taken as 100%. *b* Distribution of specific label density in the subcellular compartments of parathyroid cells analyzed by quantitative electron microscope autoradiography. Reproduced from *Habener* et al. [61].

the transit time was increased [36]. Whether the disruption of microtubules has a relative specific effect on the transit time of ProPTH or simply grossly alters the cytoarchitecture of the cell resulting secondarily in an increased transit time has not been determined at present. Decreasing the temperature of the incubation and depletion of ATP by addition of antimycin A also increased the transit time as well as decreasing the rate of synthesis [36]. On the other hand partial inhibition of protein synthesis with cycloheximide and treatment with Tris, which affects the morphology of the Golgi, did not affect the transit time [36]. On the basis of conversion of ProPTH to PTH the transport of ProPTH from the rough endoplasmic reticulum to the Golgi apparently requires energy, is temperature dependent, and may be facilitated by microtubules but is not dependent on continued protein synthesis.

Regulation of the Biosynthesis of PTH

Chronic stimulation of parathyroid tissue in vitro or in vivo results in hyperplasia of parathyroid tissue and increased synthesis and secretion of PTH [8]. The mechanisms underlying these effects are not known. It is not clear whether the increased syntheis of PTH is a specific effect or part of a general increase in the number and activity of the parathyroid cells. The development of a long-term in vitro culture system of parathyroid cells would be very useful in studying these chronic effects.

After a short exposure to low concentrations of calcium, the regulating effects on the secretion and biosynthesis of PTH are clearly dissociated. Secretion is affected within minutes and is stimulated up to 5-fold (see below). The effect on synthesis is considerably less, although there is some variance in reports from different laboratories. In recent experiments using short incubations of 15–20 min with radioactive amino acids to minimize the effects of degradation of PTH, synthesis of PTH was stimulated only 15–25% by incubation with low medium calcium concentrations for up to several hours [35, 63, 132, 134]. Simulations of 1- to 3-fold have been observed in other studies, but in most of these cases incubation with radioactive amino acids was for longer period of time or the efficiency of the extraction of radioactive proteins was not monitored [80, 82, 84, 131]. Particularly with the longer incubations degradation of PTH may have contributed to the apparent increase in synthesis [35, 63]. In similar in vitro studies the amount of PTH mRNA was not stimulated in incubation in low calcium medium [88]. The general consensus of these experiments is that the regulation of PTH synthesis by acute changes in calcium is not important quantitatively but that chronic changes in calcium lead to significant changes in the amount of PTH synthesized.

Since ProPTH is less active than PTH, conversion of ProPTH to PTH is the potential site for regulation. Acute changes in calcium concentrations, however, do not affect the conversion of ProPTH to PTH [63, 65, 134].

The production of PTH is increased in response to low calcium by an effect on the inhibition of the degradation of PTH. Both chronic hypocalcemia in vivo in rats [35] and incubation of bovine parathyroid slices [63] or dispersed cells [134] with low calcium in vitro increase the amount of newly synthesized PTH that is conserved. In the bovine in vitro system up to half of the newly synthesized PTH is degraded in 4 h in suppressed tissue while less than 10% is degraded in stimulated tissue. Parathyroid tissue, thus, appears to be able to eliminate excess PTH in the suppressed tissue by degrading it. It is not clear if this is a direct effect of extracellular calcium on the rate of intracellular degradation or if the decreased degradation is secondary to the increased secretion of newly synthesized PTH from the cell. Furthermore whether a specific degradative enzyme is involved or degradation occurs through a lysosomal mechanism has not been established.

Regulation of the Secretion of PTH

Calcium

The predominant importance of extracellular calcium in the regulation of PTH secretion has been established in numerous in vitro and in vivo studies [38]. The secretion of PTH is inversely related to the concentration of calcium in the extracellular fluid. This regulation is clearly different from the requirement for calcium for secretion by exocytosis in other systems since in these cases secretion is directly related to the calcium concentration [152]. Analysis of hormone secretion in the effluent of calf parathyroid glands in situ demonstrated that regulation of PTH secretion occurs between about 8 and 10.5 mg/100 ml of plasma calcium [70, 118]. The response to changes in calcium was sigmoidal and was greater at hypocalcemic calcium levels below 9 mg/100 ml than in the normal physiological range between 9 and 10.5 mg/ml. The magnitude of the response to calcium may be conditioned by the previous history of the gland since exposure of calves to hypercalcemia for 1 or 2 h reduced the response to hypocalcemia induced by EGTA infusion. Previous exposure to hypocalcemia did not alter the response. A low level of PTH secretion was not further suppressed by high plasma calcium levels above 11 mg/100 ml of plasma. The secretory response of parathyroid tissue is very rapid, occurring within minutes. The secretion of PTH is thus regulated by calcium on a minute to minute basis within the normal physiological range of calcium and the relative response is increased as calcium levels enter hypocalcemic levels, protecting against the serious consequences of hypocalcemia.

The mechanism of calcium action involves effects on the levels of cAMP. Early studies indicated that dibutryl cAMP and theophylline increased the secretion of PTH [172] and that release of cAMP by parathyroid tissue paralleled the release of PTH in response to calcium [1]. PTH secretion has been shown to be proportional to the log of the concentration of intracellular cAMP when any of a variety of secretagogues were employed to change cAMP concentrations in dispersed parathyroid cell preparations [23, 26]. Agents other than calcium that increase cAMP also increased the activity of a cAMP-dependent protein kinase in parathyroid tissue [29]. Substances that decrease extracellular cAMP, inhibit PTH secretion [23]. Calcium, as well as magnesium, ions also lower the level of cellular cAMP levels suggesting strongly that calcium effects on PTH secretion are mediated by cAMP [26, 133].

The effects of calcium on the secretion of PTH cannot be entirely explained by changes in the levels of cAMP. In the presence of increased extracellular concentrations of calcium, increased intracellular concentrations of cAMP are required to produce equivalent increases in PTH secretion which suggested that calcium is blunting the cAMP effect in some way [26]. Paradoxically, calcium or magnesium which reduced cAMP concentration and PTH secretion did not significantly reduce the activity of the cAMP-dependent protein kinase in cells treated or not treated with agonists that increased cAMP [29]. Calcium, thus, appears to be acting in part by mechanisms independent of cAMP. This bimodal mechanism is consistent with the observation that only one of the intracellular pools of PTH responds to increased cAMP concentrations but both respond to calcium as discussed above [133, 134].

In addition to changes in cAMP, the intracellular concentration of calcium may affect PTH secretion. Incubation of parathyroid tissue with divalent ion ionophores inhibits PTH release in the presence of extracellular calcium and increases the sensitivity to extracellular calcium concentration [25, 78, 122]. Verapamil, which inhibits calcium influx, stimulates secretion in suppressed parathyroid tissue incubated in 2.5 mM calcium, also suggesting that decreases in PTH secretion are associated with increased calcium influx [147]. In contrast in stimulated parathyroid tissue incubated in 1.0 mM calcium, verapamil inhibits secretion of PTH which may reflect minimal intracellular concentrations of calcium required for the exocytotic mechanisms. The effect of intracellular calcium may be mediated by calmodulin, the presence of which has been demonstrated in parathyroid tissue [142]. A cyclic nucleotide phosphodiesterase activity which is stimulated by calcium and calmodulin has been demonstrated and could contribute to the decrease in cAMP induced by calcium [20, 176]. Other calmodulin-activated processes could be involved in the cAMP-independent actions of calcium.

Effects of calcium on membrane function and ion transport have also been shown. *Bruce and Anderson* [30] showed that hypocalcemia caused hyperpolarization of mouse parathyroid cells. *Browne* et al. [28] showed that anion transport blockers inhibited the stimulated and basal levels of PTH secretion. Anion transport may be important in the exocytotic mechanism itself. The relationship of these membrane changes to calcium inhibition of PTH secretion is not clear.

Catecholamines

β-Adrenergic agonists stimulate PTH secretion and increase cellular cAMP in vitro and these effects are inhibited by propanolol, a β-blocker [27]. In calves, approximately physiological concentrations of epinephrine resulted in increased secretion of PTH [14, 53, 119]. Furthermore in calves [14], normal humans [110] and patients with secondary hyperparathyroidism [111], propranolol reduced PTH secretion as estimated by levels of circulating PTH. Although these catecholamine effects could be due to hemodynamic or other secondary effects, when combined with the in vitro data they certainly suggest that PTH secretion is modulated by β-adrenergic agents particularly when the gland is stimulated by hypocalcemia [14, 119].

In dispersed bovine parathyroid cells dopaminergic agonists also increase PTH secretion and cellular cAMP levels [4, 24]. This is mediated by a dopamine receptor distinct from the adrenergic receptor since blockers of dopamine action but not α- or β-adrenergic blockers will inhibit the stimulation. Agonists and antagonists of the D_1 class of dopamine receptors affect PTH secretion while D_2 modulaters have little effect [22]. The physiological significance of the dopamine effects is unknown.

Vitamin D

Since PTH and vitamin D cooperate in the regulation of calcium metabolism and PTH modulates the formation of the active metabolite of vitamin D, it is an attractive idea that vitamin D should regulate parathyroid activity forming a feedback loop. Binding proteins, presumably receptors for 1,25(OH)vitamin D have been identified in parathyroid tissue [91, 170]. In various in vivo and in vitro studies, 1,25(OH)$_2$vitamin D either inhibits, stimulates or has no effect and 24,25(OH)$_2$vitamin D either suppresses or has no effect on the secretion of PTH [60]. Additional studies will be required to resolve these conflicting studies and establish the role of vitamin D in the secretion of PTH.

Magnesium

Magnesium ions, like calcium ions, inhibit the release of PTH by the parathyroid gland, but are active at concentrations 2- to 3-fold above normal physiological levels and thus probably are not physiologically significant [71, 132, 165]. Severe hypomagnesia, however, inhibits the release of PTH in vivo [31] and this phenomenon has been confirmed in in vitro studies [132, 165]. Defective PTH secretion in hyperfunctioning human parathyroid cells is associated with decreased affinity for magnesium ion by the adenylate cyclase complex [10]. In rat parathyroid tissues in vitro, decreased PTH secretion caused by magnesium depletion was associated with reduced cAMP production [114]. Magnesium stimulated adenyl cyclase and reduced the inhibition of adenyl cyclase by calcium. The effects of hypomagnesia on PTH secretion thus probaby involve the adenyl cyclase system.

Other Hormones

Various hormones have been reported to affect the secretion of PTH but the physiological significance is not clear. Calcitonin has been reported to stimulate PTH in vitro [54] and this is consistent with studies in vivo [47, 130]. Calcitonin acts in other tissues by increasing cAMP and thus high concentrations of calcitonin may increase PTH secretion via cAMP. Cortisol has been shown in vitro to stimulate PTH secretion [5] and PTH secretion also increases after cortisol administration in vivo [56, 175]. Somatostatin has been reported to suppress secretion of PTH in vivo and in vitro, regardless of whether low calcium concentrations, dibutyryl cyclic AMP or isoproterenol was the stimulus [87, 109]. Studies with somatostatin antisera suggest endogenous somatostatin regulates PTH secretion [173]. Another study indicated that there was no effect of somatostatin on PTH secretion [46]. Secretin stimulates PTH secretion in vitro and in vivo [158, 177]. Secretin also increases cAMP concentrations in parathyroid cells [177]. The concentration of secretin obtained in the human studies were about 3-fold higher than normal postprandial levels and thus its role as a modulator of PTH secretion after a meal to conserve calcium is not established.

Miscellaneous Agents

Lithium has been reported to induce a change in the set point for PTH secretion in response to calcium [21]. Ethanol increases PTH secretion in vivo and in vitro [171]. Histamine stimulates release of PTH secretion in human but not bovine cells in vitro [23], although stimulation was reported in bovine slices [174]. In vivo the H_2 antagonist, cimetidine, decreased circulating immunoreactive PTH in normal human but not in patients with primary hyperparathyroidism [174]. Conflicting studies in both groups of patients have been reported [144, 159].

Parathyroid Secretory Protein

The secretion of undefined proteins larger than PTH has been described [6, 160] and *Kemper* et al. [98] reported the secretion of a defined larger protein, designated parathyroid secretory protein (PSP), that contained about 70% of the radiolabelled protein secreted by bovine parathyroid tissue in vivo. The synthesis, cellular location and secretion of PSP closely parallels that of PTH [43, 98, 132, 133]. PSP and PTH have been detected in the same secretory granules by immunofluorescence studies [149] and PSP appears to be present in two intracellular pools as is PTH [133]. It seems likely that PSP and PTH are present in the same secretory granules and are cosecreted by exocytosis.

Translation of PSP mRNA in a cell-free system produced multiple species of pre-PSP [116], but in the presence of microsomal membranes these were converted to a single form which corresponded to the single form detected intracellularly after a 5-min incubation of parathyroid tissue with radioactive amino acids [115]. The multiple forms of pre-PSP may be the result of different lengths of pre-sequences resulting from multiple initiation sites or mRNAs. Since such heterogeneity has not been confirmed in intact cells, the possibilities that multiple pre-PSPs are an artifact of the cell-free system cannot be ruled out. Secreted PSP is glycosylated [98, 136] and consists of at least two or three species [115, 138]. The single species of PSP observed in cells after a 5-min incubation can be 'chased' partially into several larger species of PSP, probably as a result of glycosylation [115]. These intracellular species correspond to the secreted species and their identification is consistent with other reports of multiple intracellular species of PSP [138]. Glycosylation appears to occur after transport from rough endoplasmic reticulum since tunicamycin did not

inhibit glycosylation and glycosylation did not occur in cell-free transla-
tions of PSP mRNA in the presence of canine pancreatic microsomes [115].
Therefore, the initial translational product of PSP is heterogeneous, at
least in vitro. These species are converted to a single species by proteolytic
cleavage, and heterogeneity is then reintroduced by glycosylation.

PSP has been purified to near homogeneity and radioimmunoassays
have been developed [42, 164]. Consistent with earlier studies [98], PSP
monomers have a molecular weight of about 70,000 daltons while the
purified native protein has an apparent molecular weight of 150,000 to
300,000 [42], indicating that the protein in solution contains 2–4 subunits.
PSP contains 30–35% acidic residues and has a pK of about 4.5. Estimates
of carbohydrate content ranged 2.6 [42] to 18% [164] and preparations
contained one or two major species. These differences may reflect the
preferential isolation in the different procedures of the various species of
PSP which are differentially glycosylated.

The function of PSP has not yet been defined. Its acidic nature suggests
that it might bind the basic peptide PTH or ProPTH but evidence to
support this is lacking. Its close association with PTH suggests that it must
play some role in the intracellular processing for secretion of PTH. There
is no evidence that PSP has a biological function after secretion. PTH itself
corrects the changes in calcium metabolism in the parathyroidectomized
animal although more subtle long-term effects of PSP certainly have not
yet been excluded.

Conclusion

The biosynthesis and secretion of PTH is summarized in figure 5.
Both the bovine and human PTH genes contain intervening sequences that
are present in the 5′ portion of the gene. The sequence of bovine PTH
mRNA has been determined and contains about 670 nucleotides exclusive
of polyadenylate at the 3′ terminus. Most of the sequence for the human
mRNA has been determined. Considerable homology exists between the
two mRNAs particularly in the 5′ noncoding and coding regions. The 3′
noncoding region of human mRNA contains about 125 nucleotides more
than that of the bovine mRNA.

The initial translational product of PTH mRNA is preProPTH which
is converted by two sequential proteolytic cleavage to ProPTH and then
PTH. Conversion of preProPTH to ProPTH occurs cotranslationally in

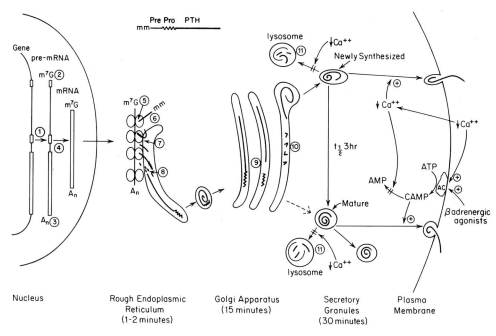

Fig. 5. Schematic summary of the biosynthesis and secretion of PTH. The times under the cytoplasmic organelles indicate the approximate time required for newly synthesized PTH to reach the organelles. In the gene and pre-mRNA the open boxes represent exons and the lines represent introns and flanking regions of the gene. AC is adenyl cyclase. The stimulation of secretion from the mature PTH pool by cAMP and from both the mature and newly synthesized pools by decreasing the extracellular calcium concentrations has thus far been observed only in porcine cells. The circled numbers indicate enzymatic activities involved in the processing of PTH or its mRNA template and are: 1=RNA polymerase II; 2=enzymes involved in capping and methylation of the 5′ terminus of mRNA; 3=polyadenylate polymerase; 4=RNA-splicing enzymes; 5=peptidyl transferase, 6=initiator methionyl aminopeptidase; 7=signalase; 8=enzymes that degrade the pre-sequence; 9=a tryptic-like enzyme that removes the pro-sequence; 10=amino and carboxyl peptidases that degrade the pro-sequence, and 11=proteolytic enzymes that degrade PTH intracellularly.

the rough endoplasmic reticulum and conversion of ProPTH to PTH occurs by a tryptic-like cleavage about 15 min after synthesis at the time when the ProPTH reaches the Golgi apparatus. The 25 amino acid presequence functions as a classical 'signal sequence' while the function of the hexapeptide prosequence is unknown. Intracellular transport of PTH appears to follow the pathway for secreted proteins proposed by *Palade* [144].

Secretion of PTH is controlled primarily by the concentration of extracellular calcium. Stimulation of PTH secretion by decreasing concentrations of calcium involves an increase in intracellular cAMP and possibly a decrease in intracellular calcium concentrations. Secretion of PTH appears to follow two pathways. One pathway involves 'old' PTH whose secretion is stimulated by agents, including calcium, that increase intracellular cAMP concentrations, and the other pathway involves newly synthesized PTH whose secretion is stimulated by decreased intracellular calcium but not by secretagogues that increase intracellular cAMP. Morphological studies and parallel secretion of PSP and plasminogen activator indicate that secretion occurs by exocytosis.

A second major secreted product of parathyroid tissue is PSP which has a molecular weight of 70,000 daltons and, by mass, accounts for 70% of the protein secreted. PSP is initially synthesized as pre-PSP; the presequence contains no homology to that of preProPTH. PSP that is secreted is heterogeneous probably because of differential extents of glycosylation. PSP has no known hormonal activity and its function, if any, in the secretion of PTH is not known.

References

1   Abe, M.; Sherwood, L. M.: Regulation of parathyroid hormone secretion by adenyl cyclase. Biochem. biophys. Res. Commun. *48:* 396 (1972).
2   Altenahr, E.; Seifert, G.: Ultrastructural comparison of human parathyroid glands in secondary hyperparathyroidism and primary parathyroid adenoma. Virchows Arch. Abt. A Path. Anat. *353:* 60 (1971).
3   Altenahr, E.: Ultrastructure of rat parathyroid glands in normo-, hyper-, and hypocalcemia. Virchows Arch. Abt. A Path. Anat. *351:* 122 (1970).
4   Attie, M. F.; Brown, E. M.; Gardner, D. G.; Spiegel, A. M.; Aurbach, G. D.: Characterization of the dopamine-responsive adenylate cyclase of bovine parathyroid cells and its relationship to parathyroid hormone secretion. Endocrinology *107:* 1776 (1980).
5   Au, W. Y. W.: Cortisol stimulation of parathyroid hormone secretion. Science *193:* 1015 (1976).
6   Au, W. Y. W.; Poland, A. P.; Stern, P. H.; Raisz, L. G.: Hormone synthesis and secretion by rat parathyroid glands in tissue culture. J. clin. Invest. *49:* 1639 (1970).
7   Aurbach, G. D.: Isolation of parathyroid hormone after extraction with phenol. J. biol. Chem. *234:* 3179 (1959).
8   Aurbach, G. D.; Marx, S. J.; Spiegel, H. M.: Parathyroid hormone, calcitonin, and the calciferols; in Williams, Textbook of endocrinology; 6th ed., p. 922 (Saunders, Philadelphia 1981).
9   Austen, B. M.; Ridd, D. H.: The signal peptide and its role in membrane penetration. Biochem. Soc. Symp. *46:* 235 (1981).

10   Bellorin-Font, E.; Martin, K.J.; Freitag, J.J.; Anderson, C.; Sicard, G.; Slatopolsky, E.; Klahr, S.: Altered adenylate cyclase kinetics in hyperfunctioning human parathyroid cells. J. clin. Endocr. Metab. *52:* 499 (1981).

11   Berman, L.: Crystalline substance from the parathyroid gland that influences the calcium content of the blood. Proc. Soc. exp. Biol. Med. *21:* 465 (1924).

12   Blobel, G.; Dobberstein, B.: Transfer of proteins across membranes. I. Presence of proteolytically processed and unprocessed nascent immunoglobulin light chains on membrane-bound ribosomes of murine myeloma. J. Cell Biol. *67:* 835 (1975).

13   Blobel, G.; Sabatini, D.D.: Ribosome membrane interaction in eukaryotic cells; in Manson, Biomembranes, vol.2, p.193 (Plenum Press, New York 1971).

14   Blum, J.W.; Fischer, J.A.; Hunziker, W.H.; Binswanger, U.; Picotti, G.B.; DaPrada, M.; Guillebeau, A.: Parathyroid hormone responses to catecholamines and to changes of extracellular calcium in cows. J. clin. Invest. *61:* 1113 (1978).

15   Boime, I.; Szczesna, E.; Smith, D.: Membrane-dependent cleavage of the human placental lactogen precursor to its native form in ascites cell-free extracts. Eur. J. Biochem. *73:* 515 (1977).

16   Breathnach, R.; Chambon, P.: Organization and expression of eukaryotic split genes coding for proteins. A. Rev. Biochem. *50:* 349 (1981).

17   Brewer, H.B.; Fairwell, T.; Ronan, R.; Rittel, W.; Arnaud, C.: Human parathyroid hormone; in Talmage, Owen, Parsons, Calcium regulating hormones, p.23 (Excerpta Medica, Amsterdam 1975).

18   Brewer, H.B., Jr.; Fairwell, T.; Ronan, R.; Sizemore, G.W.; Arnaud, C.D.: Human parathyroid hormone and amino acid sequence of the amino terminal residues 1–34. Proc. natn. Acad. Sci. USA *69:* 3585 (1972).

19   Brewer, H.B.; Ronan, R.: Bovine parathyroid hormone: amino acid sequence. Proc. natn. Acad. Sci. USA *67:* 1862 (1970).

20   Brown, E.M.: Calcium-regulated phosphodiesterase in bovine parathyroid cells. Endocrinology *107:* 1998 (1980).

21   Brown, E.: Lithium induces abnormal calcium-regulated PTH release in dispersed bovine parathyroid cells. J. clin. Endocr. Metab. *52:* 1046 (1981).

22   Brown, E.M.; Attie, M.F.; Reen, S.; Gardner, D.G.; Kebabian, J.; Aurbach, G.D.: Characterization of dopamine receptors in dispersed bovine parathyroid cells. Molec. Pharmacol. *18:* 335 (1980).

23   Brown, E.M.; Aurbach, G.D.: Role of cyclic nucleotides in secretory mechanisms and actions of parathyroid hormone and calcitonin. Vitams Horm. *38:* 206 (1980).

24   Brown, E.M.; Carroll, R.J.; Aurbach, G.D.: Dopaminergic stimulation of cyclic AMP accumulation and parathyroid hormone release from dispersed bovine parathyroid cells. Proc. natn. Acad. Sci. USA *74:* 4210 (1977).

25   Brown, E.M.; Gardner, D.G.; Aurbach, G.D.: Effects of the calcium ionophore A23187 on dispersed bovine parathyroid cells. Endocrinology *106:* 133 (1980).

26   Brown, E.M.; Gardner, D.G.; Windeck, R.A.; Aurbach, G.D.: Relationship of intracellular 3′,5′-adenosine monophosphate accumulation to parathyroid hormone release from dispersed bovine parathyroid cells. Endocrinology *103:* 2323 (1978).

27   Brown, E.M.; Hurwitz, S.; Aurbach, G.D.: Beta-adrenergic stimulation of cyclic AMP content and parathyroid hormone release from isolated bovine parathyroid cells. Endocrinology *100:* 1696 (1977).

28   Brown, E.M.; Pazoles, C.J.; Creutz, C.E.; Aurbach, G.D.; Pollard, H.B.: Role of

anions in parathyroid hormone release from dispersed bovine parathyroid cells. Proc. natn. Acad. Sci. USA 75: 876 (1978).

29 Brown, E. M.; Thatcher, J. G.: Adenosine 3'5'-monophosphate (cAMP)-dependent protein kinase and the regulation of parathyroid hormone release by divalent cations and agents elevating cellular cAMP in dispersed bovine parathyroid cells. Endocrinology 110: 1374 (1982).

30 Bruce, B. R.; Anderson, N. C., Jr.: Hyperpolariztion in mouse parathyroid cells by low calcium. Am. J. Physiol. 236: C15 (1979).

31 Chase, L. R.; Slatopolsky, E.: Secretion and metabolic efficacy of parathyroid hormone in patients with severe hypomagnesemia. J. clin. Endocr. 38: 363 (1974).

32 Chertow, B. S.; Manke, D. J.; Williams, G. A.; Baker, G. R.; Hargis, G. K.; Buschmann, R. J.: Secretory and ultrastructural responses of hyperfunctioning human parathyroid tissues to varying calcium concentration and vinblastine. Lab. Invest. 36: 198 (1977).

33 Chow, P. Y.; Fasman, G. D.: Prediction of the secondary structure of proteins from their amino acid sequence. Adv. Enzymol. 47: 45 (1978).

34 Chu, L. L. H.; Huang, D. W. Y.; Littledike, E. T.; Hamilton, J. W.; Cohn, D. V.: Porcine proparathyroid hormone: identification, biosynthesis and partial amino acid sequence. Biochemistry 14: 3631 (1975).

35 Chu, L. L. H.; MacGregor, R. R.; Anast, C. S.; Hamilton, J. W.; Cohn, D. V.: Studies on the biosynthesis of rat parathyroid hormone and proparathyroid hormone: adaptation of the parathyroid gland to dietary restriction of calcium. Endocrinology 93: 915 (1973).

36 Chu, L. L. H.; MacGregor, R. R.; Cohn, D. V.: Energy-dependent intracellular translocation of proparathormone. J. Cell Biol. 72: 1 (1977).

37 Chu, L. L. H.; MacGregor, R. R.; Hamilton, J. W.; Cohn, D. V.: Conversion of proparathyroid hormone to parathyroid hormone: the use of amines as specific inhibitors. Endocrinology 95: 1431 (1974).

38 Cohn, D. V.; MacGregor, R. R.: The biosynthesis, intracellular processing, and secretion of parathormone. Endocrine Rev. 2: 1 (1981).

39 Cohn, D. V.; MacGregor, R. R.; Chu, L. L. H.; Hamilton, J. W.: Studies on the biosynthesis in vitro of parathyroid hormone and other calcemic polypeptides of the parathyroid gland; in Talmage, Munson, Calcium parathyroid hormone and calcitonin, p.173 (Excerpta Medica, Amsterdam 1972).

40 Cohn, D. V.; MacGregor, R. R.; Chu, L. L. H.; Huang, D. W. Y.; Anast, C. S.; Hamilton, J. W.: Biosynthesis of proparathyroid hormone and parathyroid hormone. Chemistry, physiology, and role of calcium in regulation. Am. J. Med. 56: 767 (1974).

41 Cohn, D. V.; MacGregor, R. R.; Chu, L. L. H.; Kimmel, J. R.; Hamilton, J. W.: Calcemic fraction-A: biosynthetic peptide precursor of parathyroid hormone. Proc. natn. Acad. Sci. USA 69: 1521 (1972).

42 Cohn, D. V.; Morrissey, J. J.; Hamilton, J. W.; Shofstall, R. E.; Smardo, F. L.; Chu, L. L. H.: Isolation and partial characterization of secretory protein 1 from bovine parathyroid glands. Biochemistry 20: 4135 (1981).

43 Cohn, D. V.; Morrissey, J. J.; Shofstall, R. E.; Chu, L. L. H.: Cosecretion of secretory protein-1 and parathormone by dispersed bovine parathyroid cells. Endocrinology 110: 625 (1982).

44 Cohn, D. V.; Smardo, F. L.; Morrissey, J. J.: Evidence for internal homology in bovine preproparathyroid hormone. Proc. natn. Acad. Sci. USA 76: 1469 (1979).

45 Collip, J. B.: The extraction of a parathyroid hormone which will prevent or control parathyroid tetany and which regulates the level of blood calcium. J. biol. Chem. *63:* 395 (1925).

46 Deftos, L. J.; Lorenzi, M.; Bohanon, N.; Tsalakian, E.; Schneider, V.; Gerich, J. E.: Somatostatin does not suppress plasma parathyroid hormone. J. clin. Endocr. Metab. *43:* 205 (1976).

47 Deftos, L. J.; Parthemore, J. G.: Secretion of parathyroid hormone in patients with medullary thyroid carcinoma. J. clin. Invest. *54:* 416 (1974).

48 DeLuca, H. F.; Holick, M. F.: Vitamin D: biosynthesis, metabolism and mode of action; in DeGroot, Endocrinology, vol. 2, p. 653 (Grune & Stratton, New York 1979).

49 Dietel, M.; Dorn-Quint, G.: By-pass secretion of human parathyroid adenomas. Lab. Invest. *43:* 116 (1980).

50 Dietel, M.; Dorn, G.; Altenahr, E.: PTH biosynthesis and secretion: quantitative correlation of biochemistry and ultrastructure in response to substances modulating parathyroid activity (Abstract); in Cohn, Talmage, Matthews, Hormonal control of calcium metabolism, p. 341 (Excerpta Medica, Amsterdam 1981).

51 Dorner, A. J.; Kemper, B.: Conversion of pre-proparathyroid hormone to proparathyroid hormone by dog pancreatic microsomes. Biochemistry *17:* 5550 (1978).

52 Dorner, A. J.; Kemper, B.: Formation in vitro of membrane-bound polysomes containing parathyroid hormone messenger ribonucleic acid. Biochemistry *24:* 5496 (1979).

53 Fischer, J. A.; Blum, J. W.; Binswanger, U.: Acute parathyroid hormone response to epinephrine in vivo. J. clin. Invest. *52:* 2434 (1973).

54 Fischer, J. A.; Oldham, S. B.; Sizemore, G. W.; Arnaud, C. D.: Calcitonin stimulation of parathyroid hormone secretion in vivo. Hormone metabol. Res. *3:* 223 (1971).

55 Flueck, J. A.; Di Bella, F. P.; Edis, A. J.; Kehrwald, J. M.; Arnaud, C. D.: Immunoheterogeneity of parathyroid hormone in venous effluent serum from hyperfunctioning parathyroid glands. J. clin. Invest. *60:* 1367 (1977).

56 Fucik, R. F.; Kukreja, S. C.; Hargis, G. K.; Bowser, E. N.; Henderson, W. J.; Williams, G. A.: Effect of glucocorticoids on function of the parathyroid glands in man. J. clin. Endocr. Metab. *40:* 152 (1975).

57 Gley, M. E.: Sur les effets de l'extirpation du corps thyroïde. C. r. Séanc. Soc. Biol. *43:* 551 (1891).

58 Goltzman, D.; Callahan, E. N.; Tregear, G. W.; Potts, J. T., Jr.: Conversion of proparathyroid hormone to parathyroid hormone: studies in vitro with trypsin. Biochemistry *15:* 5076 (1976).

59 Gordon, D. F.; Kemper, B.: Synthesis, characterization and molecular cloning of near full length PTH cDNA. Nucl. Acids Res. *8:* 5669 (1980).

60 Habener, J. F.: Regulation of parathyroid hormone secretion and biosynthesis. A. Rev. Physiol. *43:* 211 (1981).

61 Habener, J. F.; Amherdt, M.; Ravazzola, M.; Orci, L.: Parathyroid hormone biosynthesis. Correlation of conversion of biosynthetic precursors with intracellular protein migration as determined by electron microscope autoradiography. J. Cell Biol. *80:* 715 (1979).

62 Habener, J. F.; Chang, H. T.; Potts, J. T., Jr.: Enzymic processing of proparathyroid hormone by cell-free extracts of parathyroid glands. Biochemistry *16:* 3910 (1977).

63 Habener, J. F.; Kemper, B.; Potts, J. T., Jr.: Calcium-dependent intracellular degrada-

tion of parathyroid hormone: a possible mechanism for the regulation of hormone stores. Endocrinology *97:* 431 (1975).

64   Habener, J. F.; Kemper, B.; Potts, J. T., Jr.; Rich, A.: Bovine proparathyroid hormone: structural analysis of radioactive peptides formed by limited cleavage. Endocrinology *92:* 219 (1973).

65   Habener, J. F.; Kemper, B. W.; Potts, J. T., Jr.; Rich, A.: Calcium-independent intra-cellular conversion of proparathyroid hormone to parathyroid hormone. Endocr. Res. Commun. *1:* 239 (1974).

66   Habener, J. F.; Kemper, B.; Potts, J. T., Jr.; Rich, A.: Parathyroid mRNA directs the synthesis of pre-proparathyroid hormone and proparathyroid hormone in the Krebs ascites cell-free system. Biochem. biophys. Res. Commun. *67:* 1114 (1975).

67   Habener, J. F.; Kemper, B.; Potts, J. T., Jr.; Rich, A.: Pre-proparathyroid hormone identified by cell-free translation of messenger RNA from hyperplastic human para-thyroid tisue. J. clin. Invest. *56:* 1328 (1975).

68   Habener, J. F.; Kronenberg, H. M.; Potts, J. T., Jr.; Orci, L.: Biosynthesis of pre-pro-parathyroid hormone. Methods Cell Biol. *23:* 51 (1981).

69   Habener, J. F.; Maunus, R.; Dee, P. C.; Potts, J. T., Jr.: Early events in the cellular formation of proparathyroid hormone. J. Cell Biol. *85:* 292 (1980).

70   Habener, J. F.; Potts, J. T., Jr.: Chemistry, biosynthesis, secretion and metabolism of parathyroid hormone, in Handbook of physiology. Sect. 7: Endocrinology. Vol. VII: Parathyroid gland, p. 313 (American Physiological Society, Washington 1976).

71   Habener, J. F.; Potts, J. T., Jr.: Relative effectiveness of magnesium and calcium on the secretion and biosynthesis of parathyroid hormone in vitro. Endocrinology *98:* 197 (1976).

72   Habener, J. F.; Potts, J. T., Jr.: Biosynthesis of parathyroid hormone. Part I. New Engl. J. Med. *299:* 580 (1978).

73   Habener, J. F.; Potts, J. T., Jr.: Biosynthesis of parathyroid hormone. Part II. New Engl. J. Med. *299:* 635 (1978).

74   Habener, J. F.; Potts, J. T., Jr.: Subcellular distributions of parathyroid hormone, hor-monal precursors, and parathyroid secretory protein. Endocrinology *104:* 265 (1979).

75   Habener, J. F.; Potts, J. T., Jr.; Rich, A.: Pre-proparathyroid hormone: evidence for an early biosynthetic precursor of parathyroid hormone. J. biol. Chem. *251:* 3893 (1976).

76   Habener, J. F.; Rosenblatt, M.; Dee, P. C.; Potts, J. T., Jr.: Cellular processing of pre-proparathyroid hormone involves rapid hydrolysis of the leader sequence. J. biol. Chem. *254:* 10596 (1979).

77   Habener, J. F.; Rosenblatt, M.; Kemper, B.; Kronenberg, H. M.; Rich, A.; Potts, J. T., Jr.: Preproparathyroid hormone: amino acid sequence, chemical synthesis and some biological studies of the precursor region. Proc. natn. Acad. Sci. USA *75:* 2616 (1978).

78   Habener, J. F.; Stevens, T. D.; Ravazzola, M.; Orci, L.; Potts, J. T., Jr.: Effects of calcium ionophores on the synthesis and release of parathyroid hormone. Endocrinol-ogy *101:* 1524 (1977).

79   Habener, J. F.; Tregear, G. W.; Stevens, T. D.; Dee, P. C.; Potts, J. T., Jr.: Radioim-munoassay for proparathyroid hormone. Endocr. Res. Commun. *1:* 1 (1974).

80   Hamilton, J. W.; Cohn, D. V.: Studies on the biosynthesis in vitro of parathyroid hor-
      mone. I. Synthesis of parathyroid hormone by bovine parathyroid gland slices and its
      control by calcium. J. biol. Chem. *244:* 5421 (1969).

81   Hamilton, J. W.; Huang, D. W. Y.; Chu, L. L. H.; MacGregor, R. R.; Cohn,
      D. V.: Chemical and biological properties of proparathyroid hormone; in Talmage,
      Owen, Parsons, Calcium regulating hormones, p. 40 (Excerpta Medica, Amsterdam
      1975).

82   Hamilton, J. W.; MacGregor, R. R.; Chu, L. L. H.; Cohn, D. V.: The isolation and
      partial purification of a nonparathyroid hormone calcemic fraction from bovine para-
      thyroid glands. Endocrinology *89:* 1440 (1971).

83   Hamilton, J. W.; Niall, H. D.; Jacobs, J. W.; Keutmann, H. T.; Potts, J. T., Jr.; Cohn,
      D. V.: The N-terminal amino acid sequence of bovine proparathyroid hormone. Proc.
      natn. Acad. Sci. USA *71:* 653 (1974).

84   Hamilton, J. W.; Spieto, F. W.; MacGregor, R. R.; Cohn, D. V.: Studies on the biosyn-
      thesis in vitro of parathyroid hormone. II. The effect of calcium and magnesium on
      synthesis of parathyroid hormone isolated from bovine parathyroid tissue and incu-
      bation medium. J. biol. Chem. *246:* 3224 (1971).

85   Hanley, D. A.; Takatsuki, K.; Sultan, J. M.; Schneider, A. B.; Sherwood, L. M.: Direct
      release of parathyroid hormone fragments from functioning bovine parathyroid glands
      in vitro. J. clin. Invest. *62:* 1247 (1978).

86   Hanson, A. M.: The hormone of the parathyroid gland. Proc. Soc. exp. Biol. Med. *22:*
      560 (1924).

87   Hargis, G. K.; Williams, G. A.; Reynolds, W. A.; Chertow, B. S.; Kukreja, S. C.; Bow-
      ser, E. N.; Henderson, W. J.: Effect of somatostatin on parathyroid hormone and cal-
      citonin secretion. Endocrinology *102:* 745 (1978).

88   Heinrich, G.; Kronenberg, H. M.: Effect of extracellular calcium on cellular levels of
      the mRNA encoding pre-proparathyroid hormone. Measurements with a sensitive
      radiodensitometric cDNA hybridization assay; in Program and Abstracts 63rd Annu.
      Meet. Endocr. Soc., p. 58 (Endocrine Soc., Bethesda, Md. 1981).

89   Hendy, G. N.; Kronenberg, H. M.; Potts, J. T., Jr.; Rich, A.: Nucleotide sequence of
      cloned cDNAs encoding human preproparathyroid hormone. Proc. natn. Acad. Sci.
      USA *78:* 7365 (1981).

90   Huang, D. W. Y.; Chu, L. L. H.; Hamilton, J. W.; McGregor, D. H.; Cohn, D. V.: The
      NH$_2$-terminal amino acid sequence of human proparathyroid hormone by radioisotope
      microanalysis. Archs Biochem. Biophys. *166:* 67 (1975).

91   Hughes, M. R.; Haussler, M. R.: 1,25-dihydroxyvitamin D$_3$ receptors in parathyroid
      glands. J. biol. Chem. *253:* 1065 (1978).

92   Jacobs, J. W.; Kemper, B.; Niall, H. D.; Habener, J. F.; Potts, J. T., Jr.: Structural
      analysis of human proparathyroid hormone by a new microsequencing approach.
      Nature, Lond. *249:* 155 (1974).

93   Jamieson, J. D.; Palade, G. E.: Synthesis, intracellular transport, and discharge of
      secretory proteins in stimulated pancreatic exocrine cells. J. Cell Biol. *50:* 135 (1971).

94   Kemper, B.: Inactivation of parathyroid hormone mRNA by treatment with periodate
      and aniline. Nature, Lond. *262:* 321 (1976).

95   Kemper, B.; Habener, J. F.; Ernst, M. D.; Potts, J. T., Jr.; Rich, A.: Preproparathyroid
      hormone: analysis of radioactive tryptic peptides and amino acid sequence. Biochem-
      istry *15:* 15 (1976).

96 Kemper, B.; Habener, J. F.; Mulligan, R. C.; Potts, J. T., Jr.; Rich, A.: Pre-propara-thyroid hormone: a direct translation product of parathyroid messenger RNA. Proc. natn. Acad. Sci. USA 71: 3731 (1974).

97 Kemper, B.; Habener, J. F.; Potts, J. T., Jr.; Rich, A.: Proparathyroid hormone: iden-tification of a biosynthetic precursor to parathyroid hormone. Proc. natn. Acad. Sci. USA 69: 643 (1972).

98 Kemper, B.; Habener, J. F.; Rich, A.; Potts, J. T., Jr.: Parathyroid secretion: discovery of a major calcium-dependent protein. Science 184: 167 (1974).

99 Kemper, B.; Habener, J. F.; Rich, A.; Potts, J. T., Jr.: Microtubules and the intracel-lular conversion of proparathyroid hormone to parathyroid hormone. Endocrinology 96: 903 (1975).

100 Kemper, B.; Weaver, C. A.; Gordon, D. F.: Structure and function of bovine parathy-roid hormone messenger RNA; in Cohn, Talmage, Matthews, Hormonal control of calcium metabolism, p. 19 (Excerpta Medica, Amsterdam 1981).

101 Keutmann, H. T.: Chemistry of parathyroid hormone; in DeGroot, Endocrinology, vol. 2, p. 593 (Grune & Stratton, New York 1979).

102 Keutmann, H. T.; Aurbach, G. D.; Dawson, B. F.; Niall, H. D.; Deftos, L. J.; Potts, J. T., Jr.: Isolation and characterization of the bovine parathyroid isohormones. Bio-chemistry 10: 2779 (1971).

103 Keutmann, H. T.; Dawson, B. F.; Aurbach, G. D.; Potts, J. T., Jr.: A biologically active amino-terminal fragment of bovine parathyroid hormone prepared by dilute acid hydrolysis. Biochemistry 11: 1973 (1972).

104 Keutmann, H. T.; Niall, H. D.; O'Riordan, J. L. H.; Potts, J. T., Jr.: A reinvestigation of the amino-terminal sequence of human parathyroid hormone. Biochemistry 14: 1842 (1975).

105 Keutmann, H. T.; Sauer, M. M.; Hendy, G. N.; O'Riordan, J. L. H.; Potts, J. T., Jr.: Complete amino acid sequence of human parathyroid hormone. Biochemistry 17: 5723 (1978).

106 Kozak, M.: Mechanism of mRNA recognition by eukaryotic ribosomes during initia-tion of protein synthesis. Curr. Top. Microbiol. Immunol. 93: 81 (1981).

107 Kronenberg, H. M.; McDevitt, B. E.; Majzoub, J. A.; Nathans, J.; Sharp, P. A.; Potts, J. T., Jr.; Rich, A.: Cloning and nucleotide sequence of DNA coding for bovine pre-proparathyroid hormone. Proc. natn. Acad. Sci. USA 76: 4981 (1979).

108 Kronenberg, H. M.; Vasicek, T. J.; Hendy, G. N.; Rich, A.; Potts, J. T., Jr.: Structural analysis of the human parathyroid hormone gene (Abstract). Calcif. Tissue Res. 33: 322 (1981).

109 Kukreja, S. C.; Hargis, G. K.; Bowser, E. N.; Williams, G. A.; Mode of action of so-matostatin in inhibiting parathyroid hormone secretion. Hormone metabol. Res. 12: 621 (1980).

110 Kukreja, S. C.; Williams, G. A.; Hargis, G. K.; Bowser, E. N.; Banerjee, P.; Vora, N. M.; Henderson, W. J.: Dual control of suppressibility of parathyroid hormone by calcium and by β-adrenergic blockade. Mineral Electrolyte Metab. 2: 316 (1979).

111 Kukreja, S. C.; Williams, G. A.; Vora, N. M.; Hargis, G. K.: Normal responsiveness of serum parathyroid hormone to β-adrenergic blockade in patients with secondary hyperparathyroidism. Hormone metabol. Res. 13: 233 (1981).

112 L'Heureux, M. V.; Melius, P.: Differential centrifugation of bovine parathyroid tissue. Biochim. biophys. Acta 20: 447 (1956).

113 MacCallum, S. G.; Voegtlin, C.: On the relation of tetany to the parathyroid glands and to calcium metabolism. J. exp. Med. *11:* 118 (1909).

114 Mahaffee, D. D.; Cooper, C. W.; Ramp, W. K.; Ontjes, D. A.: Magnesium promotes both parathyroid hormone secretion and adenosine 3′, 5′-monophosphate production in rat parathyroid tissues and reverses the inhibitory effects of calcium on adenylate cyclase. Endocrinology *110:* 487 (1982).

115 Majzoub, J. A.; Dee, P. C.; Habener, J. F.: Cellular and cell-free processing of parathyroid secretory proteins. J. biol. Chem. *257:* 3581 (1982).

116 Majzoub, J. A.; Kronenberg, H. M.; Potts, J. T., Jr.; Rich, A.; Habener, J. F.: Identification and cell-free translation of mRNA coding for a precursor of parathyroid secretory protein. J. biol. Chem. *254:* 7449 (1979).

117 Majzoub, J. A.; Rosenblatt, M.; Fennick, B.; Maunus, R.; Kronenberg, H. M.; Potts, J. T., Jr.; Habener, J. F.: Synthetic pre-proparathyroid hormone leader sequence inhibits cell-free processing of placental, parathyroid and pituitary prehormones. J. biol. Chem. *255:* 11478 (1980).

118 Mayer, G. P.; Hurst, J. G.: Sigmoidal relationship between parathyroid hormone secretion rate and plasma calcium concentration in calves. Endocrinology *102:* 1036 (1978).

119 Mayer, G. P.; Hurst, J. G.; Barto, J. H.; Keaton, J. A.; Moore, M. P.: Effect of epinephrine on parathyroid hormone secretion in calves. Endocrinology *104:* 1181 (1979).

120 Mayer, G. P.; Keaton, J. A.; Hurst, J. G.; Habener, J. F.: Effects of plasma calcium concentration on the relative proportion of hormone and carboxyl fragments in parathyroid venous blood. Endocrinology *104:* 1778 (1979).

121 McGregor, D. H.; Chu, L. L. H.; MacGregor, R. R.; Cohn, D. V.: Disruption of the Golgi zone and inhibition of the conversion of proparathyroid hormone to parathyroid hormone in human parathyroid tissue by tris(hydroxymethyl)aminomethane. Am. J. Path. *87:* 553 (1977).

122 McGregor, D. H.; Morrisey, J. J.; Cohn, D. V.: The effects of tris(hydroxymethyl)aminomethane and calcium ionophores on the biosynthesis of proparathyroid hormone and the formation and degradation of parathormone in bovine parathyroid tissue. Am. J. Path. *102:* 336 (1981).

123 MacGregor, R. R.; Chu, L. L. H.; Cohn, D. V.: Conversion of proparathyroid hormone to parathyroid hormone by a particulate enzyme of the parathyroid gland. J. biol. Chem. *251:* 6711 (1976).

124 MacGregor, R. R.; Chu, L. L. H.; Hamilton, J. W.; Cohn, D. V.: Studies on the subcellular localization of proparathyroid hormone and parathyroid hormone in the bovine parathyroid gland: separation of newly synthesized from mature forms. Endocrinology *93:* 1387 (1973).

125 MacGregor, R. R.; Hamilton, J. W.; Cohn, D. V.: The bypass of tissue hormone stores during the secretion of newly synthesized parathyroid hormone. Endocrinology *97:* 178 (1975).

126 MacGregor, R. R.; Hamilton, J. W.; Cohn, D. V.: The mode of conversion of proparathormone to parathormone by a particulate converting enzymic activity of the parathyroid gland. J. biol. Chem. *253:* 2012 (1978).

127 MacGregor, R. R.; Hamilton, J. W.; Kent, G. N.; Shofstall, R. E.; Cohn, D. V.: The degradation of proparathormone and parathormone by parathyroid and liver cathepsin B. J. biol. Chem. *254:* 4428 (1979).

128 MacGregor, R. R.; Hamilton, J. W.; Shofstall, R. E.; Cohn, D. V.: Isolation and characterization of porcine parathyroid cathepsin B. J. biol. Chem. *254:* 4423 (1979).

129 MacGregor, R. R.; Huang, W. Y.; Cohn, D. V.: Identification of chicken proparathormone: comparison of amino acid sequence to that of chicken parathormone (Abstract). Fed. Proc. *35:* 1695 (1976).

130 Metz, S. A.; Deftos, L. J.; Baylink, D. J.; Robertson, R. P.: Neuroendocrine modulation of calcitonin and parathyroid hormone in man. J. clin. Endocr. Metab. *47:* 151 (1978).

131 Moran, J. R.; Born, W.; Tuchschmid, C. R.; Fischer, J. A.: Calcium-regulated biosynthesis of parathyroid secretory protein, proparathyroid hormone, and parathyroid hormone in dispersed bovine parathyroid cells. Endocrinology *108:* 2264 (1981).

132 Morrissey, J. J.; Cohn, D. V.: The effects of calcium and magnesium on the secretion of parathormone and parathyroid secretory protein by isolated porcine parathyroid cells. Endocrinology *103:* 2081 (1978).

133 Morrissey, J. J.; Cohn, D. V.: Regulation of secretion of parathormone and secretory protein-I from separate intracellular pools by calcium; dibutyryl cyclic AMP, and (1)-isoproterenol. J. Cell Biol. *82:* 93 (1979).

134 Morrissey, J. J.; Cohn, D. V.: Secretion and degradation of parathormone as a function of intracellular maturation of hormone pools: modulation by calcium and dibutyryl cyclic AMP. J. Cell Biol. *83:* 521 (1979).

135 Morrissey, J. J.; Cohn, D. V.; Shofstall, R. E.; MacGregor, R. R.: The biosynthesis and secretion of parathormone (PTH), secretory protein-I (SP-I) and plasminogen activator (PA) by bovine parathyroid cells (Abstract); in Cohn, Talmage, Matthews, Hormonal control of calcium metabolism, p.352 (Excerpta Medica, Amsterdam 1981).

136 Morrissey, J. J.; Hamilton, J. W.; Cohn, D. V.: The secretion of parathormone and glycosylated proteins by parathyroid cells in culture. Biochem. biophys. Res. Commun. *82:* 1279 (1978).

137 Morrissey, J. J.; Hamilton, J. W.; MacGregor, R. R.; Cohn, D. V.: The secretion of parathormone fragments 34–84 and 37–84 by dispersed porcine parathyroid cells. Endocrinology *107:* 164 (1980).

138 Morrissey, J. J.; Shofstall, R. E.; Hamilton, J. W.; Cohn, D. V.: Synthesis, intracellular distribution and secretion of multiple forms of parathyroid secretory protein I. Proc. natn. Acad. Sci. USA *77:* 6406 (1980).

139 Nakagami, K.; Warshawsky, H.; LeBlond, C. P.: The elaboration of protein and carbohydrate by rat parathyroid cells as revealed by electron microscope autoradiography. J. Cell Biol. *51:* 596 (1971).

140 Niall, H. D.; Keutmann, H. T.; Sauer, R.; Hogan, M.; Dawson, B.; Aurbach, G. D.; Potts, J. T., Jr.: The amino acid sequence of bovine parathyroid hormone. Hoppe-Seyler's Z. physiol. Chem. *351:* 1586 (1970).

141 Nussinov, R.: Nearest neighbor nucleotide patterns. Structural and biological implications. J. biol. Chem. *256:* 8458 (1981).

142 Oldham, S. B.; Lipson, L. G.; Tietjen, G. E.: Evidence for calmodulin in parathyroid tissue (Abstract); in Cohn, Talmage, Matthews, Hormonal control of calcium metabolism, p.354 (Excerpta Medica, Amsterdam 1981).

143 Palade, G.: Intracellular aspects of the process of protein synthesis. Science *189:* 347 (1975).

144 Palmer, F.J.; Sawyers, T.M.; Wierzbinski, S.J.: Cimetidine and hyperparathyroidism. New Engl. J. Med. *302:* 692 (1980).

145 Patt, H.M.; Luckhardt, A.B.: Relationship of a low blood calcium to parathyroid secretion. Endocrinology *31:* 384 (1942).

146 Proudfoot, N.G.; Brownlee, G.G.: 3′ Non-coding region sequences in eukaryotic messenger RNA. Nature, Lond. *263:* 211 (1976).

147 Ramp, W.K.; Cooper, C.W.; Ross, A.J., III; Welk, S.A., Jr.: Effects of calcium and cyclic nucleotides on rat calcitonin and parathyroid hormone secretion. Mol. cell Endocrinol. *14:* 205 (1979).

148 Ravazzola, M.: Golgi complex alterations induced by X537A in chief cells of rat parathyroid gland. Lab. Invest. *35:* 425 (1976).

149 Ravazzola, M.; Orci, L.; Habener, J.F.; Potts, J.T., Jr.: Parathyroid secretory protein: immunocytochemical localisation within cells that contain parathyroid hormone. Lancet *ii:* 371 (1978).

150 Rosenblatt, M.; Beaudette, N.V.; Fasman, G.: Conformational studies of the synthetic precursor-specific region of preproparathyroid hormone. Proc. natn. Acad. Sci. USA *77:* 3983 (1980).

151 Roth, S.I.; Capen, C.C.: Ultrastructural and functional correlations of the parathyroid gland. Int. Rev. exp. Path. *13:* 161 (1974).

152 Rubin, R.P.: The role of calcium in the release of neurotransmitter substances and hormones. Pharmac. Rev. *22:* 389 (1970).

153 Rasmussen, H.; Craig, L.C.: Parathyroid hormone. The parathyroid polypeptides. Recent Prog. Horm. Res. *18:* 269 (1962).

154 Rasmussen, H.; Sze, Y.-L.; Young, R.: Further studies on the isolation and characterization of parathyroid polypeptides. J. biol. Chem. *239:* 2852 (1964).

155 Salser, W.: Globin mRNA sequences: analysis of base pairing and evolutionary implications. Cold Spring Harb. Symp. quant Biol., vol.42, p.985 (Cold Spring Harbor Laboratory, Cold Spring Harbor 1977).

156 Sauer, R.T.; Niall, H.D.; Hogan, M.L.; Keutmann, H.T.; O'Riordan, J.L.H.; Potts, J.T., Jr.: The amino acid sequence of porcine parathyroid hormone. Biochemistry *13:* 1994 (1974).

157 Segre, G.V.; Niall, H.D.; Sauer, R.T.; Potts, J.T., Jr.: Edman degradation of radioiodinated parathyroid hormone: application to sequence analysis and hormone metabolism in vivo. Biochemistry *16:* 2417 (1977).

158 Sethi, R.; Kukreja, S.C.; Bowser, E.N.; Hargis, G.K.; Williams, G.A.: Effect of secretin on parathyroid hormone and calcitonin secretion. J. clin. Endocr. Metab. *53:* 153 (1981).

159 Sherwood, J.K.; Ackroyd, F.W.; Garcia, M.: Effect of cimetidine on circulating parathyroid hormone in primary hyperparathyroidism. Lancet *i:* 616 (1980).

160 Sherwood, L.M.; Rodman, J.S.; Lundberg, W.B.: Evidence for a precursor to circulating parathyroid hormone. Proc. natn. Acad. Sci. USA *67:* 1631 (1970).

161 Silverman, R.; Yalow, R.S.: Heterogeneity of parathyroid hormone. Clinical and physiologic implications. J. clin. Invest. *52:* 1958 (1973).

162 Steiner, D.F.; Quinn, P.S.; Chan, S.J.; Marsh, J.; Tager, H.S.: Processing mechanisms in the biosynthesis of proteins; in Zimmerman, Mumford, Steiner, Precursor processing in the biosynthesis of proteins, p.1 (New York Academy of Sciences, New York 1980).

163 Stolarsky, L.; Kemper, B.: Characterization and partial purification of parathyroid hormone messenger RNA. J. biol. Chem. *253:* 7194 (1978).

164 Takatsuki, K.; Schneider, A. B.; Shin, K. Y.; Sherwood, L. M.: Extraction purification and partial characterization of bovine parathyroid secretory protein. J. biol. Chem. *256:* 2342 (1981).

165 Targovnik, J. H.; Rodman, J. S.; Sherwood, L. M.: Regulation of parathyroid hormone secretion in vitro: quantitative aspects of calcium and magnesium ion control. Endocrinology *88:* 1477 (1971).

166 Thiele, J.: Changes in the plasma membrane associated with endocrine activity. Virchows Arch. Abt. B ZellPath. *34:* 219 (1980).

167 Tregar, G. W.; Rietschoten, J. van; Green, E.; Keutmann, H. T.; Niall, H. D.; Reit, B.; Parsons, J. A.; Potts, J. T., Jr.: Bovine parathyroid hormone: minimum chain length of synthetic peptide required for biological activity. Endocrinology *93:* 1349 (1973).

168 Weaver, C. A.; Gordon, D. F.; Kemper, B.: Introduction by molecular cloning of artifactual inverted sequences at the 5' terminus of the sense strand of bovine parathyroid hormone cDNA. Proc. natn. Acad. Sci. USA *78:* 4073 (1981).

169 Weaver, C. A.; Gordon, D. F.; Kemper, B.: Nucleotide sequence of bovine parathyroid hormone messenger RNA. Mol. cell Endocrinol. (in press, 1982).

170 Wecksler, W. R.; Ross, F. P.; Mason, R. S.; Posen, S.; Norman, A. W.: Biochemical properties of the $1\alpha,25$-dihydroxyvitamin D_3 cytoplasmic receptors from human and chick parathyroid glands. Archs Biochem. Biophys. *201:* 95 (1980).

171 Williams, G. A.; Bowser, E. N.; Hargis, G. K.; Kukreja, S. C.; Shah, J. H.; Vora, N. M.; Henderson, W. J.: Effect of ethanol on parathyroid hormone and calcitonin secretion in man. Proc. Soc. exp. Biol. Med. *159:* 187 (1978).

172 Williams, G. A.; Hargis, G. K.; Bowser, E. N.; Henderson, W. J.; Martinez, N. J.: Evidence for a role of adenosine 3',5'-monophosphate in parathyroid hormone release. Endocrinology *92:* 687 (1973).

173 Williams, G. A.; Hargis, G. K.; Ensinck, J. W.; Kukreja, S. C.; Bowser, E. N.; Chertow, B. S.; Henderson, W. J.: Role of endogenous somatostatin in the secretion of parathyroid hormone and calcitonin. Metabolism *28:* 950 (1979).

174 Williams, G. A.; Langley, R. S.; Bowser, E. N.; Hargis, G. K.; Kukreja, S. C.; Vora, N. M.; Johnson, P. A.; Jackson, B. L.; Kawahara, W. J.; Henderson, W. J.: Parathyroid hormone secretion in normal man and in primary hyperparathyroidism: Role of histamine H_2 receptors. J. clin. Endocr. Metab. *52:* 122 (1981).

175 Williams, G. A.; Peterson, W. C.; Bowser, E. N.; Henderson, W. J.; Hargis, G. K.; Martinez, N. J.: Interrelationship of parathyroid and adrenocortical function in calcium homeostasis in the rat. Endocrinology *95:* 707 (1974).

176 Willgoss, D.; Jacobi, J. M.; deJersey, J.; Bartley, P. C.; Lloyd, H. M.: Effect of calcium on cyclic nucleotide phosphodiesterase in parathyroid tissue. Biochem. biophys. Res. Commun. *94:* 763 (1980).

177 Windek, R.; Brown, E. M.; Gardner, D. G.; Aurbach, G. D.: Effect of gastrointestinal hormones on isolated bovine parathyroid cells. Endocrinology *103:* 2020 (1978).

B. Kemper, PhD, Department of Physiology and Biophysics, School of Basic Medical Sciences, University of Illinois, Urbana, IL 61801 (USA)

Cell Biology of the Secretory Process, pp. 481–516 (Karger, Basel 1984)

Mast Cell Secretion[1]

Thomas W. Martin, David Lagunoff

Department of Pathology, St. Louis University School of Medicine,
St. Louis, Mo., USA

Introduction

Mast cells are highly specialized secretory cells of the connective tissue (fig. 1–3). It is a common belief, though by no means a proven fact, that secretion is an infrequent event in the life of a mast cell. In contrast to endocrine or exocrine secretory cells, mast cells are considered to respond to stimuli that occur at highly irregular intervals. According to this concept of mast cell function, the cells after synthesizing their granules, vegetate for relatively long periods before being called on for any major secretory act. This model for mast cell behavior is supported by the ultrastructural appearance of most tissue mast cells, packed with secretory granules, without evidence of ongoing or cyclic resynthesis of their secretory constituents. The function mast cells are accorded in the repulsion of parasites [2, 14, 42] would appear to be well served by a vigilant force of primed cells prepared to rapidly discharge their potent granule contents in response to local stimulation by a soluble product of the parasite or direct contact, both enhanced in many instances by IgE antibodies capable of binding to specific receptors on the mast cell surface. The consequences of inappropriate mast cell secretion in genetically predisposed atopic individuals take the form of hives, hay fever, and perhaps asthma.

While the life of a mast cell between secretory events may well differ from that of most secreting cells, what we know of the actual process of mast cell secretion indicates a great degree of parallelism in actual secretory activity. Accumulated evidence supports the conviction that mast cell secretion is exocytotic in character (fig. 4–8), that receptors borne on the

[1] Supported in part by grant No. HL 25049.

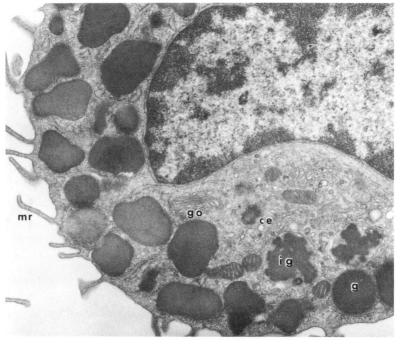

1

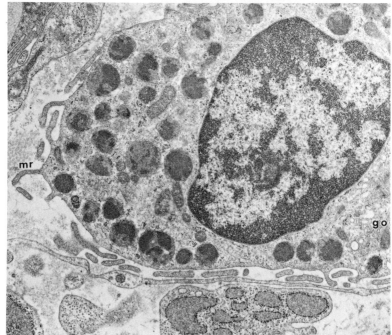

2

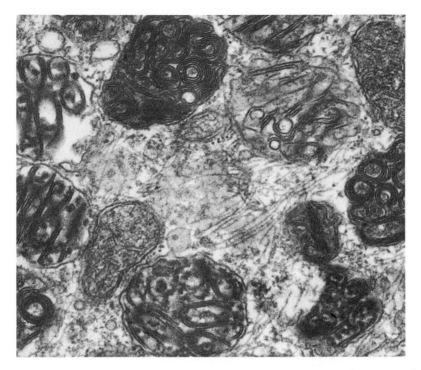

Fig. 3. Human mast cell granules. The varied structural forms of the granules in cross-section are evident in the electron micrograph of a lung mast cell from a 14-year-old girl. ×55,000.

surface of the cell combine with secretory stimuli, that an increase in free cytoplasmic Ca^{++} is a necessary presecretory event, and that ATP is required for the culmination of secretory activities [28, 84, 85]. In contrast to this set of reasonably well-established events, there is a collection of cellular activities whose relationship to mast cell secretion remains uncer-

---

Fig. 1. A portion of an untreated rat peritoneal mast cell. Microridges (mr), appearing as finger-like projections, are present at the cell surface. Most of the cytoplasm is filled with uniformly dense mast cell granules (g). Two granules (ig) are present in the region containing the Golgi apparatus (go) and a centriole (ce); these granules have the characteristic aggregated appearance of immature granules. ×38,000.

Fig. 2. A normal human mast cell from connective tissue. Prominent microridges (mr) are evident in cross-section at the cell surface. A portion of a small Golgi system (go) is present. The secretory granules exhibit inhomogeneity of structure in place of the uniform dense appearance of rat mast cell granules. ×17,000.

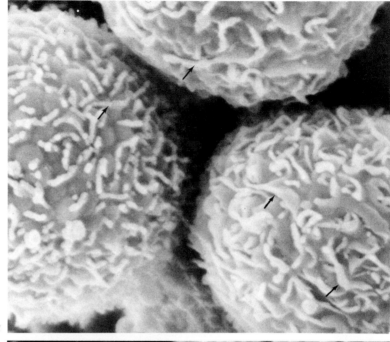

4

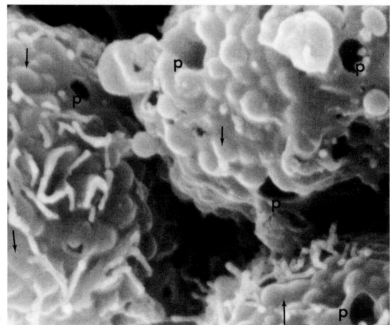

5

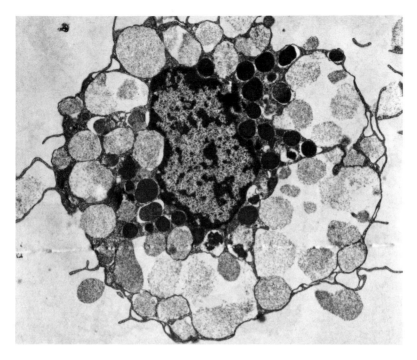

Fig. 6. A rat mast cell stimulated to secrete with polymyxin B ($2\,\mu g/\mu l$). The extensive exocytotic degranulation is evident, involving most but not all of the secretory granules. The very thin, anastomosing, cytoplasmic processes formed in the secretory process are well demonstrated. ×22,000.

tain: phospholipid turnover, phospholipase A_2 hydrolysis of membrane lipid, phospholipid methylation, cAMP generation and breakdown, protein phosphorylation, cytoskeletal perturbations, and granule swelling. Our attempt in this chapter is to critically consider the evidence for the role of the mechanisms variously proposed as part of the secretory activity of the mast cell. We have tried to avoid forcing observations into consistence

---

Fig. 4. Scanning electron micrograph of untreated isolated rat peritoneal mast cells. The surface shows that the protrusion seen in sectioned cells (fig. 1) are predominantly microridges (arrows). ×9,000.

Fig. 5. Scanning electron micrograph of rat mast cells treated with $1\,\mu g/ml$ polymyxin B for 10 min at 22 °C. Pores (p) through which granules have been extruded are visible. The microridges are markedly reduced on the surface of the secreting cell, and bulging of the granules (arrows) beneath the plasma membrane is apparent. ×11,000.

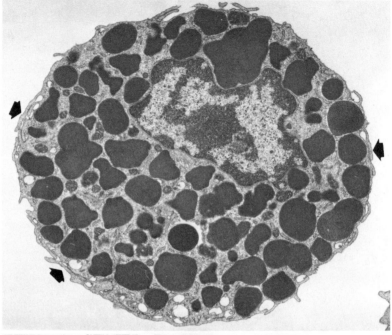

7

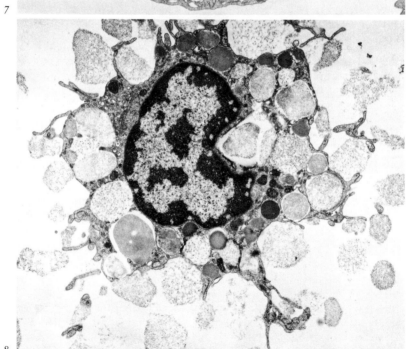

8

with any one scheme in the belief that the evidence is still too fragmentary to support such an effort.

Model Systems

Most studies of mast cell secretion have been performed with rat peritoneal mast cells. Mast cells in this species occur normally in the peritoneal cavity. In adult commercially outbred white rats (often called Sprague-Dawley, to the consternation of competitors of the firm of that name), so long as they are healthy and free of infection, between 1 and 2×10^6 mast cells can be washed from the peritoneal cavity with no more manipulation than adding and recovering 5–10 ml of appropriately buffered, heparinized, physiologic salt solution [107]. The absence of any requirement for pretreating the rats to elicit mast cells and the ease with which the mast cells can be separated from the other cells, members of the monocyte-macrophage series with a few eosinophils, by a one-step centrifugation into or through albumin [107], Ficoll [185], metrizamide [121] or Percoll [37, 141] make the rat peritoneal mast cell a favorite cell for study. A rather more cumbersome method has been developed for the isolation of mast cells from human lung [146]. In spite of the long time required to carry out the procedure and the intensive enzyme digestion of the lung that is necessary, the resulting population of mast cells does not appear any the worse for wear by electron microscopy, and they exhibit excellent secretory activity. Perhaps because of the difficulties attendant on isolating these cells, there have been few reports on the secretory activities of isolated human lung mast cells.

Studies of secretion by the basophil, a closely related but distinct cell type, have been numerous. Human basophils are quite difficult to obtain

---

Fig. 7. A mast cell exposed to Con A (100 µg/ml) but not PS. No secretion has occurred. The only evident effect of the treatment has been the adherence of microridges to one another at the intervening cell surface (arrows), presumably by the intermediation of Con A. Irregular granule forms are prevalent in this cell. These are not the result of Con A treatment but probably are created by granules fusing with one another. ×30,000.

Fig. 8. A mast cell stimulated with Con A (100 µg/ml) and PS (10 µg/ml) in the presence of 1 mM Ca^{++}. The cell exhibits extreme exocytotic secretion of granules. Only a few granules close to the nucleus have not participated in the secretory event. Note the thin ramified cell extensions formed peripherally in the process. ×27,000.

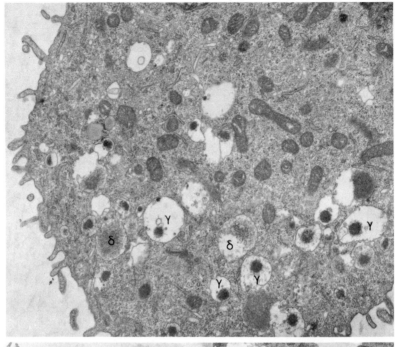

9

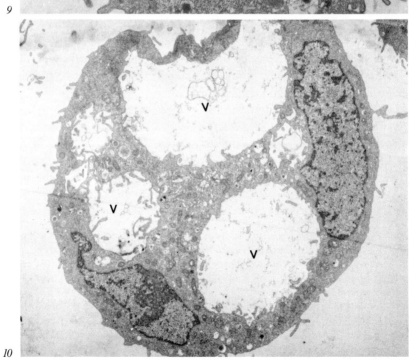

10

in the degree of purity [27, 68, 154] necessary for studies of biochemical changes in metabolites like ATP or cAMP, or enzymes like phospholiphase A_2 or protein kinase, but it has been possible to investigate the actions of an array of agents in terms of stimulation, enhancement, and inhibition of secretion of histamine by basophils. As long as the modifying agents are well behaved, the results of such experiments are useful. When unusually high concentrations of the modifiers are required or the modifiers are intrinsically not specific, considerable caution must be exercised in the interpretation of results in the absence of firm knowledge that the agent is actually performing in the manner expected. Guinea pig basophils are more readily obtained than human basophils [35, 145], but, even so, studies of these cells' secretory behavior have been largely restricted to detailed electron microscope examination [32–34, 67].

In 1974, a rat basophil leukemia cell line was established in tissue culture by *Kulczycki* et al. [101] from a transplantable leukemia which had originally arisen in a rat treated with β-chloroethylamine [36]. The original cell line did not secrete under the influence of any tested putative secretagogue, but a subline was subsequently derived under obscure circumstances that did release histamine and serotonin in response to perturbation of surface IgE receptors [161, 165, 176] or treatment with the calcium ionophore A23187 [43]. *Barsumian* et al. [4] and *McGivney* et al. [131] are involved in the clever exploitation of the rapidly growing secretory basophil leukemia cells by deriving mutant cell lines blocked at various points in the secretory process. Unfortunately, and embarrassingly for those of us who have studied the cell by electron microscopy, the morphologic events of release of histamine and serotonin by the rat basophil leukemia cells remain undefined. Even the intracellular location of the releasable histamine and serotonin has not been definitively established (fig. 9, 10).

Fig. 9. An untreated rat basophil leukemia cell. The vesicle form that occurs with the highest frequency in these cells and is the best candidate for the secretory organelle contains an eccentrically placed centrally dense body (γ). Other vesicles (δ) are more diffusely filled with amorphous material. To date, isolation and characterization of the storage vesicles of the rat basophil leukemia cell has not been achieved. ×32,000.

Fig. 10. A rat basophil leukemia cell exposed to IgE and then anti-IgE for 30 min. Most of the normal complement of granules is gone and several large vacuoles (v) have appeared. Extensive vacuolation is occasionally present in unstimulated rat basophil leukemia cells, and no actual exocytotic events involving the cell surface or the vacuoles have been identified. ×13,500.

Secretory Granule Structure

The mean volume of a peritoneal mast cell from a 6-month-old male outbred white rat is $1.08 \pm 0.08 \times 10^3 \, \mu m^3$ and the $1,220 \pm 507$ secretory granules represent 50% of that volume [66]. Detailed analysis of the distribution of measured granule areas indicates a multimodal distribution of granule size, and we propose that this pattern arises from fusion of granules with one another throughout the lifetime of the cell [66]. Within the cell, each granule is surrounded by its own perigranule membrane. Recently, it has become possible to prepare granules with their membranes [100], but there are no reported studies of the membranes.

Although it has been possible to prepare mast cell granules in high purity for almost 20 years [110], there has been little interest in characterizing their constituents, and there is still not a satisfactory complete accounting of granule components. We estimate that approximately 50% of the dry mass of the mast cell granule is protein [110]. On SDS gel electrophoresis, 6 major protein bands are present [111]. The only major granule protein that has been characterized is a protease resembling chymotrypsin in its substrate specificity and variously called chymotrypsin-like mast cell esterase, chymase, and mast cell protease I. It is a basic protein with a molecular weight of 28,500. It is a single chain and its amino acid sequence appears to be distinctive [189]. Mast cell protease I is clearly related to a second mast cell protease that is present in a restricted population of atypical mast cells but not in peritoneal mast cells [189]. While histochemical studies originally suggested that human mast cells and dog mast cells [102] contained two enzymes, one with esterase activity resembling trypsin and another enzyme with esterase activity more closely allied to chymotrypsin, examination of isolated human mast cells has revealed only a trypsin-like enzyme [158, 160]. A distinct carboxypeptidase has also been isolated from rat mast cells [108]. Its quantitative contribution to the granule proteins has not yet been measured.

Several typical lysosomal hydrolases with acidic pH optima have been identified in rat mast cell granules: N-acetyl-β-glucosaminidase [158, 159], β-glucuronidase [158, 159], and arylsulfatase A [121, 158]. Acid phosphatase, the archetypal lysosomal enzyme, is found in mast cells but not apparently in the secretory granules [158, 159]. The granule lysosomal enzymes are only trace constituents, accounting in aggregate for less than 5% of total granule protein [111].

A number of small proteins <20,000 daltons are variably present in SDS gels after electrophoresis of solubilized granules. Since their appearance can be largely suppressed by irreversible inactivation of mast cell protease prior to isolation of the granules and solubilization of the proteins, we believe that these low molecular weight proteins are products of the action of the protease on other granule proteins [108]. Whether or not the small amounts of eosinophil chemotactic tetrapeptides associated with rat mast cell granules are similarly generated is not known.

In addition to protein, the major contributor to the granule dry mass is the glycosaminoglycan anticoagulant, heparin; about 25% of the dry mass of the rat granule is heparin [110]. In the rat, heparin occurs as a high molecular weight polymer (750,000 daltons) in which many heparin chains are densely attached at their reducing ends by xylosidic bonds to a polypeptide consisting largely of repeating glycylserine units [132]. Human heparin while similarly fashioned has a lower molecular weight [169].

The granule proteins and heparin together form the granule matrix. The strictly ionic nature of the binding of most of the granule protein and heparin is attested to by the ease with which the components are separated by increasing salt concentrations. About 40% of the granule heparin is solubilized by 2.0 M KCl and the heparin is freed of all but the covalently bound heparin core polypeptide.

Histamine and a much smaller amount of serotonin in rat and mouse mast cells, are bound by virtue of their positive charges to the negatively charged granule. One of us has argued for the proposal that histamine binds to heparin negative sites that are not occupied in the heparin-protein interaction [106]. The alternative is binding to protein carboxyl groups as promulgated by *Uvnäs and Åborg* [182] and *Uvnäs* et al. [183]. While there has been agreement that histamine is not released from the granule matrix in the absence of an adequate concentration of a displacing cation, recent experiments with granules isolated with intact membranes in our laboratory lead us to believe that the binding of most of the granule histamine to the matrix is much weaker than previously thought [112]. As is the case for the chromaffin granule [89], and the lysosome [143], the pH in the mast cell granule is probably between 5 and 6 [90, 112]. This low pH may be responsible both for keeping the mast cell protease inactive and increasing the affinity of histamine for its binding sites in the matrix.

The mast cell granule matrix within its protective membrane has a density of 1.2 [100] indicating a relatively low water content. On stripping the membrane in the presence of 0.18 M NaCl, the granule matrix swells and the homogeneously dense appearance of the granule in the electron microscope gives way to a diffuse filamentous mass [104]. A similar change is characteristic of the morphologic events that ensue on granule exocytosis.

Neither the highly structured quality of some human mast cell granules nor the variation in human granule substructure has any counterpart in the rat. The most frequently described structure of the human mast cell granule is the scroll composed of tightly wound lamellae. There are also crystalline granules and amorphous granules in human mast cells [15]. Granule structure is promptly lost during the course of secretion [15]. Nothing is known of the arrangement of the macromolecular components of the human mast cell granule responsible for the ultrastructural images.

Secretory Stimuli

There is little question that in man the most common stimulus for mast cell secretion is the binding of an antigen to two molecules of IgE which are attached to their membrane receptors. Analysis of the mechanism of action of IgE-mediated secretion indicates that cross-linking of its receptors is the essential event in initiating secretory activity [16, 79, 80]. The cross-linking effector can be antigen [133], anti-IgE [78–80], anti-IgE receptor [81] or even an artificial IgE dimer [161]. A recent report indicates that multimers of IgE are more active in initiating secretion than dimers [44, 45]. Some doubt has been cast on the necessity of rigid receptor cross-linking by the observation that haptens presented in mobile form in a liposome are effective in eliciting secretion [25].

Concanavalin A (Con A) may function by cross-linking IgE or some other glycoprotein on the cell surface [95, 122, 166]. This secretagogue shares with IgE receptor-mediated stimulation and dextran, a requirement in the case of rat mast cells for exogenous phosphatidylserine (PS) to achieve maximal secretion [17, 170, 171].

The largest class of stimuli of mast cell secretion is constituted by the polyamines. 48/80 (a mixture of 4–8 unit lengths of p-methoxy-N-methyl-phenylethylamine units) [70], the topical antibiotic polymyxin B [13], polylysine [11], and a large number of polypeptides [38, 59, 82, 86, 87, 99, 139, 177] with two or more nearby free amino or guanidino groups all share the ability to act on the mast cell to cause release of histamine. The demonstration that both polymyxin B [140] and 48/80 [70] when divalently attached to Sepharose beads retain activity indicates that polyamines like the IgE-receptor-dependent stimuli act at the cell surface.

Two highly purified enzymes, α-chymotrypsin [109] and porcine pancreatic phospholipase A_2 [19], have been shown to activate mast cell secretion. The evidence is solid in the case of chymotrypsin that the catalytic activity of the enzyme is essential for stimulating secretion, and it is highly likely that phospholipase A_2 activity is required for activating secretion. Two agents, ATP [10, 24, 29] and NaF [147, 148], which cause mast cell secretion do not fall into the standard classes of secretagogues. Good evidence indicates that ATP^{-4} permeabilizes the cell membrane in general and induces secretion by letting Ca^{++} in [23, 24]. Fluoride may act in a similar fashion but not much work has been done with this agent. The effect of ionophores in causing secretion is discussed in the section on the role of Ca^{++}. Hydrogen peroxide has been shown to induce secretion [144]; this effect is markedly potentiated by eosinophil peroxidase together with halide ion [69].

Phospholipids and Mast Cell Secretion

Experiments performed in several laboratories over the past few years have provided much evidence for the involvement of phospholipids in mast cell secretion. New information has accumulated on the role of phospholipases and the effect of exogenous PS, and increases in phospholipid turnover and phospholipid methylation following stimulation of rat mast cells with certain secretagogues have been demonstrated in several laboratories.

When added exogenously to mast cells isolated from Sprague-Dawley and Wistar rats, PS activates secretion induced by IgE-dependent agents [63, 168] including Con A [126, 170, 171], dextran [17, 63], and the anaphylatoxic complement-derived polypeptides [87]; PS has no such activating effect on secretion induced with polyamines or A23187 [7, 65] and may inhibit polyamine-induced secretion by forming a less active complex [157]. No other diacyl phospholipid tested to date has been found to substitute for PS [6, 63, 136, 170].

Outside of its selectivity and specificity, not much else is known about the mechanism of action of PS. There appears to be a positive correlation between the effectiveness of PS in activating the secretory response to a given agent and the extracellular Ca^{++} dependence of the reaction. With the exception of A23187 and ATP, those secretagogues with the most stringent requirement for extracellular Ca^{++} are most affected by the presence of PS [7, 52]. Several workers have provided evidence that PS enhances secretion by decreasing the rate of desensitization following exposure of cells to the primary secretagogue, thus increasing the initial rate of secretion [5, 48]. These observations have led to the suggestion that PS acts to augment secretion by facilitating Ca^{++} influx through Ca^{++} channels initially activated by binding of the primary stimulus to the cell membrane. Direct measurements of $^{45}Ca^{++}$ uptake by rat mast cells exposed to several PS-dependent secretagogues [49] and rabbit antibody directed against IgE receptors [78] in the presence and absence of PS provide preliminary evidence to support this hypothesis.

Recent experiments performed in our laboratory have shed additional light on the chemical and physical characteristics of the interaction between PS and mast cells. Exogenous PS interacts with mast cells in a micellar state, presumably in the form of bilayer vesicles [126]. Sonication of PS dispersions containing multilamellar vesicles converts them to unilamellar vesicles and increases the potency of the PS preparation 3- to 4-fold on a molar basis [129]. PS molecules with longer hydrocarbon chains are more potent activators of mast cell secretion, but fluidity of PS dispersions appears to have little influence on their activity [129].

N-substituted derivatives of PS have been found to be totally ineffective in activating secretion [127]. Several such derivatives tested act as inhibitors of the PS effect and are potent inhibitors of secretion induced by A23187 as well as by PS-dependent secretagogues [127]. Inhibition of histamine release by one such derivative, NBD-PS, has been related to its ability to competitively block binding of radiolabeled exogenous PS to mast cells providing a potential explanation for its inhibitory effect on PS-dependent secretagogues. This cannot explain how NBD-PS inhibits secretion induced by A23187 which is unaffected by exogenous PS.

Speculation on the involvement of phospholipases in mast cell secretion dates to experiments performed by *Feldberg and Kellaway* [39, 40] who demonstrated that snake venom [39] and bee venom [40] liberated histamine from perfused lungs and lysophosphatidylcholine released histamine when injected into perfused liver [41]. *Hogberg and Uvnäs* [72, 184] found that mast cells treated with phospholipase A_2 released histamine. These experiments have since been discredited by the finding of peptide contaminants in these enzyme preparations with direct histamine-releasing activity [59, 139]. *Chi* et al. [19] have used a highly purified phospholipase A$_2$ from porcine pancreas and shown that it does cause noncytotoxic release of histamine from mast cells. Secretion was inhibited if phospholipase A$_2$ was pretreated with an enzyme inhibitor prior to incubation with mast cells. In view of these results, it would be of particular interest to know which membrane phospholipids are modified by the exogenous phospholipase A$_2$ and how these effects induce secretion, since neither lysophospholipid [128] nor fatty acids [172] added exogenously to mast cells are capable of initiating secretion.

The fact that stimulated mast cells synthesize prostaglandins and leukotrienes during secretion [116] suggests the participation of phospholipases in mast cell secretory activity. The actual enzymes involved in the generation of arachidonic acid required for the synthesis of these metabolites has not been elucidated, but by analogy with other cells, it is likely that the arachidonic acid is derived from membrane phospholipids via the direct action of phospholipase A$_2$ [186] or by the action of a diglyceride lipase [8] after initial attack by phospholipase C or phospholipase D and phosphatidate phosphohydrolase. Additional support for the involvement of a phospholipase A$_2$ in mast cell secretion comes from the demonstration that lyso-PS added exogenously to mast cells activates secretion induced by Con A [128, 167]. The effect of lyso-PS is noncytotoxic and is not exhibited by the N-substituted lyso derivative [128]. Lysophosphatidylcholine and lysophosphatidylethanolamine are incapable of activating Con-A-induced secretion, and none of the lysophospholipids including lyso-PS evokes secretion in the absence of Con A although each releases histamine from mast cells by a cytotoxic effect at high concentrations [128]. Lyso-PS was found to be 50- to 100-fold more potent than PS in activating Con-A-induced secretion [128, 167], and we have suggested that the effect of PS on secretion may involve its conversion to lyso-PS by a mast cell phospholipase A$_2$ [128].

Phospholipase A$_2$ has recently been demonstrated in purified rat peritoneal mast cells [130]. Enzyme activity is detectable in unstimulated intact cells exposed to exogenous phospholipid substrates. NBD-PS, an inhibitor of mast cell secretion, was found to be a potent inhibitor of mast cell phospholipase A$_2$, and there was a positive correlation between the extent of enzyme inhibition by NBD-PS and of secretion induced by Con A and A23187 [130]. This set of observations provides suggestive evidence for the involvement of phospholipase A$_2$ in mast cell secretion, but it must be acknowledged that the experiments do not rule out the possibility that NBD-PS acts at a site other than phospholipase A_2 to inhibit secretion. Additional experiments are necessary to evaluate the possible role of the surface mast cell phospholipase A_2 in providing arachidonic acid from endogenous membrane phospholipids during secretion.

Consistent findings in a variety of studies of phospholipid metabolism during secretion are: (1) specific ligand-receptor binding; (2) enhanced turnover of the anionic phospholipid, phosphatidylinositol (PI), and (3) Ca^{++}-dependent cell activation [113]. *Michell* [134, 135] proposed that enhanced PI turnover is in some way linked to activation of Ca^{++} channels following ligand-receptor binding. The pathway for PI turnover begins with the breakdown of PI to diglyceride. Diglyceride is then phosphorylated to phosphatidic acid (PA) which in turn combines with cytidine triphosphate to form cytidine diphosphodiglyceride. PI is reformed by incorporaton of inositol into this precursor with the release of cytidine monophosphate.

Kennerly et al. [96, 97] prelabeled mast cells with $^{32}PO_4$ and measured its incorporation into cell phospholipids after stimulation with Con A, anti-IgE, 48/80, and A23187. They observed an increased labeling of phospholipids in stimulated cells with the extent of increase dependent on the particular phospholipid class and the secretagogue. Increased labeling was most pronounced in phosphatidylcholine (PC), PI, and PA with little or no increase in other phospholipid classes. All secretagogues were capable of stimulating increased phospholipid labeling, with A23187 the least effective. Cells exposed to anti-IgE also incorporated about twice the amount of ^3H-inositol into PI as control cells. Enhanced $^{32}PO_4$ labeling of phospholipids was temperature-dependent and was increased by the presence of PS when PS-dependent secretagogues were used, but nearly 4-fold higher concentrations of Con A and anti-IgE were required for histamine release than were necessary for comparable changes in phospholipid labeling. Kinetic experiments demonstrated increased PA labeling with $^{32}PO_4$ as early as 3 s after stimulation with anti-IgE. *Cockcroft and Gomperts* [22] studied the incorporation of $^{32}PO_4$ and ^3H-inositol into PI in rat mast cells stimulated with antigen, Con A, 48/80, and chymotrypsin. All secretagogues were effective in increasing the labeling of PI, but the absolute increase varied from 2- to 4-fold depending on the secretagogue and labeled precursor employed. PS enhanced $^{32}PO_4$ incorporation into PI and histamine release in the presence of Con A, and had no effect on either response in the absence of Con A. Histamine release induced by PS and Con A required extracellular Ca^{++} whereas enhanced PI labeling occurred in the absence of added Ca^{++}. Increased $^{32}PO_4$ labeling of PI was detectable within minutes after stimulation with Con A but continued long after histamine release had stopped. It was not possible from their findings to determine if PI labeling preceded histamine release.

These studies provide evidence for enhanced phospholipid turnover in mast cells during secretion but do not establish if enhanced turnover is a requirement for secretion, an associated phenomenon, or a consequence of secretion. Although changes in PI turnover are detectable in stimulated cells, its importance in the activation of Ca^{++} channels is unclear since enhanced turnover occurs with secretagogues like 48/80 which do not require extracellular Ca^{++}. A recent observation may provide a link between phospholipid metabolism and mast cell secretion. Diacylglycerol (DAG) is one of the intermediates in the postulated pathway for phospholipid turnover which occurs during mast cell secretion, and *Kennerly* et al. [98] have demonstrated that mast cells stimulated with 48/80 accumulate DAG. They propose that the increased turnover of phospholipids observed during mast cell secretion may serve to provide an intermediate fusogen such as DAG which is necessary for secretion. Alternatively, DAG could serve as the principal immediate source of arachidonic acid through the action of DAG lipase. The data indicate that some of the DAG formed is derived from membrane phospholipids, but they have not yet reported direct evidence for the presence of either phospholipase C or D in mast cells.

Sullivan and Parker [172] have studied the role of arachidonic acid and its metabolites in mast cell secretion. When added alone to mast cells, arachidonic acid has no effect on histamine release until cytotoxic levels are reached, but preincubation of arachidonic acid with mast cells inhibits secretion induced by anti-IgE. This effect is inhibited if the preincubation with arachidonic acid is performed in the presence of indomethacin or aspirin which alone have no effect on secretion [117]. Arachidonic acid added during secretion actually potentiates secretion induced by anti-IgE and Con A. These results suggest a possible modulator role for arachidonic acid metabolites in mast cell secretion with cyclooxygenase metabolites being inhibitory. Additional experiments were performed with ETYA, an acetylenic analogue of arachidonic acid which inhibits lipoxygenase and cyclooxygenase activities in many tissues [181]. Mast cell secretion induced by anti-IgE, Con A, and A23187 was inhibited by ETYA but not by aspirin or indomethacin [172]. ETYA also inhibited basal and stimulated incorporation of $^{32}PO_4$ into phospholipids with dose-response curves which paralleled those obtained for inhibition of secretion [124]. *Nemeth and Douglas* [142] have also studied the effect of ETYA on secretion. ETYA effectively inhibited secretion induced by antigen, dextran, Con A, and A23187, but it had little or no effect on secretion induced by polyamine secretagogues. The inhibitory action of ETYA on A23187-induced secretion was attributed to interference with the ionophoretic ability of A23187 rather than an inhibitory effect mediated at the level of the cell. In view of these findings, they argue that the effect of ETYA is probably at some early step in secretion prior to Ca^{++} influx rather than at a step intrinsic to secretion; the question as to whether or not lipoxygenase metabolites are involved in an intrinsic step in secretion has yet to be answered definitively. *Chi* et al. [19] have recently demonstrated that ETYA inhibits secretion induced by treatment of mast cells with phospholipase A_2.

An entirely different series of metabolic events involving membrane phospholipids has been studied in rat mast cells and RBL cells [71]. *Hirata* et al. [71] demonstrated that rat mast cells stimulated with Con A and PS incorporated methyl groups derived from methionine into phospholipids. Kinetic experiments revealed that phospholipid methylation increased concomitant with histamine release and then decreased during the later stages of secretion. Thin-layer chromatography established that the methyl groups were incorporated into phosphatidylethanolamine (PE) forming mono-, di-, and trimethylated derivatives. Phospholipid methylation and histamine secretion were inhibited by prior treatment of cells with 3-deazaadenosine (3-DZA) which inhibits S-adenosylmethionine-dependent transmethylation. Phospholipid methylation stimulated by Con A required Mg^{++} but not extracellular Ca^{++}, whereas histamine secretion was dependent on added Ca^{++}. By varying the Mg^{++} concentration during a preincubation step with Con A and PS in the absence of Ca^{++}, it was possible to correlate the extent of phospholipid methylation during the preincubation with the amount of histamine release from washed cells subsequently exposed to Ca^{++}. Phospholipid methylation was not increased in mast cells stimulated with 48/80 or A23187. Kinetic experiments by *Ishizaka* et al. [79] have provided evidence that lipid methylation precedes $^{45}Ca^{++}$ influx and histamine release in mast cells triggered to secrete with divalent antibody directed against the IgE receptor. Monovalent Fab' fragments stimulated neither lipid methylation nor histamine secretion. Inhibition of transmethylation with S-isobutyryl-3-deazaadenosine (SIDZA) or 3-DZA resulted in inhibition of lipid methylation, $^{45}Ca^{++}$ influx, and histamine secretion. Dose-response curves indicated, however, that higher concentrations of SIDZA were required for inhibition of histamine release than were necessary for inhibition of lipid methylation. No identification of the chloroform-extracted methylated products was reported in these studies.

Work by *Morita and Siraganian* [138] provides support for the proposal that phospholipid methylation is important only in IgE-dependent mast cell secretion. Agents were used to inhibit transmethylation in mast cells and while these agents inhibited secretion induced by anti-IgE or dextran, they had no inhibitory effect when secretion was triggered with polymyxin B, 48/80, A23187, or ATP. The concentration of 3-DZA (10^{-4} *M*) required for 50% inhibition by anti-IgE was nearly two orders of magnitude higher than in earlier studies [71, 79].

A novel approach for the study of the involvement of methyltransferases in histamine secretion was employed by *McGivney* et al. [131]. Somatic genetic variants derived from secretory sublines of RBL-1 cells were selected which were deficient in methyltransferases and incapable of secretion induced by specific antigen after sensitization with IgE. The variants were still able to release histamine in the presence of A23187 and Ca^{++}. When two separately derived variants were fused, isolated hybrids expressed normal methyltransferase activity and were capable of releasing histamine after stimulation with antigen.

These experiments form a substantial basis for the hypothesis that phospholipid methylation is essential for histamine secretion induced by IgE-dependent secretagogues. However, the detailed model [71, 80] that has arisen which implicates asymmetric plasma membrane methyltransferases, transbilayer movement of methylated PE, and production of lyso-PC from the methylated derivatives via the action of phospholipase A_2 seems to go far beyond the evidence available at this time. There is no direct evidence for methyltransferase activity associated with mast cell plasma membranes, and there are no quantitative measurements of methylated products or their distribution in the mast cell membranes. As for experiments performed with the methyltransferase inhibitors, the data would be more convincing if an appropriate temporal correlation were established between inhibition of histamine secretion and phospholipid methylation. It should be noted that whereas 3-DZA and homocysteine thiolactone added together to rat platelets inhibits platelet aggregation and secretion induced by thrombin as well as phospholipid methylation [156], phospholipid methylation is completely inhibited within minutes after addition of 3-DZA and homocysteine thiolactone, but inhibition of platelet aggregation requires nearly 2 h of incubation with the inhibitors.

Calcium and Mast Cell Secretion

The advent of calcium ionophores and their application to mast cells have added considerable support to earlier ideas cogently summarized in *Douglas'* [31] proposal of a calcium-coupled stimulation-secretion mechanism analogous to the stimulation-contraction system of muscle. A23187 [21, 57, 91] and ionomycin [9] both induce mast cell secretion. The more thoroughly studied A23187-induced secretion is dependent on adequate levels of mast cell ATP [164] and proceeds by the standard exocytotic mechanism with granule extrusion. Under conventional conditions, the induction of mast cell secretion by the ionophores is dependent on the presence of extracellular calcium [21, 57, 83]. It has been possible, however, to elicit histamine release with A23187 at a subthreshold concentration of calcium by diluting cells incubated in a calcium-containing medium into a calcium-free medium in which the ionophore is present [187]. If the addition of the ionophore is delayed after the dilution step, there is a progressive loss of secretory activity with time. This result has been interpreted as evidence for a cell-associated pool of calcium that rapidly equilibrates with the medium. In other experiments, $10^{-4} M$ EDTA or EGTA in the medium enhanced histamine release by A23187 and ionomycin when calcium was absent [150]. The proposed explanation for this effect is removal of calcium from membrane regulatory sites by EDTA allowing intracellular translocations of calcium by the ionophores. A partial test of this hypothesis by substantial depletion of intracellular calcium stores prior to exposure to the ionophore plus chelator has not yet been reported. It would also be useful to know if histamine release by A23187 in the presence of chelators was secretory or dependent on some other mechanism. If a considerable portion of mast cell histamine is retained within the confines of the granules by an active granule membrane pump of some sort, then treatments that compromise pump activity could result in histamine release by leakage rather than secretion.

Several other stratagems have been employed to increase intracellular calcium concentrations; these include direct injection [94], fusion of mast cells with calcium-carrying liposomes [178], and permeabilization of the cell membrane to ionized calcium with ATP^{4-} [23]. However the calcium is introduced, secretion is induced by the increase of calcium in the mast cell cytoplasm.

All the conventional mast cell secretagogues are dependent on calcium for maximal stimulation, but the extent of secretion in the absence of calcium varies with the agent as well as the laboratory. Generally, IgE receptor-related agents are more dependent on the presence of extracellular calcium than the polyamines. In our hands, for instance, histamine release by Con A with PS is absolutely dependent on the presence of calcium, whereas absence of calcium in the medium reduces polymyxin-B-stimulated release by 33–50%. The anomalous enhancement of histamine release by the inclusion of EDTA in calcium-free medium noted above for the ionophores has been found by *Pearce* et al. [150] with all secretagogues they tested with the single exception of dextran.

Limited but convincing measurements of ^{45}Ca influx associated with exposure of cells to secretagogues [40, 50, 155] and correlation of inhibited influx with inhibited secretion [50] further encourage the assignment of a significant role to calcium entry in triggering mast cell secretory activity. Strontium can in general substitute for calcium in mast cell secretion, but 10–30 times higher concentrations are required [3, 49, 54, 55, 58, 149]. Strontium has the interesting property of being able by itself to induce secretion of histamine at sufficiently high concentration [47, 51]. Here again morphologic proof for an exocytotic secretory mechanism seems to be lacking. Barium also can induce spontaneous release of histamine. The evidence suggests that strontium acts by penetrating putative calcium channels but that barium simply displaces cell-associated calcium and thereby increases the cytoplasmic calcium concentration [3, 149]. Lanthanum [53, 56] and other lanthanides [151] block calcium-dependent histamine release.

Cyclic Nucleotides and Mast Cell Secretion

From its inception, experimentation on the role of cyclic nucleotides, predominantly cAMP, has attracted intense interest [26, 62, 64, 92]. The first observations indicated that β-adrenergic agonists and theophylline, an inhibitor of phosphodiesterase, which were expected to increase cellular cAMP levels, inhibited histamine release from human basophils [118] and chopped human lung preparations [93] stimulated by the IgE receptor-dependent mechanism. Later, measurements of cAMP levels generally confirmed the expected effects of the agents. This approach coupled with more limited evidence on enhancing effects of α-adrenergic agonists, cholinergic agonists, and β-adrenergic antagonists led to a widely promulgated concept that in contrast to the situation in many other secretory systems, mast cell secretion was inhibited by intracellular increases in cAMP. The finding of a decrease in cAMP levels on exposure of mast cells to 48/80 [61, 173] seemed to confirm the initial impression.

Careful study of the kinetics of cAMP changes in mast cells treated with Con A indicated a more complex pattern, with a rapid rise and a subsequent decrease to below resting levels [174]. Other complications that challenged the dogma were the great differences between concentrations of agents capable of increasing cell cAMP levels that were necessary to inhibit histamine secretion [88, 119, 174], the failure of some agents like IBTX which greatly increased cAMP in mast cells to inhibit histamine secretion [60, 175], and the effects of inhibitors and activators of adenyl cyclase opposite to those predicted by the dogma [1, 60, 75, 76]. Out of the ashes of the cAMP-as-inhibitor theory, a new bird has risen, turkey or phoenix remains to be determined.

An important foundation of the new theory [116] is provided by experiments on R (ribose) and P (purine) site analogues of adenosine which act on adenylate cyclase [120]. N^6-phenylisopropyladenosine (pip A), an R site analogue, elevates mast cell cAMP and enhances IgE-mediated release [74]. However, elevation of cAMP by this R site analog to levels as high or higher than those achieved in the presence of the immunologic stimulus for secretion did not result in secretion [74]. Adenosine was earlier shown to activate mast cell secretion [125], and its effects on cAMP levels and stimulation of secretion are roughly parallel to those of pipA. Interestingly, at the lowest concentrations of both agents there was a significant increase in secretory activity in the absence of any measurable increase in cell cAMP [74]. No inhibitory effects of adenosine were observed with respect to either cAMP or secretion at 10^{-4} M, a concentration at which inhibitory effects on adenylate cyclase are typically found in other systems. Basophil secretion is inhibited by adenosine at all concentrations tested without evidence of activation [123]. Dideoxyadenosine, a P site analogue, prevented the normal rise in mast cell cAMP and correspondingly inhibited mast cell secretion [74].

Protein Kinases and Mast Cell Secretion

By analogy with other cell systems, it would be expected that activation of adenyl cyclase with a transient increase in cAMP might activate a protein kinase which in turn would phosphorylate a protein or proteins critical in some phase of the secretory mechanism of cAMP-dependent phosphorylation. Exposure of mast cells to several secretagogues has been shown to induce the increased phosphorylation of the same three proteins with apparent molecular weights of 42,000, 59,000 and 68,000 daltons [162, 180]. One secretory stimulus, 48/80, also increased the incorporation of $^{32}PO_4$ into a 78,000-dalton protein in a delayed fashion, and several inhibitors of secretion, including sodium cromoglycate and quercetin, stimulated the phosphorylation of the 78,000-dalton protein and not the other three protein species [163, 179]. This group of results suggests the possibility that phosphorylation of specific proteins may be part of the common secretory mechanism and that the phosphorylation of at least one protein may be responsible for turning off or inhibiting secretion.

cAMP independent protein kinases that are activated by Ca^{++} either via calmodulin or PS are known but neither has yet been studied in mast cells. Studies of activation of two cAMP-dependent protein kinases identified in rat mast cells have been hampered by an inability to quantitatively recover the activated kinases [73, 188]. It has therefore been necessary to adopt the expedient of measuring the depletion of the activatable holoenzyme in order to assess protein kinase activation. The limited results available are consistent with activation of two such enzymes during IgE receptor-mediated secretion and dependence of activation on the elevation of cAMP [73, 188]. There is no reason to doubt that secretion is associated with changes in adenylate cyclase activity, elevated cAMP levels, protein kinase activation, and protein phosphorylation. How and where any of these metabolic events fit into the secretory mechanism remains obscure.

Cytoskeleton and Mast Cell Secretion

Another theme that permeates the literature on secretion has been the subject of considerable study in mast cells; this is the role of cytoskeletal elements. It is no surprise that mast cells have the usual complement of standard cytoskeletal components: thin actin filaments, intermediate filaments, and microtubules. Although colchicine does inhibit rat mast cell secretion at high concentrations, poor correlation between the effects of colchine on microtubule integrity and inhibition of secretion raises a serious question as to the meaning of the observed inhibition [103]. Similarly, the finding that cytochalasin B eliminates cell contractility at a much lower concentration than that required for limited interference with secretion does not support the argument for a major role of actin filaments in the secretory process. The electron-microscopic evidence for intermediate filament association with granules provides an interesting lead for further experimentation [15].

Exocytosis

Classical exocytotic secretion has been repeatedly demonstrated for mast cells [104, 105] and basophils [33, 34]; the only exception is the rat basophil leukemia cell in which exocytosis has not been identified and in which the ultrastructural correlates of histamine and serotonin release remain unclear. At an early stage prior to granule membrane association with plasma membrane and ensuing membrane fission to form a pore, bulging of the plasma membrane by the underlying granules, loss of the typical surface ridges, and a striking change in the distribution of intra-membranous particles with clearing of the particles from large portions of the bulging plasma membrane and the apposing granule membranes have been described [20, 114, 115, 137]. The absence of these modifications in cells that were rapidly frozen without prior fixation has led *Chandler and Heuser* [18] to emphasize highly localized membrane interactions in pore formation and to propose that the earlier observations on both thin section and freeze-fractured preparations resulted from an artifact produced by glutaraldehyde acting on a perturbed membrane. In the absence of kinetic studies of histamine secretion under the condition used by *Chandler and Heuser* [18], and their use of high concentrations of 48/80 (4–8 µg) at 22 °C, it is difficult to be sure that they were looking early enough to see the events reported by *Chi* and *Lagunoff* who conducted their experiments at 10 °C in order to capture early transitory changes in the membrane. In any case, important modification of membrane structure leading to pore formation and exoctytosis are critical for the secretory process, and there is a notable absence of information on the molecular character of these changes.

Granule Swelling and Mast Cell Secretion

Pollard et al. [12, 152, 153] have posited an ingenious mechanism whereby exocytosis would be dependent on osmotically driven granule swelling. They suggest that the process results from the inward movement of permeant anions in response to an actively generated proton gradient across the secretory granule membrane. They associate the influx of Cl⁻with the formation of fusion between granule membranes and plasma membrane prior to pore formation. *Caulfield* et al. [15] have observed that prior to exocytosis crystalline granules in human mast cells are transformed to an amorphous state that indicates a change in the intragranular ionic environment. A direct test of *Pollard's* model with basophils [77] has demonstrated that agents, SITS, probenecide and sulfite, which block anion transport and secretion in other cells do not inhibit histamine release, and replacement of all ions other than millimolar Ca^{++} and Cl^- or SO_4^- with any of several monosaccharides did not prevent secretion. Neither did substantial increases in osmolarity interfere with release. Hyperosmolarity can actually induce basophils to actively secrete, an observation that has been offered in explanation of the prevalence of symptoms related to histamine release in patients given hypersomolar solutions of radiocontrast material intravenously [46]. It would seem from these results that the chemisomotic swelling mechanism does not operate in basophils. Rat mast cells in contrast to human basophils do respond to hypertonic concentration of NaCl with reversibly inhibited histamine release [40]. Other tests of the anionic swelling model have not been reported for mast cells.

Coda

However strange mast cell secretory behavior may appear at times, the essentials of secretion appear to be comparable to those of other secretory cells. While the apparent variety of pathways in the mast cell for modifying cytoplasmic Ca^{++} is noteworthy, it seems highly likely that there is a set of events common to mast cell secretion stimulated by the full range of secretagogues and that an increase in cytoplasmic Ca^{++} is sufficient and probably necessary to set the events in motion. Protein phosphorylation and membrane phospholipid modification are probably included in these events, but we are still far from a clear view of the mechanism of mast cell exocytotic secretion or the details of its control.

References

1 Alm, P. E.; Bloom, G. D.: What – if any – is the role of adrenergic mechanisms in histamine release from mast cells? Agents Actions *11:* 60–66 (1981).

2 Askenase, P. W.: Immune inflammatory responses to parasites: the role of basophils, mast cells and vasoactive amines. Am. J. trop. Med. Hyg. *26:* 96–103 (1977).

3 Atkinson, G.; Ennis, M.; Pearce, F. L.: The effect of alkaline earth cations on the release of histamine from rat peritoneal mast cells treated with compound 48/80 and peptide 401. Br. J. Pharmacol. *65:* 395–402 (1979).

4 Barsumian, E. L.; Isersky, C.; Petrino, M. G.; Siraganian, R. P.: IgE-induced histamine release from rat basophilic leukemia cell lines; isolation of releasing and nonreleasing clones. Eur. J. Immunol. *11:* 317–323 (1981).

5 Baxter, J. H.; Adamik, R.: Control of histamine release: effects of various conditions on rate of release and rate of cell desensitization. J. Immun. *114:* 1034–1041 (1975).

6 Baxter, J. H.; Adamik, R.: Effects of calcium and phosphatidylserine in rat mast cell reaction to dextran. Proc. Soc. exp. Biol. Med. *152:* 266–271 (1976).

7 Baxter, J. H.; Adamik, R.: Differences in requirements and actions of various histamine-releasing agents. Biochem. Pharmac. *27:* 497–503 (1978).

8 Bell, R. L.; Kennerly, D. A.; Stanford, N.; Majerus, P. W.: Diglyceride lipase: a pathway for arachidonate release from human platelets. Proc. natn. Acad. Sci. USA *76:* 3238–3241 (1979).

9 Bennett, J. P.; Cockcroft, S.; Gomperts, B. D.: Ionomycin stimulates mast cell histamine secretion by forming a lipid-soluble calcium complex. Nature, Lond. *282:* 851–853 (1979).

10 Bennett, J. P.; Cockcroft, S.; Gomperts, B. D.: Rat mast cells permeabilized with ATP secrete histamine in response to calcium ions buffered in the micromolar range. J. Physiol., Lond. *317:* 335–345 (1981).

11 Bergmann, F.: Polylysine as histamine releaser. J. Pharm. Pharmac. *30:* 319–320 (1968).

12 Brown, E. M.; Pazoles, C. J.; Creutz, C. E.; Aurbach, G. D.; Pollard, H. B.: Regulation of parathyroid hormone release from dispersed bovine parathyroid cells by permeant anions. Proc. natn. Acad. Sci. USA *75:* 876–880 (1978).

13 Bushby, S. R. M.; Green, A. F.: The release of histamine by polymyxin B and polymyxin E. Br. J. Pharmacol. *10:* 215–218 (1955).

14 Capron, A.; Dessaint, J. P.: IgE and cells in protective immunity. Pathol. Biol. *25:* 287–290 (1977).

15 Caulfield, J. P.; Lewis, R. A.; Hein, A.; Austen, K. F.: Secretion in dissociated human pulmonary mast cells. Evidence for solubilization of granule contents before discharge. J. Cell Biol. *85:* 299–311 (1980).

16 Chabay, R.; DeLisi, C.: Receptor cross-linking and histamine release in basophils. J. biol. Chem. *255:* 4628–4635 (1980).

17 Chakravarty, N.; Goth, A.; Sen, P.: Potentiation of dextran-induced histamine release from rat mast cells by phosphatidylserine. Acta physiol. scand. *88:* 469–480 (1973).

18 Chandler, D. E.; Heuser, J. E.: Membrane fusion during exocytosis in quick-frozen mast cells. J. Cell Biol. *86:* 666–674 (1980).

19 Chi, E. Y.; Henderson, W. R.; Klebanoff, S. J.: Phospholipase A_2-induced mast

cell secretion. Role of arachidonic acid metabolites. Lab. Invest. *47:* 579-585 (1982).

20   Chi, E.Y.; Lagunoff, D.; Koehler, J.K.: Freeze-fracture study of mast cell secretion. Proc. natn. Acad. Sci. USA *73:* 2823–2827 (1976).

21   Cochrane, D.E.; Douglas, W.W.: Calcium-induces extrusion of secretory granules (exocytosis) in mast cells exposed to 48/80 or the ionophores A-23187 and X-537A. Proc. natn. Acad. Sci. USA *71:* 408–412 (1974).

22   Cockcroft, S.; Gomperts, B.D.: Evidence for a role of phosphatidylinositol turnover in stimulus-secretion coupling. Studies with rat peritoneal mast cells. Biochem. J. *178:* 681–687 (1979).

23   Cockcroft, S.; Gomperts, B.D.: Activation and inhibition of calcium-dependent histamine secretion by ATP ions applied to rat mast cells. J. Physiol., Lond. *296:* 229–243 (1979).

24   Cockcroft, S.; Gomperts, B.D.: The ATP^{4-} receptor of rat mast cells. Biochem. J. *188:* 789–798 (1979).

25   Cooper, A.D.; Balakirshnan, K.; McConnell, H.M.: Mobile haptens in liposomes stimulate serotonin release by rat basophil leukemia cells in the presence of specific immunoglobulin E. J. biol. Chem. *256:* 9379–9381 (1981).

26   Coulson, C.J.; Ford, R.E.; Marshall, S.; Walker, J.L.; Wooldridge, K.R.H.; Bowden, K.; Coombs, T.J.: Interrelationship of cyclic nucleotides and anaphylactic reactions. Nature, Lond. *265:* 545–547 (1976).

27   Day, R.P.: Basophil leucocyte separation from human peripheral blood: a technique for their isolation in high purity and higher yield. Clin. Allergy *2:* 205–212 (1972).

28   Diamant, B.: Energy production in rat mast cells and its role for histamine release. Int. Archs Allergy appl. Immun. *49:* 155–171 (1975).

29   Diamant, B.; Kruger, P.G.: Histamine release from isolated rat peritoneal mast cells induced by adenosine-5'-triphosphate. Acta physiol. scand. *71:* 291–302 (1967).

30   Diamant, B.; Patkar, S.A.: The influence of hypertonicity on histamine release from isolated rat mast cells. Agents Actions *10:* 140–144 (1979).

31   Douglas, W.W.: The role of calcium in stimulus-secretion coupling; in Cose, Goebell, Stimulus-secretion coupling in the gastrointestinal tract, pp.17–29 (University Park Press, Baltimore 1976).

32   Dvorak, A.M.: Biology and morphology of basophilic leukocytes; in Bach, Immediate hypersensitivity, modern concepts and developments (Marcel Dekker, New York 1978).

33   Dvorak, A.M.; Galli, S.J.; Morgan, E.; Galli, A.S.; Hammond, M.E.; Dvorak, M.F.: Anaphylactic degranulation of guinea pig basophilic leukocytes. Lab. Invest. *44:* 174–191 (1981).

34   Dvorak, A.M.; Newball, H.; Dvorak, H.F.; Lichtenstein, L.M.: Antigen-induced IgE-mediated degranulation of human basophils. Lab. Invest. *43:* 126–139 (1980).

35   Dvorak, H.F.; Selvaggio, S.S.; Dvorak, A.M.; Colvin, R.B.; Lean, D.B.; Rypyse, J.: Purification of basophilic leukocytes from guinea pig blood and bone marrow. J. Immun. *113:* 1694–1702 (1974).

36   Eccleston, E.; Leonard, B.J.; Lowe, J.; Welford, H.: Basophilic leukemia in the albino rat and a demonstration of the basoprotein. Nature new Biol. *244:* 73–76 (1973).

37   Enerback, L.; Svensson, I.: Isolation of rat peritoneal mast cells by centrifugation on density gradients of percoll. J. immunol. Methods *39:* 135–145 (1980).

38 Erjavec, F.; Lembeck, F.; Florjanc-Irman, T.; Skofitsch, G.; Donnerer, J.; Saria, A.; Holzer, P.: Release of histamine by substance P. Arch. Pharmacol. *317:* 67–70 (1981).

39 Feldberg, W.; Kellaway, C. H.: Liberation of histamine from the perfused lung by snake venoms. J. Physiol., Lond. *90:* 257–279 (1937).

40 Feldberg, W.; Kellaway, C. H.: Liberation of histamine from the perfused lung of the guinea pig by bee venom. J. Physiol., Lond. *91:* 2P–3P (1937).

41 Feldberg, W.; Kellaway, C. H.: Liberation of histamine and formation of lysolecithin-like substances by cobra venom. J. Physiol., Lond. *94:* 187–226 (1938).

42 Ferguson, A.; Miller, H. R. P.: Role of the mast cell in the defense against gut parasites; in Pepys, Edwards, The mast cell. Its role in health and disease, pp. 159–165 (Pitman, Tunbridge Wells 1979).

43 Fewtrell, C.; Lagunoff, D.; Metzger, H.: Secretion from rat basophilic leukemia cells induced by calcium ionophores. Effect of pH and metabolic inhibition. Biochim. biophys. Acta *644:* 363–368 (1981).

44 Fewtrell, C.; Metzger, H.: The role of aggregation in the function of the IgE Fc receptors. J. reticuloendoth. Soc. *28:* 35S–42S (1980).

45 Fewtrell, C.; Metzger, H.: Large oligomers of IgE are more effective than dimers in stimulating rat basophilic leukemia cells. J. Immun. *125:* 701–710 (1980).

46 Findlay, S. R.; Dvorak, A. M.; Kagey-Sobotka, A.; Lichtenstein, L. M.: Hyperosmolar triggering of histamine release from human basophils. J. clin. Invest. *67:* 1604–1613 (1981).

47 Foreman, J. C.: Spontaneous histamine secretion from mast cells in the presence of strontium. J. Physiol., Lond. *271:* 215–232 (1977).

48 Foreman, J. C.; Garland, L. G.: Desensitization in the process of histamine secretion induced by antigen and dextran. J. Physiol., Lond. *239:* 381–391 (1974).

49 Foreman, J. C.; Hallett, M. B.; Mongar, J. L.: The relationship between histamine secretion and ^{45}calcium uptake by mast cells. J. Physiol., Lond. *271:* 193–214 (1977).

50 Foreman, J. C.; Hallett, M. B.; Mongar, J. L.: Movement of strontium ions into mast cells and its relationship to the secretory response. J. Physiol., Lond. *271:* 233–251 (1977).

51 Foreman, J. C.; Lichtenstein, L. M.: Spontaneous histamine secretion from leukocytes in the presence of strontium. J. Pharma. exp. Ther. *210:* 75–81 (1979).

52 Foreman, J. C.; Mongar, J. L.: Effect of calcium on dextran-induced histamine release from isolated mast cells. Br. J. Pharmacol. *46:* 767–769 (1972).

53 Foreman, J. C.; Mongar, J. L.: Dual effect of lanthanum on histamine release from mast cells. Nature new Biol. *240:* 255–256 (1972).

54 Foreman, J. C.; Mongar, J. L.: The role of the alkaline earth ions in anaphylactic histamine secretion. J. Physiol., Lond. *224:* 753–769 (1972).

55 Foreman, J. C.; Mongar, J. L.: The interaction of calcium and strontium with phosphatidyl serine in the anaphylactic secretion of histamine. J. Physiol., Lond. *230:* 493–507 (1973).

56 Foreman, J. C.; Mongar, J. L.: The action of lanthanum and manganese on anaphylactic histamine secretion. Br. J. Pharmacol. *48:* 527–537 (1973).

57 Foreman, J. C.; Mongar, J. L.; Gomperts, B. D.: Calcium ionophores and movement of calcium ions following the physiological stimulus to a secretory process. Nature, Lond. *245:* 249–251 (1973).

58 Foreman, J.C.; Sobotka, A.K.; Lichtenstein, L.M.: The action of strontium on basophil leukocytes and its use to probe the relationship between immunologic stimulus and
   secretory response. J. Immun. *123:* 153–159 (1979).
59 Fredholm, B.: Studies on a mast cell degranulating factor in bee venom. Biochem.
   Pharmac. *15:* 2037–2043 (1966).
60 Fredholm, B.B.; Guschin, I.; Elwin, K.; Schwab, G.; Uvnäs, B.: Cyclic AMP independent inhibition by papaverine of histamine release induced by compound 48/80.
   Biochem. Pharmac. *25:* 1583–1588 (1975).
61 Gillespie, E.: Compound 48/80 decreases adenosine 3'5'-monophosphate formation in
   rate peritoneal mast cells. Experientia *29:* 47–448 (1972).
62 Gillespie, E.; Lichtenstein, L.M.: Histamine release from human leukocytes: relationships between cyclic nucleotide, calcium, and antigen concentrations. J. Immun. *115:*
   1572–1576 (1975).
63 Goth, A.; Adams, H.R.; Knoohuizen, M.: Phosphatidylserine: selective enhancer of
   histamine release. Science *173:* 1034–1035 (1971).
64 Goth, A.; Johnson, A.R.: Current concepts on the secretory function of mast cells. Life
   Sci. *16:* 1201–1214 (1975).
65 Grosman, N.; Diamant, B.: The influence of phosphatidylserine on the release of
   histamine from isolated rat mast cells induced by different agents. Agents Action *5:*
   296–301 (1975).
66 Hammel, I.; Lagunoff, D.: Unpublished observations.
67 Hastie, R.: A study of the ultrastructure of human basophil leukocytes. Lab. Invest. *31:*
   223–231 (1982).
68 Hempstead, B.L.; Parker, C.W.; Kulczycki, A.: Characterization of the IgE receptor
   isolated from human basophils. J. Immun. *123:* 2283–2291 (1979).
69 Henderson, W.R.; Chi, E.Y.; Klebanoff, S.J.: Eosinophil peroxidase-induced mast cell
   secretion. J. exp. med. *152:* 265–279 (1980).
70 Hino, R.H.; Lau, C.K.H.; Read, G.W.: The site of action of the histamine releaser
   compound 48/80 in causing mast cell degranulation. J. Pharmac. exp. Ther. *200:*
   658–663 (1977).
71 Hirata, F.; Axelrod, J.; Crews, F.T.: Concanavalin A stimulates phospholipid methylation and phosphatidylserine decarboxylation in rat mast cells. Proc. natn. Acad. Sci.
   USA *76:* 4813–4816 (1979).
72 Hogberg, B.; Uvnäs, B.: The mechanism of the disruption of mast cells produced by
   compound 48/80. Acta physiol. scand. *41:* 345–369 (1957).
73 Holgate, S.T.; Lewis, R.A.; Austen, K.F.: 3',5'-Cyclic adenosine monophosphate-
   dependent protein kinase of the rat serosal mast cell and its immunologic activation.
   J. Immun. *124:* 2093–2099 (1980).
74 Holgate, S.T.; Lewis, R.A.; Austen, K.F.: Role of adenylate cyclase in immunologic
   release of mediators from rat mast cells: agonist and antagonist effects of purine- and
   ribose-modified adenosine analogs. Proc. natn. Acad. Sci. USA *77:* 6800–6804
   (1980).
75 Holgate, S.T.; Lewis, R.A.; Maguire, J.F.; Roberts, L.J.; Oates, J.A.; Austen,
   K.F.: Effects of prostaglandin D_2 on rat serosal mast cells: discordance between immunologic mediator release and cyclic AMP levels. J. Immun. *125:* 1367–1373
   (1980).
76 Holgate, S.T.; Winslow, C.M.; Lewis, R.A.; Austen, K.F.: Effects of prostaglandin D_2

and theophylline on rat serosal mast cells: discordance between increased cellular levels of cyclic AMP and activation of cyclic AMP-dependent protein kinase. J. Immun. *127:* 1520–1533 (1981).

77 Hook, W. A.; Siraganian, R. P.: Influence of anions, cations and osmolarity on IgE-mediated histamine from human basophils. Immunology *43:* 723–731 (1981).

78 Ishizaka, T.; Foreman, J. C.; Sterk, A. R.; Ishizaka, K.: Induction of calcium flux across the rat mast cell membrane by bridging IgE receptors. Proc. natn. Acad. Sci. USA *76:* 5858–5862 (1979).

79 Ishizaka, T.; Hirata, F.; Ishizaka, K.; Axelrod, J.: Stimulation of phospholipid methylation, Ca^{+2} influx, and histamine release by bridging of IgE receptors on rat mast cells. Proc. natn. Acad. Sci. USA *77:* 1903–1906 (1980).

80 Ishizaka, T.; Hirata, F.; Ihiszaka, K.; Axelrod, J.: Transmission and regulation of triggering signals induced by bridging of IgE receptors on rat mast cells; in Becker, Simon, Austen, Biochemistry of the acute allergic reations, pp. 213–227 (Liss, New York 1981).

81 Ishizaka, T.; Ishizaka, K.: Triggering of histamine release from rat mast cells by diavalent antibodies against IgE-receptors. J. Immun. *120:* 800–805 (1978).

82 Jasani, B.; Dreil, G.; Mackler, B. F.; Stanworth, D. R.: Further studies on the structural requirements for polypeptide-mediated histamine release from rat mast cells. Biochem. J. *181:* 623–632 (1979).

83 Johansen, T.: Mechanism of histamine release from rat mast cells induced by the ionophore A23817: effects of calcium and temperature. Br. J. Pharmacol. *63:* 643–649 (1978).

84 Johansen, T.: Adenosine triphosphate levels during anaphylactic histamine release in rat mast cells in vitro. Effects of glycolytic and respiratory inhibitors. Eur. J. Pharmacol. *58:* 107–115 (1979).

85 Johansen, T.: Dependence of anaphylactic histamine releasse from rat mast cells on cellular energy metabolism. Eur. J. Pharmacol. *72:* 281–286 (1981).

86 Johnson, A. R.; Erdos, E. G.: Release of histamine from mast cells by vasoactive peptides. Proc. Soc. exp. Biol. Med. *142:* 1252–1256 (1973).

87 Johnson, A. R.; Hügli, T. E.; Müller-Eberhard, H. J.: Release of histamine from rat mast cells by the complement peptides C3a and C5a. Immunology *28:* 1067–1080 (1975).

88 Johnson, A. R.; Moran, N. C.; Mayer, S. E.: Cyclic AMP content and histamine release in rat mast cells. J. Immun. *112:* 511–519 (1973).

89 Johnson, R. G.; Carlson, N. J.; Scarpa, A.: pH and catecholamine distribution in isolated chromaffin granules. J. biol. Chem. *253:* 1512–1521 (1978).

90 Johnson, R. G.; Carty, S. E.; Fingerhood, B. J.; Scarpa, A.: The internal pH of mast cell granules. FEBS Lett. *120:* 75–79 (1980).

91 Kagayama, M.; Douglas, W. W.: Electron microscope evidence of calcium-induced exocytosis in mast cells treated with 48/80 or the ionophores A-23187 and X-537A. J. Cell Biol. *62:* 519–526 (1980).

92 Kaliner, M.; Austen, K. F.: Cyclic AMP, ATP, and reversed anaphylactic histamine release from rat mast cells. J. Immun. *112:* 664–674 (1973).

93 Kaliner, M.; Austen, K. F.: Cyclic nucleotides and modulation of effector systems of inflammation. Biochem. Pharmac. *23:* 763–771 (1974).

94 Kanno, T.; Cochrane, D. E.; Douglas, W. W.: Exocytosis (sectretory granule extrusion)

induced by injection of calcium into mast cells. Can. J. Physiol. Pharamcol. *51:* 1001–1004 (1973).

95 Keller, R.: Concanavalin A, a model 'antigen' for the in vitro detection of cell-bound reaginic antibody in the rat. Clin. exp. Immunol. *13:* 139–147 (1973).

96 Kennerly, D.A.; Secosan, C.J.; Parker, C.W.; Sullivan, T.J.: Modulation of stimulated phospholipid metabolism in mast cells by pharmacologic agents that increase cyclic 3′,5′ adenosine monophosphate levels. J. Immun. *123:* 1519–1524 (1979).

97 Kennerly, D.A.; Sullivan, T.J.; Parker, C.W.: Activation of phospholipid metabolism during mediator release from stimulated rat mast cells. J. Immun. *122:* 152–159 (1979).

98 Kennerly, D.A.; Sullivan, T.J.; Sylwester, P.; Parker, C.W.: Diacylglycerol metabolism in mast cells: a potential role in membrane fusion and arachidonic acid release. J. exp. Med. *150:* 1039–1044 (1979).

99 Kitada, C.; Ashida, Y.; Maki, Y.; Fujino, M.; Hirai, Y.; Yasuhara, T.; Nakagima, T.; Takeyama, M.; Koyama, K.; Yajima, H.: Synthesis of granuliberin-R and various fragment peptides and their histamine-releasing activities. Chem. pharm. Bull., Tokyo *28:* 887–892 (1980).

100 Krüger, P.G.; Lagunoff, D.; Wan, H.: Isolation of rat mast cell granules with intact membranes. Expl Cell Res. *129:* 83–93 (1980).

101 Kulczycki, A.J.; Isersky, C.; Metzger, M.: The interaction of IgE with rat basophil leukemia cells. I. Evidence for specific binding of IgE. J. exp. Med. *139:* 600–616 (1974).

102 Lagunoff, D.: Histochemistry of proteolytic enzymes. Meth. Achiev. exp. Pathol., vol. 2, pp. 55–77 (Karger, Basel 1967).

103 Lagunoff, D.; Chi, E.: Effect of colchicine on rat mast cells. J. Cell Biol. *71:* 182–195 (1976).

104 Lagunoff, D.: Contributions of electron microscopy to the study of mast cells. J. invest. Derm. *58:* 296–311 (1972).

105 Lagunoff, D.: Membrane fusion during mast cell secretion. J. Cell Biol. *57:* 252–259 (1973).

106 Lagunoff, D.: Analysis of dye binding sites in mast cell granules. Biochemistry *13:* 3982–3986 (1974).

107 Lagunoff, D.: Localization of histamine in cells; Glick, Rosenbaum, Techniques of biochemical and biophysical morphology, vol. 2, pp. 283–305 (Wiley Interscience, New York 1975).

108 Lagunoff, D.: Neutral proteases of the mast cell; in Becker, Simon, Austen, Biochemistry of the acute allergic reaction. Fourth Int. Symp. Kroc Foundation Series, vol. 14, pp. 89–101 (Liss, New York 1981).

109 Lagunoff, D.; Chi, E.Y.; Wan, H.: Effects of chymotrypsin and trypsin on rat peritoneal mast cells. Biochem. Pharmac. *24:* 1573–1578 (1975).

110 Lagunoff, D.; Phillips, M.T.; Iseri, O.A.; Benditt, E.P.: Isolation and preliminary characterization of rat mast cell granules. Lab. Invest. *13:* 1331–1344 (1964).

111 Lagunoff, D.; Pritzl, P.: Characterization of rat mast cell granule proteins. Archs Biochem. Biophys. *173:* 554–563 (1975).

112 Lagunoff, D.; Rickard, A.: unpublished observations.

113 Lapetina, D.; Michell, R.H.: Phosphatidylinositol metabolism in cells receiving extracellular stimulation. FEBS Lett. *31:* 1–10 (1973).

114 Lawson, D.; Fewtrell, C.; Gomperts, B.; Raff, M.C.: Anti-immunoglobulin-induced histamine secretion by rat peritoneal mast cells studied by immunoferritin electron microscopy. J. exp. Med. *142:* 391–402 (1975).

115 Lawson, D.; Raff, M.C.; Gomperts, B.; Fewtrell, C.; Gilula, N.B.: Molecular events during membrane fusion. A study of exocytosis in rat peritoneal mast cells. J. Cell Biol. *72:* 242–259 (1977).

116 Lewis, R.A.; Austen, K.F.: Mediation of local homeostasis and inflammation by leukotrienes and other mast cell-dependent compounds. Nature, Lond. *293:* 103–108 (1981).

117 Lewis, R.A.; Holgate, S.T.; Roberts, L.J.; Maguire, J.F.; Oates, J.A.; Austen, K.F.: Effects of indomethacin on cyclic nucleotide levels and histamine release from rat serosal mast cells. J. Immun. *123:* 1663–1668 (1979).

118 Lichtenstein, L.M.; Margolis, S.: Histamine release in vitro: inhibition by catecholamines and methylxanthines. Science *161:* 902–903 (1968).

119 Loeffler, L.J.; Lovenberg, W.; Sjoerdsma, A.: Effects of dibutyryl-3',5'-cyclic adenosine monophsophate, phosphodiesterase inhibitors and prostaglandin E$_1$ on the compound 48/80-induced histamine release from rat peritoneal mast cells in vitro. Biochem. Pharmac. *20:* 2287–2297 (1971).

120 Londos, C.; Wolff, J.: Two distinct adenosine-sensitive sites on adenylate cyclase. Proc. natn. Acad. Sci. USA *74:* 5482–5486 (1977).

121 Lynch, S.M.; Austen, K.F.; Wasserman, S.I.: Release of arylsulfatase A but not B from rat mast cells by noncytolytic secretory stimuli. J. Immun. *121:* 1394–1399 (1978).

122 Magro, A.M.: Involvement of IgE in Con A-induced histamine release from human basophils. Nature, Lond. *249:* 572–573 (1974).

123 Marone, G.; Findlay, S.R.; Lichtenstein, L.M.: Adenosine receptor on human basophils: modulation of histamine release. J. Immun. *123:* 1473–1477 (1979).

124 Marquardt, D.L.; Nicolotti, R.A.; Kennerly, D.A.; Sullivan, T.J.: Lipid metabolism during mediator release from mast cells: studies of the role of arachidonic acid metabolism in the control of phospholipid metabolism. J. Immun. *127:* 845–849 (1981).

125 Marquardt, D.L.; Parker, C.W.; Sullivan, T.J.: Potentiation of mast cell mediator release by adenosine. J. Immun. *120:* 871–878 (1978).

126 Martin, T.W.; Lagunoff, D.: Interaction of phosphatidylserine with mast cells. Proc. natn. Acad. Sci. USA *75:* 4997–5000 (1978).

127 Martin, T.W.; Lagunoff, D.: Inhibition of mast cell histamine secretion by N-substituted derivatives of phosphatidylserine. Science *204:* 631–633 (1979).

128 Martin, T.W.; Lagunoff, D.: Interactions of lysophospholipids and mast cells. Nature, Lond. *279:* 250–252 (1979).

129 Martin, T.W.; Lagunoff, D.: Activation of histamine secretion from rat mast cells by aqueous dispersions of phosphatidylserine. Biochemistry *19:* 3106–3113 (1980).

130 Martin, T.W.; Lagunoff, D.: Rat mast cell phospholipase A$_2$: activity towards exogenous phosphatidylserine and inhibition by N-7-Nitro-2,1,3-benzoxadiazol-4yl phosphatidylserine. Biochemistry *21:* 1254–1260 (1982).

131 McGivney, A.; Crews, F.T.; Hirata, F.; Axelrod, J.; Siraganian, R.P.: Rat basophilic leukemia cell lines defective in phospholipid methyltransferase enzymes, Ca^{+2} influx, and histamine release: reconstitution by hybridization. Proc. natn. Acad. Sci. USA *78:* 6176–6180 (1981).

132 Metcalfe, D.D.; Smith, J.A.; Austen, K.F.: Polydispersity of rat mast cell heparin. J. biol. Chem. *255:* 11753–11758 (1980).

133 Metzger, H.: The IgE-mast cell system as a paradigm for the study of antibody mechanisms. Immunol. Rev. *41:* 186–199 (1978).

134 Michell, R.H.: Inositol phospholipids and cell surface receptor function. Biochim. biophys. Acta *415:* 81–147 (1975).

135 Michell, R.H.: Inositol phospholipids in membrane function. Trends biochem. Sci. *4:* 128–131 (1979).

136 Mongar, J.L.; Svec, P.: The effect of phospholipids on anaphylactic histamine release. Br. J. Pharmacol. *46:* 741–752 (1972).

137 Montesano, R.; Vassalli, P.; Perrelet, A.; Orci, L.: Distribution of filipin-cholesterol complexes at sites of exocytosis – a freeze-fracture study of degranulating mast cells. Cell Biol. int. Rep. *4:* 975–984 (1980).

138 Morita, Y.; Siraganian, R.P.: Inhibition of IgE-mediated histamine release from rat basophilic leukemia cells and rat mast cells by inhibitors of transmethylation. J. Immun. *127:* 1339–1344 (1981).

139 Morrison, D.C.; Roser, J.F.; Henson, P.M.; Cochrane, C.G.: Isolation and characterization of a non-cytotoxic mast cell activator from cobra venom. Inflammation *1:* 103–115 (1975).

140 Morrison, D.C.; Roser, J.F.; Cochrane, C.G.; Henson, P.M.: The initiation of mast cell degranulation: activation at the cell membrane. J. Immun. *114:* 966–970 (1975).

141 Nemeth, A.: Rapid separation of rat peritoneal mast cells with Percoll. Eur. J. Cell Biol. *20:* 272–275 (1980).

142 Nemeth, E.F.; Douglas, W.W.: Differential inhibitory effects of the arachidonic acid analog ETYA on rat mast cell exocytosis evoked by secretagogues utilizing cellular or extracellular calcium. Eur. J. Pharmacol. *67:* 439–450 (1980).

143 Ohkuma, S.; Poole, B.: Fluorescence probe measurement of the intralysosomal pH in living cells and the perturbation of pH by various agents. Proc. natn. Acad. Sci. USA *75:* 3327–3331 (1978).

144 Ohmori, H.; Yammaoto, I.; Akagi, M.; Tasaka, T.: Properties of hydrogen peroxide-induced histamine release from rat mast cells. Biomed. Pharmacol. *29:* 741–745 (1980).

145 Orenstein, N.S.; Galli, S.J.; Dvorak, A.M.; Dvorak, H.G.: Glycosaminoglycans and proteases of guinea pig basophilic leukocytes; in Becker, Simon, Austen, Biochemistry of the acute allergic reaction. Fourth Int. Symp. Kroc Foundation Series, vol.14, pp.123–143 (Liss, New York 1981).

146 Paterson, N.A.M.; Wasserman, S.I.; Said, J.W.; Austen, K.F.: Release of chemical mediators from partially purified human lung mast cells. J. Immun. *117:* 1356–1362 (1976).

147 Patkar, S.A.; Kazimierczak, W.; Diamant, B.: Sodium fluoride – a stimulus for a calcium-triggered secretory process. Int. Archs Allergy appl. Immun. *55:* 193–200 (1977).

148 Patkar, S.A.; Kazimierczak, W.; Diamant, B.: Histamine release by calcium from sodium fluoride-activated rat mast cells. Further evidence for a secretory process. Int. Archs Allergy appl. Immun. *57:* 146–154 (1978).

149 Payne, A.N.; Garland, L.G.: Interaction between barium, strontium and calcium

in histamine release by compound 48/80. Eur. J. Pharmacol. *52:* 329–334 (1978).

150 Pearce, F. L.; Ennis, M.; Truneh, A.; White, J. R.: Role of intra- and extracellular calcium in histamine release from rat peritoneal mast cells. Agent Action *11:* 51–54 (1981).

151 Pearce, F. L.; White, J. R.: Effect of lanthanide ions on histamine secretion from rat peritoneal mast cells. Br. J. Pharmacol. *72:* 341–347 (1981).

152 Pollard, H. B.; Pazoles, C. J.; Creutz, C. F.; Ramu, A.; Strott, C. A.; Ray, P.; Brown, E.; Aurbach, G. D.; Tack-Goldman, K. M.; Shulman, N. R.: A role for anion transport in the regulation of release from chromaffin granules and exocytosis from cells. J. supramol. Struct. *7:* 277–285 (1977).

153 Pollard, H. B.; Tack-Goldman, K.; Pazoles, C. J.; Cruetz, C. E.; Shulman, R.: Evidence for control of serotonin secretion from human platelets. Proc. natn. Acad. Sci. USA *74:* 5295–5299 (1977).

154 Pruzansky, J. J.; Patterson, R.: Enrichment of human basophils. J. immunol. Methods *44:* 183–190 (1981).

155 Ranadive, N. R.; Dhanani, N.: Movement of calcium ions and release of histamine from rat mast cells. Int. Archs Allery appl. Immun. *61:* 9–18 (1980).

156 Randon, J.; Lecompte, T.; Chignard, M.; Siess, W.; Marlas, G.; Dray, F.; Vargaftig, B. B.: Dissociation of platelet activation from transmethylation of their membrane phospholipids. Nature, Lond. *293:* 660–662 (1981).

157 Read, G. W.; Knoohuizen, M.; Goth, A.: Relationship between phosphatidylserine and cromolyn in histamine release. Eur. J. Pharmacol. *42:* 171–177 (1977).

158 Schwartz, L. B.; Austen, K. F.: Acid hydrolases and other enzymes of rat and human mast cell secretory granules; in: Becker, Simon, Austen, Biochemistry of the acute allergic reaction. Fourth Int. Symp. Kroc Foundation Series, vol. 14, pp. 103–121 (Liss, New York 1981).

159 Schwartz, L. B.; Austen, K. F.; Wasserman, S. I.: Immunologic release of β-hexosaminidase and β-glucuronidase from purified rat serosal mast cells. J. Immun. *123:* 1445–1450 (1979).

160 Schwartz, L. B.; Lewis, R. A.; Seldin, D.; Austen, K. F.: Acid hydrolases and tryptase from secretory granules of dispersed human lung mast cells. J. Immun. *126:* 1290–1294 (1981).

161 Segal, D. M.; Taurog, J. D.; Metzger, H.: Dimeric immunoglobulin E serves as a unit signal for mast cell degranulation. Proc. natn. Acad. Sci. USA *74:* 2993–2997 (1977).

162 Sieghart, W.; Theoharides, T. C.; Alper, S. L.; Douglas, W. W.; Greengard, P.: Calcium-dependent protein phosphorylation during secretion by exocytosis in the mast cell. Nature, Lond. *275:* 329–331 (1978).

163 Sieghart, W.; Theoharides, T. C.; Douglas, W. W.; Greengard, P.: Phosphorylation of a single mast cell protein in response to drugs that inhibit secretion. Biochem. Pharmac. *30:* 2737–2738 (1981).

164 Siraganian, R. P.; Kulczycki, A.; Mendoza, G.; Metzger, H.: Ionophore A-23187 induced histamine release from rat mast cells and rat basophil leukemia (RBL-1) cells. J. Immun. *115:* 1599–1602 (1975).

165 Siraganian, R. P.; Metzger, H.: Evidence that the 'mouse mastocytoma' cell line (MCT-1) is of rat origin. J. Immun. *121:* 2584–2585 (1978).

166 Siraganian, R. P.; Siraganian, P. A.: Mechnism of action of concanavalin A on human basophils. J. Immun. *114:* 886–893 (1975).

167 Smith, G. A.; Hesketh, T. R.; Plumb, R. W.; Metcalfe, J. C.: The exogenous lipid requirement for histamine release from rat peritoneal mast cells stimulated by concanavalin A. FEBS Lett. *105:* 58–62 (1979).

168 Stechschulte, D. J.; Austen, K. F.: Phosphatidylserine enhancement of antigen-induced mediator release from rat mast cells. J. Immun. *112:* 970–978 (1974).

169 Stevens, R. L.; Austen, K. F.: Proteoglycans of the mast cell; in Becker, Simon, Austen, Biochemistry of the acute allergic reaction. Fourth Int. Symp. Kroc Foundation Series, vol. 14, pp. 69–88 (Liss, New York 1981).

170 Sugiyama, K.; Sasaki, J.; Yamasaki, H.: Potentiation by phosphatidylserine of calcium-dependent histamine release from rat mast cells induced by concanavalin A. Jap. J. Pharmacol. *25:* 485–487 (1975).

171 Sullivan, T. J.; Greene, W. C.; Parker, C. W.: Concanavalin A-induced histamine release from normal rat mast cells. J. Immun. *115:* 278–282 (1975).

172 Sullivan, T. J.; Parker, C. W.: Possible role of arachidonic acid and its metabolites in mediator release from rat mast cells. J. Immun. *122:* 431–436 (1979).

173 Sullivan, T. J.; Parker, K. L.; Eisen, S. A.; Parker, C. W.: Modulation of cyclic AMP in purified rat mast cells. II. Studies on the relationship between intracellular cyclic AMP concentrations and histamine release. J. Immun. *114:* 1480–1485 (1975).

174 Sullivan, T. J.; Parker, K. L.; Kulczycki, A., Jr.; Parker, C. W.: Modulation of cyclic AMP in purified rat mast cells. III. Studies on the effects on concanavalin A and anti-IgE or cyclic AMP concentrations during histamine release. J. Immun. *117:* 713–716 (1976).

175 Sydbom, A.; Fredholm, B.; Uvnäs, B.: Evidence against a role of cyclic nucleotides in the regulation of anaphylactic histamine release in isolated red mast cells. Acta physiol. scand. *112:* 47–56 (1981).

176 Taurog, J. D.; Mendoza, G. R.; Hook, W. A.; Siraganian, R. P.; Metzger, H.: Noncytotoxic IgE-mediated release of histamine and serotonin from murine mastocytoma cells. J. Immun. *119:* 1757–1761 (1977).

177 Theoharides, T. C.; Betchaku, T.; Douglas, W. W.: Somatostatin-induced histamine secretion in mast cells. Characterization of the effect. Eur. J. Pharmacol. *69:* 127–137 (1981).

178 Theoharides, T. C.; Douglas, W. W.: Secretion in mast cells induced by calcium entrapped within phospholipid vesicles. Science *201:* 1143–1145 (1978).

179 Theoharides, T. C.; Sieghart, W.; Greengard, P.; Douglas, W. W.: Antiallergic drug cromolyn may inhibit histamine secretion by regulating phosphorylation of a mast cell protein. Science *207:* 80–82 (1980).

180 Theoharides, T. C.; Sieghart, W.; Greengard, P.; Douglas, W. W.: Somatostatin-induced phosphorylation of mast cell proteins. Biochem. Pharmac. *30:* 2735–2736 (1981).

181 Tobias, L. D.; Hamilton, J. G.: The effect of 5, 8, 11, 14-eicosatetraynoic acid on lipid metabolism. Lipids *14:* 181–193 (1979).

182 Uvnäs, B.; Åborg, C.-H.: On the cation exchanger properties of rat mast cell granules and their storage of histamine. Acta physiol. scand. *100:* 309–314 (1976).

183 Uvnäs, B.; Åborg, C.-H.; Bergendorff, A.: Storage of histamine in mast cells. Evidence

for an ionic binding of histamine to protein carboxyls in the granule heparin-protein complex. Acta physiol. scand. *78:* 3–16 (1970).

184 Uvnäs, B.; Diamant, B.; Hogberg, B.: Trigger action of phospholipase A on mast cells. Archs int. Pharmacodyn. Thér. *140:* 577–580 (1962).

185 Uvnäs, B.; Thon I.-L.: Evidence for enzymatic release from isolated rat mast cells. Expl. Cell Res. *23:* 45–57 (1961).

186 Vogt, W.: Role of phospholipase A_2 in prostaglandin formation. Adv. Prostaglandin Thromboxane Res. *3:* 89–95 (1978).

187 White, J.R.; Pearce, F.L.: Role of membrane bound calcium in histamine secretion from rat peritoneal mast cells. Agents Actions *11:* 324–329 (1981).

188 Winslow, C.M.; Lewis, R.A.; Austen, K.F.: Mast cell mediator release as a function of cyclic AMP-dependent protein kinase activation. J. exp. Med. *154:* 1125–1133 (1981).

189 Woodbury, R.G.; Neurath, H.: Structure, specificity and localization of the serine proteases of connective tissue. FEBS Lett. *114:* 189–106

Thomas W. Martin, PhD, Department of Pathology, St. Louis Unversity School of Medicine, St. Louis, Mo. 63104 (USA)

Cell Biology of the Secretory Process, pp. 517–545 (Karger, Basel 1984)

Stimulus-Secretion Coupling in Human Polymorphonuclear Leukocytes

James E. Smolen, Helen M. Korchak, Gerald Weissmann

Division of Rheumatology, Department of Medicine, New York University Medical Center, School of Medicine, New York, N.Y., USA

Introduction

When a mature polymorphonuclear leukocyte (PMN) encounters a suitable ingestible particle, the cell's surface membrane invaginates and surrounds it. The vesicle containing the particle, now called a phagosome or phagocytic vesicle, pinches off from the surface and is introduced into the cytoplasm of the cell [46, 51, 106]. Digestion of the engulfed particle is initiated by fusion of the membranes of granules with the phagosome membrane. The lysosomes subsequently discharge their contents into the phagosome, a process called degranulation [13, 31], and the vesicle is now a 'phagolysosome'.

The primary role of PMN is bactercidal, to which end the contents of the lysosomes are tailored (fig. 1). However, some granule contents (particularly those enzymes designed for scavenging as well as microbicidal roles) are also capable of damaging host tissues. This would not be a problem were the phagocytic scheme presented above neatly followed in practice: dangerous granule contents would remain sequestered inside the PMN, concentrated in the phagolysosome as required. However, under pathological conditions, massive release of granule contents can occur, promoting tissue injury and exacerbating existing inflammatory conditions.

Metabolic Responses

When PMN encounter ingestible particles or certain soluble stimuli, prompt metabolic changes are observed. These responses include increased oxygen consumption [5, 69], increased hexose monophosphate

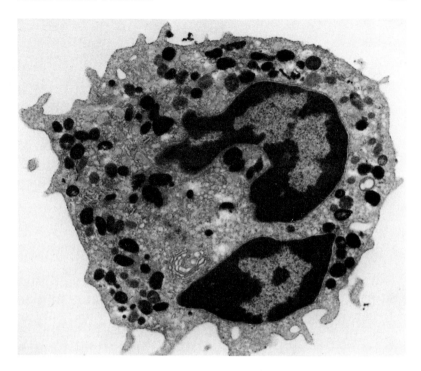

Fig. 1. Morphology of a human PMN. A human peripheral blood PMN reacted for myeloperoxidase to distinguish azurophil or primary granules from specific or secondary granules. The peroxidase-positive (black) azurophil granules tend to be larger than the peroxidase-negative or specific granules but the two populations overlap. Two of the cell's nuclear lobes are shown and illustrate the high proportion of heterochromatin compared to euchromatin present in these cells. × 11,000 [courtesy of Dr. *S. T. Hoffstein*].

shunt activity [17], increased hydrogen peroxide [36] and superoxide anion [2] production, and increased lipid turnover [41]. A critical first step in these events is the reduction of molecular oxygen to superoxide anion (O_2^-) by an oxidase linked either to NADH or NADPH.

Superoxide anion is enzymatically (by superoxide dismutase) or spontaneously converted to hydrogen peroxide, which may in turn react with additional O_2^- to form hydroxyl radicals (OH·). All of these species of oxygen are highly reactive and possess various degrees of bactericidal activity. Hydrogen peroxide is particularly important in that it is responsible for stimulation of hexose monophosphate shunt activity, the required oxidation of NADPH by H_2O_2 being mediated by glutathione-linked reactions [64]. Consumption of hydrogen peroxide by this mechanism and by cytoplasmic catalase provides the PMN with the means to protect itself against this noxious compound.

There is evidence that some, if not all, of the neutrophil's reactive oxygen moieties are originally generated by enzymes on the plasma membrane [9, 19, 21, 66]. This idea is appealing since generation of superoxide and other reactive derivatives at the plasma membrane level would be advantageous; toxic bactericidal materials would thus be concentrated around ingested organisms (and minimized in the cell cytoplasm and extracellular space) once the membrane was formed into phagosomes.

Cytoskeletal Structures

Microtubules and microfilaments, the most prominent cytoskeletal structures of PMN, appear to be intimately involved in the processes of cell motility, adhesion and ingestion [1, 63]. These structures not only generate motive power for the PMN, but could conceivably mediate a transfer of information between plasma membrane and the cell interior.

In PMN, microtubules seem to be of particular importance. The number of these structures increases when cells are exposed to various surface stimuli [33] (fig. 2), including the chemotactic factor C5a [22]. The fact that colchicine can inhibit particle ingestion (in some phagocytic systems), chemotaxis, migration, surface adhesion, and degranulation suggests that microtubules are important in all of these processes. Microtubules have been implicated in other more subtle phenomena. There is evidence that these structures can govern the planar disposition of membrane components during phagocytosis; microtubules appear to be able to prohibit [100] or promote [56] the incorporation of various membrane markers into phagosomes.

Microfilaments seem to constitute the PMN contractile system. These structures have been identified as actin polymers and are prominent in areas of the cell involved in adhesion and particle ingestion [63]. Strong similarities between the PMN contractile system and skeletal muscle have recently become apparent. Actin, myosin (with actin-activated Mg^{++}-ATPase activity), actin-binding protein, and a cofactor which allows actin to activate the above-mentioned ATPase have all been isolated from phagocytic cells [8]. Appropriate extracts of rabbit alveolar macrophages can be induced to form gels and contract in vitro, and significantly more actin-binding protein can be obtained from phagocytizing than from resting cells [94].

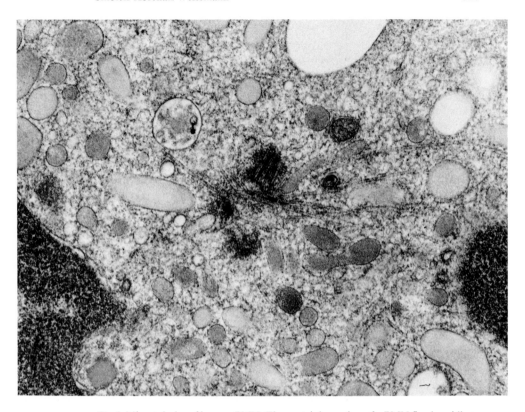

Fig. 2. Microtubules of human PMN. The centriolar region of a PMN fixed rapidly to preserve microtubule organization. Portions of two centrioles and two centriole-associated microtubule-organizing sites are visible. Most of the microtubules emerge from the microtubule-organizing sites situated between the centrioles. These microtubules radiate from the cytocenter to the cell periphery and appear to maintain the organization of the cytocenter. × 40,000 [courtesy of Dr. *S. T. Hoffstein*].

The microfilament system can be inhibited by interference with production of metabolic energy (ATP). More specific disruption of the contractile system can be achieved using cytochalasin B. The cytochalasins are fungal metabolites which at low concentrations interfere with the function of actin-binding protein [25]. While the effect of cytochalasin B is usually assumed to be specific against microfilaments, this drug is also known to inhibit hexose transport [105]. The fact that cytochalasin B is a powerful inhibitor of PMN migration and phagocytosis, in conjunction with ultrastructural data, suggests a vital role for microfilaments in these active processes [1].

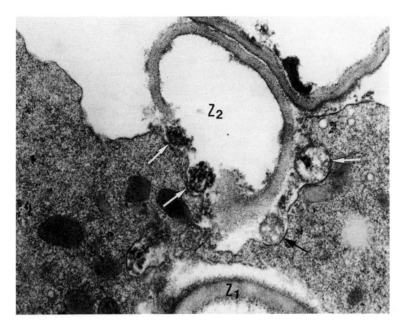

Fig. 3. Phagocytosis and degranulation. A higher magnification view of a PMN exposed to opsonized zymosan (Z). One particle in the lower portion of the figure (Z_1) has been completely ingested and another particle, Z_2, has been only partially engulfed. Degranulation has already begun adjacent to Z_2 (arrows) and these granule contents are free to diffuse to the extracellular space. × 39,000 [courtesy of Dr. *S. T. Hoffstein*].

Degranulation

Under normal conditions, degranulation is a direct accompaniment of phagocytosis, and the close relationship between these events suggests common, or at least similar, triggering mechanisms [95]. During degranulation, lysosomes fuse with phagosomes containing ingested particles, thus discharging their contents in concentrated form in the immediate vicinity of their target (fig. 3). Ideally, both the extracellular milieu and the PMN cytoplasm are spared the deleterious effects of the granule enzymes. In practice, some lysosomal contents are released outside the cell during particle ingestion, usually because of premature degranulation into those phagosomes which are still open at the cell surface.

Extensive extracellular degranulation can be stimulated in some pathological or experimental systems, giving rise to secretion. For example, a PMN might try to ingest an oversized particle such as an opsonin-coated surface; phagocytosis cannot be accomplished but degranulation is stimulated, resulting in extracellular discharge of lysosomal contents (a process called 'reverse endocytosis'). An experimental system for study of these processes is provided by PMN whose ability to ingest particles is blocked by the presence of cytochalasin B. The surface of such cells can be stimulated by a variety of agents such as particles, opsonins, or lectins, leading to extrusion of granule (but not cytoplasmic) contents. Here, cytochalasin B converts PMN from phagocytic cells into model secretory cells, consequently making it possible to monitor *extracellularly* (after fusion of lysosomes with plasma membranes) processes that ordinarily occur *intracellularly* (fusion of lysosomes with phagocytic vacuoles).

Much previous work has been devoted to the study of both the normal and pathologic responses of human PMN to surface stimulation, with the result that many of the mechanisms which govern these responses are becoming clear. However, only recently has attention been paid to the initiation of these responses and to biochemical and physiological events which take place during the first minute of granulocyte stimulation. Measurements of these earliest events and determinations of the temporal order in which they occur have already contributed to an understanding of the controlling mechanisms and promise to be similarly productive in the future. Determinations of the temporal order of events allows a separation of ends from means and causes from effects. This work has placed us on the verge of unravelling the 'secretory code' of human PMN.

Early Events in PMN Stimulation

Receptor-Ligand Interactions
The initial step in PMN stimulation, a process leading ultimately to secretory events such as degranulation and superoxide anion (O_2^-) generation, must involve either a ligand-receptor interaction at the cell surface or some other direct perturbation of the plasma membrane. The kinetics of receptor-ligand interactions appear to be rapid, when defined stimuli are examined; binding of the chemotactic peptide N-formyl-norleucyl-leucyl-phenylalanine (FNLP) to rabbit PMN or rabbit PMN plasma membranes was both time-dependent and saturable [77]. Saturating concentrations of FNLP occupied 50% of the available receptors of intact PMN within 30 s. Binding was even more rapid on the isolated plasma membrane.

Membrane Perturbations

Membrane perturbations produced either by direct interactions of the stimulus with the plasmalemma or, indirectly, by ligand-receptor interactions, would be expected to follow surface stimulation. The fluorescent probe 1-anilino-8-naphthalene sulfonate (ANS), when added to guinea pig PMN, exhibited an enhancement and blue shift of fluorescence, suggestive of plasma membrane binding [65]. When the ANS-labelled PMN were exposed to polystyrene latex particles, there was a prompt increase in fluorescent intensity. This rapid response, taking place within 2 s of particle addition, was indicative of a conformational change in the cell membrane; the exact nature of this conformational change is unknown.

Changes in Membrane Potential

Surface responses in secretory cells are usually accompanied by changes in membrane potential. To quantify this in the PMN, which are too small for the insertion of microelectrodes, the lipophilic triphenyl-methylphosphonium cation (TPMP+) has been used [44]. The distribution of this molecule across cell membranes is determined by the strength and polarity of the existing electrochemical gradient. Changes in the ratio of $TPMP^+_{in}$ to $TPMP^+_{out}$ therefore reflected changes in the membrane potential which could be calculated from the Nernst equation. Using an experimentally determined intracellular volume, it was possible to calculate that the resting potential of human PMN was approximately -26 mV.

The effects of secretory concentrations of concanavalin A and immune complexes on the surface potential of human PMN (measured as TPMP+ uptake) were also studied [44]. When purified granulocytes were exposed to concanavalin A, they responded within 10 s by a sharp hyperpolarization followed by depolarization followed in turn by long slow hyperpolarization. When endocytosis of concanavalin A was blocked by cytochalasin B, similar results were seen. In experiments in which the cells were exposed to immune complexes of BSA/anti-BSA, to the chemotactic peptide N-formyl-methionyl-leucyl-phenylalanine (FMLP), and to latex beads, a similar hyperpolarization response was observed. In contrast, cells exposed to ionophore A23187 did not exhibit a hyperpolarization response.

Changes in PMN membrane potential have also been measured using the fluorescent dye dipentyloxacarbocyanine (Di-O-C_5-(3)) [74]. Exposure of dye-loaded cells to the chemotactic peptide FMLP produced a biphasic change in fluorescent intensity, seemingly similar to that seen by the

TPMP+ method [44]. However, the fluorescence response indicated an initial depolarization rather than a hyperpolarization, was inhibited by cytochalasin B, and EGTA blocked the second phase. Furthermore, calcium ionophore A23187 produced fluorescence changes. These latter three observations are at variance with results obtained by TPMP+ method. Although carbocyanine dyes have been used extensively, it is also clear that these dyes interfere with cellular metabolism and have some toxic effects [35]. On balance, however, both methods (despite their unique problems) agree that membrane potential changes occur in stimulated PMN and that these changes take place as rapidly as they can be measured (<5 s).

Several other laboratories have reported apparent depolarization responses to a variety of stimuli using $Di-S-C_3-(5)$ as a fluorescent probe [39, 41, 97]. However, *Seligmann and Gallin* [73] recently reported that secretory concentrations of the tumor promotor phorbol myristate acetate (PMA) and calcium ionophore A23187 evoked transient hyperpolarization followed by a large depolarization. Several reports have suggested that membrane potential changes, measured either by the TPMP+ [72] or cyanine dye [40, 72, 103] techniques, are required for subsequent generation of O_2^- by human PMN. This correlation is bolstered by the observation that PMN from some patients with chronic granulomatous disease (CGD), which do not produce O_2^-, also have marked deficiencies in stimulated transmembrane potential changes [72, 103].

Loss of Membrane Calcium

Chlortetracycline (CTC) is a probe which forms highly fluorescent complexes with membrane-bound Ca^{++} or Mg^{++}; the complexes with these two divalent cations can be distinguished on the basis of their spectral properties. CTC-loaded rabbit PMN showed an immediate loss of fluorescence when exposed to FMLP or C5a [54, 55]. This loss was very rapid (taking place within $2-5$ s) and was independent of the presence or absence of extracellular Ca^{++}, suggesting that it was attendant to the earliest events following receptor-ligand interaction and not to the later Ca^{++} influx. The loss of fluorescence was also accompanied by a spectral shift indicating that CTC was now reporting membrane-bound Mg^{++}; Ca^{++} had apparently been liberated into the intracellular medium. Changes in CTC fluorescence also displayed desensitization: subsequent exposure of cells to the same stimulus provoked little or no response.

Takeshige et al. [96] reported that guinea pig PMN responded to *Escherichia coli* and cytochalasin D with similar losses of CTC fluorescence. These CTC responses were blocked by TMB-8, an inhibitor of intracellular calcium mobilization. Human PMN also showed prompt decrements in CTC fluorescence when exposed to FMLP, PMA, and Con A [92]. However, the details of the response varied with the stimulus employed. Neither TMB-8 nor inhibitors of calmodulin (W-7 and trifluperazine) affected the response in human PMN. The properties of the CTC response and its correlations with enzyme secretion [85] suggest that this probe is monitoring a 'trigger pool' of intracellular calcium, the mobilization of which is a critical first step in PMN stimulation.

Membrane-bound Ca^{++} of human PMN has been observed by means of electron microscopy [32]. Cations were precipitated in situ by fixing cells in aqueous solutions of osmium hydroxide and calcium pyroantimonate. Cells were permitted to ingest zymosan particles and were then immediately stained by the pyroantimonate method. Whereas the plasma membrane stained heavily for calcium, the membranes of the phagocytic vacuoles were entirely devoid of the precipitates (fig. 4). Those areas of the plasma membrane of the PMN not in proximity to the zymosan particles still retained calcium at the surface. However, at sites of the PMN membrane which were at the region of contact with the zymosan particles, calcium had been lost. Similar results were obtained when PMN were exposed to concanavalin A-Sepharose beads.

Oxidative Metabolism

Stimulated PMN consume far more molecular oxygen than resting PMN. Molecular oxygen is reduced by an NADH- and/or NADPH-dependent oxidase, resulting in the formation of superoxide anion (O_2^-), H_2O_2, hydroxyl radicals, and singlet oxygen. The subcellular localization and pyridine nucleotide specificity of this critical oxidase has long been in dispute [3]. However, a cell surface localization of the oxidase(s) has been gaining support in recent years [9, 19, 21, 66]. NADH oxidase was demonstrated on the surface of human PMN [9]; phagocytizing PMN generated H_2O_2, visualized by precipitates deposited in the presence of cerous ions, on the cell surface and on phagosome membranes when NADH was provided. Treatment of intact PMN with the poorly-penetrating reagent diazotized sulfanilic acid resulted in inhibition of O_2^- generation without concomitant inhibition of intracellular lactate dehydrogenase [21]. Recent cell fractionation studies, in which marker enzymes were used appropri-

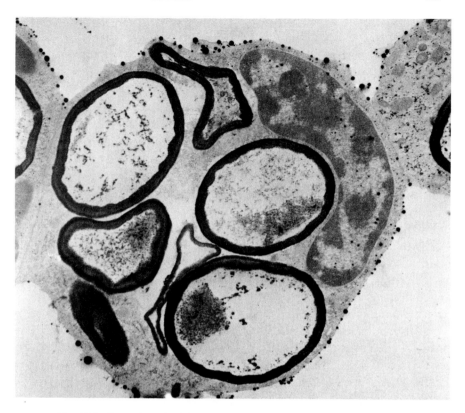

Fig. 4. Membrane-bound calcium deposits. A portion of a PMN aggregate that had been exposed to zymosan, then fixed and reacted with potassium pyroantimonate plus osmium tetroxide to reveal cell-associated divalent cations. Calcium containing pyroantimonate precipitates are found over the free cell membrane but not where the PMN membranes are adherent and not on phagosome membrane. × 11,500 [courtesy of Dr. *S. T. Hoffstein*].

ately to assess membrane contamination of 'granule' fractions, indicated that 60–70% of the O_2^- generating activity of human PMN was located in the plasma membrane [15]. NADPH oxidase activity was found in the same fractions. Superoxide permeates the erythrocyte membrane by way of the anion 'channel' and the diffusion of O_2^- can be inhibited with the anion channel blockers SITS and DIDS [47]. If O_2^- generation by PMN was taking place intracellularly, it would be expected that efflux of this anion radical would take place via the anion channel and that its extracellular detection would be inhibited by SITS and DIDS. In fact, concentrations of SITS and DIDS sufficient to block anion fluxes did not inhibit extracellular detection of O_2^- [43]. All of these reports strongly suggest that the pyridine nucleotide-dependent oxidase which produces O_2^- and H_2O_2 is located on the PMN cell surface.

The induction of oxidative metabolism may begin with an allosteric transition in the oxidase. Particulate fractions from resting PMN (containing substantial amounts of plasma membrane) had NADPH oxidase activity which displayed sigmoidal kinetics with respect to NADPH [14]. Similar fractions from cells which had ingested zymosan particles contained an oxidase with hyperbolic kinetics. Thus, activation of the enzyme by phagocytosis seemed to involve an allosteric change in the oxidase which permitted substantial NADPH oxidation to take place at the low concentrations customary for endogenous NADPH. NADH oxidase activity displayed the same kinetic behavior. The PMN from patients with chronic granulomatous disease, which do not have a stimulated oxidative burst, did not display the transition from sigmoidal to hyperbolic kinetics.

Oxygen consumption begins within seconds of cell-particle contact. However, a close examination of the kinetics of this process showed that oxygen consumption by PMN did not begin until approximately 25 s after the addition of latex particles [70]. Thus, unlike changes in membrane potential [39, 40, 50, 72–74, 97, 103], increases in ANS fluorescence [65], and losses in membrane-bound Ca^{++} [32, 54, 85, 92, 96], the onset of oxidative metabolism has a distinct, measurable lag phase. The lag period seemed to be intrinsic to the activation process and was not dependent upon the frequency or number of particle-cell contacts (i.e., the stimulus concentration).

If oxygen consumption, the seminal event in the burst of oxidative metabolism, displayed a lag period, then production of reduced oxygen derivatives should have similar or longer lag periods. *Root* et al. [67] devised a continuous assay for H_2O_2 production. H_2O_2 liberated into the extracellular medium by stimulated PMN was reduced by horseradish peroxidase using scopoletin as an electron donor. When PMN were stimulated by latex particles, opsonized yeast, or *Staphylococcus aureus*, a lag period of 10–15 s elapsed between addition of the stimuli and the onset of fluorescence extinction. Thus, H_2O_2 production, like oxygen consumption, was initiated after a distinct lag period in response to surface stimuli.

Activation of the O_2^- generating system of PMN is also not immediate. Using a simple technique whereby O_2^- formation was continuously monitored as extracellular cytochrome c reduction, *Cohen and Chovaniec* [11, 12] demonstrated that 30–60 s elapsed before guinea pig PMN produced O_2^- in response to digitonin. The lag period was decreased by increasing the concentration of the stimulus or by elevation of the temperature. The lag period was lengthened by N-ethylmalemide and 2-deoxyglucose. They further confirmed the observation of *Root* et al. [67] that H_2O_2 production had a lag time of 25–30 s, which under the conditions employed in those experiments, coincided with that obtained for O_2^- production. This would be expected since H_2O_2 is primarily the product of spontaneous or enzymatic dismutation of O_2^- to H_2O_2 and O_2.

These results were verified in human PMN by *Korchak and Weissmann* [44]. Generation of O_2^- was assayed continuously in the presence of cytochalasin B, used to maximize production and detection of the anion radical. Exposure of PMN to the lectin concanavalin A or to an immune complex consisting of bovine serum albumin and antibovine serum albumin ('BSA/anti-BSA') gave rise to O_2^- generation after distinct lag periods. Stimulation by the immune complex resulted in a lag period of approximately 30 s while Con A induced O_2^- generation after 42 s. Generation of O_2^- was clearly shown to follow the earlier and apparently immediate hyperpolarization responses. These results demonstrated that lag periods for O_2^- production are not species-specific, are not dependent upon the presence or absence of cytochalasin B, and that their durations are dependent upon the stimulus presented.

Degranulation and Secretion

Perhaps the earliest study of the initial kinetics of degranulation was provided by the classic work of *Bainton* [4]. Granule discharge into phagosomes of rabbit PMN was monitored by electron microscopy. Cell preparations were viewed shortly after exposure to bacteria (*E. coli* or *S. aureus*) and subsequent cytochemical reaction for alkaline phosphatase, a specific granule marker, and peroxidase, an azurophil granule marker. She found that alkaline phosphatase was present in phagocytic vacuoles within 30 s of the initiation of phagocytosis. Peroxidase was discharged after 1–3 min. Thus, specific granule contents were released into the phagosome before the contents of azurophil granules, a concept called 'sequential degranulation'. This order of discharge was in accord with the pH optima of the enzymes contained within these types of lysosomes. Specific granule enzymes having neutral or alkaline pH optima, were discharged when the phagosome was at neutral pH; azurophil granule enzymes, predominantly acid hydrolases, were introduced into the phagosome later, when the pH was substantially lower [38]. Since earlier time points were not examined in this work, it was clear whether or not degranulation displayed a lag period. More recent microscopic studies, using rhodamine and fluorescein-labelled antisera to myeloperoxidase and lactoferrin have not only verified sequential degranulation but also suggested that some granule discharge occurs within 5 s of exposure to bacteria [60].

It is clear that degranulation is a rapid event. However, a quantitative examination of the kinetics of this process has only recently appeared in the literature. The relative kinetics of degranulation into phagosomes was measured by *Segal* et al. [71]. These investigators assayed the enzymatic contents of phagosomes isolated from PMN which had invested latex particles. While the stopping procedure employed (dilution in ice-cold buffer) may not have been adequate for determining *absolute* kinetics of degranulation, the experimental technique was suitable for measuring relative kinetics. They reported that lactoferrin (a specific granule enzyme) and myeloperoxidase (a reputed azurophil granule marker) appeared in phagosomes promptly, with no apparent lag periods. However, azurophil granule enzymes such as β-glucuronidase and acid phosphatase appeared much later, with lag periods of 1–2 min. Thus, discharge of azurophil granule enzymes was significantly delayed in both a relative (compared to degranulation of specific granule enzymes) and an absolute (compared to changes in membrane-bound calcium and electrochemical potential) sense. The authors also suggest that the kinetics indicate that myeloperoxidase is not contained in either specific or azurophil granules.

As discussed earlier, suitable continuous assay techniques had existed for monitoring consumption of O_2 and production of H_2O_2 and O_2^-. The temporal resolution of the continuous methods was such that it was readily determined that these processes did not commence until after lag periods of 15–60 s following stimulation. Similar continuous techniques for measurement of extracellular lysosomal enzyme secretion were not available. The time resolution of conventional centrifugation methods for obtaining cell-free supernatants was limited by, among other factors, the period required for centrifugation. It is not surprising that several investigators concluded that degranulation was virtually an immediate consequence of PMN stimulation, having no discernible lag period. Cytochemical techniques, which potentially possessed adequate time resolution, were not quantitative [4].

To solve this problem, a semicontinuous flow dialysis system was devised to monitor enzyme secretion [87]. The cytochalasin B-treated PMN were placed in the upper chamber of a modified flow-dialysis cell to which a bolus containing a concentrated stimulus and [^{14}C]-inulin was added. The inulin served as the extracellular space marker; it immediately began to cross a Millipore filter into the lower chamber and thus served to mark the moment of stimulation and to calibrate the entire apparatus. The appearance of [^{14}C]-inulin and lysosomal enzymes in collected fractions could then be used to obtain an absolute time-course for secretion. It was possible to show that lag periods did exist for lysosomal enzyme release and that the lengths of these lag periods were stimulus-dependent. Values for these lag periods were also comparable to those seen for O_2^- generation. The lag periods for both responses were not dependent upon the dose of the stimulus. A variety of agents such as corticosteroids, colchicine, 2-deoxyglucose, and N-ethylmalemide, affected the magnitudes of the responses, but not the lag periods, when FMLP was the stimulus. When BSA/anti-BSA was used as the stimulus, 2-deoxglucose and N-ethylmalemide increased the lag period for O_2^- generation but not for lysosomal enzyme release. These latter observations suggested that the initial events leading to the two responses were similar and possibly parallel, but not tightly coupled.

This semicontinuous flow-dialysis technique was compared to a modern centrifugation method wherein cells were rapidly separated from the medium by short (20 s) high-speed centrifugation through silicone oil. Measurement of O_2^- generation by this latter method revealed the existence of lag periods following stimulation of PMN by FMLP or A23187. However, these lag periods were 10–20 s shorter than those obtained by the continuous method, indicating that even this modern centrifugation technique was inadequate for the purpose of absolute timing. Lysosomal enzyme release also had lag periods when centrifugation was used; once again, the values for these lags were 10–20 s too short. Thus, these two independent techniques confirmed that lag periods existed for lysosomal enzyme release. Furthermore, the two techniques provide comparable values for the duration of the lag periods when the 10–20 s underestimation by the centrifugation method was taken into account.

In spite of the fact that centrifugation was not suitable for purposes of absolute timing, it was better than flow dialysis for *relative* timing and for the direct determination of quantities of enzyme release. When both β-glucuronidase and lysozyme were measured in the same cell supernatants, it was clear that lysozyme release preceded that of β-glucuronidase in response to A23187 (i.e., sequential degranulation was observed). The time courses of secretion of these two enzymes were virtually identical when FMLP was the stimulus.

The advantages of the centrifugation technique for relative timing were exploited by *Bentwood and Henson* [7]. These investigators measured the release of lactoferrin, β-glucuronidase, and myeloperoxidase from cytochalasin B-treated human PMN exposed to opsonized zymosan, aggregated immunoglobulin, C5a, FMLP, PMA and A23187. Lag periods were stimulus-dependent; as expected, FMLP provoked secretion with little or no apparent lag period, probably due to the uncertainties with respect to absolute timing discussed above. However, for all stimuli the specific granule marker lactoferrin was secreted before release of the two azurophil granule enzymes. This was apparent not only in the greater initial rate of release of lactoferrin but also in a distinctly shorter lag period. Thus, this report provided a quantitative basis for sequential extracellular degranulation.

Ion Requirements

The ionic composition of the extracellular medium is critical in PMN stimulation. In particular, the fact that the calcium ionophore A23187 is alone sufficient to cause lysosomal enzyme release and O_2^- generation attests to a critical role of Ca^{++} and suggests that fluxes of this divalent cation underlie the stimulatory process. This discussion will focus upon both ionic requirements for PMN stimulation and upon the measurement of ionic fluxes.

Degranulation measured in either phagocytic or exocytic systems is Ca^{++}-dependent [6, 25, 35, 84, 104]. Mg^{++} is relatively less important. For O_2^- generation, removal of Ca^{++} was greatly inhibitory [44]; in the presence of Ca^{++}, removal of Mg^{++} actually increased O_2^- generation. High concentrations of Ca^{++} per se were sufficient to provoke lysozyme release from PMN prepared in a divalent cation-free medium [23]; the azurophil granule enzyme β-glucuronidase was not released by this treatment. Lysozyme secretion was potentiated by the presence of phosphate ions or A23187.

The effect of monovalent cations on enzyme secretion by PMN was reported by *Showell* et al. [78]. Degranulation by cytochalasin B-treated rabbit PMN in response to FMLP or A23187 (in the presence of Ca^{++}) was enhanced by extracellular K^+ (5.5 mM). This enhancement was not prevented by ouabain, suggesting that activation of Na^+, K^+-ATPase was not involved. The K^+ ionophore valinomycin (but not nigeracin) provoked enzyme release by itself. Neither K^+ ionophore effected degranulation induced by chemotactic factors, suggesting that maintenance of a transmembrane K^+ gradient was not essential for this response. Enzyme release induced by FMLP in the presence of Ca^{++} was diminished when Na^+ was replaced with K^+ or choline$^+$. La^{3+}, a Ca^{++} channel blocker, inhibited degranulation at low concentrations; at high concentrations, La^{3+} sustained degranulation, apparently by substituting for Ca^{++}. The authors suggested that both cytochalasin B and K^+ facilitate Ca^{++} influxes induced by secretagogues, thereby enhancing lysosomal enzyme release. Removal of Na^+ could suppress degranulation as a result of membrane depolarization or by depression of a putative Na^+-Ca^{++} exchange mechanism.

In human PMN, the role of monovalent ions has only scarcely been studied. Removal of Na^+ from the medium (with replacement by choline$^+$ or K^+) resulted in greatly diminished O_2^- generation, lysosomal enzyme release and transmembrane potential changes [45]. A possible role for anions in the degranulation process has also been suggested [43]. Anion channel blockers such as SITS and DIDS inhibited lysosomal enzyme release from human PMN in response to immune complexes; however, O_2^- generation was unaffected. The role of anion fluxes per se is still unclear since no crucial species of anion has yet been identified.

The fact that secretion can be induced by ionophores such as A23187 and valinomycin or by high concentrations of Ca^{++} (presented to cells prepared without divalent cations) has suggested that fluxes of cations, particularly Ca^{++}, are intimately involved in the mechanism of stimulus-secretion coupling. Both A23187 and zymosan-activated serum provoked large rapid increases in Ca^{++} associated with human PMN [84]. Increased cellular Ca^{++}, presumably the result of Ca^{++} influx, was seen within 1 min, and was accompanied by cGMP accumulation and β-glucuronidase release. The fact that verapamil blocks O_2^- generation suggests that this response also requires a Ca^{++} influx [81].

In cytochalasin B-treated rabbit PMN, FMLP induced no change in total cellular Ca^{++} in the presence of 1 mM extracellular Ca^{++} [59]. In the absence of extracellular Ca^{++}, levels diminished in stimulated cells. When ^{45}Ca was used as a tracer, the specific activity of cell-associated Ca^{++} increased following exposure to FMLP, far more in the presence than in the absence of extracellular Ca^{++}. The authors suggested that while it might first appear that the changes in cell-associated Ca^{++} were merely due to greater permeability of this cation in stimulated PMN, nonexchangeable Ca^{++} could be lost during degranulation, resulting in artificially high specific activities for $^{45}Ca^{++}$.

A number of studies have shown that stimulation of PMN produces enhanced Ca^{++} influx, Ca^{++} efflux, and Na^+ influx. Probably the most valuable study of cation fluxes in PMN is that of *Naccache* et al. [53]. These researchers used rapid centrifugation of PMN through silicone oil as a means of separating cells from the medium; this technique provided prompt termination of ion fluxes. Cell-associated Ca^{++} increased rapidly upon exposure of rabbit PMN to FMLP and cytochalasin B; nearly maximal Ca^{++} influx was obtained within 30 s. FMLP or cytochalasin B alone produced only slight increments. Maximal Ca^{++} influx in response to FMLP or A23187 required extracellular K^+ (5 mM). The increments in Ca^{++} uptake closely paralleled lysozyme release. Ca^{++} efflux, on the other hand, was increased to a far lesser extent by FMLP; the further addition of cytochalasin B produced no significant enhancement. FMLP produced a substantial increase in Na^+ influx, which was greatly enhanced by the presence of cytochalasin B. In the presence of both agents, maximum Na^+ uptake was achieved within 30 s. Na^+ influx was not Ca^{++}-dependent when FMLP was the stimulus. K^+ influx was not substantially changed by FMLP and/or cytochalasin B. K^+ efflux, on the other hand, was increased by FMLP plus cytochalasin B (FMLP alone had no effect) or A23187 plus cytochalasin B. K^+ efflux due to A23187 appeared to be delayed approximately 1 min relative to K^+ efflux induced by FMLP. Both K^+ efflux responses appeared to have lag periods and were delayed relative to the prompt Na^+ and Ca^{++} influxes; near maximal K^+ efflux was achieved at 1 min by FMLP and at 2 min by A23187.

In summary, then, stimulation of PMN resulted in prompt Na^+ and Ca^{++} influxes, relatively slower K^+ efflux, and a slightly enhanced Ca^{++} efflux. *Naccache* et al. [53] suggested that ligand-receptor interaction produced transient increases in membrane permeability to Na^+ and K^+. This would cause membrane depolarization and a local increase in free intracellular Ca^{++}, which could be enhanced by the presence of cytochalasin B. A putative Na^+-Ca^{++} exchange mechanism would enhance Na^+ influx and produce the observed Ca^{++} efflux. The same mechanism would provide an additional Ca^{++} influx if the intracellular Na^+ concentration was sufficiently high. Since Na^+ influx can be obtained in the absence of Ca^{++} and consequently in the absence of degranulation, it appears that Ca^{++}, not Na^+, is the crucial cation in stimulus-secretion coupling. K^+ efflux, which could give rise to membrane hyperpolarization, did not appear to be crucial in the coupling mechanism and might have been a delayed response to elevations of intracellular Ca^{++}.

In spite of the fact that Ca^{++} influxes increase during stimulation of PMN, this does not mean that a Ca^{++} influx is the trigger for secretion: enhanced ion fluxes could be the result rather than the cause of stimulation. The use of inhibitors of Ca^{++} influx, such as La^{3+} [78], and verapamil [81] suffers from uncertainty with regard to the specificity of these agents. Indeed, the fact that human PMN can secrete lysozyme in response to PMA in the absence of extracellular Ca^{++} [82] strongly suggests that extracellular Ca^{++} and a Ca^{++} influx are not absolute requirements. This latter observation has been made with a wide variety of stimuli; lysosomal enzyme secretion and O_2^- generation induced by FMLP, PMA, immune complexes, opsonized zymosan, activated complement, Con A and A23187 were not completely blocked by saturating concentrations of EGTA [88].

If a Ca^{++} influx is not an absolute requirement for stimulation of PMN, then perhaps elevated stimulatory levels of intracellular Ca^{++} could be provided by the mobilization of intracellular Ca^{++} stores. Evidence for such a mechanism is provided by the observation that TMB-8, an inhibitor of the mobilization of intracellular Ca^{++}, completely blocks PMN responses [82, 88]. Similarly, the calmodulin inhibitors W-7 and trifluoperazine completely block enzyme secretion and O_2^- generation [52, 88], without inhibiting stimulated changes in CTC fluorescence [52, 92]. All of these data suggest that the mobilization of intracellular Ca^{++}, which can perhaps be monitored by CTC fluorescence [54, 55, 92, 95, 96], is a sine qua non for secretion by PMN. Calmodulin-dependent reactions, distal to this mobilization are also required. On the other hand, influxes of extracellular Ca^{++}, while not absolutely required, can be important in the modulation and maximization of PMN responses.

Changes in Cyclic Nucleotide Levels

In spite of the well-documented fact that cAMP and agents which elevate intracellular levels of cAMP *diminish,* whereas exogenous cGMP and agents which elevate intracellular levels of cGMP *enhance* both degranulation and other stimulated responses of PMN, changes in endogenous cellular cyclic nucleotide levels resulting from stimulation have rarely been reported.

A number of reports have stated that cAMP levels did not change following stimulation of PMN [27, 34], seemingly eliminating this cyclic nucleotide from mechanistic considerations. However, *Herlin* et al. [29] have recently shown that exposure of human PMN to latex particles provoked a rapid two-fold increase in cAMP levels. This increment was prompt (maximal within 15 s) and brief. The fact that basal cAMP levels were restored within 1–2 min explained why previous investigators, who had assayed the cell suspensions after 2–5 min, were unable to detect this response. These investigators suggested that the brief increment in cAMP might have a regulatory role in glycogen metabolism [29, 58]. In any case, it is clear that such a rapid response, the timing of which is comparable to changes in membrane potential and release of membrane-bound Ca^{++}, could possibly play a role in stimulus-secretion coupling.

Stimulated increments in cAMP levels have since been observed in both rabbit [37] and human PMN [48, 80, 86]. In all of these recent reports, the increments peaked at 10–30 s and were transient, returning to baseline after 2–5 min. During the same early time interval, no changes in cGMP levels were observed [80, 86]. Stimulus-dependent lag periods could be observed with some secretagogues [86]. Changes in cAMP levels induced by FMLP were not affected by EGTA, cytochalasin B, or the absence of extracellular Na^+ [86]. The fact that changes in cAMP levels display specific desensitization after repeated exposures to the same stimulus and that this desensitization parallels that for O_2^- generation [79] suggests that changes in cyclic nucleotide levels may be of mechanistic significance.

However, two lines of evidence suggest that increments of cAMP are not significant in stimulation of PMN. First, cAMP increments induced by $10^{-9} M$ FMLP are only slightly lower than those induced by $10^{-7} M$ FMLP [48, 86]. Since virtually no secretion of lysosomal enzymes or generation of O_2^- occurs at the lower dose of stimulus (but is optimal at the higher level of $10^{-7} M$), the observed increments do not seem to be *sufficient* for the later secretory events. However, these dose-response data have been disputed [53, 79]. Second, generation of O_2^- and secretion of lysozyme, but not β-glucuronidase, can be induced by Con A and PMA, stimuli which provoke no changes in cAMP levels [91]. Thus, increments in cAMP do not seem to be necessary for discharge of specific granules and O_2^- production. The findings that increments in cAMP levels may be neither necessary nor sufficient for secretion of at least specific granules seems to leave Ca^{++} in sole possession of the second messenger role (see previous discussion).

Nonetheless, redistribution of cAMP (without concomitant elevated levels) could still be of mechanistic significance. Indeed, immunochemical observations of cAMP show that this nucleotide (but not cGMP) is localized near forming phagosomes and thus may serve as a trigger for the initiation of phagocytosis [61].

Lipid Metabolism

A recent active area of investigation in PMN stimulation is that of lipid metabolism, particularly as it relates to changes in membrane composition. The membrane phospholipids are not merely an inert matrix for membrane proteins and a barrier to water-soluble substances. Rather there is a dynamic interaction between membrane proteins and lipids. Perturbation of the membrane by treatment with surface-active agents such as deoxycholate, digitonin and saponin, stimulates a respiratory burst in a manner similar to that of phagocytosis [25, 68]. Phospholipase C treatment of neutrophils results in a similar respiratory burst [57]. A change in the lipoprotein structure of the plasma membrane was proposed as the stimulus for this cell activation. Indeed, changes in membrane lipid composition accompany phagocytosis [49, 89].

In common with other secretory cells, neutrophils utilize Ca^{++} as a second messenger to mediate cellular responses. Thus, the mobilization of calcium is a central event in the activation sequence. The means by which this is accomplished in cells is at present imperfectly understood. However, a role for lipid metabolism has been suggested through the mechanism of a phosphatidylinositol/phosphatidic acid (PI/PA) cycle leading to Ca^{++} gating [10]. In addition, a role for prostanoids in Ca^{++} translocation has been suggested. Since phosphatidic acid demonstrated Ca^{++} ionophore activity in the liposome assay [75], it is of interest to further examine the putative role of the PI/PA cycle in Ca^{++} translocation during the earliest part of the activation sequence. There are, however, other candidates for an endogenous ionophore. On the basis of inhibitor studies, a role for lipoxygenase pathway mediator(s) has been proposed for Ca^{++} translocation in the neutrophil [101]. In support of this concept, it has been demonstrated in liposomes that a polymeric prostaglandin derivative PGB_x translocates Ca^{++} [75] and that PGB_x can also activate O_2^- generation and degranulation [76]. Although PGB_x is not a natural derivative and an unlikely candidate for a naturally occurring ionophore, it does serve as a useful model.

Stimulated neutrophils actively generate oxygenated arachidonic acid derivatives which can serve to modulate cell function. Since the size of the free arachidonic acid pool serves to regulate the rate of production of these products, it is of considerable interest to define the source of the free arachidonate and the metabolic path leading to its release. The scheme initially proposed for the release of arachidonate involved the action of a phospholipase A_2 [16, 18, 102] to cleave arachidonate from the C-2 position of a phosphatide. A phospholipase A_2 activity has been demonstrated in neutrophils [16, 18]; however, the substrate phosphatide for release of arachidonate has not been rigorously established [16] nor have free lyso-phosphatides been demonstrated, suggesting that if this pathway is active, a rapid reacylation must occur. Indeed, endogenous lysolecithin is converted to lecithin in the presence of neutrophils [18], indicating the possibility of an active deacylation-reacylation cycle. This type of recycling could explain the fatty acid remodeling observed in phagocytosing neutrophils [49, 89]. In addition, the calcium requirement of both phospholipase A_2 and the generation of arachidonate make this a plausible pathway.

Neutrophils exposed to phagocytic stimuli generate stable prostaglandins of the E and F series, as well as the lipoxygenase derivatives hydroxyeicosatetraenoic acids (HETEs), hydroperoxyeicosatetraenoic acids (HPETEs) and leukotrienes. *Goldstein* et al. [24] demonstrated that human neutrophils exposed to serum-treated zymosan generated thromboxane B_2 in a time- and concentration-dependent fashion. Conversion by neutrophils of [^{14}C]-arachidonic acid to [^{14}C]-thromboxane B_2 was confirmed by thin-layer chromatography and mass spectrometry. Thromboxane B_2 generation was inhibited by the presence of indomethacin but was not affected by the presence of cytochalasin B. Thromboxane B_2 is the stable end-product of thromboxane A_2, a biologically active compound which causes rapid irreversible platelet aggregation as well as the contraction of vascular and tracheal smooth muscle.

The role of the lipoxygenase and cyclooxygenase products is at present not understood. Studies with inhibitors of arachidonate metabolism, i.e., inhibitors of phospholipases, lipoxygenase and cyclooxygenase, suggested that arachidonate metabolite(s) might play a role in mediating and enhancing the activation sequence in the neutrophil [90]. In FMLP-treated neutrophils, the phospholipase inhibitor bromophenacyl bromide inhibited degranulation and O_2^- generation. Similarly, indomethacin (inhibits cyclooxygenase and phospholipase) and ETYA (inhibits cyclooxygenase and lipoxygenase) inhibited degranulation and O_2^- generation [83, 90]. In contrast, the thromboxane synthetase inhibitor imidazole did not inhibit degranulation, O_2^- generation or aggregation [*Korchak and Friedman,* unpublished observations]. Thus, products of both the lipoxygenase and cyclooxygenase pathways might play a role in the mediation and/or modulation of neutrophil activation. Other workers have shown that exogenous arachidonate can cause degranulation [101] in the rabbit neutrophil and that this response was inhibited by ETYA. In the human neutrophil, however, exogenous arachidonate was *not* capable of stimulating these responses [16]. HETEs, in addition to being chemotactic for neutrophils, can also induce release of specific granule contents [20, 93, 99]. Prostanoids do not always play an enhancing role in the activation sequence since prostaglandins of the E series and prostacyclin (PGI$_2$) have been shown to inhibit degranulation and O_2^- generation [86, 91]. These prostaglandins interact with cAMP metabolism in the resting and activated neutrophil. Concentrations of PGE$_1$ and prostacyclin (PGI$_2$) sufficient to inhibit degranulation and O_2^- generation produced sustained elevated levels of cAMP. Stimulation of pretreated cells evoked very large increments in cAMP (8- to 10-fold basal level) [91]. Thus, exposure to these prostaglandins 'primes' the cells in such a manner that the usual modest stimulus-elicited increments in cAMP are magnified into much higher, probably inhibitory levels [41, 86].

It thus seems probable that different arachidonate metabolites can serve either to amplify or to inhibit the neutrophil's activity. The sites of action at which the arachidonic acid metabolites act to either enhance or inhibit the activation sequence have not been established. A role for prostanoids in Ca^{++} mobilization and translocation has been suggested and preliminary data support this role. There is also evidence for the role of prostanoids, in conjunction with cAMP, as an inhibitory feedback mechanism in neutrophil activation [42]. A close interrelationship between calcium and cyclic nucleotides in cell activation was shown in many cell systems including the platelet, salivary gland, parotid gland and pancreatic B cell [62]. The neutrophil provides an ideal system in which to investigate the interrelationship of the arachidonic cascade with Ca^{++} translocation and cAMP metabolism and how this interrelationship serves to mediate and regulate cellular activation. A more complete temporal analysis correlating onset of prostanoid synthesis with the onset of secretion and the changes in Ca^{++} translocation and changes in cAMP should give more insight into these relationships.

References

1 Allison, A. C.; Davies, P.; DePetris, J.: Role of contractile microfilaments in macrophages movement and endocytosis. Nature new Biol. *232:* 153–155 (1971).

2 Babior, B. M.; Kipnes, R. S.; Curnutte, J. T.: Biological defense mechanisms: the production by leukocytes of superoxide , a potential bactericidal agent. J. clin. Invest. *52:* 741–744 (1973).

3 Badwey, J. A.; Curnutte, J. T.; Karnovsky, M. L.: The enzyme of granulocytes that produces superoxide and peroxide. An elusive pimpernel. New Engl. J. Med. *300:* 1157–1159 (1979).

4 Bainton, D. F.: Sequential degranulation of the two types of polymorphonuclear leukocytes granules during phagocytosis of microorganisms. J. Cell Biol. *58:* 249–264 (1973).

5 Baldridge, C. W.; Gerard, R. W.: The extra respiration of phagocytosis. Am. J. Physiol. *103:* 236–236 (1933).

6 Becker, E. L.; Showell, H. J.: The ability of chemotactic factors to induce lysosomal enzyme release. II. The mechanism of release. J. Immun. *112:* 2055–2062 (1974).

7 Bentwood, B. J.; Henson, P. M.: The sequential release of granule constituents from human neutrophils. J. Immun. *124:* 855–862 (1980).

8 Boxer, L. A.; Stossel, T. P.: Interactions of actin, myosin, and an actin-binding protein of chronic myelogenous leukemia leukocytes. J. clin. Invest. *57:* 964–976 (1976).

9 Briggs, R. T.; Drath, D. B.; Karnovsky, M. L.; Karnovsky, M. J.: Localization of NADH oxidase on the surface of human polymorphonuclear leukocytes by a new cytochemical method. J. Cell Biol. *67:* 566–586 (1975).

10 Cockcroft, S.; Bennett, J. P.; Gomperts, B. D.: F-Met-Leu-Phe-induced phosphatidyl inositol turnover in rabbit neutrophils is dependent on extracellular calcium. FEBS Lett. *110:* 115–118 (1980).

11 Cohen, H. J.; Chovaniec, M. E.: Superoxide generation by digitonin-stimulated guinea pig granulocytes, a basis for a continuous assay for maintaining superoxide production and for the study of the activation of the generating system. J. clin. Invest. *61:* 1081–1087 (1978).

12 Cohen, H. J.; Chovaniec, M. E.: Superoxide production by digitonin-stimulated guinea pig granulocytes. The effects of N-ethyl malemide, divalent cations, and glycolytic and mitochondrial inhibitors on the activation of the superoxide generating system. J. clin. Invest. *61:* 1088–1096 (1978).

13 Cohn, Z. A.; Hirsch, J. G.: The influence of phagocytosis on the intracellular distribution of granule-associated components of polymorphonuclear leukocytes. J. exp. Med. *112:* 1015–1022 (1960).

14 DeChatelet, L. R.; Shirley, P. S.; McPhail, L. C.; Iverson, D. B.; Doellgast, G. J.: Allosteric transformation of reduced nicotinamide adenine dinucleotide (phosphate) oxidase induced by phagocytosis in human polymorphonuclear leukocytes. Infect. Immun. *20:* 398–405 (1978).

15 DeWald, B.; Baggioloni, M.; Curnutte, J. T.; Babior, B. M.: Subcellular localization of the superoxide-forming enzyme in human neutrophils. J. clin. Invest. *63:* 21–29 (1979).

16   Elsbach, P.; Weiss, J.: Lipid metabolism by phagocytic cells; in Sbarra, Strauss, The reticuloendothelial system. A comprehensive treatise, pp. 91–119 (1980).

17   Evans, W. H.; Karnovsky, M. L.: The biochemical basis of phagocytosis. IV. Some aspects of carbohydrate metabolism during phagocytosis. Biochemistry *1:* 159–166 (1962).

18   Franson, R.; Weiss, J.; Martin, L.; Spitznagel, J. K.; Elsbach, P.: Phospholipase A activity associated with the membranes of human polymorphonuclear leukocytes. Biochem. J. *167:* 839–841 (1977).

19   Goetzl, E. J.; Austen, K. F.: Stimulation of the human neutrophil hexose monophosphate shunt (HMPS) by purified chemotactic factors. Fed. Proc. *32:* 973 (1973).

20   Goetzl, E. J.; Brash, A. R.; Tauber, A. I.; Oates, J. A.; Hubbard, W. D.: Modulation of human neutrophil function by monohydroxyeicosatetraenoic acids. Immunology *39:* 491–501 (1980).

21   Goldstein, I. M.; Cerqueira, M.; Lind, S.; Kaplan, H. B.: Evidence that the superoxide-generating system of human leukocytes is associated with the cell surface. J. clin. Invest. *59:* 249–254 (1977).

22   Goldstein, I.; Hoffstein, S.; Gallin, J.; Weissmann, G.: Mechanisms of lysosomal enzyme release from human leukocytes: microtubule assembly and membrane fusion induced by a component of complement. Proc. natl. Acad. Sci. USA *70:* 2916–2920 (1973).

23   Goldstein, I. M.; Horn, J. K.; Kaplan, H. B.; Weissmann, G.: Calcium-induced lysozyme secretion from human polymorphonuclear leukocytes. Biochem. biophys. Res. Commun. *60:* 807–812 (1974).

24   Goldstein, I. M.; Malmsten, C. L.; Kindahl, H.; Kaplan, H. B.; Radmark, O.; Samuelsson, B.; Weissmann, G.: Thromboxane generation by human peripheral blood polymorphonuclear leukocytes. J. exp. Med. *148:* 787–792 (1978).

25   Graham, R. C.; Karnovsky, M. J.; Shafer, A. W.; Glass, E. A.; Karnovsky, M. L.: Metabolic and morphological observations on the effect of surface-active agents on leukocytes. J. Cell Biol. *32:* 629–647 (1967).

26   Hartwig, J. H.; Stossel, T. P.: Interactions of actin, myosin and an actin-binding protein of rabbit pulmonary macrophages. III. Effects of cytochalasin B. J. Cell Biol. *71:* 295–303 (1976).

27   Hatch, G. E.; Nichols, W. K.; Hill, H. R.: Cyclic nucleotide changes in human neutrophils induced by chemoattractants and chemotactic peptides. J. Immun. *119:* 450–456 (1977).

28   Henson, P. M.; Oades, Z. G.: Stimulation of human neutrophils by soluble and insoluble immunoglobulin aggregates. Secretion of granule constituents and increased oxidation of glucose. J. clin. Invest. *56:* 1053–1061 (1975).

29   Herlin, T.; Petersen, C. S.; Esmann, V.: The role of calcium and cyclic adenosine-3′,5′-monophosphate in the regulation of glycogen metabolism in phagocytozing human polymorphonuclear leukocytes. Biochim. biophys. Acta *542:* 63–76 (1978).

30   Hirsch, J. G.: Cinemicrophotographic observations on granule lysis in polymorphonuclear leukocytes during phagocytosis. J. exp. Med. *116:* 827–834 (1962).

31   Hirsch, J. G.; Cohn, Z. A.: Degranulation of polymorphonuclear leukocytes following phagocytosis of microorganism. J. exp. Med. *112:* 1005–1014 (1960).

32   Hoffstein, S. T.: Ultrastructural demonstration of calcium loss from local regions of the

plasma membrane of surface stimulated human granulocytes. J. Immun. *123:* 1395–1402 (1979).

33 Hoffstein, S.; Soberman, R.; Goldstein, I.; Weissmann, G.: Concanavalin A induces microtubules assembly and specific granule discharge in human polymorphonuclear leukocytes. J. Cell Biol. *68:* 781–787 (1976).

34 Ignarro, L.J.; George, W.J.: Hormonal control of lysosomal enzyme release from human neutrophils: elevation of cyclic nucleotide levels by autonomic neurohormones. Proc. natn. Acad. Sci. USA *71:* 2027–2031 (1974).

35 Ignarro, L.J.; George, W.J.: Mediation of immunologic discharge of lysosomal enzymes from human neutrophils by guanosine-3′,5′-monophosphate. Requirement of calcium, and inhibition by adenosine-3′-5′-monophosphate. J. exp. Med. *140:* 225–238 (1974).

36 Iyer, G.Y.N.; Islam, D.F.M.; Quastel, J.H.: Biochemical aspects of phagocytosis. Nature, Lond. *192:* 535–541 (1961).

37 Jackowski, S.; Sha'afi, R.I.: Response of adenosine cyclic 3′,5′-monophosphate level in rabbit neutrophils to the chemotactic peptide formyl-methionyl-leucyl-phenylalanine. Molec. Pharmacol. *16:* 473–481 (1979).

38 Jenson, M.S.; Bainton, D.F.: Temporal changes in pH within the phagocytic vacuole of the polymorphonuclear neutrophilic leukocyte. J. Cell Biol. *56:* 379–388 (1973).

39 Jones, G.S.; Van Dyke, K.; Castronova, V.: Purification of human neutrophils by centrifugal elutriation and measurement of transmembrane potential. J. cell. Physiol. *104:* 425–431 (1980).

40 Jones, G.S.; Van Dyke, K.; Castronova, V.: Transmembrane potential changes associated with superoxide release from human granulocytes. J. cell. Physiol. *106:* 75–83 (1981).

41 Karnovsky, M.L.; Wallach, D.F.H.: Metabolic basis of phagocytosis. III. Incorporation of inorganic phosphate into various classes of phosphatides during phagocytosis. J. biol. Chem. *236:* 1895–1901 (1961).

42 Kennerly, D.A.; Secojan, C.J.; Parker, C.W.; Sullivan, T.J.: Modulation of stimulated phospholipid metabolism in mast cells by pharmacologic agents that increase cyclic 3′,5′-adenosine monophosphate levels. J. Immun. *123:* 1519–1524 (1979).

43 Korchak, H.M.; Eisenstat, B.A.; Hoffstein, S.T.; Dunham, P.B.; Weissmann, G.: Anion channel blockers inhibit lysosomal enzyme secretion from human neutrophils without affecting generation of superoxide anion. Proc. natn. Acad. Sci. USA *77:* 2721–2725 (1980).

44 Korchak, H.M.; Weissmann, G.: Changes in membrane potential of human granulocytes antecede the metabolic responses to surface stimulation. Proc. natn. Acad. Sci. USA *75:* 3818–3822 (1978).

45 Korchak, H.M.; Weissmann, G.: Stimulus-response coupling in the human neutrophil. Membrane potential changes and the role of extracellular Na^+. Biochim. biophys. Acta *601:* 180–194 (1980).

46 Korn, E.D.; Weisman, R.A.: Phagocytosis of latex beads by *Acanthamoeba.* II. Electron microscope study of the initial events. J. Cell Biol. *34:* 219–227 (1967).

47 Lynch, R.E.; Fridovich, I.: Permeation of the erythrocyte stroma by superoxide radical. J. biol. Chem. *253:* 4697–4699 (1978).

48 Marx, R.J.; McCall, C.E.; Bass, D.A.: Chemotaxis-induced changes in cyclic adenosine monophosphate levels in human neutrophils. Infect. Immun. *29:* 284–286 (1980).

49   Mason, R.Z.; Stossel, T.P.; Vaughan, M.: Lipids of alveolar macrophages, polymor-
      phonuclear leukocytes and their phagocytic vesicles. J. clin. Invest. *51:* 2399–2407
      (1972).

50   Montecucco, C.; Pozzan, T.; Rink, T.: Dicarbocyanine fluorescent probes of mem-
      brane potential block lymphocyte capping, deplete cellular ATP and inhibit respira-
      tion of isolated mitochondria. Biochim. biophys. Acta *552:* 552–567 (1979).

51   Mudd, J.; McCutcheon, M.; Lucke, B.: Phagocytosis. Physiol. Rev. *14:* 210–275
      (1934).

52   Naccache, P.H.; Molski, T.F.P.; Alobaidi, T.; Becker, E.L.; Showell, H.J.; Sha'afi,
      R.I.: Calmodulin inhibitors block neutrophil degranulation at a step distal from the
      mobilization of calcium. Biochem. biophys. Res. Commun. *97:* 62–68 (1980).

53   Naccache, P.H.; Showell, H.J.; Becker, E.L.; Sha'afi, R.I.: Changes in ionic move-
      ments across rabbit polymorphonuclear leukocyte membranes during lysosomal en-
      zyme release. Possible ionic basis for lysosomal enzyme release. J. Cell Biol. *75:*
      635–649 (1977).

54   Naccache, P.H.; Showell, H.J.; Becker, E.L.; Sha'afi, R.I.: Involvement of membrane
      calcium in the response of rabbit neutrophils to chemotactic factors as evidenced by the
      fluorescence of chlortetracycline. J. Cell Biol. *83:* 179–186 (1979).

55   Naccache, P.H.; Volpi, M.; Showell, H.J.; Becker, E.L.; Sha'afi, R.I.: Chemotactic
      factor-induced release of membrane calcium in rabbit neutrophils. Science *203:*
      461–463 (1979).

56   Oliver, J.M.; Ukena, T.C.; Berlin, R.D.: Effects of phagocytosis and colchicine on the
      distribution of lectin-binding sites on cell surfaces. Proc. natn. Acad. Sci. USA *71:*
      394–398 (1974).

57   Patriarca, P.; Zatti, M.; Cramer, R.; Rossi, F.: Stimulation of the respiration of poly-
      morphonuclear leukocytes by phospholipase C. Life Sci. *9:* 841–849 (1970).

58   Petersen, C.S.; Herlin, T.; Esmann, V.: Effects of catecholamines and glucagon on
      glycogen metabolism in human polymorphonuclear leukocytes. Biochim. biophys.
      Acta *542:* 77–87 (1978).

59   Petroski, R.J.; Naccache, P.H.; Becker, E.L.; Sha'afi, R.I.: Effect of the chemotactic
      factor formyl-methionyl-leucyl-phenylalanine and cytochalasin B on the cellular levels
      of calcium in rabbit neutrophils. FEBS Lett. *100:* 161–165 (1979).

60   Pryzwansky, K.B.; MacRae, E.K.; Cooney, M.H.; Spitznagel, N.J.: Early degranula-
      tion of human polymorphonuclear neutrophils: immunocytochemical studies of sur-
      face and intracellular phagocytic events. Fed. Proc. *38:* 1023 (1979).

61   Przwansky, K.B.; Steiner, A.L.; Spitznagel, J.K.; Kapoor, C.L.: Compartmentaliza-
      tion of cyclic AMP during phagocytosis by human neutrophil granulocytes. Science
      211: 407–410 (1981).

62   Rasmussen, H.; Goodman, D.B.P.: Relationships between calcium and cyclic nucleo-
      tides in cell activation. Physiol. Rev. *57:* 421–509 (1977).

63   Reaven, E.P.; Axline, S.G.: Subplasmalemmal microfilaments and microtubules in
      resting and phagocytizing cultivated macrophages. J. Cell Biol. *59:* 12–27 (1973).

64   Reed, P.W.: Glutathione and the hexose monophosphate shunt in phagocytizing and
      hydrogen peroxide-treated rat leukocytes. J. biol. Chem. *244:* 2459–2464 (1969).

65   Romeo, D.; Cramer, R.; Rossi, F.: Use of 1-anilino-8-naphthalene sulfonate to study
      structural transitions in cell membrane of PMN leukocytes. Biochem. biophys. Res.
      Commun. *41:* 582–588 (1970).

66 Romeo, D.; Zabucchi, G.; Rossi, F.: Reversible metabolic stimulation of polymorphonuclear leukocytes and marcrophages by concanavalin A. Nature new Biol. *243:*
111–112 (1973).

67 Root, R.K.; Metcalf, J.; Oshino, N.; Chance, B.: H_2O_2 release from human granulocytes during phagocytosis. I. Documentation, quantitation, and some regulating factors. J. clin. Invest. *55:* 945–955 (1975).

68 Root, F.; Zatti, M.: Mechanism of the respiratory stimulation in saponin-treated leukocytes. Biochim. biophys. Acta *153:* 296–299 (1968).

69 Sbarra, A.J.; Karnovsky, M.L.: The biochemical basis of phagocytosis. I. Metabolic
changes during the ingestion of particles by polymorphonuclear leukocytes. J. biol.
Chem. *234:* 1355–1362 (1959).

70 Segal, A.W.; Coade, S.B.: Kinetics of oxygen consumption by phagocytosing human
neutrophils. Biochem. biophys. Res. Commun. *84:* 611–617 (1978).

71 Segal, A.W.; Dorling, J.; Coades, S.: Kinetics of fusion of the cytoplasmic granules with
phagocytic vacuoles in human polymorphonuclear leukocytes. J. Cell Biol. *85:* 42–59
(1980).

72 Seligmann, B.E.; Gallin, J.I.: Use of lipophilic probes of membrane potential to assess
human neutrophil activation of abnormality in chronic granulomatous disease. J. clin.
Invest. *6:* 493–503 (1980).

73 Seligmann, B.; Gallin, J.I.: Secretagogue modulation of the response of human neutrophils to chemoattractants: studies with a membrane potential sensitive cyanine dye.
Molec. Immunol. *17:* 191–200 (1980).

74 Seligmann, B.E.; Gallin, E.K.; Martin, D.L.; Shain, W.; Gallin, J.E.: Interaction of
chemotactic factors with human polymorphonuclear leukocytes: studies using a membrane potential sensitive cyanine dye. J. Membr. Biol. *52:* 257–272 (1980).

75 Serhan, C.; Anderson, P.; Goodman, E.; Dunham, P.; Weissmann, G.: Phosphatidate
and oxidized fatty acids are calcium ionophores. Studies employing arsenazo III in
liposomes. J. biol. Chem. *256:* 2736–2741 (1981).

76 Serhan, C.N.; Korchak, H.M.; Weissmann, G.: PGB_x, a prostaglandin derivative,
mimics the actions of the calcium ionophore A23187 on human neutrophils. J. Immun.
(in press, 1982).

77 Sha'afi, R.I.; Williams, K.; Wacholtz, M.C.; Becker, E.L.: Binding of the chemotactic
synthetic peptide [^{3}H]-formyl-Nor-Leu-Leu-Phe to plasma membrane of rabbit neutrophils. FEBS Lett. *91:* 305–309 (1978).

78 Showell, H.J.; Naccache, P.H.; Sha'afi, R.I.; Becker, E.L.: The effects of extra-cellular
K^+, Na^+, and Ca^{++} on lysosomal enzyme secretion from polymorphonuclear leukocytes.
J. Immun. *119:* 804–811 (1977).

79 Simchowitz, L.; Atkinson, J.P.; Spilberg, I.: Stimulus-dependent deactivation of chemotactic factor-induced cyclic AMP response and superoxide generation by human
neutrophils. J. clin. Invest. *66:* 736–747 (1980).

80 Simchowitz, L.; Fischbein, L.C.; Spilberg, I.; Atkinson, J.P.: Induction of a transient
elevation in intracellular levels of adenosine-3′,5′-cyclic monophosphate by chemotactic factors: an early event in human neutrophil activation. J. Immun. *124:* 1482–1491
(1980).

81 Simchowitz, L.; Spilberg, I.: Generation of superoxide radicals by human peripheral
neutrophils activated by chemotactic factor. Evidence for the role of calcium. J. Lab.
clin. Med. *93:* 583–593 (1979).

82   Smith, R. J.; Iden, S. S.: Phorbol myristate acetate-induced release of granule enzymes from human neutrophils: inhibition by the calcium antagonist 8-(N, N-diethyl-amino)-octyl-3, 4, 5,-trimethoxybenzoate hydrochloride. Biochem. biophys. Res. Commun. *91:* 263–271 (1979).

83   Smith, R. J.; Iden, S. S.: Pharmacological modulation of chemotactic factor elicited release of granule-associated enzymes from human neutrophils. Effects of prostaglandins, non-steroid anti-inflammatory agents and corticosteroid. Biochem. Pharmac. *29:* 2389–2395 (1980).

84   Smith, R. J.; Ignarro, L. J.: Bioregulation of lysosomal enzyme secretion from human neutrophils: roles of guanosine 3′, 5′-monophosphate and calcium in stimulus-secretion coupling. Proc. natn. Acad. Sci. USA *72:* 103–112 (1975).

85   Smolen, J. E.; Eisenstat, B. A.; Weissmann, G.: The fluorescence response of chlorotetracycline-loaded human neutrophils. II. Correlations with lysosomal enzyme release and evidence for a 'trigger pool' of calcium Biochim. biophys. Acta *717:* 422–431 (1982).

86   Smolen, J. E.; Korchak, H. M.; Weissmann, G.: Increased levels of cyclic adenosine-3′, 5′-monophosphate in human polymorphonuclear leukocytes after surface stimulation. J. clin. Invest. *65:* 1077–1085 (1980).

87   Smolen, J. E.; Korchak, H. M.; Weissmann, G.: Initial kinetics of lysosomal enzyme secretion and superoxide anion generation by human polymorphonuclear leukocytes. Inflammation *4:* 145–163 (1980).

88   Smolen, J. E.; Korchak, H. M.; Weissmann, G.: The roles of extracellular and intracellular calcium in lysosomal enzyme release and superoxide anion generation by human polymorphonuclear leukocytes. Biochim. biophys. Acta *677:* 512–520 (1981).

89   Smolen, J. E.; Shohet, S. B.: Remodeling of granulocyte membrane fatty acids during phagocytosis. J. clin. Invest. *53:* 726–734 (1974).

90   Smolen, J. E.; Weissmann, G.: Effects of indomethacin, 5, 8, 11, 14-eicosatetraynoic acid, and *p*-bromophenacyl bromide on lysosomal enzyme release and superoxide anion generation by human polymorphonuclear leukocytes. Biochem. Pharmac. *29:* 533–538 (1979).

91   Smolen, J. E.; Weissmann, G.: Stimuli which provoke secretion of azurophil granules induce increments in adenosine cyclic 3′, 5′-monophosphate. Biochim. biophys. Acta *672:* 197–206 (1981).

92   Smolen, J. E.; Weissmann, G.: The fluorescence response of chlortetracycline-loaded human neutrophils. I. The effect of various stimuli and calcium antagonists Biochim. biophys. Acta *720:* 172–180 (1982).

93   Stenson, W. F.; Parker, C. W.: Monohydroxytetraenoic acids (HETEs) induce degranulation of human neutrophils. J. Immun. *124:* 2100–2104 (1980).

94   Stossel, T. P.; Hartwig, J. H.: Interactions of actin, myosin, and a new actin-binding protein of rabbit pulmonary macrophages. II. Role of actin-binding protein in cytoplasmic movement and phagocytosis. J. Cell Biol. *68:* 602–619 (1976).

95   Stossel, T. P.; Pollard, T. D.; Mason, R. J.; Vaughan, M.: Isolation and properties of phagocytic vesicles from polymorphonuclear leukocytes. J. clin. Invest. *50:* 1745–1757 (1971).

96   Takeshige, K.; Nagi, Z. G.; Tatscheck, B.; Minakami, S.: Release of calcium from membranes and its relation to phagocytotic metabolic changes: a fluorescence study of

leukocytes loaded with chlortetracycline. Biochem. biophys. Res. Commun. *95:* 410–415 (1980).

97 Tatham, P.E.R.; Delues, P.J.; Shen, L.; Roitt, I.M.: Chemotactic factor-induced membrane potential changes in rabbit neutrophils monitored by the fluorescent dye 3,3′-dipropylthiadicarbocyanine iodide. Biochim. biophys. Acta *602:* 285–298 (1980).

98 Tou, J-S.; Stjernholm, R.L.: Stimulation of the incorporation of ^{32}Pi and Myo[2-^{3}H]-inositol into the phosphinositides in polymorphonuclear leukocytes during phagocytosis. Archs Biochem. *160:* 487–494 (1974).

99 Turner, S.R.; Campbell, J.A.; Lynn, W.J.: Polymorphonuclear leukocyte chemotaxis toward oxidized components of cell membranes. J. exp. Med. *141:* 1437–1441 (1975).

100 Ukena, T.E.; Berlin, R.D.: Effect of colchicine and vinblastine on the topographical separation of membrane functions. J. exp. Med. *136:* 1–7 (1972).

101 Volpi, M.; Naccache, P.H.; Sha'afi, R.I.: Arachidonate metabolite(s) increase the permeability of the plasma membrane of the neutrophils to calcium. Biochem. biophys. Res. Commun. *92:* 1231–1237 (1980).

102 Waite, M.; DeChatelet, L.R.; King, L.; Shirley, P.S.: Phagocytosis-induced release of arachidonic acid from human neutrophils. Biochem. biophys. Res. Commun. *90:* 984–992 (1978).

103 Whitin, J.C.; Chapman, C.E.; Simons, E.R.; Chovaniec, M.E.; Cohen, H.J.: Correlation between membrane potential changes and superoxide production in human granulocytes stimulated by phorbol myristate acetate. J. biol. Chem. *255:* 1874–1878 (1980).

104 Woodin, A.M.; Wieneke, A.A.: The accumulation of calcium by the polymorphonuclear leukocyte treated with staphylococcal leucocidin and its significance in the extrusion of protein. Biochem. J. *87:* 487–495 (1963).

105 Zigmond, S.H.; Hirsch, J.G.: Effects of cytochalasin B on polymorphonuclear leukocyte locomotion, phagocytosis, and glycolysis. Expl Cell Res. *73:* 383–393 (1972).

106 Zucker-Franklin, D.; Hirsch, J.G.: Microscopic studies on the degranulation of rabbit peritoneal leukocytes during phagocytosis. J. exp. Med. *120:* 569–575 (1964).

G. Weissmann, Professor and Director, Division of Rheumatology, Department of Medicine, New York University Medical Center, School of Medicine, 550 First Avenue, New York, NY 10076 (USA)

Cell Biology of the Secretory Process, pp. 546–569 (Karger, Basel 1984)

The Secretory Process in Platelets

James G. White

Department of Pediatrics, University of Minnesota, Minneapolis, Minn., USA

Introduction

The concept of platelet secretion was introduced in early morphological studies of platelet activity in blood coagulation prior to 1900 [3, 18, 26, 28]. 50 years later, *Taniguchi* et al. [37] made the following observation after evaluating platelet-bacterial interaction in vivo: 'In contrast to the phagocytosis of leukocytes, which consume living as well as dead bacteria and cellular debris by intracellular digestion after engulfing them, the blood platelets first stick to the foreign substance and digest it by secreting ferment extracellularly.'

Despite the importance of these early observations, it was *Grette* [16] who first characterized platelets as secretory cells. He found that platelets stimulated by thrombin extruded in parallel a variety of chemical constituents, including serotonin, adenine nucleotides, amino acids and fibrinogen. *Grette* [16] referred to this process as the 'release reaction', and noted that other chemical constituents which might suggest lysis or cell damage were retained by activated platelets. This point was crucial, for it demonstrated that platelets were secretory cells and not merely fragile elements whose destruction led to functional expression in hemostasis.

Grette's [16] thesis on thrombin catalyzed reactions of blood platelets was published in 1963. Two subsequent decades of effort by many investigators have only begun to explore the full potential of his monumental contribution. Interest in the topic of platelet secretion has been intense, and the matter has been examined in detail in several recent reviews [4, 9, 20, 21, 24, 27, 34, 36, 38, 47, 48]. In this chapter the subject will be framed in the context of platelet ultrastructural anatomy. Newer aspects of the release reaction will be discussed and the importance of secretion in the irreversible response of platelets will be evaluated in the light of current investigations.

Platelet Ultrastructure – General Features

Platelets in circulating blood or from samples of properly prepared platelet-rich plasma have a characteristic discoid form [44]. Thin sections of such platelets preserved in glutaraldehyde and osmic acid reveal that the lentiform appearance is supported by a circumferential bundle of micro-tubules (fig. 1, 2). The band of tubules lies just under the surface membrane along its greatest circumference in the equatorial plane. Cross sections reveal that the bundles of tubules are groups of hollow circular profiles at polar ends of the discoid cells. A variety of formed organelles and other structures are randomly distributed in the amorphous matrix of the cyto-plasm. Granules are the most common (fig. 3, 4). They are generally round or oval in shape, although rod or club forms are often observed. Occasionally the matrix has a periodic substructure, but ordinarily is amorphous with two zones of differing electron opacity. The variety of forms and internal organization manifested by granules has made it difficult to separate them into subpopulations on the basis of morphology alone [44].

Small numbers of mitochondria are present in platelets (fig. 2). They are usually oval in form and smaller than granules, but can be long or irregular. Plication of their internal membrane into cristae distinguishes mitochondria from other organelles. Electron-dense bodies are the least common of the formed structures in human platelets [43]. While as many as 7–10 have been found in single thin sections, they usually occur in a frequency of 1–1.4 per platelet (fig. 2–4). Many dense bodies appear as black spots surrounded by clear zones enclosed within a unit membrane, the so-called 'bull's eye' form. Others have a reticular substructure or, in a few, opaque substance fills the interior of the membrane. In human plate-lets the nucleoid of some granules is identical in appearance and opacity to dense bodies (fig. 4). The appearance suggests that some dense bodies may derive from a subpopulation of granules [40].

Glycogen is present in platelet cytoplasm as single particles or large masses (fig. 2). Occasional platelets contain centrioles, nuclear remnants or stacks of membranes similar to Golgi zones in the parent megakaryo-

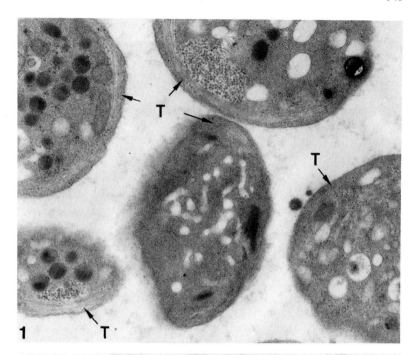

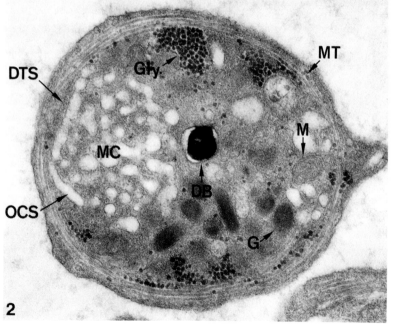

cytes. The latter structures are vestigal remnants unrelated to platelet function [44].

Elements of two-channel systems are dispersed at random in the cytoplasm [1]. Channels of surface connected open canalicular system (OCS) communicate directly with the cell surface (fig. 2, 5, 6). The openings are similar in diameter and may be clustered or randomly dispersed. Each channel follows a tortuous pathway into the cytoplasm. Alternating dilated and narrow segments give the apperance of sausage links. Channels may reconnect to other sites on the surface membrane shortly after entering the cytoplasm. More commonly, they fuse with other canaliculi forming a spiderweb of channels stretching from one side of the platelet to the other. The labyrinth of channels formed by the OCS has the appearance of a fenestrated membrane system [52].

The dense tubular system (DTS) of channels is closed (fig. 2, 5, 7). It does not communicate with the cell surface or elements of the OCS. An amorphous substance similar in density to the surrounding cytoplasm fills DTS channels, distinguishing them from canaliculi of the OCS which are clear. DTS channels are narrow and do not have the alternating dilated and narrow segments evident in elements of the OCS. Cytochemical studies have shown that the DTS is rich in divalent cation binding sites and contains a peroxidase activity associated with the enzyme, cyclo-oxygenase, involved in conversion of arachidonic acid into endoperoxides and, ultimately, thromboxane A_2 [15].

The two-channel systems of the platelet are not totally independent. In one or two areas of the cytoplasm in each cell canaliculi of the OCS and DTS are woven into membrane complexes [46] (fig. 8, 9). The close asso-

---

Fig. 1. Discoid platelets. The lentiform shape of circulating platelets is supported by a circumferential bundle of microtubules (T) lying just under the cell membrane in the equatorial plane of each cell. ×18,000.

Fig. 2. Discoid platelet. The cell in this example has been sectioned in the equatorial plane. A circumferential band of microtubules (MT) supports the discoid configuration. Numerous granules (G), a few mitochondria (M) and occasional electron-dense bodies (DB) are randomly dispersed in the cytoplasmic matrix. Glycogen (Gly) is concentrated in masses or occur as single particles. Clear channels of the open canalicular system (OCS) follow tortuous courses through the cytoplasm. Elements of the dense tubular system (DTS) of channels are often associated with the circumferential bundle of microtubules. Interactions between the two-channel systems result in formation of a twisted mass of membranous elements referred to as the membrane complex (MC). ×30,500.

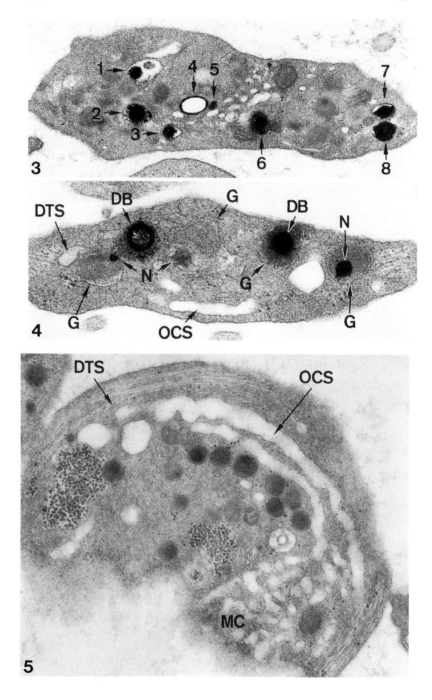

ciation of the two-channel systems in membrane complexes brings elements of the DTS very close to the inner surface of the OCS membranes. Relationships similar to this are also found between the transverse tubular system and sarcotubular system in embryonic muscle cells [49].

Secretory Organelles

The capacity to extrude specific chemical substances from intracellular storage sites to surrounding plasma without losing cytoplasmic constituents suggestive of cell damage is now accepted as an integral feature of the blood platelet response to agents which stimulate irreversible aggregation in vitro and hemostatic plug formation in vivo [16]. Products secreted by platelets are stored in the organelles described briefly above. It was pointed out that granules, though somewhat variable in ultrastructural appearance, could not be divided into subpopulations on the basis of morphology alone. However, cytochemical and biochemical techniques, together with studies of abnormal platelets deficient in specific types of organelles, have made it possible to define at least three types within the granule population. The smallest of the granules contains catalase, which can be detected by ultrastructural cytochemistry [5]. Two or three oval or round structures are specifically stained by catalase reaction product in each platelet (fig. 10). Because of their similarity to catalase containing organelles in other cell types, the stained organelles in platelets are referred

---

Fig. 3, 4. Platelet storage organelles. Granules are the most numerous organelles in platelets, but dense bodies are also present in significant number. The cell in figure 3 contains at least eight dense bodies which are considered the principle storage sites for serotonin and the nonmetabolic pool or adenine nucleotides. Dense bodies vary in size, shape, and content and occasionally appear to contain substance (2) resembling the matrix material of granules. The platelet in figure 4 also demonstrates the similarity of dense bodies and granules. Granules (G) in this cell contain nucleotides (N) with an opacity similar to that of dense bodies. Dense bodies (DB) are surrounded by material resembling granule matrix. Elements of the open canalicular system (OCS) and dense tubular system (DTS) lie in proximity to both types of secretory organelles. $3 \times 19,500$; $4 \times 35,000$.

Fig. 5. Platelet membrane systems. This example is unusual in that a channel of the OCS, though tortuous, can be followed for a long distance. Elements of the dense tubular system (DTS) are closely associated with the marginal band of microtubules. In one area of the cytoplasm components of the two-channel system are interwoven to form a membrane complex (MC). $\times 24,000$.

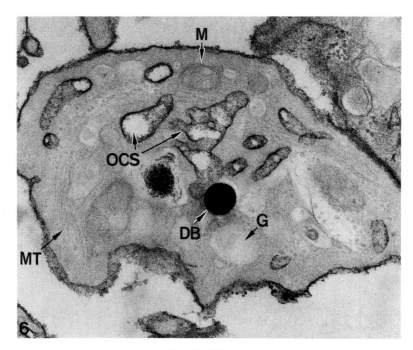

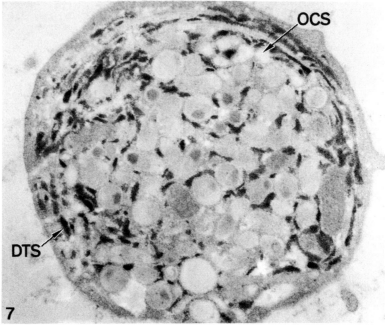

to as peroxisomes. Platelet peroxisomes and their stored products do not appear to be secreted during the release reaction.

Biochemical studies have shown that platelets contain substantial quantities of hydrolytic enzymes [8]. Early studies employing subcellular fractionation and centrifugation on continuous gradients suggested that hydrolytic enzymes were associated with vesicular constituents separate from the granules [33]. One study employing ultrastructural cytochemistry appeared to support localization of certain hydrolytic enzymes in platelet vesicles [2]. However, early [45] and more recent [50] investigations in our laboratory demonstrated that reaction products of acid phosphatase and β-glucuronidase are localized in granules (fig. 11). *Parmley* et al. [29] have recently confirmed this localization in their study of giant granules in platelets from patients with the Chédiak-Higashi syndrome. Granules storing hydrolytic enzymes in other cells and tissues are referred to as lysosomes and this term seems appropriate for similar organelles in platelets. It is not certain, however, whether platelets contain more than one kind of lysosome. The number of different enzymes found in platelets has increased rapidly in recent years, and not all of them are secreted at the same rate or to the same extent during the release reaction [24]. Further studies will be required to determine whether there are one or several kinds of platelet lysosomes.

The subpopulation of granules known as the α granules are the largest group of organelles in the platelet. Many chemical substances were found associated with α granules in studies of subcellular fractions [8, 24]. As indicated above, the granules are heterogeneous in structure, but a specific cytochemical test to distinguish different types of α granules is not yet available. Therefore, it is uncertain whether or not α granules consist of several subpopulations. Investigations of patients with the gray platelet syndrome (GPS) suggest that most, if not all, α granule constituents are

---

Fig. 6. Cytochemistry of membrane systems. The platelet was fixed in glutaraldehyde and osmium solution containing lanthanum. Electron dense tracer coats the surface of the cell and lines each channel of the surface-connected canalicular systems (OCS). Microtubules (MT), a mitochondrion (M) and granules (G) are barely visible in this unstained section, but a dense body (DB) is prominent. ×40,500.

Fig. 7. Cytochemistry of membrane systems. This platelet is from a sample of PRP incubated for peroxidase activity. Enzyme reaction product is specifically localized to channels of the dense tubular system (DTS) and none is present in the surface-connected canalicular system (OCS). ×31,000.

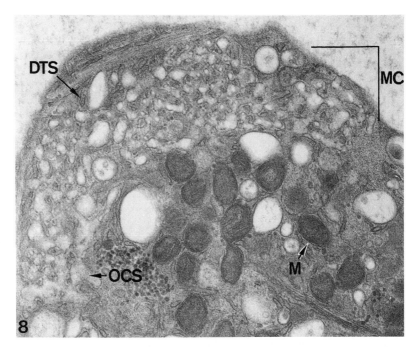

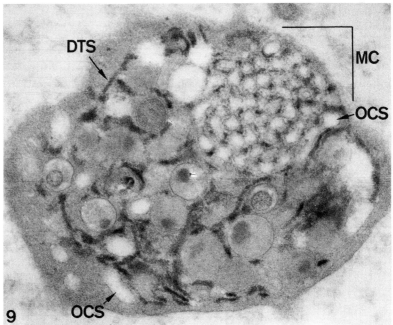

confined to a single population of organelles [50]. Platelets from patients with this disorder contain normal numbers of mitochondria, peroxisomes and lysosomes, but are very deficient in total granules. Biochemical studies [11] have shown that their platelets also lack all of the constituents associated with α granules in previous reports on subcellular fractions. It would be unlikely for the megakaryocyte to be congenitally unable to form several types of organelles. As a result, the α granule population, despite its structural variations, appears to be a homogeneous subpopulation in regard to chemical constituents.

In addition to the mitochondria, peroxisomes, lysosomes and the granules, platelets contain one other major type of organelle, the dense body [57]. The storage pool of adenine nucleotides and serotonin are localized in the electron-opaque structures, as are calcium and pyrophosphate [7]. Similar to the granule population, dense bodies are strikingly variable in size, shape and internal appearance [40, 43]. The organelles are relatively unstable, however, and easily extracted. As a result, some of the variability may be due to fixation problems. The chemical composition of dense bodies is uniform, suggesting that structural differences are probably not related to alterations in their content of stored substances.

Structural Physiology

Sequential physical alterations develop in platelets following exposure to a variety of agents capable of stimulating secretion and irreversible aggregation [44]. It is uncertain whether or not the earliest change develops in the cell surface or in the cytoplasm, for both appear to be affected from the onset of response [41]. Platelets lose their discoid form, develop long, thin pseudopods and bulky surface projections, and undergo a process of internal transformation (fig. 12, 13). Randomly dispersed organelles be-

---

Fig. 8. Membrane complex. Spongelike appearance of peripheral cytoplasm in this cell is due to interwoven associations between elements of surface-connected open canalicular system (OCS) and dense tubular system (DTS). Resulting membrane complex (MC) strongly resembles associations between transverse tubules and sarcotubules in muscle cells. This platelet is from patient with gray platelet syndrome. ×32,000.

Fig. 9. Membrane complex, platelet from sample reacted for endogenous peroxidase activity. Reaction product is localized to channels of dense tubular system (DTS). Clear channels of surface-connected open canalicular system (OCS) form very close associations with dense tubular system (DTS) in membrane complexes (MC). ×38,000.

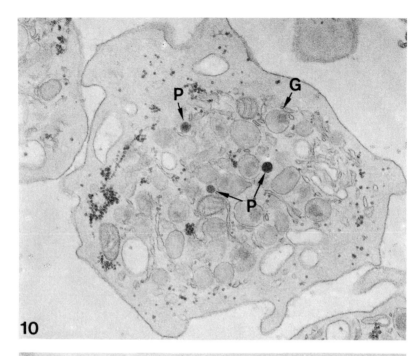

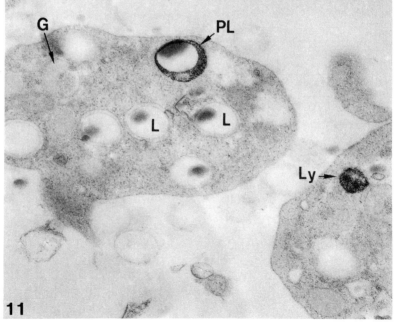

come concentrated together in the center of the cytoplasm. The closely grouped storage organelles are enclosed within a tight ring of microtubules and microfilaments. During early stages of this process the granule population and dense bodies remain intact.

If stimulation is caused by an agent that does not trigger secretion or by a low concentration of a release-inducing agent, internal transformation and shape change can recover completely [44, 47, 48]. The circumferential bundle of microtubules returns to its position just under the membrane along its greatest circumference and storage organelles are restored to random locations in the cytoplasmic matrix. Even after the stage of irreversible aggregation has been reached and secretion begun, significant numbers of transformed platelets can be dissociated and restored to an unaltered appearance [32].

Clearly, the extent of shape change and internal transformation are dependent on the state of the platelet, the conditions under which activation is stimulated and the concentration, potency and duration of exposure to the agonist. The population of platelets exposed to a stimulus varies in age from those just delivered to the circulation to cells just about ready to leave it. Considerable variation exists in the sensitivity of young versus old platelets. As a result, no two platelets are at exactly the same stage of activation at any one moment in time, and it is possible to observe nearly all stages of physical transformation in every activated sample. This has helped to identify the sequential stages of platelet activation, but made it difficult to relate any specific change to secretion.

Potent agents causing irreversible aggregation and release stimulate additional physical changes in platelet internal appearance. Granules crushed together in cell centers gradually lose their definition and are

---

Fig. 10. Organelle zone. Platelet from a sample incubated at high pH according to the procedure of Breton-Gorius to demonstrate catalase activity. Peroxisomes (P) containing the reaction product of catalase are apparent in the cytoplasm. Granules (G) are unreactive. ×25,500.

Fig. 11. Organelle zone. Platelet from a sample of citrate platelet-rich plasma mixed with latex (L) particles and left unstirred for 30 min. Sample was then fixed in glutaraldehyde and incubated for acid phosphatase activity. Organelles containing dense reaction product – lead phosphate – are smaller than most platelet granules (G) and resemble in some respects dense bodies. As a result it is difficult to distinguish lysosomes (Ly) from DB without the use of latex particles. Latex spherules taken up by platelets interact with some lysosomes and form phagolysosomes (PL), which are easy to distinguish from dense bodies. ×37,500.

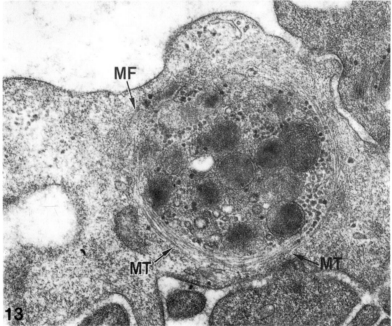

replaced in many examples by dense masses of contracted microfilaments. Some of the swollen granule membranes remaining in the platelet retain variable amounts of matrix material. At this stage microtubules are less prominent as constricted rings in cell centers and appear more frequently in pseudopodia. It is uncertain whether platelets which have reached this degree of transformation can still recover an unaltered appearance [44].

The similarity of events described above to a process of internal contraction or a contraction-relaxation cycle has been noted previously [6, 13]. It was suggested that movement of circumferential microtubules into a tight circle around centrally grouped organelles was due to contraction of the actomyosin network. However, others have suggested that microtubules do not remain intact following platelet activation [25, 35]. Colchicine-binding assays suggested that microtubules are almost completely disassembled immediately after exposure to aggregating agents. 1–4 min following activation tubulin reassembles into tubular polymers in new locations near the cell center or extending into pseudopods. Clearly these findings argue against the concept that constriction of circumferential microtubules is due to a wave of contraction.

Recently this controversy has been reexamined with the aid of taxol, a microtubule-stabilizing agent [51]. Taxol stabilized the discoid shape of unaltered platelets for long periods and blocked disassembly of microtubules by low temperature or vincristine. However, the agent had no effect on the response of platelets to aggregating agents nor did it interfere with the process of clot retraction. Examination of taxol-treated platelets after exposure to aggregating agents revealed exactly the same internal changes described above. Organelles were concentrated in cell centers and encircled by tight-fitting bands of microtubules and microfilaments. The only difference between taxol-treated as compared to control platelets following activation was the presence of nearly twice as many microtubules encircling centrally clumped granules in drug-treated cells. These findings

---

Fig. 12. Shape change in platelets fixed after exposure to adenosine diphosphate and prepared for study by scanning electron microscope. Loss of discoid shape to form irregular spheres and extension of long pseudopods are well demonstrated by this technic. ×12,000.

Fig. 13. Shape change and internal contraction. Platelet from sample exposed to adenosine diphosphate before fixation. Aggregated cell has undergone shape change and internal transformation. Organelles are clumped in cell center and enclosed by ring of microtubules (MT) and microfilaments (MF). ×41,000.

demonstrate that microtubule disassembly is not an essential event in platelet activation and that the wave of contraction is powerful enough inside platelets to drive stabilized microtubules toward cell centers even when there are twice as many as in control cells.

Secretory Process

Transfer of chemical products from storage organelles inside platelets to surrounding plasma is complex. The process involves a sequence of stimulus-activation-contraction-secretion coupling similar in many respects to discharge of stored products from other cell types. However, there are many significant differences. Most cells that secrete substances to the exterior discharge products from a single type of organelle, or, at the most, two varieties. Platelets, though much smaller than most secretory cells, release chemical substances from at least three different kinds of storage organelles. The nearly universal mechanism of granule discharge in secretory cells involves movement of organelles from more central locations to the periphery where they fuse with the cell membrane and release products to the exterior [36]. Platelets represent a rare exception to this rule. Following stimulation secretory organelles are contracted toward cell centers where they are discharged from channels to the exterior [44, 47, 48].

The organelles which discharge when platelets are stimulated and the extent of product release depend on the nature of the aggregating agent, its concentration and the duration of platelet exposure to the agent. For many years it was believed that most of the products secreted by platelets were stored in just two types of organelles, dense bodies and α granules [8]. *Grette* [16] and others noted that thrombin and collagen caused the rapid release of serotonin, adenine nucleotides and calcium, all confined within dense bodies [17, 19, 22]. The release of acid hydrolases, at that time believed to be constituents of α granules, occurred at a slower rate after thrombin addition [20]. The findings suggested two distinct phases of platelet secretion, one occurring early and the other developing later following activation. *Day and Holmsen* [9] termed the early phase 'release I' and the slower component 'release II'. Under the influence of ADP, epinephrine and small amounts of thrombin or collagen only release I was triggered, resulting in discharge of serotonin, the nonmetabolic pool of adenine nucleotides and calcium. Products stored in α granules were not released. Intermediate concentrations of thrombin and collagen stimulated release I and, to some extent, release II. Rapid, nearly complete extrusion of all chemical substances subject to secretion in release I and release II were discharged from platelets following activation by higher concentrations of collagen and thrombin and other potent aggregating agents. Thus, acid hydrolases and other granule-associated substances were believed to be extruded more slowly than products confined to dense bodies and reached peak levels outside platelets after release I had plateaued.

Electron microscopic studies, however, failed to support the patterns observed biochemically [44]. Examination of thin sections from platelet samples fixed at intervals throughout release I and release II failed to show complete discharge of dense bodies, even after exposure to large concentrations of thrombin and collagen and irreversible aggregation [55]. Granules were also retained long after release II had been temporally completed.

Reasons for the discrepancy between biochemical studies of secretion and ultrastructural observations remained obscure until the recent investigations by *Kaplan* [23] and *Kaplan* et al. [24]. Their findings indicated that there are three types of platelet release, rather than two. Depending on the nature of the stimulus and its concentration, products stored in α granules, dense bodies and lysosomes could all be released. However, α granule proteins were secreted at lower concentrations of a stimulating agent than are required for release of dense granule constituents. Products of dense bodies were secreted concomitantly with products of α granules if the stimulus was strong enough, but not as an isolated or separate phase. These findings were essentially the reverse of what had been proposed earlier [9].

ADP and epinephrine, found earlier to cause only release I, also promoted discharge of PF-4 and BTG, but not acid hydrolases. Thus, ADP and epinephrine were capable of causing release from two but not three varieties of storage organelles. Low concentrations of thrombin also caused secretion of dense body and α granule products, but not lysosomal enzymes. Higher amounts of thrombin caused release of constituents from all types of storage organelles. *Kaplan's* [23] study revealed no significant separation in time of secretion of substances from dense bodies or α granules. Furthermore, the percentage of total product secreted by α granules was nearly always higher than the percentage released from dense bodies, no matter which stimulus was used. Thus, release I involves simultaneous release of products from α granules and dense bodies, whereas release II should be used to connote release from lysosomes.

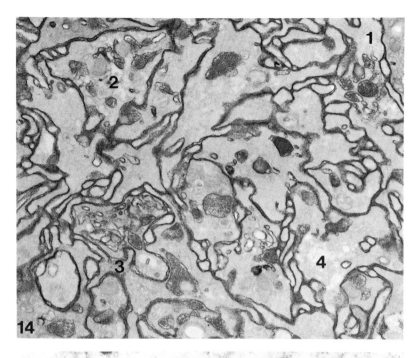

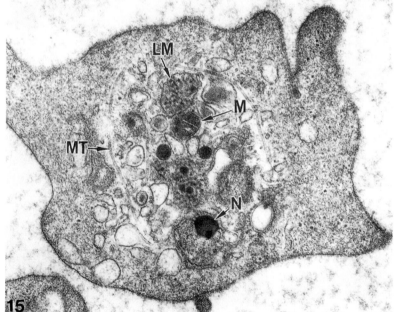

The Secretory Pathway

In a strict sense, secretion merely involves the transfer of substances stored within cells to the exterior environment. However, secretion is not a simple process. The nature of the chemical and physical agonists, the trigger mechanisms, amplifiers, modulators and inhibitors, the metabolic and structural responses, the time course of events and the disposition of storage organelles following stimulation vary greatly from one secretory cell type to another. Not all of the constituents leaving cells after stimulation are products of secretion. For example, endoperoxides and thromboxanes generated during platelet prostaglandin synthesis appear outside the cells shortly after activation, but are not extruded from secretory organelles [14]. Thus, secretion is a complex process, as variable as the types of cells capable of discharging products in a physiologic manner from storage granules.

It is not surprising, therefore, that the secretory pathway in platelets is quite different from that observed in other secretory cells. In most types the storage granules move from random positions to the surface where they fuse with the cell membrane and discharge their contents to the outside. Platelets appear to respond in exactly the opposite way. Randomly dispersed storage granules are moved to the platelet centers where they are enclosed within a tight-fitting web of microtubules and microfilaments [44]. Clearly the pathway for discharge of secretory products from organelles crushed in the cell center must be different than in cells where granules move to and fuse with the surface membranes.

A search for the secretory pathway in platelets revealed that channels of the surface-connected OCS served as canaliculi for the discharge of secretory products [42] (fig. 14). At some point following activation plate-

---

Fig. 14. Secretory pathway. Platelets in this example were aggregated by epinephrine and fixed in glutaraldehyde and osmic acid containing lanthanum citrate. The stain outlines the surfaces of closely associated platelets, and fills channels of the open canalicular system. Uptake of lanthanum particles by open channels is seen to advantage in the cells numbered 1–4. ×16,500.

Fig. 15. Secretory pathway. Platelet from a sample of C-PRP incubated with the cationic polyelectrolyte, polylysine. The agent is taken up and selectively deposited in platelet secretory organelles. The light matrix (LM) of the platelet granules is polymerized into a lattice-like structure. Granule nucleoids (N) became dense and fragmented. After prolonged incubation the cells lose their discoid form and develop internal transformation. M=Microfilaments; MT=microtubules. ×41,500.

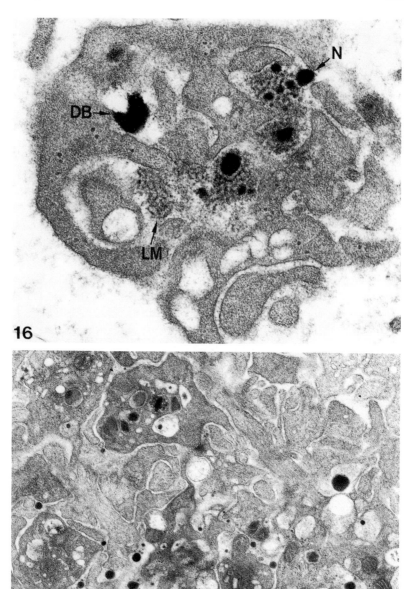

let organelles develop direct communication with OCS channels (fig. 15–17). However, it is possible that a relationship exists between at least some storage organelles and the OCS in unaltered platelets [48]. Platelets incubated with thorium dioxide or latex spherules take up the foreign particulates into channels of the OCS without losing discoid shape [39, 53]. In some of the unaltered platelets latex spherules or thorium dioxide particles appear in apparently intact granules. Examination of platelets ruptured following exposure to distilled water by the negative stain whole mount technique also suggests that granules in unstimulated cells are directly connected to elements of the OCS. The nature of the association between channels and granules in unaltered platelets, however, remains uncertain.

Movement of organelles to cell centers during shape change and internal contraction is a normal feature of the secretory response. However, communications between channels of the OCS and storage granules can be established without centripetal movement. Phorbol myristate acetate (PMA) labilizes platelet granules causing them to swell and lose some of their content [10]. If the cells are treated with cytochalasin B (CB) before exposure to PMA, communication between canaliculi of the OCS and all platelet granules is established without loss of discoid shape [54]. Under these conditions the amount of substance lost from the storage organelles is slight, even though they appear virtually empty. Exposure of the cytochalasin B treated cells to trypsin will also establish communication between granules and the OCS and cause degranulation of discoid platelets [48]. Thus, the development of openings between channels and storage organelles is an important step in secretion, but not sufficient to result in discharge of major amounts of stored products.

The platelet muscle system drives the process of secretion. Platelets exposed to a small amount of PMA will secrete less than 10% of their storage pool labeled with ^{14}C-serotonin, despite developing a 'Swiss cheese' appearance [56]. One of the endoperoxides, PGG_2, can be titered to

---

Fig. 16. Secretory pathway. After prolonged incubation with cationic polypeptides the altered platelets extrude their storage organelles. All stages in the extrusion of granule light matrix (LM) fragmented nucleoids (N) and dense body (DB) substance are apparent in this example. ×35,000.

Fig. 17. Secretory pathway. Platelet sample fixed 5 min after exposure to collagen on the platelet aggregometer. Polylysine was added to the sample just prior to collagen and the stirring rod. Storage organelles in all stages of extrusion are easily identified. ×13,500.

a level low enough so that it will also induce release of less than 10% of labeled storage pool [12]. When half of the concentrations of PMA and PGG$_2$ that caused release of 10% of the labelled storage pool were combined together they stimulated release of over 40% of the pool of ^{14}C-5HT. The study demonstrated that PGG$_2$, an agent that causes contraction, could trigger extensive release in platelets whose granules had been swollen by PMA, a chemical that labilizes storage organelles without activating contraction. Platelet secretion, then, is at least a two-stage process, involving labilization of storage organelles and contraction which drives the secreted products from the cells.

Conclusions

Platelets are secretory cells. Following exposure to agents which cause the release reaction, discoid platelets become irregular and roughly spherical with spiky surface projections. Storage organelles are moved from random positions to the centers of the cells where they are closely encircled by webs of microtubules and microfilaments. Labilization results in development of direct communication between granules and the OCS, and internal contraction drives the stored products from labilized organelles through the channels of the OCS to the platelet exterior. Secreted substances amplify the response of platelets to agonists, sustain the reaction and recruit additional platelets to the site of vascular damage. Platelet secretion, therefore, is important in the physiology of hemostasis and a factor in thrombotic disease. However, secretion is ancillary, just as are the synthesis of endoperoxides and thromboxanes. Recent work has shown that human platelets can develop irreversible aggregation in response to physiological agents by a mechanism of membrane modulation completely independent of prostaglandin synthesis and secretion [30, 31]. Thus, secretion does not appear to be the predominant mechanism regulating platelet function, but an important cofactor in the total response of the cell to vascular injury.

References

1   Behnke, O.: The morphology of blood platelet membrane systems. Ser. haematol. *3:* 3–16 (1970).
2   Bentfield, M. E.; Bainton, D. F.: Cytochemical localization of lysosomal enzymes in rat megakaryocytes and platelets. J. clin. Invest. *56:* 1635–1639 (1975).

3   Bizzozero, G.: Über einen neuen Formbestandtheil des Blutes und dessen Rolle bei der Thrombose und der Blutgerinnung. Virchows Arch. path. Anat. *90:* 261–332 (1881).

4   Born, G. V. R.: Current ideas on the mechanism of platelet aggregation. Ann. N.Y. Acad. Sci. *201:* 4 (1972).

5   Breton-Gorius, J.; Guichard, G.: Two different types of granules in megakaryocytes and platelets as revealed by the diaminobenzidine reaction. J. Microsc. Biol. cell. *23:* 197–202 (1975).

6   Cohen, I.; Gerrard, J. M.; White, J. G.: The role of contractile filaments in platelet activation; in Peeters, Protides in the biological fluids, pp. 555–566 (Pergamon Press, Oxford 1979).

7   DaPrada, M.; Richards, J. G.; Kettler, R.: Amine storage organelles in platelets; in Gordon, Platelets in biology and pathology, vol. 2, pp. 107–145 (Elsevier/North-Holland Biomedical Press, Amsterdam 1981).

8   Day, H. J.; Holmsen, H.; Hovig, T.: Subcellular particles of human platelets. Scand. J. Haematol., suppl. 7 (1969).

9   Day, H. J.; Holmsen, H.: Concepts of the blood platelet release reaction. Ser. haematol. *4:* 3 (1971).

10   Estensen, R. D.; White, J. G.: Ultrastructural features of the platelet response to phorbol myristate acetate. Am. J. Path. *74:* 441–452 (1974).

11   Gerrard, J. M.; Phillips, D. R.; Rao, G. H. R.; Plow, E. F.; Walz, D. A.; Ross, R.; Harker, L. A.; White, J. G.: Biochemical studies of two patients with the gray platelet syndrome: selective deficiency of platelet alpha granules. J. clin. Invest. *66:* 102–109 (1980).

12   Gerrard, J. M.; Townsend, D.; Stoddard, S.; Witkop, C. J., Jr.; White, J. G.: The influence of prostaglandin G_2 on platelet ultrastructure and platelet secretion. Am. J. Path. *86:* 99–116 (1977).

13   Gerrard, J. M.; White, J. G.: The structure and function of platelets, with emphasis on their contractile nature; in Iochim, Pathobiology annual, vol. 6, pp. 31–58 (Appleton Century Crofts, New York 1976).

14   Gerrard, J. M.; White, J. G.: Prostaglandins and thromboxanes: 'middlemen' modulating platelet function in hemostasis and thrombosis. Prog. Hemost. Thromb. *4:* 87–125 (1978).

15   Gerrard, J. M.; White, J. G.; Rao, G. H. R.; Townsend, D.: Localization of platelet prostaglandin production in the platelet dense tubular system. Am. J. Path. *83:* 283–298 (1976).

16   Grette, K.: Relaxing factor in extracts of blood platelets and its function in the cells. Nature, Lond. *198:* 488–489 (1963).

17   Haslam, R. J.: Role of adenosine diphosphate in the aggregation of human blood platelets by thrombin and fatty acids. Nature, Lond. *202:* 765 (1964).

18   Hayem, G.: Sur le mechanism de l'arrêt des hemorrhagies. C. r. hebd. Séanc. Acad. Sci., Paris *95:* 18–36 (1882).

19   Holmsen, H.: Collagen-induced release of adenosine diphosphate from blood platelets incubated with radioactive phosphate in vitro. Scand. J. clin. Invest. *17:* 239 (1965).

20   Holmsen, H.; Day, H. J.: The selectivity of the thrombin induced platelet release reaction. J. Lab. clin. Med. *75:* 840 (1970).

21   Holmsen, H.; Day, H. J.; Stormorken, H.: The blood platelet release reaction. Scand. J. Haemat., suppl. 8, pp. 3–26 (1969).

22 Hovig, T.: Release of a platelet aggregating substance (adenosine diphosphage) from rabbit blood platelets induced by saline 'extract' of tendons. Thromb. Diath. haemorrh. 9: 264 (1963).

23 Kaplan, K. L.: Platelet granule proteins: localization and secretion; in Gordon, Platelets in biology and pathology, vol. 2, pp. 77–90 (Elsevier/North-Holland Biomedical Press, Amsterdam 1981).

24 Kaplan, K. L.; Broekman, M. J.; Chernoff, A.; Lesnik, G. R.; Drillings, M.: Platelet α granule proteins: studies on release and subcellular localization. Blood 53: 604–618 (1979).

25 Kenney, D. M.; Chao, F. C.: Ionophore-induced disassembly of blood platelet microtubules: effect of cyclic-AMP and indomethacin. J. cell. Physiol. 103: 289–298 (1980).

26 Muller, H. F.: Über einen bisher nicht beachteten Formbestandtheil des Blutes. Zentbl. allg. path. Anat. 7: 529 (1896).

27 Mustard, J. F.; Packham, M. A.: Factors influencing platelet function: adhesion, release and aggregation. Pharmac. Rev. 22: 97 (1970).

28 Osler, W.: An account of certain organisms occurring in the liquor sanquinis. Proc. R. Soc. 22: 391–398 (1874).

29 Parmley, R. T.; Poom, M. C.; Gist, W. M.; Malluh, A.: Giant platelet granules in a child with the Chédiak-Higashi syndrome. Am. J. Hematol. 6: 51–60 (1979).

30 Rao, G. H. R.; Gerrard, J. M.; Witkop, C. J.; White, J. G.: Platelet aggregation independent of ADP release or prostaglandin synthesis in patients with Hermansky-Pudlak syndrome. Prostaglandins and Medicine 6: 459 (1981).

31 Rao, G. H. R.; Reddy, K. R.; White, J. G.: Modification of human platelet response to sodium arachidonate by membrane modulation. Prostaglandins and Medicine 6: 75–90 (1981).

32 Reimers, H. J.; Packham, M. A.; Kinlough-Rathbone, R. L.; Mustard, J. F.: Effects of repeated treatment of rabbit platelets with low concentrations of thrombin on their function, metabolism and survival. Br. J. Haemat. 25: 657–689 (1973).

33 Siegel, A.; Luscher, E. F.: Non-identity of the granules of human blood platelets with typical lysosomes. Nature, Lond. 215: 745 (1967).

34 Skaer, R. J.: Platelet degranulation; in Gordon, Platelets in biology and pathology, vol. 2, pp. 321–248 (Elsevier/North-Holland Biomedical Press, Amsterdam 1981).

35 Steiner, M.; Ikeda, Y.: Quantitation assessment of polymerized and depolymerized platelet microtubules. Changes caused by aggregating agents. J. clin. Invest. 3: 443–448 (1979).

36 Stormorken, H.: The release of secretion. Scand. J. Haematol., suppl. 9 (1969).

37 Taniguchi, T.; Joogetsu, M.; Kasahara, T.: A role of blood platelets against infection. Jap. J. exp. Med. 8: 55 (1930).

38 Weiss, H. J.: Abnormalities in platelet function due to defects in the release reaction. Ann. N.Y. Acad. Sci. 201: 161 (1972).

39 White, J. G.: The transfer of thorium particles from plasma to platelets and platelet granules. Am. J. Path. 53: 567 (1968).

40 White, J. G.: The dense bodies of human platelets: origin of serotonin storage particles from platelet granules. Am. J. Path. 53: 791 (1968).

41 White, J. G.: Fine structural alterations induced in platelets by adenosine diphosphate. Blood 31: 604 (1968).

42   White, J.G.: A search for the platelet secretory pathway using electron dense tracers. Am. J. Path. *58:* 31 (1970).

43   White, J.G.: Origin and function of platelet dense bodies. Ser. Haematol. *3:* 17 (1970).

44   White, J.G.: Platelet morphology; in Johnson, The circulating platelet, p.45 (Academic Press, New York 1971).

45   White, J.G.: The ultrastructural cytochemistry and physiology of blood platelets; in Mostofi, Brinkhous, The platelet physiology of blood platelets, pp.83–115 (Williams & Wilkins, Baltimore 1971).

46   White, J.G.: Interaction of membrane systems in blood platelets. Am. J. Path. *66:* 295–312 (1972).

47   White, J.G.: Identification of platelet secretion in the electron microscope. Ser. haematol. *6:* 429–459 (1973).

48   White, J.G.: Electron microscopic studies of platelet secretion. Prog. Hemost. Thromb. *2:* 49–98 (1974).

49   White, J.G.: Is the canalicular system the equivalent of the muscle sarcoplasmic reticulum? Hemostasis *4:* 185–191 (1975).

50   White, J.G.: Ultrastructural studies of the gray platelet syndrome. Am. J. Path. *95:* 445–462 (1979).

51   White, J.G.: Influence of a microtubule stabilizing agent on platelet physiology (in press, 1982).

52   White, J.G.; Clawson, C.C.: The surface-connected canalicular system of blood platelets – a fenestrated membrane system. Am. J. Path. *101:* 353–364 (1980).

53   White, J.G.; Clawson, C.C.: Effects of large latex particle uptake on the surface connected canalicular system of blood platelets: a freeze-fracture and cytochemical study. Ultrastruct. Path. *2:* 277–287 (1981).

54   White, J.G.; Estensen, R.D.: Degranulation of discoid platelets. Am. J. Path. *68:* 289 (1972).

55   White, J.G.; Gerrard, J.M.: The cell biology of platelets; in Weissman, Handbook of inflammation (the cell biology of inflammation), p.83 (Elsevier/North-Holland, New York 1980).

56   White, J.G.; Rao, G.H.R.; Estensen, R.D.: Investigation of the release reaction in platelets exposed to phorbol myristate acetate. Am. J. Path. *75:* 301 (1974).

57   Wood, J.G.: Electron microscopic localization of 5-hydroxytryptamine (5-HT). Tex. Rep. Biol. Med. *23:* 828–837 (1965).

J.G. White, MD, Department of Pediatrics, Box 490 Mayo Memorial Building, University of Minnesota, Minneapolis, MN 55455 (USA)

Cell Biology of the Secretory Process, pp. 570–589 (Karger, Basel 1984)

Secretory Process in Normal and Malignant Trophoblast

Robert O. Hussa

Medical College of Wisconsin, Department of Gynecology and Obstetrics, Milwaukee, Wisc., USA

Introduction

The trophoblast is a rich source of many secretory products including chorionic gonadotropin [12, 13, 29, 30, 32–36, 61], placental lactogen [12, 13, 24, 29, 30], chorionic corticotropin [12, 29], chorionic thyrotropin [12, 29], chorionic follicle-stimulating hormone [42], chorionic gonadotropin-releasing hormone [12, 27, 32, 34], chorionic thyrotropin-releasing hormone [12, 27], heat-stable alkaline phosphatase [12], oxytocinase [12], histaminase [12], pregnancy-specific β_1-glycoprotein [12, 28], pregnancy-associated plasma proteins A, B and C [12, 28], placental protein 5 [12, 28], and many steroids including progesterone, 17β-estradiol, estrone, and estriol [30]. This chapter reviews the secretory mechanisms for the two most extensively studied proteins, human chorionic gonadotropin (hCG) and human placental lactogen (hPL), in both normal and malignant trophoblast.

Trophoblast Cell Types

The trophoblast layer of the placenta contains villous (or Langhans') cytotrophoblast cells in the inner layer, syncytial trophoblast in the outer layer, and transitional trophoblast cell types [62]. Langhans' cells, which are comparatively undifferentiated, are proliferative, mononucleated, and of moderate size. The syncytial trophoblast layer lines the placental villi which are in contact with the maternal circulation in the intervillous space. The syncytiotrophoblast layer is a syncytium of nonproliferative, multinucleated giant cells, and arises by differentiation of cytotrophoblast. The syncytiotrophoblast contains abundant rough-surfaced endoplasmic reticulum, Golgi complexes, mitochondria, lysosomes, and osmiophilic lipid droplets. The synthesis of steroid and peptide hormones occurs predominantly in the syncytiotrophoblast [23, 45, 62], although recent immunocytochemical evidence suggests a role for cytotrophoblast in production of α-subunit of hCG (reviewed below). The relative concentration of cytotrophoblast cells is greatest during the first trimester of pregnancy. Toward the end of the first trimester, the proportion of cytotrophoblast cells begins to decrease, although cytotrophoblast cells persist in reduced numbers throughout gestation. The syncytial trophoblast gradually becomes more dominant beginning in the second trimester of pregnancy [62]. In other regions of the placenta there are cytotrophoblastic variants (e.g., basal plate and cell-column cytotrophoblasts) that are more complex than Langhans' cells. These cytotrophoblast cells, like the transitional cells in the villi, may approach or equal the mature syncytial trophoblast in ultrastructural complexity [57, 62].

Choriocarcinomas are malignancies containing trophoblastic cells, and usually occur postgestationally. Like the normal placenta, choriocarcinoma tumors contain the entire spectrum of trophoblast cells, from undifferentiated cytotrophoblast to syncytium [63]. Continuous cell lines derived from choriocarcinoma contain predominantly cytotrophoblast-like cells (96–99% of the total population). The remaining cells are syncytiotrophoblast-like in that they are giant, multinucleated, and nonproliferative [23]. Transplantation of such cells to the hamster cheek pouch or into nude mice leads to trophoblastic tumors that contain a considerably greater population of syncytiotrophoblast cells.

The Process of hCG Secretion in the Trophoblast

hCG is a major secretory product of the first-trimester placenta. The presence of hCG is detectable in serum within the first week after fertilization. Serum hCG levels rise rapidly in early gestation, with a doubling time of about 2 days. The concentration of hCG reaches a peak during the 8–11th week of pregnancy. Subsequently, circulating levels decrease to 10% of the peak levels for the remaining two trimesters of pregnancy [13, 29, 32, 61]. Like the pituitary glycoprotein hormones luteinizing hormone (LH), follicle-stimulating hormone (FSH), and thyrotropin (TSH), this hormone

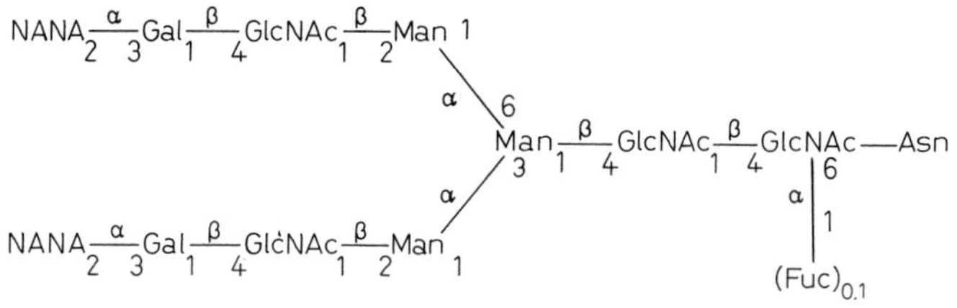

NANA$\frac{\alpha}{2\quad3}$Gal$\frac{\beta}{1\quad4}$GlcNAc$\frac{\beta}{1\quad2}$Man 1

α \\ 6

Man$\frac{\beta}{3\quad1}$GlcNAc$\frac{\beta}{4}$GlcNAc——Asn

α / 6

α | 1

NANA$\frac{\alpha}{2\quad3}$Gal$\frac{\beta}{1\quad4}$GlcNAc$\frac{\beta}{1\quad2}$Man 1

(Fuc)$_{0.1}$

Fig. 1. Structure of N-glycosidic oligosaccharide units attached to asparagine residues 52 and 78 of hCGα, and 13 and 30 of hCGβ. Fucose is found on the oligosaccharide unit of hCGβ only [references in 32]. Gal=D-galactose; GlcNAc=N-acetyl-D-glucosamine; Man=D-mannose; Fuc=L-fucose; NANA=N-acetylneuraminic acid.

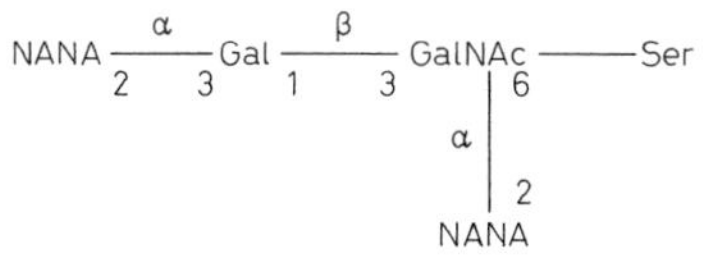

NANA$\frac{\alpha}{2\quad3}$Gal$\frac{\beta}{1\quad3}$GalNAc——Ser

α | 6

α | 2

NANA

Fig. 2. Structure of O-glycosidic oligosaccharide units attached to serine residues 121, 127, 132, and 138 of hCGβ [references in 32]. GalNAc=N-acetyl-D-galactosamine.

(molecular weight 37,000) is composed of two nonidentical subunits, designated α and β. The 92-amino acid α-subunit is common to all four glycoprotein hormones, whereas the β-subunits differ and confer specificity to each hormone. The β-subunit of hCG contains 145 amino acids [32, 34]. Each subunit of hCG contains two asparagine-linked complex-type oligosaccharides (fig. 1). In addition, the β-subunit contains four simple oligosaccharide units linked O-glycosidically to serine residues in the unique 30-amino acid C-terminal portion of the molecule (fig. 2).

Separate Synthesis of α- and β-Subunits

Several different lines of evidence indicate that the α- and β-subunits of hCG are synthesized from two separate mRNAs. *First,* many instances of isolated or unbalanced production of either the α- or β-subunit of hCG (usually α-subunit) have been described [32, 34]. For example, in early pregnancy the placenta secretes predominantly complete hCG but very little of either subunit. After the peak of hCG at weeks 8–11 of pregnancy, the levels of complete hCG and β-subunit decrease, while α-subunit increases until term [29, 32, 61]. A number of human malignant tissues and cell lines in tissue culture secrete predominantly either the α- or β-subunit [references in 32, 34]. *Second,* the cell-free translational activities for hCGα and hCGβ sedimented at different positions in sucrose density gradients [14]. *Third,* separate complementary DNA (cDNA) molecules for the two subunits have been synthesized and sequenced by *Fiddes and Goodman* [20, 21]. The amino acid sequences of the α- and β-subunits, deduced from the nucleotide sequence of the respective cDNAs, corresponded to the previously determined amino acid sequences. Furthermore, signal peptide sequences of 24 and 20 amino acids were predicted from the cDNA nucleotide sequences of pre-hCGα [20] and pre-hCGβ [21], respectively. Both presequences agreed with the amino acid sequences determined by *Birken* et al. [8] on pre-hCGα and pre-hCGβ synthesized in wheat germ cell-free systems and sequenced by Edman degradation. *Lastly,* a mechanism involving separate mRNAs for the α- and β-subunits is supported by the genetic hybrid data of *Bordelon-Riser* et al. [11]. Using clones from the fusion of hCG-producing choriocarcinoma cells with mouse cells, these investigators found that two human chromosomes, 10 and 18, were needed in the same cell for the production of hCG.

Translation of the mRNAs occurs on the ribosomes in the rough endoplasmic reticulum. The nascent α- and β-subunits contain amino-terminal hydrophobic signal peptides of 24 and 20 amino acids, respectively. Once the nascent peptide has grown to a size of 60–80 amino acids, the signal peptide is cleaved by a microsomal endopeptidase [32, 34]. In membrane-depleted cell-free incubations, the prepeptide is not cleaved from the peptide, and the apoprotein containing its signal peptide accumulates. However, it is important to note that such products are artifacts of cell-free translation systems and have never been detected in whole-cell or tissue incubations.

A number of reports of high molecular weight species of hCG have appeared [43; other refs. in 32, 34]. In spite of evidence for a large molecular species of hCG in placenta, and incorporation of labeled amino acids into the large form of hCG in placental incubations [43], the physiological significance of this material remains obscure. *Maruo* et al. [43] have speculated that it may be an intermediate component of hCG biosynthesis, or a product of posttranslational modification, probably of the carbohydrate moieties of hCG. The clarification of these findings must await further research.

Glycosylation

In the membranes of the endoplasmic reticulum, high-mannose oligo-saccharides are attached en bloc in N-glycosidic linkage to asparagine residues 52 and 78 of the α-subunit, and to asparagine residues 13 and 30 of the β-subunit [32]. The attachments are probably sequential [55], but the order of attachment to the two positions of each subunit is not known. The oligosaccharides are carried by dolichol phosphate in the membranes, and face the lumen of the endoplasmic reticulum [34]. Cleavage of the pre-peptide and glycosylation probably occur in tandem, but it has not yet been established if cleavage of the signal peptide occurs first. However, glycosylation is not obligatory for cleavage of the signal peptide to occur, since the apoproteins accumulate in incubations containing membranes from cells incubated with tunicamycin [7, 55]. The antibiotic tunicamycin prevents the formation of the oligosaccharide-lipid molecules which serve as donors for the synthesis of mannose-rich oligosaccharide units [32]. Glycosylation may occur just prior to release of the polypeptide from the ribosome [6, 55].

High-Mannose Intermediates

High-mannose intermediates of both subunits accumulate in cell-free incubations containing normal microsomal membranes [6, 32] and in intact trophoblast cells [53, 55]. *Ruddon* et al. [53–55] carried out pulse-chase experiments with cultures of human choriocarcinoma cells. Radioactive α- and β-subunits of hCG were isolated by immune precipitation and analyzed by electrophoresis on sodium dodecyl sulfate gels. Partially glycosylated precursors for α-subunit (apparent molecular weight 18,000) and β-subunit (24,000) were identified in the cell lysates. Both precursors had intracellular half-lives of at least 1 h and contained the high-mannose core but not the terminal carbohydrate sequences as judged by sensitivity to endoglycosidase H (specific for high-mannose oligosaccharides) but resistance to digestion by endoglycosidase D. The latter enzyme cleaves side chain-free complex-type glycopeptides but not high-mannose glyco-peptides [53]. To explain the observation of two major precursor forms of each subunit in the malignant cell lines, *Ruddon* et al. [54, 55] hypothe-sized the stepwise addition of N-linked high-mannose oligosaccharides to the apoprotein core of each subunit. This mechanism would be analogous to the processing pathway followed by other glycoproteins [44]. When ^{35}S-methionine was used as tracer, the authors could not detect any accu-mulation of fully processed α- or β-subunit (and therefore complete hCG) intracellularly in the choriocarcinoma cells, suggesting that there is a rate-limiting step in the processing of the subunits and that once the mature forms of the subunits are generated, secretion occurs rapidly. Significant amounts of mature subunits were observed, however, when ^{3}H-glucosamine was used as tracer, perhaps reflecting a greater sensitivity of detection since there are more glucosamine residues than methionine in each subunit [54].

Although the structure of the high-mannose oligosaccharide intermediates have not been determined for any of the glycoprotein hormones, a structure containing 9 mannosyl residues and up to 3 glucosyl residues would be consistent with the evidence thus far and is the same as the intermediate involved in the synthesis of other N-linked glycoproteins [32, 34].

Trimming of High-Mannose Intermediates

Glucose residues are probably removed first, by glucosidase action. Trimming of excess mannose residues requires at least two different mannosidases and occurs in a different intracellular compartment than the initial glycosylation reactions [5, 32]. In choriocarcinoma cell lines, trimming reactions and later glycosylation and assembly reactions must occur rapidly since subsequent intermediates do not accumulate. It has not been determined if all trimming steps are completed before peripheral glycosylation begins or if the first N-acetylglucosaminyl transferase uses a partially trimmed intermediate as substrate.

Glycosyl Transfer to N-Linked Core

Separate glycosyl transferases probably catalyze the sequential transfer of GlcNAc, Gal, and NANA from appropriate sugar nucleotides to nonreducing terminal mannosyl residues of the inner core oligosaccharides of the trimmed intermediates [32]. The intracellular location of the peripheral glycosyl transferases has not yet been determined but is likely to be in the Golgi apparatus and/or the plasma membrane. Final trimming of excess mannose residues may also occur during this time.

Glycosyl Transfer to Serine Residues of hCGβ

The four unique O-glycosidic oligosaccharides are synthesized on serine residues 121, 127, 132, and 138 of the β-subunit. Transfer of GalNAc, Gal, and NANA residues probably occurs sequentially by the action of separate glycosyl transferases, and without the involvement of lipid-linked intermediates [32]. The order of glycosylation of the four serine residues, or the intracellular location of glycosyl transfer, are not known. *Elting* [18] studied the order of *N*- and *O*-glycosylation of hCGβ in choriocarcinoma cells in time-course experiments with ^{14}C-amino acids and ^{3}H-glucosamine (which serves as precursor of ^{3}H-GlcNAc in *N*-linked oligosaccharides and of ^{3}H-GalNAc in *O*-linked oligosaccharides). The labeled subunit was immune precipitated. Whereas ^{14}C-amino acid-labeled hCG began to appear in the medium 1.5 h following addition of label (reflecting the biosynthetic lag during synthesis in the rough endoplasmic reticulum), little lag was seen before the appearance of ^{3}H-hCG. The ^{3}H-hexosamine label in hCG secreted prior to 1 h was found predominantly in GalNAc of the *O*-linked oligosaccharides isolated from hCG. Only after 1 h was a significant fraction of label found in GlcNAc of the *N*-linked oligosaccharides isolated from hCG. These results suggested that attachment of *N*-linked oligosaccharides occurs close to the time of protein synthesis, but that *O*-glycosylation occurs at a later stage, perhaps in the Golgi, in the biosynthesis of hCGβ.

Mature Subunits

The disulfide bridges (5 in hCGα, 6 in hCGβ) are complete in the mature subunits, although the point in biosynthesis that this occurs is not known. The fully glycosylated α-subunit (molecular weight 22,000 by SDS-PAGE) possesses in its sequence all the information needed for correct disulfide bond formation. This is apparently not the case for the 6 disulfide bridges of the mature β-subunit (molecular weight 34,000), where the oligosaccharides may interfere with refolding of the denatured mature β-subunit [references in 32, 34].

Subunit Assembly

The mature α- and β-subunits of hCG are paired into a loose complex. The assembly reaction is second-order and therefore requires a high concentration of each subunit in a single compartment [25]. The loose αβ complex is inactive, but provides the necessary information for slow refolding into the active dimer in a first-order reaction [25]. The active conformation stabilizes the dimer. It is in this conformation that complete hCG is secreted and expresses its biological activity. Whether assembly of subunits occurs in Golgi, secretory granules, or plasma membrane has not been determined for hCG. Recent immunohistochemical evidence has led to the hypothesis [1, 9, 10, 13, 23, 26, 37, 57] that the secretion of hCG may be coupled to differentiation of the trophoblast. As reviewed below, several workers have suggested that α-subunit is synthesized by the undifferentiated cytotrophoblast, whereas synthesis of hCGβ, assembly of subunits, and secretion of complete hCG occur only in the syncytiotrophoblast, or at least after the cytotrophoblast cells have coalesced and have begun to differentiate (i.e., transitional cells).

Continuous Secretion versus Storage of hCG

Normal placental extracts contain intact hCG and α-subunit, but not β-subunit, when analyzed by gel filtration and radioimmunoassay (RIA) [1, 61]. Nevertheless, *Heap* et al. [30] estimate that the concentration of hCG stored in placental tissue is generally low and represents less than a 5-min supply of hormone. In extracts from trophoblastic tumors, several workers have noted predominantly intact hCG and abnormal levels and altered forms of subunits [1, 61]. Malignant trophoblast cells in continuous culture do not store appreciable quantities of complete hCG. Instead, the intracellular profile consists of subunits of apparently smaller molecular weight than the secreted forms [3, 15, 32, 34, 35, 51, 53, 55]. *Folman* et al. [22] used tissue slices and organ culture of first trimester placenta to study hCG synthesis and secretion. There was a 2 h lag between incorporation of labeled amino acids and appearance of labeled hCG in the medium. The capacity of the placental tissue to synthesize and secrete hCG was proportional to its hCG content and probably reflected the relative amounts of mRNA in the tissue. Puromycin and cycloheximide inhibited hCG secretion, but synthesis and secretion were renewed upon removal of the inhibitor. We have reported similar results with inhibitors of both RNA synthesis [50] and protein synthesis [58] in malignant trophoblastic cell lines. *Patrito* et al. [48] used pulse-chase incubations and radioimmunoprecipitation techniques to show that newly synthesized hCG was not stored by the early placenta, but was released as it was synthesized. The above findings suggest that hCG is secreted continuously and never accumulates to a significant extent in its cell of origin. As discussed later, this is in contrast to the discontinuous secretion of proteins which are stored in secretory granules or zymogen granules and released into the bloodstream after the cell has received an appropriate stimulus [47].

Plasma Membrane Involvement

Immunohistochemical studies at the light microscopic level have suggested a localization of hCG and/or its subunits on or near the plasma

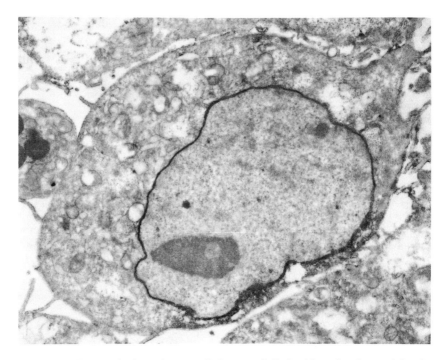

Fig. 3. Human choriocarcinoma cell (BeWo cell line) with perinuclear staining for anti-hCGβ. The immunoperoxidase procedure gives a dark reaction product. The BeWo cells had been incubated for 24 h with 1 m*M* dibutyryl cyclic AMP and 1 m*M* theophylline. ×7,500 [reproduced from ref. 64 with permission].

membrane of normal placenta [16, 64]. Using antibody to hCG and immunoperoxidase staining techniques, *Dreskin* et al. [17] studied the ultrastructural localization of hCG in human term placenta. Staining was found in surface microvilli and in cisternae of the rough endoplasmic reticulum, but was absent from the Golgi, mitochondria, and basement membrane. The authors suggested that neither the Golgi apparatus nor secretory granules may be required for export of hCG (an alternate interpretation is that the Golgi complex and secretory granules contain a modified form of hCG that is not recognized by the antibody). The occasional staining of cytoplasmic granules observed in both control and experimental sections was attributed to peroxidase intrinsic to the granules. The results of that study were confirmed by *Martin* et al. [41] in staining experiments with the lectin, Con A, on early and term placenta. In

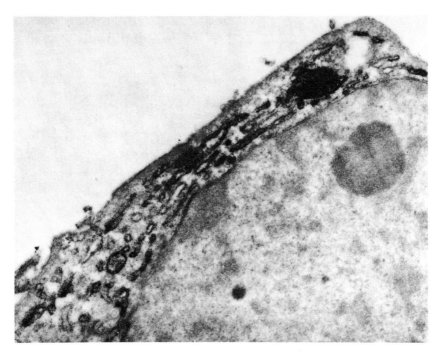

Fig. 4. Immunoperoxidase localization of anti-hCGβ-reactive material in the cisternae of BeWo cell endoplasmic reticulum. ×20,000 [reproduced from ref. 64 with permission].

immunoperoxidase ultrastructural localization studies on choriocarcinoma cells, we observed staining with anti-hCGβ localized to the perinuclear space (fig. 3), cisternae of the rough endoplasmic reticulum (fig. 4), and plasma membrane (fig. 5) [64]. Incubation of the cultures with dibutyryl cyclic AMP increased the amount of staining on the plasma membrane.

Mechanism of hCG Secretion

The results obtained in most protein hormone secretory systems, including pituitary secretion of LH and FSH, are consistent with the cisternal packaging-exocytosis hypothesis [19, 38, 39, 47; other references in 32, 34]. In this pathway, newly synthesized proteins are segregated in the cisternal space of the rough endoplasmic reticulum and packaged into

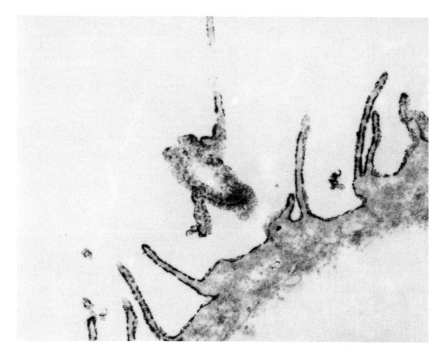

Fig. 5. Anti-hCGβ-reactive material on the microvillous surface of a stimulated BeWo cell. ×12,700 [reproduced from ref. 64 with permission].

secretory granules in the condensing vacuoles of the Golgi apparatus. The hormone is discharged by exocytosis, in which the membrane of the secretory granule fuses with the cell membrane and ruptures, allowing release of the hormone [47].

Distinctive membrane-enclosed secretory granules have been described for insulin, anterior and posterior pituitary hormones, catecholamines of the adrenal medulla, and other secretory proteins [19, 38, 39, 47, 60]. Furthermore, the size and density of secretory granules change according to the physiological demand for the stored hormone. For example, castration results in an increase in secretory granules in the pituitary gonadotroph, while release of LH and FSH in response to gonadotropin-releasing hormone leads to a decrease in size, and migration of the secretory granules to the plasma membrane [60]. In contrast to these observations, secretory granules are sparse in placenta and other hCG-producing trophoblastic tissues [17, 63]. *Sideri* et al. [57] used transmission and freeze-fracture electron microscopic techniques to examine first trimester and term placenta. These workers found no evidence of condensation of protein material in the Golgi apparatus. Furthermore, freeze-fracture study showed no evidence of exocytosis into the intervillous space.

A second characteristic of secretory systems involving exocytosis is the rapid release of stored hormone following a stimulus. The glucose-induced release of insulin from pancreatic islets occurs within a few minutes [19]. Likewise, release of anterior pituitary hormones by hypothalamic releasing hormones or by cyclic nucleotides is rapid [4, 19, 32, 34, 38, 52, 60]. In the case of hCG secretion, however, no immediate release of hormone has been demonstrated with any stimulus. In placental explant cultures and choriocarcinoma cell lines, several hours of incubation with dibutyryl cyclic AMP were required before stimulation of hCG secretion was observed [32, 34, 50]. The increase in secreted hCG was preceded by an increase in intracellular hormone, and could be prevented by inhibitors of RNA synthesis [2, 32, 34, 50]. On the other hand, the stimulated release of stored LH from rat pituitary glands in vitro was not affected by inhibition of protein synthesis [56]. These findings suggested that, in contrast to pituitary LH secretion, the early stimulation of hCG secretion required the synthesis of new RNA and protein, and that there was no significant release of prestored hCG [2, 32, 34, 50].

In choriocarcinoma cell cultures, the plant lectin Con A stimulated the secretion of hCG and α-subunit [3]. The stimulation was dose-dependent and was prevented by α-methyl mannoside. On the other hand, Con A inhibited basal and stimulated release of FSH, TSH, LH, and other hormones in pituitary cell cultures. The dose-dependent inhibition was partially reversed by α-methyl mannoside. The authors concluded that Con A interferes either with exocytosis or with the pituitary binding site for hypothalamic-releasing factors [52].

The divalent cation ionophore, A23187, stimulates secretory responses in a number of tissues. This carboxylic acid antibiotic apparently alters calcium transport across biological membranes, causing a shift in the distribution of intracellular calcium [4, 32]. This agent stimulated the secretion of insulin and amylase from the pancreas, vasopressin from the posterior pituitary, catecholamines from the adrenal medulla, and LH from the pituitary [4]. In contrast, hCG secretion in choriocarcinoma cells was not enhanced by A23187 when tested under a wide range of conditions [36]. Since calcium is required for secretion by exocytosis [4, 39, 47] as well

Table I. Comparison of hCG secretion and exocytosis

Parameter	Exocytosis	hCG secretion
Intact intracellular hormone	+	±
Secretory granules	+	−
Earliest stimulation	minutes	hours
Ionophore A23187	stimulation	inhibition
Con A	inhibition	stimulation
High potassium	stimulation	no effect
Calcium dependence	+	+

as for hCG secretion [36], the disparate findings with A23187 may reflect basic differences in secretory mechanisms between hCG secretion and other endocrine tissues.

High concentrations of potassium in the incubation medium stimulate the release of prestored hormone, perhaps by depolarization of the plasma membrane resulting in increased permeability of the membrane to calcium [4, 38, 39]. High potassium potentiated the basal or stimulated release of TSH, LH, FSH, growth hormone, prolactin, and ACTH from the anterior pituitary, of growth hormone and prolactin from pituitary tumor cells in culture, and of insulin from the pancreas [4, 38; other references in 32]. Furthermore, high potassium caused depletion of cellular stores of hormone. In contrast to these results, high potassium was without effect on basal or stimulated hCG secretion in choriocarcinoma cultures, nor was there any reduction of cellular hCG levels [33].

The inconsistencies between the mechanism of exocytosis and hCG secretion are summarized in table I. It appears that secretory granules may not play a significant role in hCG secretion. Furthermore, accumulating evidence suggests a major role for the plasma membrane in the secretion of hCG [17, 32, 33, 36, 41, 55, 64]. Even if hCG is not condensed in the Golgi apparatus nor stored in secretory granules, however, it is possible that the Golgi complex is intimately involved in hCG secretion, analogous to the secretion of immunoglobulin by plasma cells [40, 46]. It is well known that, while inner core glycosylation and processing of glycoproteins (including hCGα and hCGβ) occur in the membranes of the endoplasmic reticulum, the Golgi apparatus is rich in glycosyl transferases for peripheral sugars (i.e., GlcNAc, Gal, NANA).

Role of Trophoblast Cell Type in hCG Secretion

The availability in recent years of specific antisera for hCG and its subunits has led to immunocytochemical and immunohistochemical findings that have provocative implications with respect to the process of hCG secretion. A number of investigators have localized hCG, hCGβ, and hCGα immunoreactivity in the syncytiotrophoblast (and in some transitional cells), but only hCGα in the cytotrophoblast [1, 26, 31, 37]. Based on these observations, several workers [9, 13, 26, 31, 34, 37, 57] have proposed that the differentiated syncytiotrophoblast is the source of the β-subunit (and of assembly of subunits into complete hCG), while the relatively undifferentiated cytotrophoblast is the site of synthesis for α-subunit.

The hypothesis would predict an increase in secretion of complete hCG in conditions that cause differentiation of cytotrophoblast to syncytiotrophoblast. Furthermore, this hypothesis would explain a number of experimental observations. For example, the unbalanced overproduction of α-subunit in malignant trophoblast cell lines comprised mainly of cytotrophoblast cells is well known [references in 32, 34]. Cytotrophoblast cells are notably more resistant to adverse tissue culture conditions, whereas syncytiotrophoblast cells are rapidly damaged [26]. In placental explants incubated in vitro for 5 days, the syncytiotrophoblast became progressively more damaged while the cytotrophoblast thrived [37]. In the same cultures the secretion of free α-subunit increased markedly, while the secretion of free β-subunit rose only moderately and secretion of intact hCG remained constant. In clinical conditions usually associated with severe placental hypoxia (inevitable abortion, ectopic pregnancy, pre-eclampsia and eclampsia) – where there is preferential preservation of cytotrophoblast rather than of syncytiotrophoblast cells – there is an increase in maternal serum levels of hCGα and in the α/hCG ratio [26]. In hydatidiform mole, there is a much greater α/hCG ratio in the relatively hypoxic molar vesicles than in maternal serum [1]. *Friedman and Skehan* [23] found that concentrations of methotrexate too low to inhibit DNA or protein synthesis led to differentiation of cytotrophoblastic choriocarcinoma cells into syncytiotrophoblast cells. The methotrexate-induced cells were multinucleated and contained increased surface microvilli, Golgi, and endoplasmic reticulum. Since comparable concentrations of methotrexate enhanced hCG secretion in these cells, the possibility is raised that trophoblast cell differentiation may be part of the mechanism by which some agents stimulate hCG secretion [reviewed in 32, 34]. *Chatterjee and*

Munro [13] have speculated that the cytotrophoblast cells may be programmed to develop releasing factors at different times during gestation in order to regulate synthesis of the corresponding peptide hormone in the syncytiotrophoblast. We can anticipate much exciting research activity in the area of trophoblast differentiation and hormone secretion in the near future.

The Process of Human Placental Lactogen Secretion in the Trophoblast

Human placental lactogen (hPL) is a single-chain polypeptide with 191 amino acid residues (molecular weight 21,500). The hPL concentrations parallel the increase in placental growth during pregnancy [13, 24, 29]. At term, hPL represents 7–10% of the total peptides made by the placenta, and is the major protein released from incubations of placental slices [13]. Malignant trophoblastic cells in continuous culture also produce hPL, but in very low amounts [49]. The low values correlate with the relative amount of syncytial cells in the choriocarcinoma cell cultures, and with the low order of production of hPL in patients with choriocarcinoma.

Synthesis of hPL is carried out on membrane-bound polyribosomes in the placenta. The form of the polypeptide initially synthesized on the ribosomes contains an additional sequence of 25 amino acids at its amino-terminal end [13, 29, 59], which facilitates passage of the peptide across the membrane of the endoplasmic reticulum. This hydrophobic signal peptide is hydrolyzed in the membranes of the endoplasmic reticulum. *McWilliams* et al. [44] used DNA complementary to the isolated hPL mRNA to quantitate specific mRNA sequences in the total RNA of early and late placentas, and found 4 times greater abundance of hPL mRNA in the RNA of full-term placenta. This appears to be due to increased transcription, since the number of gene copies for hPL remains unchanged.

Immunohistochemical and immunocytochemical studies on placenta have localized hPL to the syncytiotrophoblast, where the hormone is probably synthesized, with greater staining intensity in term placenta than early placenta [16, 26, 29]. Except for the electron microscopic and morphologic studies, which apply to all products of the trophoblast, there is little other experimental evidence relating to the secretory mechanism for hPL. While it seems likely that the secretion process for hPL may be the same as for hCG, further speculation at this time is not warranted.

Acknowledgement

The author is grateful to Dr. *Donald E. Yorde,* Department of Anatomy, Medical College of Wisconsin, for providing the photomicrographs of the immunoperoxidase studies.

References

1   Ashitaka, Y.; Nishimura, R.; Takemori, M.; Tojo, S.: Production and secretion of HCG and HCG subunits by trophoblastic tissue; in Segal, Chorionic gonadotropin, pp. 147–175 (Plenum Pres, New York 1980).

2   Azizkhan, J. C.: Stimulation of human chorionic gonadotropin synthesis in choriocarcinoma cells by inhibition of DNA synthesis and by adenosine 3′,5′-cyclic monophosphate; PhD Diss. University of Maryland (1978).

3   Benveniste, R.; Speeg, K. V., Jr.; Long, A.; Rabinowitz, D.: Concanavalin-A stimulates human chorionic gonadotropin (hCG) and hCG-alpha secretion by human choriocarcinoma cells. Biochem. biophys. Res. Commun. *84:* 1082–1087 (1978).

4   Berridge, M. J.: in Greengard, Robison, Advances in cyclic nucleotide research, vol. 6, pp. 1–98 (Raven Press, New York 1975).

5   Bielinska, M.; Boime, I.: Glycosylation of human chorionic gonadotropin in mRNA-dependent cel-free extracts: posttranslational processing of an asparagine-linked mannose-rich oligosaccharide. Proc. natn. Acad. Sci. USA *76:* 1208–1212 (1979).

6   Bielinska, M.; Boime, I.: mRNA-dependent synthesis of a glycosylated subunit of human chorionic gonadotropin in cell-free extracts derived from ascites tumor cells. Proc. natn. Acad. Sci. USA *75:* 1768–1772 (1978).

7   Bielinska, M.; Grant, G. A.; Boime, I.: Processing of placental peptide hormones synthesized in lysates containing membranes derived from tunicamycin-treated ascites tumor cells. J. biol. Chem. *253:* 7117–7119 (1978).

8   Birken, S.; Fetherston, J.; Canfield, R.; Boime, I.: The amino acid sequences of the prepeptides contained in the α and β subunits of human choriogonadotropin. J. biol. Chem. *256:* 1816–1823 (1981).

9   Boime, I.; Landefeld, T.; McQueen, S.; McWilliams, D.: The biosynthesis of chorionic gonadotropin and placental lactogen in first- and third-trimester human placenta; in McKerns, Structure and function of the gonadotropins, pp. 235–267 (Plenum Press, New York 1978).

10   Boothby, M.; Daniels-McQueen, S.; McWilliams, D.; Zernik, M.; Boime, I.: Human chorionic gonadotropin alpha and beta subunit mRNAs: translatable levels during pregnancy and molecular cloning of DNA sequences complementary to hCG; in Segal, Chorionic gonadotropin, pp. 253–275 (Plenum Press, New York 1980).

11   Bordelon-Riser, M. E.; Siciliano, M. J.; Kohler, P. O.: Necessity for two human chromosomes for human chorionic gonadotropin production in human-mouse hybrids. Somatic Cell Genet. *5:* 597–613 (1979).

12   Chard, T.; Grudzinskas, J. G.: New placental proteins – biology and clinical applications; in Klopper, Genazzani, Crosignani, The human placenta: proteins and hormones, pp. 3–16 (Academic Press, New York 1980).

13 Chatterjee, M.; Munro, H.N.: Structure and biosynthesis of human placental peptide hormones. Vitams Horm. *35:* 149–208 (1977).

14 Daniels-McQueen, S.; McWilliams, D.; Birken, S.; Canfield, R.; Landefeld, T.; Boime, I.: Identification of mRNAs encoding the α and β subunits of human choriogonadotropin. J. biol. Chem. *253:* 7109–7114 (1978).

15 Dean, D.J.; Weintraub, B.D.; Rosen, S.W.: De novo synthesis and secretion of heterogeneous forms of human chorionic gonadotropin and its free alpha subunit in the human choriocarcinoma clonal cell line, JEG-3. Endocrinology *106:* 849–858 (1980).

16 De Ikonikoff, L.K.; Cedard, L.: Localization of human chorionic gonadotropic and somatomammotropic hormones by the peroxidase immunohistoenzymologic method in villi and amniotic epithelium of human placentas (from six weeks to term). Am. J. Obstet. Gynec. *116:* 1124–1132 (1973).

17 Dreskin, R.B.; Spicer, S.S.; Greene, W.B.: Ultrastructural localization of chorionic gonadotropin in human term placenta. J. Histochem. Cytochem. *18:* 862–874 (1970).

18 Elting, J.: N- and O-glycosylation of human chorionic gonadotropin (hCG) (Abstract). Fed. Proc. *40:* 1882 (1981).

19 Farquhar, M.G.; Skutelsky, E.H.; Hopkins, C.R.: Structure and function of the anterior pituitary and dispersed pituitary cells. In vitro studies; in Tixier-Vidal, Farquhar, The anterior pituitary, pp. 84–135 (Academic Press, New York 1975).

20 Fiddes, J.C.; Goodman, H.M.: Isolation, cloning and sequence analysis of the cDNA for the α-subunit of human chorionic gonadotropin. Nature, Lond. *281:* 351–356 (1979).

21 Fiddes, J.C.; Goodman, H.M.: The cDNA for the β-subunit of human chorionic gonadotropin suggests evolution of a gene by readthrough into the 3′-untranslated region. Nature, Lond. *286:* 684–687 (1980).

22 Folman, R.; Ilan, J.; de Groot, N.; Hochberg, A.A.: Human chorionic gonadotropin: studies on the mechanism of secretion; in Segal, Chorionic gonadotropin, pp. 277–294 (Plenum Press, New York 1980).

23 Friedman, S.J.; Skehan, P.: Morphological differentiation of human choriocarcinoma cells induced by methotrexate. Cancer Res. *39:* 1960–1967 (1979).

24 Friesen, H.G.: Placental protein and polypeptide hormones; in Greep, Astwood, Handbook of physiology, Section 7: Endocrinology, vol. II: Female reproductive system, part 2, pp. 295–309 (American Physiological Society, Washington 1973).

25 Garnier, J.: Molecular aspects of the subunit assembly of glycoprotein hormones; in McKerns, Structure and function of the gonadotropins, pp. 381–414 (Plenum Press, New York 1978).

26 Gaspard, U.J.; Hustin, J.; Reuter, A.M.; Lambotte, R.; Franchimont, P.: Immunofluorescent localization of placental lactogen, chorionic gonadotropin and its alpha and beta subunits in organ cultures of human placenta. Placenta *1:* 135–144 (1980).

27 Gibbons, J.M.; Mitnick, M.; Chieffo, V.: In vitro biosynthesis of TSH- and LH-releasing factors by the human placenta. Am. J. Obstet. Gynec. *121:* 127–131 (1975).

28 Grudzinskas, J.G.; Lee, J.N.; Teisner, B.; Chard, T.: Synthesis and secretion of placental proteins; in Klopper, Genazzani, Crosignani, The human placenta: proteins and hormones, pp. 87–92 (Academic Press, New York 1980).

29 Handwerger, S.; Hurley, T.W.; Golander, A.: Placental and decidual polypeptide

hormones; in Iffy, Kaminetzky, Principles and practice of obstetrics and perinatology, pp. 243–260 (Wiley, New York 1981).

30   Heap, R. B.; Perry, J. S.; Challis, J. R. G.: Hormonal maintenance of pregnancy; in Greep, Astwood, Handbook of physiology, Section 7: Endocrinology, vol. II: Female reproductive system, part 2, pp. 217–260 (American Physiological Society, Washington 1973).

31   Hoshina, M.; Ashitaka, Y.; Tojo, S.: Immunohistochemical localization of HCG and its subunits in chorionic tissue. Abstr. 6th Int. Congr. of Endocrinology, Melbourne 1980, No. 545.

32   Hussa, R. O.: Biosynthesis of human chorionic gonadotropin. Endocrine Rev. *1:* 268–294 (1980).

33   Hussa, R. O.: Effects of antimicrotubule agents, potassium, and inhibitors of energy production on hCG secretion. In Vitro *15:* 237–245 (1979).

34   Hussa, R. O.: Human chorionic gonadotropin, a clinical marker: review of its biosynthesis. Ligand Rev. *3:* suppl. 2 (1981).

35   Hussa, R. O.: Immunologic and physical characterization of human chorionic gonadotropin and its subunits in cultures of human malignant trophoblast. J. clin. Endocr. Metab. *44:* 1154–1162 (1977).

36   Hussa, R. O.: Studies on human chorionic gonadotropin secretion: effects of EGTA, lanthanum, and the ionophore A23187. J. clin. Endocr. Metab. *44:* 520–529 (1977).

37   Hustin, J.; Gaspard, U.; Reuter, A.; Hendrick, J. C.; Franchimont, P.: Protein synthesis and release by the human placenta in organ culture; in Klopper, Genazzani, Crosignani, The human placenta: proteins and hormones, pp. 259–266 (Academic Press, New York 1980).

38   Kraicer, J.: Mechanisms involved in the release of adenohypophysial hormones; in Tixier-Vidal, Farquhar, The anterior pituitary, pp. 21–43 (Academic Press, New York 1975).

39   Lacy, P. E.: Endocrine secretory mechanisms. A review. Am. J. Path. *79:* 170–187 (1975).

40   Lin, C. T.; Chang, J. P.; Chen, J. P.: Ultrastructural localization of γ-chain and immunoglobulin G in human lymphocytes using enzyme-labeled Fab fragment. J. Histochem. Cytochem. *23:* 624–631 (1975).

41   Martin, B. J.; Spicer, S. S.; Smythe, N. M.: Cytochemical studies of the maternal surface of the syncytiotrophoblast of human early and term placenta. Anat. Rec. *178:* 769–786 (1974).

42   Maruo, T.: In vitro biosynthesis of human chorionic follicle-stimulating hormone. Endocr. jap. *23:* 65–73 (1976).

43   Maruo, T.; Segal, S. J.; Koide, S. S.: Large molecular species of human chorionic gonadotropin; in Segal, Chorionic gonadotropin, pp. 177–197 (Plenum Press, New York 1980).

44   McWilliams, D.; Callahan, R. C.; Boime, I.: Human placental lactogen mRNA and its structural genes during pregnancy: quantitation with a complementary DNA. Proc. natn. Acad. Sci. USA *74:* 1024–1027 (1977).

45   Midgley, A. R., Jr.; Pierce, G. V., Jr.: Immunohistochemical localization of human chorionic gonadotropin. J. exp. Med. *115:* 289–294 (1962).

46   Murphy, M. J.; Hay, J. B.; Morris, B.; Bessis, M. C.: Ultrastructural analysis of antibody synthesis in cells from lymph and lymph nodes. Am. J. Path. *66:* 25–42 (1972).

47 Palade, G.: Intracellular aspects of the process of protein synthesis. Science *189:* 347–358 (1975).

48 Patrito, L.C.; Flury, A.; Rosato, J.; Martin, A.: Biosynthesis in vitro of chorionic gonadotropin from human placenta. Hoppe-Seyler's Z. physiol. Chem. *354:* 1129–1132 (1973).

49 Pattillo, R.A.; Gey, G.O.; Delfs, E.; Huang, W.Y.; Hause, L.; Garancis, J.; Knoth, M.; Amatruda, J.; Bertino, J.; Friesen, H.G.; Mattingly, R.F.: The hormone-synthesizing trophoblastic cell in vitro: a model for cancer research and placental hormone synthesis. Ann. N.Y. Acad. Sci. *172:* 288–298 (1971).

50 Pattillo, R.A.; Hussa, R.O.: Early stimulation of human chorionic gonadotropin secretion by dibutyryl cyclic AMP and theophylline in human malignant trophoblast cells in vitro: inhibition by actinomycin D, α-amanitin, and cordycepin. Gynecol. Invest. *6:* 365–377 (1975).

51 Pattillo, R.A.; Hussa, R.O.; Ruckert, A.C.F.; Kurtz, J.W.; Cade, J.M.; Rinke, M.L.: hCG in BeWo trophoblastic cells after twelve years in continuous culture: retention of intact hCG secretion in mechanical versus enzyme dispersed cells. Endocrinology *105:* 967–974 (1979).

52 Ponsin, G.; Khar, A.; Kunert-Radek, Y.; Jutisz, M.: Inhibition of hormone release from pituitary cell cultures by concanavalin A. Abstr. 6th Int. Congr. of Endocrinology, Melbourne 1980, No.60.

53 Ruddon, R.W.; Hanson, C.A.; Addison, N.J.: Synthesis and processing of human chorionic gonadotropin subunits in cultured choriocarcinoma cells. Proc. natn. Acad. Sci. USA *76:* 5143–5147 (1979).

54 Ruddon, R.W.; Hanson, C.A.; Bryan, A.H.; Anderson, C.: Synthesis, processing, and secretion of human chorionic gonadotropin subunits by cultured human cells; in Segal, Chorionic gonadotropin, pp.295–315 (Plenum Press, New York 1980).

55 Ruddon, R.W.; Hanson, C.A.; Bryan, A.H.; Putterman, G.J.; White, E.L.; Perini, F.; Meade, K.S.; Aldenderfer, P.H.: Synthesis and secretion of human chorionic gonadotropin subunits by cultured human malignant cells. J. biol. Chem. *255:* 1000–1007 (1980).

56 Samli, M.H.; Geschwind, I.I.: Some effects of energy-transfer inhibitors and of Ca^{++}-free or K^+-enhanced media on the release of luteinizing hormone (LH) from the rat pituitary gland in vitro. Endocrinology *82:* 225–231 (1968).

57 Sideri, M.; Fumagalli, G.; De Virgiliis, G.; Remotti, G.: The morphological basis of placental endocrine activity; in Klopper, Genazzani, Crosignani, The human placenta: proteins and hormones, pp.339–346 (Academic Press, New York 1980).

58 Story, M.T.; Hussa, R.O.; Pattillo, R.A.: Independent dibutyryl cyclic adenosine monophosphate stimulation of human chorionic gonadotropin and estrogen secretion by malignant trophoblast cells in vitro. J. clin. Endocr. Metab. *39:* 877–881 (1974).

59 Szczesna, E.; Boime, I.: mRNA-dependent synthesis of authentic precursor to human placental lactogen: conversion to its mature hormone form in ascites cell-free extracts. Proc. natn. Acad. Sci. USA *73:* 1179–1183 (1976).

60 Tixier-Vidal, A.: Ultrastructure of anterior pituitary cells in culture; in Tixier-Vidal, Farquhar, The anterior pituitary, pp.181–229 (Academic Press, New York 1975).

61 Vaitukaitis, J.L.: Glycoprotein hormones and their subunits – immunological and biological characterization; in McKerns, Structure and function of the gonadotropins, pp.339–360 (Plenum Press, New York 1978).

62 Wynn, R. M.: Fine structure of the placenta; in Greep, Astwood, Handbook of physiology, Section 7: Endocrinology, vol. II: Female reproductive system part 2, pp. 261–276 (American Physiological Society, Washington 1973).

63 Wynn, R. M.; Davies, J.: Ultrastructure of transplanted choriocarcinoma and its endocrine implications. Am. J. Obstet. Gynec. *88:* 618–633 (1964).

64 Yorde, D. E.; Hussa, R. O.; Garancis, J. C.; Pattillo, R. A.: Immunocytochemical localization of human choriogonadotropin (hCG) in human malignant trophoblast: model for hCG secretion. Lab. Invest. *40:* 391–398 (1979).

Robert O. Hussa, PhD, Professor of Gynecology and Obstetrics, Associate Professor of Biochemistry, The Medical College of Wisconsin, Department of Gynecology and Obstetrics, 8700 W. Wisconsin Avenue, Milwaukee, WI 53226 (USA)

Cell Biology of the Secretory Process, pp. 590–609 (Karger, Basel 1984)

Regulation of Immunoglobulin Synthesis and Secretion in B Lymphocytes

Abul K. Abbas

Departments of Pathology, Harvard Medical School and Brigham and Women's Hospital, Boston, Mass., USA

Introduction

The fully differentiated antibody-secreting cell, identified by morphologists as the plasma cell, has long been considered a prototype of cells whose primary function is to synthesize and secrete large quantities of a single class of protein. Plasma cells secrete about 2,00 molecules of immunoglobulin (Ig)/minute/cell; 5–40% of the total protein synthesized and 75–90% of the protein secreted by these cells consists of antibody molecules. The ultrastructural demonstration of abundant rough endoplasmic reticulum (RER) and a prominent Golgi apparatus in plasma cells further attests to their commitment to the processes of protein (antibody) synthesis and secretion. During the last decade, progress in our understanding of antibody structure, the genes coding for Ig, and the differentiation of cells destined to secrete antibody has made this system uniquely valuable for studying the biochemical and structural basis of protein synthesis and secretion, and the regulation of these processes in mammalian cells.

It has been recognized for many years that the immune system serves a limited number of physiologic functions in response to an almost infinite variety of extrinsic or endogenous stimuli (antigens). The basis of this intriguing paradigm was elucidated only when the structure of antibodies was defined. Antibody molecules are composed of two pairs of disulfide-linked chains – non-glycosylated light chains ($\varkappa$ or λ) and glycosylated heavy chains ($\mu, \gamma, \alpha, \varepsilon, \delta$, making up IgM, IgG, IgA, IgE or IgD molecules, respectively). Each chain is a bifunctional polypeptide, comprised of an invariant or constant (C) region and a highly variable (V) region; the

Table I. Stages in B lymphocyte differentiation

Stage of differentiation	Pattern of immunoglobulin expression
Stem cell	no Ig synthesis; presence inferred from tumors and bone marrow reconstitution experiments
Pre-B cell	cytoplasmic μ^+, no light-chain synthesis, surface Ig
Immature B lymphocyte	surface IgM$^+$ (μ and light-chain synthesis)
Mature B lymphocyte	surface IgM$^+$ IgD$^+$
Activated B lymphocyte	expression of other Ig isotypes (heavy chain class switch); surface Ig$^+$, low rate of Ig secretion
Antibody secreting cell	high rate of Ig secretion, reduced density of membrane Ig

former is responsible for the effector functions of antibodies, and the enormous number of V regions makes up the repertoire of combining sites specific for different antigens that is present within each individual. The molecular biologic basis for producing single protein chains which consist of highly variable and highly conserved segments has been elucidated only during the last 5 years or so (vide infra). The complexity of the immune system is now known to be even greater than previously suspected because of the demonstrated changes that occur in the constituent cells during the processes of ontogeny and antigen-stimulated differentiation (table I). Thus, the virgin, antigen-reactive cell of humoral immunity is the B ('bone-marrow'- or 'bursa'-derived) lymphocyte which expresses surface IgM and/or IgD molecules that bind specific antigens. Precursors of B lymphocytes are pre-B cells which synthesize μ heavy chains only, do not express surface Ig and are, therefore, incapable of recognizing and responding to antigens [18]. Mature B lymphocytes are stimulated by antigens to divide and differentiate into cells that (i) express other heavy chain classes, i.e. γ, α, ε, with the same antigenic specificity as the original lymphocyte (and, therefore, possessing the same V regions in association with different C regions), and (ii) begin to secrete Ig molecules and gradually reduce their density of membrane-bound receptor Ig. These proliferative and differentiative events usually require cooperative signals from regulatory thymus-derived (T) lymphocytes in addition to specific antigens, and culminate in the development of cells that synthesize and secrete large amounts of the antibodies that constitute the humoral immune response.

There are several reasons why the system of Ig biosynthesis and secretion offers particular advantages as a model for understanding the secretory process and its regulation in eukaryotic cells. First, B lymphocyte differentiation occurs along a precisely programmed sequence of steps, each of which is characterized by a specific pattern of membrane and secretory Ig expression (table I). Although the vast heterogeneity of the immune system and the fact that very few cells respond to any one antigen make it difficult to study the regulation of Ig secretion by antigens, two strategies are available to overcome these problems. The first is the use of polyclonal activators such as bacterial lipopolysaccharide ('endotoxin') and anti-Ig antibodies, which mimic the effects of antigens in many respects but do so in most or all B lymphocytes and not only the specific antigen-responsive clones. Secondly, murine and human tumor lines are available that not only represent various stages of B cell differentiation but can be specifically induced, much like normal non-neoplastic cells, to undergo sequential differentiative steps [3]. These homogeneous, autonomous tumor populations are amenable to far more detailed biochemical and morphologic analyses than normal lymphoid cells. As far as can be determined, many tumors are valid models for normal lymphocytes; perhaps equally importantly, aberrations associated with the neoplastic state have also provided useful information on the regulation of physiologic antibody production [3]. Finally, with the recent application of recombinant DNA technology to studies of antibody-producing cells, specific probes are available to define the genetic and molecular basis of Ig production [21].

In this chapter, the process of Ig biosynthesis, assembly and secretion will be described, with an emphasis on the mechanisms that regulate them. Although the detailed molecular basis of this regulation is not yet defined, it is clear that the experimental systems are available for doing so. It should be emphasized that in many instances, a change in the pattern of Ig expression, e.g. membrane to secreted, or IgM to IgG, occurs not in the same cell but in the progeny of a particular B lymphocyte. Since such changes reflect alterations in differentiative stage, they are regulated by factors that control cellular differentiation as a whole rather than Ig production alone. Nevertheless, analyses of antibody production in normal and neoplastic B lymphocytes and their progeny have raised issues that are of central importance to an understanding of protein secretion in all cells.

Biosynthesis of Antibodies

The biochemical basis of Ig biosynthesis was poorly understood until the application of recombinant DNA technology led to the delineation of the genes encoding Ig heavy and light chains and the regulatory events that occur at the level of DNA rearrangement and RNA splicing. This rapidly expanding field has been the subject of numerous recent reviews [6, 16, 17, 21]; in the following section, some salient features that are most relevant to subsequent discussions will be highlighted without any attempt to detail the experimental evidence.

Regulation of Ig Biosynthesis by DNA Rearrangement and RNA Splicing

Genes coding for Ig are prototypes of 'split genes'. In the embryo, multiple DNA segments encoding different V regions ('exons') are arranged in tandem, followed by 'joining' (J) and 'diversity' (D) segments (the latter only for heavy chains), and the constant region(s) genes. The V, J and C exons for light chains, or V, D, J and C exons for heavy chains are located on different chromosomes, and the segments are separated from one another by untranslated intervening sequences ('introns') of different lengths. In the differentiated antibody-producing cell, the segment of chromosome coding for heavy or light chain undergoes DNA rearrangement such that one V region DNA in each cell joins one D and/or J segment, the introns are deleted and the rearranged V-D-J-C (heavy chain) or V-J-C (light chain) gene is transcribed to form the primary RNA transcript (fig. 1). The C region is then spliced on to the 3' end of the J segment, to generate a messenger RNA in which the V, (D), J and C regions are contiguous (fig. 1). The combinatorial associations between different V, D and J genes generate the large diversity of antibody-combining sites in each individual; moreover, this process of rearrangement explains how highly variable and invariable segments of the same proteins (Ig heavy or light chains) can be generated, since in fact, each chain is encoded by more than one DNA segment. The available evidence indicates that during ontogeny, μ-chain DNA is rearranged first, so that pre-B cells synthesize only μ chains. This is followed by joining of ϰ light chain DNA sequences, and λ-DNA rearrangement only occurs if the ϰ chain is abortively or aberrantly rearranged. Similarly, it is believed that formation of productive Ig

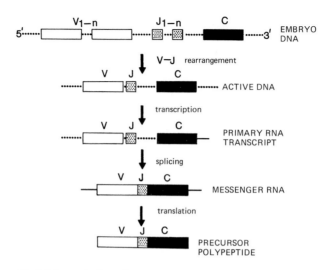

Fig. 1. Schematic view of gene rearrangement and RNA processing for an Ig light chain. One V and one J region (separated by introns; dotted lines, in germline gene) are rearranged to form the primary transcript, and the mature mRNA is produced by splicing. Not shown are the leader peptide sequences at the 5′ (NH_2 terminal) ends of each V region, and the poly-A sequence which identifies the RNA transcript as a messenger RNA. A similar sequence of events occurs with Ig heavy chains, except that additional D region exons are involved in the gene rearrangement, and each heavy chain C region may consist of 3 or more exons at the DNA level.

heavy and light chain genes on one set of chromosomes prevents rearrangement of Ig DNA on the other chromosomes, thus accounting for the expression of only one Ig allele in each cell ('allelic exclusion'). The mechanisms underlying these sequential gene tanslocation events, and the regulation of these processes by extrinsic and internal factors are intriguing issues that remain to be resolved. The identification of numerous possible mechanisms for joining of exons and deletion of introns [6, 16, 17] clearly indicates that experimental systems for analyzing Ig gene rearrangement are now available.

Three additional changes in Ig expression that occur during B lymphocyte differentiation have been extensively investigated, and can be explained by DNA rearrangements or alternative RNA-splicing mechanisms.

(i) Simultaneous expression of membrane IgM and IgD: immature B lymphocytes express membrane IgM, the antigenic specificity of which is determined by the V regions of the μ heavy chain and the ϰ or λ chain.

Mature B cells express IgD in addition to IgM, both membrane receptors having the same specificity. The simultaneous expression of μ and δ C regions with the same V region has been shown to result from alternative processing (splicing) of a primary RNA transcript that contains the V, D, J, Cμ and Cδ sequences [27].

(ii) Switch from membrane to secreted Ig: mature B lymphocytes express membrane IgM (and, sometimes IgD) but do not secrete antibody prior to stimulation; the major antibody released into the extracellular compartment is 8S IgM that is slowly turned over along with other membrane proteins [41]. Following activation by antigens or polyclonal stimuli, B cells proliferate and differentiate and begin to actively secrete 19S IgM while exhibiting a progressive, concomitant decrease in the density of membrane Ig. Numerous studies of the biosynthesis, structure and, more recently, of the molecular biology of membrane and secreted IgM [7, 11] have established that these two forms have different μ chains. The Cμ segment of membrane IgM (Cμmem) has a sequence of 26 hydrophobic amino acids at its C terminal with a short segment at the tail which is probably intracytoplasmic; this hydrophobic region is presumably responsible for insertion into the lipid bilayer of the plasma membrane, and, as expected, it is lacking in secreted IgM (Cμsec). The DNA coding for the Cμmem fragment is located 3′ to the coding domains (separated by intervening sequences of non-informational DNA) that make up the Cμ gene. The primary RNA transcript of this gene includes both exons and introns, and the coding sequences are spliced onto one another to generate the mature message (mRNA). If the primary transcript includes the Cμmem segment, it is spliced onto the main Cμ gene; this is the situation in non-secreting, membrane IgM-positive B cells. If, on the other hand, the primary RNA transcript lacks the Cμmem segment, a stop codon preceding this segment is recognized, and the mature mRNA is translated into the secretory form of the μ heavy chain. Thus, the control of membrane vs. secreted IgM is largely at the level of splicing of the primary RNA transcript [7, 11, 21].

(iii) Heavy chain class switching: further differentiation of B lymphocytes leads to the generation of cells that secrete antibodies of other classes, i.e. IgG, IgA, IgE, with the original antigenic specificity (V regions). In the IgM-producing cell, the recombined V-D-J gene joins with the Cμ gene. It is believed that a major mechanism for heavy-chain switching is deletion of the Cμ gene, which leads to joining of the V-D-J segment to C-region genes located 3′ to Cμ, i.e. the Cγ genes; similar deletion of Cγ DNA will generate antibodies of other heavy chain classes [16]. Putative 'switch regions' that have been identified adjacent to Cγ and Cα genes may play a critical role in these events. The control of heavy-chain switching by genetic, immunologic and other factors is a fascinating question, and better understanding of this phenomenon may explain several presently unresolved issues, e.g. why different antigens preferentially induce antibodies of different classes, why allergic individuals produce disproportionate amounts of IgE, and why B cells located at epithelial surfaces tend to differentiate into cells expressing the secretory immunoglobulin, IgA.

This brief outline has been attempted to highlight the roles played by DNA recombination and RNA splicing in the pattern of expression of Ig. It should be emphasized that the majority of studies to date have been done with B-lymphocyte-derived tumors at various stages of differentiation, and extrapolated to normal Ig-producing cells [3]. The subsequent steps of translation of Ig mRNA and synthesis of the polypeptide chains have been studied with tumors as well as normal, polyclonally stimulated B lymphocytes.

Synthesis of Ig Polypeptide Chains

The bulk of Ig nascent polypeptides and mRNA activity are associated with membrane-bound polyribosomes. This conclusion is based on immunoelectron-microscopic studies, analysis of nascent chains copurified with polyribosomes, and analysis of in vitro translation products using fractionated polysomes [reviewed in ref. 20]. Recently, *Mechler* [23] has presented evidence indicating that translation of Ig heavy and light chain mRNA is initiated in the free ribosomal fraction, and the amino terminal of nascent polypeptide chains provides a 'signal' for attachment of polysomes to the endoplasmic reticulum membrane. This NH_2 terminal sequence presumably corresponds to the 'leader' peptide, which has been recognized for some years as a hydrophobic sequence of about 20 amino acids that guides the growing polypeptide chain to a receptor on the membrane of the endoplasmic reticulum and is responsible for transfer of the chain into the cisternae of the endoplasmic reticulum. Shortly thereafter, the leader is cleaved by a membrane enzyme from the nascent polypeptide chain, which then undergoes post-translational modifications and assembly. Similar leader peptides have been demonstrated in other proteins destined for secretion and their role in vectorial transfer of proteins across membranes has been reviewed recently [30].

Biosynthetic studies done by *Sidman* [34] demonstrate that resting B lymphocytes synthesize non-glycosylated precursor forms of both membrane and secreted μ (molecular weights 62,000 and 60,000 daltons, respectively), each having a cleavable leader peptide. In secreting cells, the 60,000-dalton precursor undergoes a series of post-translational modification steps, being converted to a 73,000 and then a 70,000-dalton glycosylated form; in non-secreting B cells, the final step appears blocked, so that the 70,000-dalton secretory μ is not present. The 62,000-dalton membrane μ is converted to a 73,000-dalton protein, and finally to a 78,000-dalton form that is externally expressed; cells that rapidly secrete Ig but have little or no membrane IgM appear to shut off membrane μ expression by pre-translational mechanisms involving RNA processing (see above). The role of glycosylation and other factors in Ig assembly and expression are discussed below. The studies of *Sidman* [34] elegantly highlight the different synthetic and post-translational pathways followed by membrane and secreted μ heavy chains. It is also worth noting that in cells expressing both forms of IgM, the turnover rate of the secretory form is considerably higher, since secretion can be blocked by 4–6 h of exposure to cycloheximide without affecting membrane Ig [8]. Direct measurements using pulse-chase protocols indicate that the turnover time of membrane Ig in resting B cells is in the order of 20 h, whereas the half-life of secreted Ig is less than 1 h in normal plasma cells and 2–4 h in plasma cell tumors (myelomas) [20].

The biosynthesis of Ig heavy and light chains is under several other, as yet poorly understood, control mechanisms. Most neoplastic antibody-producing cells synthesize and secrete a significant molar excess of light chains, although in other cell lines either balanced (equimolar) synthesis of the two chains or intracellular degradation of excessive light chains has been demonstrated [9]. Secondly, since antibody-secreting cells do not contain significant pools of stored Ig, synthesis is apparently coordinately regulated with the rate of active secretion. In fact, the rate of Ig synthesis may be inversely related to the amount of Ig within an individual cell [36]. Finally, a number of studies suggest that the rate of Ig synthesis in B-cell-derived human tumor lines varies during the cell cycle, with the peak occurring during the late Gl or S phase. However, experiments with more rigorously cloned murine myeloma lines have generally failed to demonstrate a clear influence of the cell cycle on secreted or membrane Ig [3, 20]. So far there is no direct evidence to indicate whether these different regulatory mechanisms operate at the level of transcription, translation or post-translational modifications.

Assembly and Secretion of Antibodies

In spite of substantial numbers of ultrastructural and biochemical studies, the precise pathways from translation of mRNA to exteriorization of membrane or secretory Ig are not yet definitively established. It is generally believed that in most Ig-producing cells, the sequence of events is as follows: (i) heavy and light chains are covalently linked by interchain disulfide bond formation in the RER; (ii) core sugars are added to nascent polypeptide chains in the RER and/or to the completed chains in Golgi vesicles; (iii) the addition of terminal sugars and polymerization of secretory IgA (into dimers) and IgM (into pentamers) occurs very shortly before secretion, and (iv) secretion occurs by a process of reverse pinocytosis of Ig-containing vesicles [20, 31]. Several aspects of this scheme are noteworthy. First, most of the intracellular Ig in normal plasma cells and myeloma cells is located in the RER, no storage or secretory granules are present in such cells, and the average time for an Ig polypeptide chain to traverse the RER-SER/Golgi cisternae and be secreted is in the order of 20 min only. These kinetics are essentially identical for all Ig classes. Secondly, Ig is specifically selected for secretion among the proteins synthesized by antibody-producing cells. The presence of the signal 'leader' peptide is one factor which ensures entry of nascent Ig chains into the endoplasmic reticulum and thus selects them for secretion [30]; the role of other factors such as glycosylation, is discussed below. Finally, it is unclear whether the intracellular pathways of membrane and secreted Ig are compartmentalized in cells producing both forms, or whether random associations of the two forms of heavy chains can occur producing 'hybrid' Ig molecules [12]. In the latter instance, the pattern of predominant Ig expression (i.e. membrane or secreted) would be controlled largely by the rates of synthesis of the two polypeptide forms.

Glycosylation of Ig Chains
A large number of studies using inhibitors of glycosylation or incorporation of radioactive sugars into proteins have emphasized the role of carbohydrate addition in the secretion of Ig, and other glycoproteins [24, 25, 29, 40]. The core sugars, mannose and N-acetylglycosamine, are added to asparagine residues on growing, nascent Ig polypeptide chains in the RER from a pool of preassembled dolichol-oligosaccharide intermediates which are present in the RER [38]. It appears that specific glycosyl transferases (i.e. specific for each monosaccharide but able to work on different

glycoproteins), glycosidases and/or recognition structures for different carbohydrates are involved in this process, and thus control the transitions of secretory Ig precursors to the different molecular weight sizes that have been demonstrated by electrophoretic analysis of intracellular μ chain precursors [34]. The terminal sugars, fucose, sialic acid and, in some cases, galactose, are added to assembled Ig chains just prior to their secretion, at a site(s) between the Golgi complex and the plasma membrane [24, 25].

The role of glycosylation in transport and secretion of Ig remains controversial. Tunicamycin, which inhibits asparagine-linked glycosylation and therefore the addition of core sugars to polypeptide chains, has been shown to block secretion of Ig from normal plasma cells and myeloma cells without markedly reducing the expression of membrane Ig [14, 15]. Ultrastructural studies demonstrate the accumulation of Ig in dilated cisternae of the RER in tunicamycin-treated myeloma cells, and biosynthetic labeling experiments have shown that in μ-secreting B lymphocytes and myeloma cells, tunicamycin does not affect synthesis of early μ precursors but does block the later processing events [14, 38]. 2-deoxy-D-glucose has been shown to prevent the transport of Ig from polysomes to RER and subsequently into smooth endoplasmic reticulum (SER) vesicles but does not block the secretion of synthesized Ig molecules from the SER. Since this drug probably acts by reducing the pool of oligosaccharide intermediates that are available for addition of sugars to polypeptides [32], these experiments suggest that glycosylation of Ig (or of an unidentified non-Ig protein) is important for the transport of Ig from polysomes to SER, but addition of terminal sugars is not critical for the secretory process per se [24, 25]. Finally, myeloma cell variants that have lost their ability to secrete Ig and show abnormal patterns of glycosylation have been identified [43], although it is unclear whether the glycosylation defect causes the failure of secretion or whether both are the result of another primary structural or regulatory abnormality.

In contrast to these experiments suggesting that glycosylation is essential for correct intracellular transport and secretion of Ig, several observations support the converse view. For instance, most murine Ig light chains are non-glycosylated but are obviously secreted actively. In some instances, failure of light chains to be secreted by myeloma variants has been attributed to structural mutations which do not affect the formation or cleavage of the leader peptide but do block subsequent assembly and secretion [28]. Parenthetically, this result also demonstrates that transfer of the polypeptide across the endoplasmic reticulum membrane, which is initiated by the leader peptide, is necessary but not sufficient for secretion. Analyses of large numbers of spontaneous or mutagen-derived myeloma variants have demonstrated structural changes in Ig heavy and/or light chains that can sometimes lead to a failure of secretion, but in other instances grossly abnormal proteins are secreted at normal rates [9, 31, 33]. Comparisons between such mutant proteins might enable one to identify specific 'signal' sequences of amino acids that are critical for intracellular transport and assembly and for secretion. In the final analysis, it is apparent that glycosylation of Ig, and possibly non-Ig regulatory protein(s), is important in the secretion process but is not an essential step in all instances.

Assembly of Ig Chains

From the synthesis of Ig heavy and light chains, which are encoded by genes on different chromosomes, to their exteriorization as functional glycoproteins, two types of chain assemblies are required. The first involves formation of H_2L_2 units from two pairs of heavy (H) and light (L) chains. Noncovalent association of H and L chains probably occurs in the RER even before completion of H chain synthesis, and covalent disulfide linkage rapidly follows chain completion. Different myeloma lines show H-H or H-L precursors that are subsequently linked to form the complete H_2L_2 molecules; the factors, if any, that determine pathways of assembly are unknown [9]. Glycosylation of Ig polypeptides does not appear to be essential for H-L or H-H chain assembly [20, 38]. The second assembly reaction occurs with IgA and IgM, which are normally secreted as dimers and pentamers, respectively. In these polymeric antibodies, individual complete Ig molecules are linked by an acidic peptide called the 'J chain', which is believed to be involved in the polymerization process [19, 29]. However, recent studies showing J chain biosynthesis by B cell lines that represent early precursors, well before the acquisition of secretory function, suggests that J chains may play a role in Ig production early in differentiation rather than in the polymerization of IgA and IgM only [22]. The failure to detect large amounts of intracellular polymeric (19S) IgM in IgM-producing myelomas indicates that polymerization occurs just prior to secretion [29], although other studies suggest that J-chain-containing pentamers may be found in the RER immediately after assembly of H and L chains into Ig monomers [38]. It is of interest that the rates of biosynthesis of J chain and Ig heavy chains appear to be precisely coordinated, although neither the underlying mechanism nor the significance of this physiologic control process is understood.

Intracellular Traffic Pathway of Ig

The early events in Ig transport, namely biosynthesis on membrane-bound polysomes, segregation, assembly and glycosylation in the cisternae of the RER, and transport into Golgi vesicles, are better characterized than the subsequent intracellular traffic of Ig chains, largely because antibody-secreting cells do not store or concentrate Ig before secretion. In regulated secretory cells which possess storage compartments, secretion is known to be influenced by divalent cations (particularly Ca^{++}), intracellular cyclic nucleotide levels and cytoskeletal structures (microtubules, microfilaments), and is known to occur by an energy-dependent process of vesicle to plasma membrane fusion and exocytosis. In contrast, Ig secretion by plasma cells is unaffected by cyclic nucleotides and cytoskeletal alterations, these cells being essentially non-regulated [37]. Analysis of the Ig secretory process by pharmacologic inhibition suggests that it occurs by constant traffic of small, Ig-containing vesicles from the Golgi system to the plasma membrane, and different drugs can inhibit secretion by blocking traffic at different steps [37]. For instance, lowering intracellular Ca^{++} levels with an ionophore prevents migration of Ig-containing vesicles from the Golgi complex, where they tend to aggregate. Monensin and nigericin, which are carboxylic ionophores that promote Na^{+}-K^{+} proton exchange, appear to cause excessive fusion of Ig-containing Golgi-derived vesicles either with one another or with other, pre-existing vesicles, and these fused vesicles fail to migrate to and fuse with the plasma membrane. Finally, inhibitors of energy metabolism probably block Ig transport at a site before entry into the Golgi complex.

From such studies, a general scheme of post-translational modification and traffic of antibody molecules can be constructed (table II). In the above discussion, an attempt has been made to point out areas where knowledge is incomplete and exceptions to the general scheme might exist. Nevertheless, morphologic and biochemical techniques are readily available to dissect the sequence of events in greater detail. Perhaps the most exciting studies will answer how and at what steps the processes of Ig biosynthesis assembly and secretion are affected by external regulatory stimuli. In the following section, the stimuli capable of specifically regulating Ig production will be outlined; any discussion of operative mechanisms is necessarily largely speculative, so that emphasis will be placed

Table II. Sequence of events in assembly and secretion of immunoglobulins

Intracellular organelle	Step in Ig assembly secretion
Membrane-bound polyribosomes (RER)	translation of Ig mRNA (? initiated on free ribosomes)
↓	
Cisterna of RER	cleavage of leader peptide
↓	noncovalent assembly of H-L, H-H chains
Cisterna of Golgi complex	addition of core sugars disulfide linkage of H-L, H-H chains addition of terminal sugars
	polymerization of IgA, IgM
Ig-containing vesicles that migrate to and fuse with plasma membrane	(steps blocked by lowering Ca^{++} level; monensin, nigericin)
	secretion by reverse pinocytosis

on the potential for using these phenomena to define the cellular consequences of ligand-receptor interactions occurring at the cell surface.

Regulation of Immunoglobulin Expression by External Stimuli

Specific patterns of positive and negative responses to antigens, polyclonal stimuli and regulatory cells (such as macrophages and helper or
suppressor T lymphocytes) are the sine qua non of the immune system.
Such responses are most thoroughly studied with normal and neoplastic B
lymphocytes, and in general they consist of proliferation, differentiation
and changes in Ig expression. The *positive responses* are induced by
immunogenic antigens and polyclonal B cell activators (PBA), frequently
require cooperative stimuli from helper T cells and macrophages, and
consist of: (i) proliferation and (ii) differentiation, e.g. from membrane-
Ig-bearing non-secreting B cells to cells actively secreting Ig, or from IgM-
expressing cells to cells producing IgG, IgA or IgE. *Negative responses* are
induced by tolerogenic (suppressive) forms of antigens and by suppressor
T lymphocytes, and are composed of: (i) blocks in Ig synthesis or the
expression of membrane and/or secretory Ig, and (ii) possible inhibition of
proliferation and/or differentiation, for which the evidence is largely circumstantial and inconclusive. The ability to induce these events with
polyclonal stimuli and in homogeneous B lymphocyte-derived neoplasms

has proved critical to analyses of operative mechanisms. Although the relevance of mechanistic studies employing tumor cells to the physiologic situations of antigen-specific immune induction and suppression is difficult to establish conclusively, the available evidence does indicate that many tumors are valid models for normal immunocompetent cells [3].

In the following sections, current concepts of B lymphocyte activation and paralysis will be outlined. Studies with tolerogenic antigens and suppressor T cells are the most relevant to regulation of the secretory process. The control of cellular proliferation and differentiation are complex issues beyond the scope of this chapter, and only selected features will be described without any attempt to present the experimental evidence.

B Lymphocyte Activation by Antigens, Polyclonal Activators, and Helper T Lymphocytes

The simplified view of B cell activation that is supported by the bulk of currently available data is that most resting (G_0) B lymphocytes are triggered by binding of multivalent antigen (or anti-Ig antibody) to membrane Ig receptors, and move into the G_1 (pre-S) phase of the cell cycle and also develop surface receptors for helper-T-cell-derived factors. Helper T cells, which are activated by antigen on appropriate presenting cells, liberate a number of as yet poorly defined B cell growth and differentiation factors, which induce expansion and maturation of the antigen-stimulated B lymphocyte clone(s). Maturation is reflected by two basic changes in Ig expression, i.e. a membrane to secreted Ig switch, and heavy chain class switch. These differentiative events occur in the progeny of the original antigen-responsive clone(s). The molecular biologic basis for changes in Ig expression has been discussed above, but the mechanism of signal transduction, i.e. how antigen- and T-cell-derived substances induce specific alterations in DNA arrangement and/or RNA splicing, is entirely unknown. Two other points are noteworthy. Some (thymus-independent) antigens are capable of inducing all the above proliferative and differentiative events without the overt participation of T lymphocytes. Secondly, various experiments suggest that macrophage-derived factors may also play a role in B cell responses, and some B cells require direct contact with helper T lymphocytes (rather than T-cell-derived factors) to be activated.

The ability to reproduce these events in murine and human B-lymphocyte-derived tumors with PBAs and T-cell-derived factors provides a powerful tool for dissecting the biochemical basis of lymphocyte activation [3]. For instance, non-secreting, membrane IgM-positive B-lympho-cyte-derived, cloned tumor cells cultured with PBA and helper T cell factors differentiate into IgM-secreting cells and show a gradual reduction in the density of membrane Ig. Preliminary studies using Northern blot analysis of Cμsec and Cμmem mRNA have shown a concomitant increase in the intracellular levels of both forms of mRNA, indicating that the induction of IgM secretion and decline in membrane Ig expression is probably due to post-translational control [quoted in ref. 3]. Such analyses should not only lead to a clearer definition of the molecular biology and biochemistry of cellular differentiation in terms of Ig expression, but might also establish the fate of various ligands (antigens, PBAs, T-cell-derived factors) following their interactions with specific membrane receptors as well as the nature of such receptors.

Suppression of B Lymphocytes by Antigens

The ability of certain forms of antigens to paralyze ('tolerize') rather than induce humoral immune responses is well-established [42]. A large body of experimental work has led to the formulation of some key con-cepts. First, tolerogenic antigens can act on any differentiative stage of B cells, from the unstimulated lymphocyte to the antibody-secreting cell, provided that membrane Ig receptors for antigen are present. Secondly, immature (IgM+, IgD–) B cells are exquisitely susceptible to tolerance induction by virtually all antigens examined. In contrast, mature B lym-phocytes and antibody-secreting cells can be readily inactivated only by polymeric antigens that are degraded poorly or not at all by cellular enzymes [2]. Finally, γ-globulins and antigen-antibody complexes may induce lymphocyte unresponsiveness by a special mechanism in which the ligands bind simultaneously to membrane Ig and to receptors for the Fc portion of the γ-globulin or the antibody present in the complex, and the 'negative signal' is delivered via the Fc receptors or both the membrane Ig and Fc receptors [35]. Some recent experiments suggest that B lymphocyte suppression by antigens is an active process requiring energy metabolism and protein synthesis [39]. Any putative 'second messenger' or 'repressor' type of molecule(s) that might be actively synthesized and might block further B cell stimulation has not been identified, to date. However, in the better analyzed systems, it appears that tolerogenic ligands act on mem-brane Ig-bearing B cells and block the expression of this Ig, so that the relevant B cells are incapable of subsequent antigen recognition and acti-vation [2, 10]. The internalization of ligand-receptor complexes, their intracellular dissociation or degradation, and the re-expression or recycling of receptors have been shown to be of critical importance in the responses of many cell types to peptide hormones and growth factors [13]. As an extension of these studies to B lymphocytes, it is possible that internalized complexes of tolerogenic ligands and membrane Ig receptors may not be degraded if the ligand is insusceptible to hydrolytic enzymes (e.g. bacterial polysaccharide antigens or polypeptides composed of *D*-amino acids) or if the B cells are deficient in lysosomal functions, which might be true of neonatal lymphocytes [2]. In these situations, the ligand-receptor complexes would persist within the cells; this might not only block subsequent re-expression of receptors but could also inhibit any responses that are dependent on receptor recycling or re-expression. Clearly, then, one has a system in which it should be possible to explore the effects of intracellular ligand on membrane Ig expression and cellular functions. So far, however, it has proved difficult to carry out mechanistic studies using normal neonatal or adult B cells, which have a limited life-span in vitro.

A parallel situation exists in a mouse myeloma line, MOPC 315, which binds di- and trinitrophenol (DNP, TNP). Culture of MOPC 315 cells with DNP-coupled globulins but not a number of other DNP proteins leads to a block in antibody synthesis and secretion. Further analysis shows that DNP globulins are endocytosed by the tumor cells in larger amounts than non-suppressive hapten proteins, and they remain undergraded within the cells for prolonged periods of time [1]. In other words, intracellular persistence of antigen blocks the synthesis and secretion of secretory Ig, similar to the inhibition of membrane Ig expression in normal B cells by tolerogenic antigens. Again, the level at which Ig production in the myeloma cells is blocked is not known. It is also not known whether the reduction of Ig biosynthesis is a primary event or is secondary to a secretory block that itself is due to abnormal post-translational modification or transport or enhanced endogenous degradation.

Inhibition of Antibody Production by Suppressor T Lymphocytes

Suppressor T lymphocytes (Ts) are known to inhibit a variety of humoral and cell-mediated immune responses, and have been shown to act on different types of immunocompetent cells. Most relevant to the control of Ig secretion are the experimental systems in which antibody production by myeloma cells can be suppressed by co-culture with Ts specific for the myeloma Ig [5, 26]. This suppression is of particular interest because it is exquisitely selective in three respects. Brief (1–2 day) culture of MOPC 315 cells with Ts that recognize receptor Ig on the myeloma targets blocks the synthesis and secretion of secretory Ig without affecting membrane Ig expression or the production of non-Ig proteins [5, 26]. Moreover, somatic cell hybrids that produce two unrelated antibodies, when cultured with Ts specific for one, show inhibition of secretion of only that Ig [4]. Clearly, the selectivity of this suppression implies a compartmentalization of signal delivery that may have important implications for signal transduction effects, in general. The inhibition of secretory but not membrane Ig resembles the previously demonstrated effect of tunicamycin on antibody-secreting cells [15]. Thus, it is possible that Ts inhibit glycosylation and/or assembly of Ig chains, and the inhibition of synthesis is a feedback response to the block in secretion. Alternatively, the primary effect of Ts may be on Ig biosynthesis. In any event, the cellular and biochemical tools for analyzing such suppressive phenomena are now available, and it appears likely that definitive answers will soon be forthcoming.

Conclusions

The processes of Ig biosynthesis and secretion have already provided a wealth of information that is relevant to an understanding of secretory processes, in general. Genes coding for antibody molecules are examples of split genes that undergo highly specific rearrangements during ontogeny; such recombination events in various cellular genes may be important in the production of many diverse proteins that are markers of differentiation. The unusual structural composition of antibodies, i.e. the presence of separately synthesized but covalently linked heavy and light chains, each consisting of functionally and biochemically distinct segments, implies the need for mechanisms of polypeptide chain assembly that are applicable to many proteins composed of two or more chains. The production of two distinct forms of antibodies, i.e. membrane and secretory Ig, by the same cell has been utilized to study the biosynthetic and intracellular pathways that lead to expression of proteins that are largely identical but differ in a critical segment at the C terminal. The presence of characteristic and readily identifiable oligosaccharide residues on Ig molecules makes it feasible to examine the role of glycosylation in the secretion of glycoproteins. Antibody-secreting plasma cells are also the most typical examples of secretory cells that lack storage compartments and are, therefore, insusceptible to short-acting regulatory stimuli. All these analyses can be done because of the ready availability of highly specific immunochemical and molecular biologic probes for detecting Ig molecules and studying their expression. Even more significant is the fact that large numbers of homogeneous, cloned tumor lines that produce antibodies, and variants of such lines with specific aberrations in Ig biosynthesis or secretion, can be used as models for examining antibody production in detail. Finally, and perhaps most important, during the last 5 years it has been clearly established that Ig expression by normal and neoplastic B lymphocytes can be specifically and selectively altered by a variety of extrinsic agents, such as antigens, polyclonal activators and regulatory T lymphocytes. These experimental systems provide a unique opportunity for analyzing the mechanisms that regulate protein synthesis and secretion in eukaryotic cells. Ultimately, studies of the regulation of antibody production might provide answers to what is probably the central issue in cell biology, namely the mechanisms whereby ligand-receptor interactions at the cell surface lead to changes in gene expression and cell function.

Acknowledgements

I thank Ms. *Valerie Sherman* for her valuable assistance in the preparation of the manuscript.

The author's research is supported by NIH grant AI 16349, USPHS.

References

1   Abbas, A.K.: Antigen and T lymphocyte mediated suppression of myeloma cells: model systems for regulation of lymphocyte function. Immunol. Rev. *48:* 245–264 (1979).

2   Abbas, A.K.: Two distinct mechanisms of B-lymphocyte tolerance. Cell. Immunol. *46:* 178–183 (1979).

3   Abbas, A.K.: Immunologic regulation of lymphoid tumor cells: model systems for lymphocyte function. Adv. Immunol. *32:* 301–368 (1982).

4   Abbas, A.K.; Burakoff, S.J.; Gefter, M.L.; Greene, M.I.: T lymphocyte mediated suppression of myeloma function in vitro. J. exp. Med. *152:* 969–978 (1980).

5   Abbas, A.K.; Perry, L.L.; Bach, B.A.; Greene, M.I.: Idiotype specific T cell immunity. J. Immun. *124:* 1160–1166 (1980).

6   Adams, J.M.; Kemp, D.J.; Bernard, O.; Gough, N.; Webb, E.; Tyler, B.; Gerondakis, S.; Cory, S.: Organization and expression of murine immunoglobulin genes. Immunol. Rev. *59:* 5–32 (1981).

7   Alt, F. W.; Bothwell, A. L. M.; Knapp, M.; Siden, E.; Mather, E.; Koshland, M.; Baltimore, D.: Synthesis of secreted and membrane-bound immunoglobulin μ heavy chains is directed by mRNAs that differ at their 3' ends. Cell *20:* 293–301 (1980).

8   Bankert, R. B.; Mayers, G. L.; Pressman, D.: Clearance and re-expression of a myeloma cell's antigen-binding receptors by ligands known to be immunogenic or tolerogenic for normal B lymphocytes: a model to study membrane events associated with B cell tolerance. Eur. J. Immunol. *8:* 512–519 (1978).

9   Baumal, R.; Scharff, M. D.: Synthesis, assembly and secretion of mouse immunoglobulin. Transplant. Rev. *14:* 163–183 (1973).

10 Braun, J.; Unanue, E. R.: B lymphocyte biology studied with anti-Ig antibodies. Immunol. Rev. *52:* 3–28 (1980).

11 Early, P.; Rogers, J.; David, M.; Calame, K.; Bond, M.; Wall, R.; Hood, L.: Two mRNA's can be produced from a single immunoglobulin μ gene by alternative RNA processing pathways. Cell *20:* 313–319 (1980).

12 Goding, J. W.: Asymmetrical surface IgG on MOPC-21 plasmacytoma cells contains one membrane heavy chain and one secretory heavy chain. J. Immun. *128:* 2416–2421 (1982).

13 Goldstein, J. L.; Anderson, R. G. W.; Brown, M. S.: Coated pits, coated vesicles and receptor-mediated endocytosis. Nature, Lond. *279:* 679–685 (1979).

14 Hickman, S.; Kulczycki, A.; Lynch, R. G.; Kornfeld, S.: Studies on the mechanism of tunicamycin inhibition of IgA and IgE secretion by plasma cells. J. biol. Chem. *252:* 4402–4408 (1977).

15 Hickman, S.; Wong-Yip, Y. P.: Re-expression of nonglycosylated surface IgA in trypsin-treated MOPC 315 plasmacytoma cells. J. Immun. *123:* 389–395 (1979).

16 Honjo, T.; Nakai, S.; Nishida, Y.; Kataoka, T.; Yamawaki, Kataoka, Y.; Takashashi, N.; Obata, M.; Shimizu, A.; Yaoita, Y.; Nikaido, T.; Ishida, N.: Rearrangements of immunoglobulin genes during differentiation and evolution. Immunol. Rev. *59:* 33–68 (1981).

17 Johnson, N.; Douglas, R.; Hood, L.: Nucleic acid rearrangements in the differentiation of mouse B cells; in Klinman, Mosier, Scher, Vitetta, B lymphocytes in the immune response, pp. 3–18 (Elsevier/North-Holland, New York 1982).

18 Kincade, P. W.: Formation of B lymphocytes in fetal and adult life. Adv. Immunol. *31:* 177–245 (1981).

19 Koshland, M. E.: Structure and function of the J chain. Adv. Immunol. *20:* 41–70 (1975).

20 Kuehl, W. M.: Synthesis of immunoglobulin in myeloma cells. Curr. Top. Microbiol. Immunol. *76:* 3–47 (1977).

21 Leder, P.: The genetics of antibody diversity. Scient. Am. *244:* 102–115 (1982).

22 McCune, J. M.; Fu, S. M.; Kunkel, H. G.: J chain biosynthesis in pre-B cells and other possible precursor B cells. J. exp. Med. *154:* 138–145 (1981).

23 Mechler, B.: Membrane-bound ribosomes of myeloma cells. VI. Initiation of immunoglobulin mRNA translation occurs on free ribosomes. J. Cell Biol. *88:* 42–50 (1981).

24 Melchers, F.; Andersson, J.: Synthesis, surface deposition and secretion of immunoglobulin M in bone marrow-derived lymphocytes before and after mitogenic stimulation. Transplant. Rev. *14:* 76–130 (1973).

25   Melchers, F.; Andersson, J.: Secretion of immunoglobulins. Adv. Cytopharmacol. *2:* 225–235 (1974).

26   Milburn, G.L.; Lynch, R.G.: Immunoregulation of murine myeloma in vitro. J. exp. Med. *155:* 852–862 (1982).

27   Moore, K.W.; Rogers, J.; Hunkapiller, T.; Early, P.; Nottenburg, C.; Weissman, I.; Bazin, H.; Wall, R.; Hood, L.E.: The expression of immunoglobulin D may employ both DNA rearrangement and RNA splicing mechanisms. Proc. natn. Acad. Sci. USA *78:* 1800–1806 (1981).

28   Mosmann, T.R.; Williamson, A.R.: Structural mutations in a mouse immunoglobulin light chain resulting in failure to be secreted. Cell *20:* 283–292 (1980).

29   Parkhouse, R.M.E.: Assembly and secretion of immunoglobulin M (IgM) by plasma cells and lymphocytes. Transplant. Rev. *14:* 131–145 (1973).

30   Sabatini, D.D.; Kreibich, D.; Morimoto, T.; Adesnik, M.: Mechanisms for the incorporation of proteins in membranes and organelles. J. Cell Biol. *92:* 1–22 (1982).

31   Scharff, M.D.: The synthesis, assembly and secretion of immunoglobulins: a biochemical and genetic approach. Harvey Lect. *69:* 125–142 (1975).

32   Scholtissek, C.: Inhibition of the multiplication of enveloped viruses by glucose derivatives. Curr. Top. Microbiol. Immunol. *70:* 101–119 (1975).

33   Secher, D.S.; Milstein, C.; Adetugbo, K.: Somatic mutants and antibody diversity. Immunol. Rev. *36:* 51–72 (1977).

34   Sidman, C.L.: B lymphocyte differentiation and the control of IgM μ chain expression. Cell *23:* 379–389 (1981).

35   Sidman, C.L.; Unanue, E.R.: Control of B lymphocyte function. J. exp. Med. *144:* 882–896 (1976).

36   Stevens, R.H.; Williamson, A.R.: Translational control of immunoglobulin synthesis. I. Repression of heavy chain synthesis. J. molec. Biol. *78:* 505–516 (1973).

37   Tartakoff, A.; Vassalli, P.: Plasma cell immunoglobulin secretion. J. exp. Med. *146:* 1332–1345 (1977).

38   Tartakoff, A.; Vassalli, P.: Plasma cell immunoglobulin M molecules: their biosynthesis, assembly and intracellular transport. J. Cell Biol. *83:* 284–299 (1979).

39   Teale, J.M.; Klinman, N.R.: Tolerance as an active process. Nature, Lond. *288:* 385–387 (1980).

40   Uhr, J.W.; Schenkein, I.: Immunoglobulin synthesis and secretion: sites of incorporation of sugars as determined by subcellular fractionation. Proc. natn. Acad. Sci. USA *66:* 952–958 (1970).

41   Vitetta, E.S.; Uhr, J.W.: Synthesis, transport, dynamics and fate of cell surface Ig and alloantigens in murine lymphocytes. Transplant. Rev. *14:* 50–75 (1973).

42   Weigle, W.O.: Immunological unresponsiveness. Adv. Immunol. *16:* 61–123 (1973).

43   Weitzman, S.; Scharff, M.D.: Mouse myeloma mutants blocked in the assembly, glycosylation and secretion of immunoglobulin. J. molec. Biol. *102:* 237–252 (1976).

Abul K. Abbas, MD, Departments of Pathology, Harvard Medical School and Brigham and Women's Hospital, Boston, MA 02115 (USA)

Subject Index